ENVIRONMENTAL FACTORS IN NEURODEVELOPMENTAL AND NEURODEGENERATIVE DISORDERS

ENVIRONMENTAL FACTORS IN NEURODEVELOPMENTAL AND NEURODEGENERATIVE DISORDERS

Edited by

MICHAEL ASCHNER, PhD
Professor, Department of Molecular Pharmacology,
Albert Einstein College of Medicine, Bronx, NY, USA

LUCIO G. COSTA, PhD
Professor, Department of Environmental and Occupational Health Sciences,
University of Washington, Seattle, WA, USA; Professsor, Department of
Neuroscience, University of Parma Medical School, Parma, Italy

AMSTERDAM • BOSTON • HEIDELBERG • LONDON
NEW YORK • OXFORD • PARIS • SAN DIEGO
SAN FRANCISCO • SINGAPORE • SYDNEY • TOKYO

Academic Press is an imprint of Elsevier

Academic Press is an imprint of Elsevier
125 London Wall, London EC2Y 5AS, UK
525 B Street, Suite 1800, San Diego, CA 92101-4495, USA
225 Wyman Street, Waltham, MA 02451, USA
The Boulevard, Langford Lane, Kidlington, Oxford OX5 1GB, UK

Notices
Knowledge and best practice in this field are constantly changing. As new research and experience broaden our
understanding, changes in research methods, professional practices, or medical treatment may become necessary.

Practitioners and researchers may always rely on their own experience and knowledge in evaluating and using
any information, methods, compounds, or experiments described herein. In using such information or methods
they should be mindful of their own safety and the safety of others, including parties for whom they have a
professional responsibility.

To the fullest extent of the law, neither the Publisher nor the authors, contributors, or editors, assume any
liability for any injury and/or damage to persons or property as a matter of products liability, negligence or
otherwise, or from any use or operation of any methods, products, instructions, or ideas contained in the
material herein.

Library of Congress Cataloging-in-Publication Data
A catalog record for this book is available from the Library of Congress

British Library Cataloguing-in-Publication Data
A catalogue record for this book is available from the British Library

ISBN: 978-0-12-800228-5

For information on all Academic Press publications
visit our website at http://store.elsevier.com/

Working together
to grow libraries in
developing countries

www.elsevier.com • www.bookaid.org

Publisher: Mica Haley
Acquisition Editor: Erin Hill-Parks
Editorial Project Manager: Molly McLaughlin
Production Project Manager: Lucía Pérez
Designer: Greg Harris

Typeset by TNQ Books and Journals
www.tnq.co.in

Contents

11. Environmental Neurotoxins Linked to a Prototypical Neurodegenerative Disease

PETER S. SPENCER, C. EDWIN GARNER,
VALERIE S. PALMER, GLEN E. KISBY

12. Environmental Exposures and Risks for Parkinson's Disease

HARVEY CHECKOWAY, SUSAN SEARLES NIELSEN,
BRAD A. RACETTE

13. Parkinson's Disease: Mechanisms, Models, and Biological Plausibility

ERIC E. BEIER, JASON R. RICHARDSON

14. Genetic Models of Parkinson's Disease: Behavior, Signaling, and Pathological Features

AMY M. PALUBINSKY, BRITNEY N. LIZAMA-MANIBUSAN,
DANA MILLER, BethAnn McLAUGHLIN

15. Alzheimer's Disease and the Search for Environmental Risk Factors

WALTER A. KUKULL

16. Environmental Factors and Amyotrophic Lateral Sclerosis: What Do We Know?

PAM FACTOR-LIVAK

Preface

This book aims to impart new light on the environmental origins of neurodevelopmental and neurodegenerative diseases. Neurological disorders are characterized by degeneration and cell loss, accompanied by a reduction in cell numbers, and a consequence aberrant brain function. Remarkably, even after the enormous investment in recent research, best exemplified by the "Decade of the Brain," most neurodevelopmental and neurodegenerative diseases remain of unknown origin or idiopathic in nature.

While genetics clearly plays a role in neurodevelopmental and neurodegenerative diseases, occupational, iatrogenic, medical, and environmental exposures clearly contribute to human morbidity. In most instances exposures are to xenobiotics, but remarkably even essential compounds may cause clinical signs secondary to morphological lesions and neurodegeneration, upon exposures to exceedingly high levels or secondary to genetic susceptibility.

Neurodegeneration may occur at all life stages. Neurotoxins such as mercury, lead, or alcohol, to name a few, may cause neurodegeneration during brain development or adulthood. Furthermore, susceptibility may vary, and most often is most pronounced during developmental stages, given the dynamic processes inherent to this life stage (cell division, differentiation, migration). Neurodegeneration at early life stages may also be silent for years if not decades, unmasking itself well into adult life. Thus, there is a clear continuum between neurodevelopmental and neurodegenerative disorders.

For the majority of the neurotoxins, support for causality is limited and characterized by lack of confirmation of case diagnoses and inadequate exposure data. Keeping these limits in mind, the book was designed to provide state-of-the-art information on environmental links to neurodevelopmental and neurodegenerative disorders. In the first part, several chapters highlight the vulnerability of the central nervous system (CNS) to neurotoxins, highlighting the role of genetics and inflammatory processes in disease etiology. In the second part, focus is directed at neurodegenerative diseases with adult onset. Our primary goals were to provide the reader with information on mechanistic events leading to clinical disease, and facilitate the understanding on mechanisms associated with neurodevelopmental and neurodegenerative disorders, whether genetically or environmentally linked (or both).

We assembled a series of chapters that advance the latest developments and scientific breakthroughs in this fast-paced research area, and provide information that should be of interest to risk assessors, neurobiologists, clinicians, and neurotoxicologists. We are hopeful that the book offers the reader appreciation and renewed sense on contemporary issues in this topic. We are indebted to the authors for their contributions, and hope that as a reader, whether you are novice or a seasoned expert in the topic, the knowledge amassed herein will stimulate and transform your thinking in this contemporary health issue.

Michael Aschner and Lucio G. Costa

Foreword

The field of neurotoxicology, the study of the effect of toxic exposures on the nervous system, has expanded tremendously in the last decades from a new but descriptive science, to one in which highly state of the art, mechanistic research is being done. As a result, it is becoming clearer that neurotoxic exposures have consequences for the nervous system and that the greatest periods of vulnerabilities are the developing nervous system and the aging nervous system. While it is still a challenge to find an ideal nonhuman model of exposure to explore this research, the combination of human, mammalian, and nonmammalian studies have contributed greatly to overcoming the deficits of any one model. And it is this combination of efforts that has contributed to public policy decisions in a meaningful way and to the association of some neuronal dysfunction with exposure.

Research has suggested that exposures can affect the quality of life across the lifespan and that a combination of genetic susceptibility and acute and/or chronic exposure can contribute to the problem. Because of the more sophisticated tools now available, we have gleaned an understanding of not only which mutations contribute to one's vulnerability, but also to how epigenetic changes to genes due to these exposures can also accomplish this. We also have better evidence for the assumption that early exposure can alter neuronal development that may not be evident until much later in life, that is the long-latency period leading to neurodegeneration. And while the problems have gotten more complex, so are the tools and approaches being used to solve them. We are now more capable of appreciating how exposure to mixtures can be more detrimental to neuronal health either in an additive or synergistic manner and how heavy metals, industrial and agrochemicals, air pollutants either by themselves or combined with life style choices can magnify the challenge of maintaining neuronal health while living in a complex environment.

The series of reviews in this book *Environmental Factors in Neurodevelopmental and Neurodegenerative Disorders*, edited by Lucio Costa and Michael Aschner, is a timely, selective compendium of the state of the field of neurotoxicology on outcomes, some of which were not even associated with neurotoxic exposure until only a decade or two ago. It will provide the state-of-the-science for many of the neuronal outcomes of current concern in which the role of exposure had been underappreciated, such as autism, attention deficit-hyperactivity disorder, schizophrenia, Parkinson's, amyotrophic lateral sclerosis, Alzheimer's, and Huntington's disease. It will also comprehensively discuss gene–environment interactions and the potential for long-latency periods in these outcomes, as well as the contribution of alcohol and infection to neuronal development in the fetus.

It will inform us of the role that selective exposures—lead, endocrine disruptors, methyl mercury—have played in establishing the importance of the resulting neuronal disorders contribution to public health. This book is highly recommended to students and researchers in the field as well as public health advocates and policy makers.

Annette Kirshner, PhD
Program Director
National Institute of Environmental Health Sciences
National Institutes of Health
Research Triangle Park, North Carolina

Contributors

Michael Aschner Department of Molecular Pharmacology, Albert Einstein College of Medicine, Bronx, NY, USA

Paul Barrett Department of Neurology, University of Pittsburgh, Pittsburgh, PA, USA

Eric E. Beier Department of Environmental and Occupational Medicine, Rutgers Robert Wood Johnson Medical School and Environmental and Occupational Health Sciences Institute, Piscataway, NJ, USA

David C. Bellinger Department of Neurology, Boston Children's Hospital, Boston, MA, USA; Department of Psychiatry, Boston Children's Hospital, Boston, MA, USA; Department of Neurology, Harvard Medical School, Boston, MA, USA; Department of Environmental Health, Harvard School of Public Health, Boston, MA, USA

Terry Jo Bichell Vanderbilt Brain Institute, Vanderbilt University, Nashville, TN, USA

Aaron B. Bowman Department of Neurology, School of Medicine, Vanderbilt University, Nashville, TN, USA

Emma Bradley Department of Neurology, Vanderbilt University, Nashville, TN, USA

Thomas M. Burbacher Department of Environmental and Occupational Health Sciences, School of Public Health, University of Washington, Seattle, WA, USA; Center on Human Development and Disability, University of Washington, Seattle, WA, USA; Washington National Primate Research Center, University of Washington, Seattle, WA, USA; Amgen, Seattle, WA, USA

Harvey Checkoway Family and Preventive Medicine, University of California, San Diego, La Jolla, CA, USA

Shih-Heng Chen Laboratory of Toxicology and Pharmacology, National Institute of Environmental Health Sciences, National Institutes of Health, NC, USA

Deborah A. Cory-Slechta Department of Environmental Medicine, University of Rochester School of Medicine, Rochester, NY, USA

Lucio G. Costa Department of Environmental and Occupational Health Sciences, University of Washington, Seattle, WA, USA; Department of Neuroscience, University of Parma Medical School, Parma, Italy

Donato A. Di Monte German Center for Neurodegenerative Diseases (DZNE), Bonn, Germany

Pam Factor-Livak Department of Epidemiology, Mailman School of Public Health, Columbia University, New York, NY, USA

C. Edwin Garner Department of Internal Medicine, University of New Mexico, Albuquerque, NM, USA

Kimberly S. Grant Department of Environmental and Occupational Health Sciences, School of Public Health, University of Washington, Seattle, WA, USA; Center on Human Development and Disability, University of Washington, Seattle, WA, USA; Washington National Primate Research Center, University of Washington, Seattle, WA, USA

John T. Greenamyre Department of Neurology, School of Medicine; Pittsburgh Institute for Neurodegenerative Diseases (PIND), University of Pittsburgh, Pittsburgh, PA, USA

Marina Guizzetti Department of Behavioral Neuroscience, Oregon Health & Science University, and Research and Development Service (R&D39), Department of Veterans Affairs Medical Center, Portland, OR, USA

Jau-Shyong Hong Laboratory of Toxicology and Pharmacology, National Institute of Environmental Health Sciences, National Institutes of Health, NC, USA

Sarah A. Jewell German Center for Neurodegenerative Diseases (DZNE), Bonn, Germany

Lulu Jiang Laboratory of Toxicology and Pharmacology, National Institute of Environmental Health Sciences, National Institutes of Health, NC, USA

Michal Kicinski Centre for Environmental Sciences, Hasselt University, Belgium

Glen E. Kisby Department of Basic Medical Sciences, Western University of Health Sciences, Lebanon, OR, USA

Irene Knuesel University of Zurich, Zurich, Switzerland

Walter A. Kukull Department of Epidemiology, University of Washington, Seattle, WA, USA; National Alzheimer Coordinating Center, Department of Epidemiology, University of Washington, Seattle, WA, USA

Pamela J. Lein Department of Molecular Biosciences, School of Veterinary Medicine, University of California, Davis, CA, USA

Britney N. Lizama-Manibusan Neuroscience Graduate Program, School of Medicine, Vanderbilt University, Nashville, TN, USA; Vanderbilt Kennedy Center for Research on Human Development, Nashville, TN, USA

Maitreyi Mazumdar Department of Neurology, Boston Children's Hospital, Boston, MA, USA; Department of Neurology, Harvard Medical School, Boston, MA, USA; Department of Environmental Health, Harvard School of Public Health, Boston, MA, USA

BethAnn McLaughlin Department of Neurology, School of Medicine, Vanderbilt University, Nashville, TN, USA; Department of Pharmacology, School of Medicine, Vanderbilt University, Nashville, TN, USA; Vanderbilt Kennedy Center for Research on Human Development, Nashville, TN, USA

Dana Miller Department of Biochemistry, University of Washington, Seattle, WA, USA

Ruth E. Musgrove German Center for Neurodegenerative Diseases (DZNE), Bonn, Germany

Tim S. Nawrot Centre for Environmental Sciences, Hasselt University, Belgium; Department of Public Health & Primary Care, Leuven University, Belgium

Susan Searles Nielsen Department of Environmental and Occupational Health Sciences, University of Washington, Seattle, WA, USA

Esteban A. Oyarzabal Laboratory of Toxicology and Pharmacology, National Institute of Environmental Health Sciences, National Institutes of Health, NC, USA

Valerie S. Palmer Oregon Health & Science University, Portland, OR, USA

Amy M. Palubinsky Neuroscience Graduate Program, School of Medicine, Vanderbilt University, Nashville, TN, USA; Vanderbilt Kennedy Center for Research on Human Development, Nashville, TN, USA

Rafael Ponce Department of Environmental and Occupational Health Sciences, School of Public Health, University of Washington, Seattle, WA, USA; Amgen, Seattle, WA, USA

Brad A. Racette Department of Neurology, Washington University, St Louis, MO, USA; School of Public Health, Faculty of Health Sciences, University of the Witwatersrand, Parktown, South Africa

Jason R. Richardson Department of Environmental and Occupational Medicine, Rutgers Robert Wood Johnson Medical School and Environmental and Occupational Health Sciences Institute, Piscataway, NJ, USA

Janine Santos Laboratory of Molecular Carcinogenesis, National Institute of Environmental Health Sciences, National Institutes of Health, NC, USA

David S. Sharlin Department of Biological Sciences, Minnesota State University, Mankato, Mankato, MN, USA

Peter S. Spencer Department of Neurology, Center for Research on Occupational & Environmental Toxicology, and Global Health Center; Oregon Health & Science University, Portland, OR, USA

Michael Uhouse Department of Neurology, Vanderbilt University, Nashville, TN, USA

Qingshan Wang Laboratory of Toxicology and Pharmacology, National Institute of Environmental Health Sciences, National Institutes of Health, NC, USA

NEURODEVELOPMENTAL DISORDERS

Overview of the Role of Environmental Factors in Neurodevelopmental Disorders

Pamela J. Lein

INTRODUCTION

Neurodevelopmental disabilities, including autism spectrum disorders (ASD), attention-deficit hyperactivity disorder (ADHD), schizophrenia, learning disabilities, intellectual disability (also known as mental retardation), and sensory impairments, affect 10–15% of all births in the United States [1,2]. ADHD and ASD are among the most common of the neurodevelopmental disorders. In 2007, the worldwide prevalence of ADHD was estimated to be 5.3% of children and adolescents [3], and this prevalence is thought to be increasing worldwide [2]. The prevalence of ASD has increased dramatically from an estimated 1:150 children reported by the United States Center for Disease Control (CDC) in 2007 based on 2000 and 2002 data to the current estimate of 1:68 children or 1:42 boys [4]. Subclinical decrements in brain function are even more common than either of these neurodevelopmental disorders [5]. When considered in the context of the tremendous costs, which these disorders and disabilities exact on the affected individual, their families, and society [6–8], these statistics of their prevalence underscore the urgent need to identify factors that confer risk for neurodevelopmental disabilities.

Until recently, research on the etiology of neurodevelopmental disorders has focused largely on genetic causes [2,9,10]. However, this research has clearly shown that single genetic anomalies can account for only a small proportion of cases [2,11] and that, overall, genetic factors seem to account for at most 30–40% of all cases of neurodevelopmental disorders [5]. Credible evidence now exists that many neurodevelopmental disorders are the result of complex gene–environment interactions [11–15]. In contrast to genetic risks, which are currently irreversible, environmental factors are modifiable risk factors. Therefore, identifying specific environmental factors that increase risk for neurodevelopmental disorders may provide rational approaches for the primary prevention of the symptoms associated with these disorders. However, to date, the identities of environmental factors that influence susceptibility to and/or severity of neurodevelopmental disorders and intellectual disabilities, and the mechanisms by which environmental factors interact with genetic factors to determine individual risk, remain critical gaps in our understanding. This chapter will provide an overview of the evidence implicating environmental factors in determining risk of neurodevelopmental disorders, using autism and ADHD as prime examples. This chapter will also briefly discuss proposed mechanisms by which environmental factors might influence risk, and summarize the challenges and opportunities in this important area of research.

EVIDENCE IMPLICATING ENVIRONMENTAL FACTORS

The most compelling evidence in support of the hypothesis that environmental factors contribute to risk of neurodevelopmental disorders is the rapid increase in the prevalence of ASD and ADHD over the past several decades [2], which seems unlikely to have been caused entirely by significant shifts in the human genome. Concerns have been raised as to how much of the increased prevalence of these disorders represents true increases in the numbers of affected children. It has been suggested that diagnostic substitution, for example, the labeling of individuals as autistic or ADHD who previously would not have been so labeled, because of broadening of diagnostic criteria, coupled with increased awareness and improved detection of these disorders explain their increased prevalence [16–18]. However, several studies that have investigated these concerns in the context of ASD concluded that the frequency of this disorder has truly increased, with factors other than broadening of diagnostic criteria and increased awareness likely accounting for more than half of all new cases [19–21].

Studies of the genetic causes of neurodevelopmental disorders also support a role for environmental factors in determining the risk for many neurodevelopmental deficits. Autism is considered to be one of the most heritable complex neurodevelopmental disorders [22,23]; however, genes linked to ASD rarely segregate in a simple Mendelian manner [22]. This widespread observation has led many to posit that multiple genetic etiologies, including rare, private (de novo) single-gene mutations that are highly penetrant, inherited common functional variants of multiple genes with small to moderate effects on ASD, or copy number variation (CNV), occur in combination to determine ASD risk [24–28]. An alternative interpretation of the genetic findings is that environmental factors act as risk modifiers [11,29]. The consistent finding of incomplete monozygotic concordance in twin studies of both autism and ADHD [11,30] as well as data demonstrating that even in genetic syndromes highly associated with ASD, a significant percentage of carriers do not express autistic phenotypes [11,25], are consistent with a model in which environmental factors interact with genetic susceptibilities to influence ASD/ADHD risk, clinical phenotype, and/or treatment outcome [11,29]. Given the extensive literature documenting chemical-induced mutagenesis, the identification of de novo gene mutations associated with clinical diagnosis of ASD [11,22,25] is also consistent with the idea of environmental risk modifiers for ASD and other neurodevelopmental disorders.

More definitive human evidence corroborating a role for environmental factors has recently been reported. In one of the largest twin studies conducted to date, 192 monozygotic- and dizygotic twin pairs were analyzed to quantify the relative contributions of genetic heritability vs. the shared environment. The findings from these analyses suggested that 38% of ASD cases are attributable to genetic causes, whereas 58% are linked to the shared in utero environment [12]. The model used in this study had a number of inherent biases (e.g., it was assumed that gene × environment interactions did not occur, monozygotic and dizygotic twins were assumed to share the environment to the same extent, and questions arise regarding the validity of the values used for the prevalence for autism and ASD). However, similar conclusions were recently reached in an independent study of over 14,000 children with autism in Sweden that demonstrated a heritability of 50%, supporting an equally strong role for environmental risk factors [31]. Collectively, these studies suggest that environmental factors can significantly influence susceptibility and the variable expression of traits related to neurodevelopmental disorders, thereby providing a plausible explanation for both the dramatic increase in the prevalence of complex neurodevelopmental disorders and the significant clinical heterogeneity, which is a hallmark of both autism and ADHD [11,32].

ENVIRONMENTAL FACTORS ASSOCIATED WITH INCREASED RISK FOR NEURODEVELOPMENTAL DISORDERS

Early indications of an environmental contribution to neurodevelopmental disorders came from observations of a high incidence of autism associated with congenital rubella [33]. Subsequent studies linked prenatal infections to increased risk for not only ASD but also other neurodevelopmental disorders, particularly schizophrenia [34–37], and expanded the range of nongenetic risk factors for neurodevelopmental disorders to include other intrauterine stresses [38–42], paternal age [42–45] (but see [46]), maternal nutrition and metabolic status [11,47–49] and hormones, including sex hormones that contribute to the well-documented sex differences in autism susceptibility [50] via gene-specific epigenetic modifications of DNA and histones [51].

Early reports that in utero exposure to valproic acid or thalidomide during critical periods of development was associated with increased expression of autism-related traits in children

[52] raised significant interest regarding the role of chemical exposures in neurodevelopmental disorders. Subsequent epidemiologic studies reported an increased risk for ASD and other neurodevelopmental disorders associated with maternal use of various medications and drugs of abuse, including alcohol [52–58], as well as prenatal or early postnatal exposure to diverse environmental chemicals. Environmental chemicals postulated to confer risk for neurodevelopmental disorders include legacy chemicals known to be toxic to the developing human nervous system, such as lead, mercury, and polychlorinated biphenyls (PCBs) [2,59,60], as well as more contemporary contaminants, such as pesticides, including organophosphorus (OP) and organochlorine (OC) pesticides, neonicotinoids and pyrethroids, flame retardants including the polybrominated diphenyl ethers (PBDEs), plasticizers such as phthalates and bisphenol A (BPA), and complex environmental mixtures such as air pollution and cigarette smoke [11,14,32,61–63].

The epidemiologic data linking environmental chemicals to increased risk for ADHD [32] or ASD [62] were recently critically reviewed, and the major conclusions from these reviews are briefly summarized here. Polanska and colleagues [32] reviewed epidemiologic studies of environmental chemicals and ADHD published in peer-reviewed English language journals since the year 2000. Out of 72 articles identified in their initial literature search, 40 met their inclusion criteria. Many of these studies focused on exposure to tobacco smoke, and were largely consistent in identifying a positive association between tobacco smoke exposure and a diagnosis of ADHD or expression of ADHD symptoms in children [32]. Although the strength of the association differed between studies, the authors concluded that it generally appeared that the children of smokers are approximately 1.5–3 times more likely to have ADHD or ADHD symptoms than the children of nonsmokers. Other environmental chemicals addressed in the articles included in their review were ethanol, phthalates, BPA, polyfluoroalkyl chemicals (PFCs), and polycyclic aromatic hydrocarbons (PAHs). Interestingly, a number of studies in experimental animal and cell-based models suggest that developmental exposure to these chemicals elicits outcomes with face validity to at least a subset of ADHD-like symptoms (reviewed in [32]). Nevertheless, the authors found no consistent findings in the epidemiologic literature of an association between early life exposure to any of these environmental chemicals and increased risk for ADHD [32]. However, the authors concluded that the number of studies was insufficient regarding the impact of phthalates, BPA, PFCs, PAHs, and alcohol on ADHD to allow for meaningful conclusions as to whether these chemicals may indeed be environmental modifiers of ADHD risk.

Polanska and colleagues did not comment on the literature implicating lead and PCBs as risk factors for ADHD, although a recent review by Eubig, Aguiar, and Schantz [64] presents a compelling case for such associations. The analysis by Eubig and colleagues demonstrated that studies of lead exposure in children provide evidence for impairments in executive function and attention that parallel those observed in children diagnosed with ADHD. Animal studies of developmental lead exposure indicate behavioral and neurochemical changes with face validity to ADHD [64]. Importantly, an association between blood lead levels and a diagnosis of ADHD has been reported in a number of epidemiologic studies (reviewed in [64]). Similarly, the epidemiologic evidence of PCB effects on children shows parallels to behavioral domains affected in ADHD, although these were subtly different from the deficits noted in lead-exposed children [64]. The findings from animal studies of developmental PCB exposures reveal deficits on behavioral tasks that assess many of the same cognitive domains that are impaired in children with ADHD, as well as many changes in dopamine neurotransmission, which are reflective of neurochemical changes associated with ADHD [64]. However, the relationship between PCB exposure and ADHD in children remains largely unexplored.

The epidemiologic evidence linking environmental chemical exposures and ASD was recently reviewed by Kalkbrenner and colleagues [62]. These authors identified 58 relevant articles in the peer-reviewed literature published prior to March 1, 2014, of which 32 met their inclusion criteria (individual-level data on autism, exposure measures pertaining to pregnancy or the first year of life, valid comparison groups, control for confounders and adequate sample sizes). These 32 articles included studies of autism and estimates of exposure to tobacco, air pollutants, volatile organic compounds and solvents, metals, pesticides, PCBs, PBDEs, PFCs, BPA, and phthalates. As with the studies of these chemicals in relation to ADHD risk, many of these chemicals have been shown to cause effects in experimental animal or cell-based models that have face validity to behavioral outcomes or cellular phenotypes associated with autism (reviewed in [62,65]). Some of these environmental exposures also showed associations with autism in the analysis of the epidemiologic literature by Kalkbrenner and colleagues. The most strongly and consistently associated were traffic-related air pollutants, some metals, and the OP and OC pesticides, with suggestive trends for some volatile organic compounds and phthalates [62]. Interestingly, neither tobacco smoke, which was strongly associated with increased ADHD risk [32], nor alcohol showed a consistent association with increased risk for ASD.

All three of these comprehensive analyses of the human literature reached similar broad conclusions: the relevant publications that are currently available, with the possible exception of studies of tobacco and alcohol, are too limited in scope to either infer causality or rule out the possibility that these or additional environmental chemicals confer risk for ADHD or ASD [32,62,64]. All three sets of authors identified similar challenges facing epidemiologic studies of environmental chemicals and neurodevelopmental disorders. These include difficulties inherent in most epidemiologic studies, such as obtaining accurate measures of exposure, particularly for chemicals with short half-lives, such as some of the pesticides, phthalates, and BPA, controlling for confounding factors, especially socioeconomic stressors that tend to covary with environmental exposures, and dealing with coexposures, a not insignificant issue in light of reports that 250 xenobiotic chemicals were detected in biological samples from a 2013 representative sample of the United States in the National Health and Nutrition Examination Survey. However, epidemiologic studies of environmental factors in complex neurodevelopmental disorders face an additional challenge in that these disorders are phenotypically heterogeneous. Children are diagnosed as having ASD or ADHD based on observable symptom clustering, not causal pathways, a phenomenon known as etiological heterogeneity [66–68]. The complexity of heritable risk factors contributing to neurodevelopmental disorders like ADHD and ASD likely creates a range of sensitivities to environmental risk factors [29], which masks clear associations between exposure and diagnosis (see Figure 1.1). It has been proposed that a more nuanced "phenome" approach that identifies populations based on distinctive phenotypes may aid in the further discovery of modifiable environmental chemical risk factors that impact neurodevelopment until mechanism-based biomarkers for distinctive and discrete etiologies are available [62].

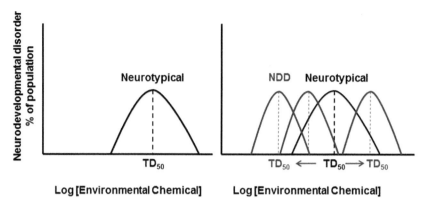

FIGURE 1.1   **The challenge of using epidemiologic approaches to identify environmental risk factors for neurodevelopmental disorders (NDDs).** In any population, including the neurotypical population, variation in the genetic makeup, timing of exposure, nutritional status, and/or coincidental illnesses will create a range of vulnerabilities to the developmental neurotoxicity of an environmental chemical. The TD_{50} is the exposure dose of the environmental chemical (or chemical mixture), which produces a phenotypic trait of relevance to an NDD. The genetic heterogeneity associated with risk for NDDs like ADHD and ASD likely creates a range of sensitivities to environmental risk factors; however, stratification of NDDs based on distinctive phenotypes may result in definable subpopulations with different sensitivity to environmental modifiers of NDD-related deficits, severity, and/or treatment outcomes.

MECHANISMS BY WHICH ENVIRONMENTAL FACTORS INFLUENCE RISK OF NEURODEVELOPMENTAL DISORDERS

Genetic, histologic, in vivo imaging, and functional data are converging on altered patterns of neuronal connectivity as the biological basis underlying the behavioral and cognitive abnormalities associated with many neurodevelopmental disorders and intellectual disabilities [69–74]. The candidate genes most strongly implicated in the causation of neurodevelopmental disorders encode proteins that regulate the patterning of neuronal networks during development and influence the balance of excitatory to inhibitory synapses [65,69,75,76]. Histological studies report increased spine density in Golgi-impregnated pyramidal neurons of Layer II in the frontal, temporal, and parietal lobes and Layer V of the temporal lobe in ASD patients [77], suggesting skewed cortical synaptic activity toward excitation, although functional evidence of excitation is needed to further support this hypothesis. Abnormal spine morphology is also found in the temporal and visual cortices of patients with Fragile-X, a syndrome with a high incidence of ASD [78] and in patients with schizophrenia [79,80]. Neuroimaging studies of children with ASD suggest a pattern of early brain overgrowth in the first few years of life followed by aberrant maturation in adolescence, whereas functional analyses suggest impaired long-range connectivity but increased local and/or subcortical connectivity (reviewed in [81]). In contrast, in childhood-onset schizophrenia, deficits in cerebral volume, cortical thickness, and white matter maturation seem most pronounced in childhood and adolescence and may level off in adulthood, and functional analyses suggest deficits in local connectivity, with increased long-range connectivity (reviewed in [81]). Thus, even though ASD and childhood-onset schizophrenia are thought to share an underlying genetic architecture [73,82–84], the dynamic changes that occur in the developing brain of children diagnosed with these neurodevelopmental disorders vary across space and time. These differences perhaps reflect the influence of differential environmental modifiers acting on a shared genetic substrate. Alternatively, these differences may reflect differences in the timing of exposure, a factor well known to influence neurodevelopmental outcomes following exposure to ethanol [85].

It has been proposed that these inherent imbalances in synaptic connectivity in children at risk for neurodevelopmental disorders likely provide the biological substrate for enhanced susceptibility to environmental triggers that modulate neurodevelopmental processes that determine synaptic connectivity, including neuronal migration, apoptosis, elaboration of axons and dendrites, and activity-dependent refinements of synaptic connections [65,76]. Perturbations of the spatiotemporal patterns or magnitude of any of these events can interfere with the formation of functionally meaningful networks and give rise to the deficits in functional connectivity associated with neurodevelopmental disorders [76,86]. Research findings from studies of transgenic animal models expressing neurodevelopmental disorder risk genes further support the hypothesis that the changes in synaptic connectivity observed in children at risk for neurodevelopmental disorders arise from perturbations of dendritic growth, synapse formation, and synapse stabilization [69,82].

Diverse pathophysiologic mechanisms have been hypothesized to mediate interactions between environmental chemicals and genetic risk factors that influence neuronal connectivity. These include (1) mutations in genes that encode proteins involved in detoxification of endogenous toxins and xenobiotics, such as the enzyme paraoxonase 1 (PON1), which hydrolyzes OP pesticides and has been associated with increased risk of autism (reviewed in [11]); (2) endocrine disruption, based on the rationale that ASD and ADHD disproportionately affect males [87] and the requirement for thyroid hormone and other hormones for normal neurodevelopment [88,89]; (3) chemical perturbation of the microbiome, derived from evidence suggesting that microbiota may regulate responses to xenobiotics and influence brain function, and that the microbiota may differ in children with neurodevelopmental disorders vs. typically developing children [90,91]; and (4) mitochondrial dysfunction and oxidative stress [11,92–94]. Although experimental evidence demonstrates that environmental chemicals can disrupt endocrine signaling, alter the microbiome, and cause oxidative stress, and neurodevelopmental disorders have been associated with endocrine disruption, altered gut microbiota, and biomarkers of oxidative stress in the central nervous system (CNS), causal relationships between chemical effects on these endpoints, and increased risk of neurodevelopmental disorders are largely lacking. Additional hypotheses involving immune dysregulation, epigenetics, and the convergence of environmental factors and genetic susceptibilities on neurotransmitters or signaling pathways that critically influence neuronal connectivity are discussed in more detail in the following paragraphs (Table 1.1).

Epigenetic mechanisms. Perhaps the most theoretically straightforward mechanism of interaction between environmental factors and susceptibility genes are physical interactions between chemicals and DNA (e.g., chemically induced de novo mutations) or chromatin (e.g., epigenetic changes in miRNA expression, DNA methylation patterns, and/or histone acetylation) [95,96]. Although experimental and clinical studies clearly demonstrate that epigenetic mechanisms are critically involved in regulating normal neurodevelopment, synaptic plasticity, and cognitive function, and that epigenetic changes in the brain are associated with

TABLE 1.1   Mechanisms Hypothesized to Mediate Gene–Environment Interactions that Influence the Risk and/or Severity of Neurodevelopmental Disorders (NDDs)

Mechanism	Rationale	Key references
Heritable deficits in xenobiotic metabolism	Decreased ability to detoxify environmental chemicals might effectively increase the neurotoxic potential of an environmental chemical	[11]
Endocrine disruption	Many NDDs are predominant in one sex and many hormones are required for normal neurodevelopment (sex steroids and thyroid hormones) or have significant effects on neurodevelopment (glucocorticoids)	[87–89]
Disruption of the gut microbiome	Emerging evidence indicates that the gut microbiome regulates host response to pathogenic microbial or xenobiotic exposures, and the gut microbiota in children with autism differs from that of neurotypical children	[11,92–94]
Epigenetic	Environmental chemicals have been demonstrated to alter DNA methylation, histone acetylation, and miRNA expression profiles, and these parameters are altered in at least some children with NDDs	[95–102,111]
Immune dysregulation	Cross talk between the nervous and immune systems is essential for normal neurodevelopment, environmental chemicals can alter immune function, and evidence of immune dysregulation in NDDs is increasing	[115–117,130–133]
Convergence of environmental and genetic factors on signaling pathways	Many genetic risk factors for NDDs converge on several major signaling pathways that critically regulate synaptic connectivity in the developing brain, and environmental chemicals have been identified that target these same signaling pathways	[13,29,65]

neurodevelopmental disorders [97–101], little data as yet causally link environmental chemicals to neurodevelopmental disorders via epigenetic mechanisms. One of the few potentially relevant examples is the recent demonstration that the nondioxin-like (NDL) PCB congener, PCB 95, increased synaptogenesis via upregulated expression of miR132 [102]. miR132 represses expression of the transcriptional repressor, methyl-CpG-binding protein-2 (MeCP2) [103]. MeCP2 dysfunction is linked to dendritic and synaptic dysregulation [104], and mutations in MeCP2 are the genetic cause of Rett syndrome and are associated with increased ASD risk [105–107]. miR132 also interacts with Fragile X mental retardation protein to regulate synapse formation [108], and more recently, expression of miR132 was shown to be dysregulated in schizophrenia [109,110]. Intriguingly, a recent analysis of persistent organic pollutants in children's brains found significantly higher levels of PCB 95, the same congener found to influence miR132 levels, in postmortem brains of children with a syndromic form of autism, as compared to neurotypical controls [111]. In contrast, levels of other PCBs and PBDEs were comparable across groups. The samples with detectable PCB 95 levels were almost exclusively those with maternal 15q11-q13 duplication (Dup15q) or deletion in Prader–Willi syndrome. When sorted by birth year, Dup15q samples represented five of six samples with detectable

PCB 95 levels and a known genetic cause of ASD born after the 1976 ban on PCB production. Dup15q was the strongest predictor of PCB 95 exposure across age, gender, or year of birth. Dup15q brain samples had lower levels of repetitive DNA methylation as measured by LINE-1 pyrosequencing, although methylation levels were confounded by year of birth. Collectively, these results suggest the possibility that NDL PCBs contribute to an increased risk of neurodevelopmental disorders via epigenetic mechanisms. This is of potential significance because recent studies clearly demonstrate that NDL PCBs remain a current and significant risk to the developing human brain. First, contemporary unintentional sources of NDL PCBs have been identified, most notably commercial paint pigments [112]. Second, in the past year it was reported that PCB levels in the indoor air of elementary schools in the United States exceed the EPA's 2009 public health guidelines, and the NDL congener PCB 95 was the second most abundant congener identified in indoor school environments [113]. Third, the latest NHANES study confirmed widespread exposure to PCBs among women of childbearing age living in the United States [114].

Immune dysregulation. Cross talk between the nervous and immune systems is essential for normal neurodevelopment [115–117]. Multiple cytokines are normally produced in the healthy brain where they play critical roles in most aspects of neurodevelopment, including neuronal differentiation, synapse formation, and structural plasticity [117,118]. Two cytokine classes of particular interest in neurodevelopmental disorders include the neuropoietic cytokines, exemplified by IL-6 [115,119], and the type II interferon, gamma interferon (IFNγ) [120]. The neuropoietic cytokines interact directly with diverse neuronal cell types in vitro to modulate dendritic growth and synapse formation [121–123], and in vivo to regulate synapse elimination in the developing rodent brain [124]. Chronic IL-6 overexpression reduces neuronal expression of glutamate receptors and L-type Ca^{2+} channels in vitro and in vivo [125]. IFNγ also modulates dendritic growth via effects on signal transducer and activator of transcription (STAT)1 [126] and mitogen-activated protein kinase (MAPK) signaling [127]. Increased IFNγ skews the balance of excitatory to inhibitory synapses toward increased excitatory signaling in cultured hippocampal neurons [128], whereas decreased IFNγ triggers decreased density of presynaptic terminals in the spinal cord of adult mice [129]. Based on these observations, it is perhaps not surprising that a growing body of evidence supports a key role for immune system dysregulation in neurodevelopmental disorders [120,130–133]. As discussed previously, maternal infection during pregnancy is associated with increased risk of neurodevelopmental disorders [115], and experimental evidence confirms that many cytokines can cross the placenta and blood–brain barrier [119]. Collectively, such studies illustrate the potential for altered cytokine profiles to adversely impact neurodevelopment during gestation. Whether immune and nervous system dysfunction in neurodevelopmental disorders reflect parallel outcomes or whether deficits in one system contribute to deficits in the other remain open to speculation [115]. Another unanswered question is whether environmental chemicals known to influence immune status can cause adverse neurodevelopmental outcomes of relevance to ASD and other disorders by altering cytokine profiles in the developing brain.

Convergence of environmental factors and genes on critical signaling pathways. One fundamental way in which heritable genetic vulnerabilities might amplify the adverse effects triggered by environmental exposures is if both factors (genes × environment) converge to dysregulate the same neurotransmitter and/or signaling systems at critical times during development [29,65].

Evidence from both human genetics studies and experimental models expressing genetic risk factors has yielded the startling fact that many ASD risk genes converge on several major signaling pathways, which play key roles in regulating neuronal connectivity in the developing brain [25,69,82,134,135]. Data emerging from studies of syndromic ASD and rare highly penetrant mutations or CNVs in ASD have identified distinct adhesion proteins needed to maintain and modify synapses in response to experience, as well as intracellular signaling pathways that control dendritic arborization and/or synaptogenesis signaling systems that appear to represent convergent molecular mechanisms in ASD. The former includes neuroligins, neurexins, contactin-associated protein-2 (CNTNAP2), and SH3 and multiple ankyrin repeat domains (SHANK) proteins [24,26,69,136]; the latter, ERK and PI3K signaling [25], as well as Ca^{2+}-dependent signaling [134]. Many of these same signaling pathways have also been implicated in schizophrenia risk [73,82–84].

Dendritic arborization and synaptic connectivity are determined by the interplay between genetic and environmental factors [137–139]. During normal brain development, neuronal activity is a predominant environmental factor in shaping dendritic arbors and synaptic connectivity. The importance of neuronal activity is evidenced by the remarkable effect of experience on the development and refinement of synaptic connections, which not only patterns neural circuitry during development [140,141] but also underlies associative learning throughout life [142]. The dynamic structural remodeling of dendrites, which occurs during development, is driven in large part by Ca^{2+}-dependent signaling pathways triggered by N-methyl-D-aspartate (NMDA) receptor activation and other extrinsic cues, such as hormones and neurotrophins [143,144]. NMDA receptor-mediated Ca^{2+}-dependent signaling couples neuronal activity to dendritic arborization, and this is mediated by sequential activation of calmodulin-dependent protein kinase kinase (CaMKK), CaMKI, and MAPK ERK kinases/extracellular signal-regulated kinase (MEK/ERK) to enhance cAMP response element-binding protein (CREB)-mediated transcription of Wnt2 [145]. A number of environmental chemicals modulate calcium signaling [13], but thus far, only NDL PCBs that bind to the ryanodine receptor (RyR) have been found to trigger this same Ca^{2+}-dependent signaling pathway to enhance dendritic growth in primary cultures of rat hippocampal neurons [146]. These effects of PCB 95 map onto signaling pathways implicated in neurodevelopmental disorders. For example, Timothy syndrome, which is caused by a gain-of-function mutation in the L-type Ca^{2+} channel CaV1.2, has a 100% incidence of neurodevelopmental disorders and a 60% rate of autism [147]. Neurons differentiated from induced pluripotent stem cells derived from Timothy syndrome patients exhibit increased Ca^{2+} oscillations and upregulated expression of genes linked to Ca^{2+}-dependent regulation of CREB, including CaMK [148]. Collectively, these studies support the hypothesis that NDL PCBs amplify the risk and/or severity of neurodevelopmental disorder by converging on signaling pathways targeted by heritable defects in Ca^{2+}-dependent signaling pathways that regulate neuronal connectivity (Figure 1.2).

The hypothesis of an imbalance between excitation and inhibition within developing CNS circuits as a common etiological factor contributing to autism, Fragile X, and schizophrenia is supported by several reports of gamma-aminobutyric acid (GABA) receptor abnormalities in these neurodevelopmental disorders [149]. A significantly lower level of GABRβ3 expression has been documented in postmortem brain samples obtained from children diagnosed with autism, Rett syndrome, and Angelman syndrome [150]. The gene encoding for

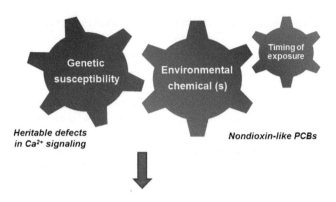

FIGURE 1.2 **Interactions between genetic susceptibilities and environmental factors likely determine the risk and/or phenotypic expression of many neurodevelopmental disorders.** If the environmental factor is a chemical, then the timing of exposure relative to the stage of neurodevelopment may influence the impact of the environmental chemical. The mechanism(s) by which environmental chemicals modify genetic risks to influence neurodevelopmental outcome are largely unknown. One possibility illustrated in this figure is that the genetic factor (heritable defects in Ca^{2+} signaling) and environmental factor (nondioxin-like PCBs) converge on the same critical signaling pathway (Ca^{2+}-dependent activation of the CaMK–CREB–Wnt signaling pathway) to amplify adverse effects on a neurodevelopmental endpoint (synaptic connectivity) of relevance to neurodevelopmental disorders.

GABRβ$_3$ is located in a nonimprinting region of chromosome 15q11-13 suggesting an overlapping pathway of gene dysregulation common to all three disorders that may stem from abnormal regulation of DNA methylation by the MeCP2 [151]. Quantitative immunoblot analyses of the brains of MeCP2 deficient mice, a model of human Rett syndrome, are deficient in the GABRβ$_3$ subunit [150], further supporting the working hypothesis that epigenetic mechanisms conferring gene dysregulation within 15q11-q13 in Angelman syndrome and autism may significantly contribute to heightened susceptibility to compounds that potently block the chloride channel of GABRβ$_3$. Autism has been linked to polymorphisms within additional genes that encode GABR, including GABRγ$_1$ located on 15q11-13 [152,153]. Complex epistatic interactions between genes that encode GABRα$_4$ and GABRβ$_1$ within 4q12 have also been reported in autistic probands [154].

Collectively, these data implicate GABA receptor dysregulation as a major contributor to imbalances between excitation and inhibition in children whose brains may be vulnerable to environmental modulation of epigenetic mechanisms and/or GABA receptor expression and/or function. Relevant to the latter are the OC pesticides that possess polychloroalkane structures and are known to bind to GABR in the mammalian brain, many with nanomolar affinity, to potently block Cl$^-$ conductance through the receptor [14]. In the United States, toxaphene represents one of the most heavily used OC pesticides but it was banned in 1990; the OC pesticides heptachlor and dieldrin were banned in the late 1980s. Examples of more recently used polychloroalkane insecticides include endosulfan and lindane. Dicofol, an OC structure similar to DDT, is currently registered for use on agricultural crops. Because of their chemical stability, global distribution from countries that continue to use these compounds, and their propensity to bioaccumulate, exposures to OC insecticides continue to be a significant concern to human health worldwide. Another class of insecticide that interferes with GABA neurotransmission and has attained broad domestic and commercial use is the 4-alkyl-1-phenylpyrazoles. Although these compounds do not persist in the environment to the same extent as OC pesticides, they are heavily used within the home and for commercial pest control. For example, fipronil is formulated as a topical for control of fleas and ticks on pets. Fipronil's insecticidal activity is mediated primarily through its actions as a noncompetitive antagonist of GABR. In this regard, fipronil was initially developed as an insecticide because of its higher selectivity for insect vs. mammalian GABR [155]. However, results from recent studies indicate that fipronil, like endosulfan and lindane, is indeed a high-affinity noncompetitive antagonist for the mammalian β_3-homopentameric GABR [156,157]. In fact, fipronil binds the GABRβ_3 with nanomolar affinity in a manner indistinguishable from its interaction with insect GABA receptors. These findings raise questions about the high degree of selectivity originally attributed to 4-alkyl-1-phenylpyrazoles. More importantly, the finding that a diverse group of widely used insecticides converges on a common molecular target—the mammalian GABRβ_3—has important implications for human risk, especially in those individuals with heritable impairments in GABA receptor signaling pathways.

CONCLUSIONS

A significant contribution from environmental factors in determining the risk for neurodevelopmental disorders is consistent with both the rapid increase in prevalence of these disorders and the clinical heterogeneity, which is the hallmark of the two most common complex disorders, ASD and ADHD. However, this phenotypic heterogeneity together with the complex multigenic etiologies of these disorders significantly increases the difficulty of identifying specific environmental factors that confer increased risk. If the complexity of heritable factors contributing to susceptibility to neurodevelopmental disorders creates a range of sensitivities of the developing brain to the adverse effects of environmental factors [11,25,29], then establishing clear associations between exposure to environmental factors and diagnosis of specific neurodevelopmental disorders will be challenging.

Several signaling pathways and developmental processes have emerged from genetic studies as convergent molecular mechanisms or targets in neurodevelopmental disorders, and data are already accumulating of environmental chemicals that potentially interact with those mechanisms or targets. These emerging data suggest a biological framework for not only studying mechanisms by which specific environmental and genetic factors interact to confer risk for neurodevelopmental orders but also for informing epidemiologic studies and for setting up in vitro and simple model systems to screen environmental chemicals for their potential to increase risk of neurodevelopmental disorders.

Clearly, work is urgently needed to better predict which combination of defective genes and environmental exposures poses the greatest risk for neurodevelopmental disorders. The fact that chemical exposures are more readily controlled than genetic factors to prevent or mitigate the expression of phenotypic traits related to neurodevelopmental disorders coupled with the significant toll that neurodevelopmental disorders and disabilities exact on the affected individual, their families, and society, provide compelling reasons to engage in this endeavor.

References

[1] Bloom B, Cohen RA, Freeman G. Summary health statistics for U.S. children: national health interview survey, 2009. Vital Health Stat 2010;10:1–82.

[2] Landrigan PJ, Lambertini L, Birnbaum LS. A research strategy to discover the environmental causes of autism and neurodevelopmental disabilities. Environ Health Perspect 2012;120:a258–60.

[3] Polanczyk G, de Lima MS, Horta BL, Biederman J, Rohde LA. The worldwide prevalence of ADHD: a systematic review and metaregression analysis. Am J Psychiatry 2007;164:942–8.

[4] CDC. In: Prevalence of autism spectrum disorder among children aged 8 years — autism and developmental disabilities monitoring network, 11 sites, United States, 2010. C.f.D.C.a. Prevention; 2014.

[5] Grandjean P, Landrigan PJ. Neurobehavioural effects of developmental toxicity. Lancet Neurol 2014;13: 330–8.

[6] Trasande L, Liu Y. Reducing the staggering costs of environmental disease in children, estimated at $76.6 billion in 2008. Health Aff (Millwood) 2011;30:863–70.

[7] Bellinger DC. A strategy for comparing the contributions of environmental chemicals and other risk factors to neurodevelopment of children. Environ Health Perspect 2012;120:501–7.

[8] Grandjean P, Pichery C, Bellanger M, Budtz-Jorgensen E. Calculation of mercury's effects on neurodevelopment. Environ Health Perspect 2012;120:A452. author reply A52.

[9] Autism Genome Project C, Szatmari P, Paterson AD, Zwaigenbaum L, Roberts W, Brian J, et al. Mapping autism risk loci using genetic linkage and chromosomal rearrangements. Nat Genet 2007;39:319–28.

[10] Buxbaum JD, Hof PR. The emerging neuroscience of autism spectrum disorders. Brain Res 2011;1380:1–2.

[11] Herbert MR. Contributions of the environment and environmentally vulnerable physiology to autism spectrum disorders. Curr Opin Neurol 2010;23:103–10.

[12] Hallmayer J, Cleveland S, Torres A, Phillips J, Cohen B, Torigoe T, et al. Genetic heritability and shared environmental factors among twin pairs with autism. Arch Gen Psychiatry 2011;68:1095–102.

[13] Pessah IN, Cherednichenko G, Lein PJ. Minding the calcium store: ryanodine receptor activation as a convergent mechanism of PCB toxicity. Pharmacol Ther 2010;125:260–85.

[14] Shelton JF, Hertz-Picciotto I, Pessah IN. Tipping the balance of autism risk: potential mechanisms linking pesticides and autism. Environ Health Perspect 2012;120:944–51.

[15] Zuk O, Hechter E, Sunyaev SR, Lander ES. The mystery of missing heritability: genetic interactions create phantom heritability. Proc Natl Acad Sci USA 2012;109:1193–8.

[16] Coo H, Ouellette-Kuntz H, Lloyd JE, Kasmara L, Holden JJ, Lewis ME. Trends in autism prevalence: diagnostic substitution revisited. J Autism Dev Disord 2008;38:1036–46.

[17] Bishop DV, Whitehouse AJ, Watt HJ, Line EA. Autism and diagnostic substitution: evidence from a study of adults with a history of developmental language disorder. Dev Med Child Neurol 2008;50:341–5.

[18] Parner ET, Schendel DE, Thorsen P. Autism prevalence trends over time in Denmark: changes in prevalence and age at diagnosis. Arch Pediatr Adolesc Med 2008;162:1150–6.

[19] Hertz-Picciotto I, Delwiche L. The rise in autism and the role of age at diagnosis. Epidemiology 2009;20:84–90.

[20] Grether JK, Rosen NJ, Smith KS, Croen LA. Investigation of shifts in autism reporting in the California department of developmental services. J Autism Dev Disord 2009;39:1412–9.

[21] King M, Bearman P. Diagnostic change and the increased prevalence of autism. Int J Epidemiol 2009;38:1224–34.

[22] El-Fishawy P, State MW. The genetics of autism: key issues, recent findings, and clinical implications. Psychiatr Clin North Am 2010;33:83–105.

[23] Geschwind DH. Genetics of autism spectrum disorders. Trends Cogn Sci 2011;15:409–16.

[24] Abrahams BS, Geschwind DH. Advances in autism genetics: on the threshold of a new neurobiology. Nat Rev Genet 2008;9:341–55.

[25] Levitt P, Campbell DB. The genetic and neurobiologic compass points toward common signaling dysfunctions in autism spectrum disorders. J Clin Invest 2009;119:747–54.

[26] O'Roak BJ, State MW. Autism genetics: strategies, challenges, and opportunities. Autism Res 2008;1:4–17.

[27] Veenstra-Vanderweele J, Christian SL, Cook Jr EH. Autism as a paradigmatic complex genetic disorder. Annu Rev Genomics Hum Genet 2004;5:379–405.

[28] Judson MC, Eagleson KL, Levitt P. A new synaptic player leading to autism risk: met receptor tyrosine kinase. J Neurodev Disord 2011;3:282–92.

[29] Pessah IN, Lein PJ. Evidence for environmental susceptibility in autism: what we need to know about gene×environment interactions. In: Zimmerman A, editor. Autism: current theories and evidence. Totowa (NJ): Humana Press; 2008. pp. 409–28.

[30] Smith AK, Mick E, Faraone SV. Advances in genetic studies of attention-deficit/hyperactivity disorder. Curr Psychiatry Rep 2009;11:143–8.

[31] Sandin S, Lichtenstein P, Kuja-Halkola R, Larsson H, Hultman CM, Reichenberg A. The familial risk of autism. JAMA 2014;311:1770–7.

[32] Polanska K, Jurewicz J, Hanke W. Exposure to environmental and lifestyle factors and attention-deficit/hyperactivity disorder in children – a review of epidemiological studies. Int J Occup Med Environ Health 2012;25:330–55.

[33] Chess S, Fernandez P, Korn S. Behavioral consequences of congenital rubella. J Pediatr 1978;93:699–703.

[34] Brown AS. Epidemiologic studies of exposure to prenatal infection and risk of schizophrenia and autism. Dev Neurobiol 2012;72:1272–6.

[35] Canetta SE, Brown AS. Prenatal infection, maternal immune activation, and risk for schizophrenia. Transl Neurosci 2012;3:320–7.

[36] Zerbo O, Iosif AM, Walker C, Ozonoff S, Hansen RL, Hertz-Picciotto I. Is maternal influenza or fever during pregnancy associated with autism or developmental delays? Results from the charge (childhood autism risks from genetics and environment) study. J Autism Dev Disord 2013;43:25–33.

[37] Ohkawara T, Katsuyama T, Ida-Eto M, Narita N, Narita M. Maternal viral infection during pregnancy impairs development of fetal serotonergic neurons. Brain Dev 2015;37:88–93.

[38] Bale TL, Baram TZ, Brown AS, Goldstein JM, Insel TR, McCarthy MM, et al. Early life programming and neurodevelopmental disorders. Biol Psychiatry 2010;68:314–9.

[39] Lupien SJ, McEwen BS, Gunnar MR, Heim C. Effects of stress throughout the lifespan on the brain, behaviour and cognition. Nat Rev Neurosci 2009;10:434–45.

[40] Elovitz MA, Brown AG, Breen K, Anton L, Maubert M, Burd I. Intrauterine inflammation, insufficient to induce parturition, still evokes fetal and neonatal brain injury. Int J Dev Neurosci 2011;29:663–71.

[41] Uher R. Gene-environment interactions in severe mental illness. Front Psychiatry 2014;5:48.

[42] Hamlyn J, Duhig M, McGrath J, Scott J. Modifiable risk factors for schizophrenia and autism–shared risk factors impacting on brain development. Neurobiol Dis 2013;53:3–9.

[43] Idring S, Magnusson C, Lundberg M, Ek M, Rai D, Svensson AC, et al. Parental age and the risk of autism spectrum disorders: findings from a Swedish population-based cohort. Int J Epidemiol 2014;43:107–15.

[44] Parner ET, Baron-Cohen S, Lauritsen MB, Jorgensen M, Schieve LA, Yeargin-Allsopp M, et al. Parental age and autism spectrum disorders. Ann Epidemiol 2012;22:143–50.

[45] Shelton JF, Tancredi DJ, Hertz-Picciotto I. Independent and dependent contributions of advanced maternal and paternal ages to autism risk. Autism Res 2010;3:30–9.

[46] Geier DA, Hooker BS, Kern JK, Sykes LK, Geier MR. An evaluation of the effect of increasing parental age on the phenotypic severity of autism spectrum disorder. J Child Neurol 2014 Aug 27: pii: 0883073814541478.

[47] Krakowiak P, Walker CK, Bremer AA, Baker AS, Ozonoff S, Hansen RL, et al. Maternal metabolic conditions and risk for autism and other neurodevelopmental disorders. Pediatrics 2012;129:e1121–8.

[48] Schmidt RJ, Hansen RL, Hartiala J, Allayee H, Schmidt LC, Tancredi DJ, et al. Prenatal vitamins, one-carbon metabolism gene variants, and risk for autism. Epidemiology 2011;22:476–85.

[49] Schmidt RJ, Tancredi DJ, Ozonoff S, Hansen RL, Hartiala J, Allayee H, et al. Maternal periconceptional folic acid intake and risk of autism spectrum disorders and developmental delay in the charge (childhood autism risks from genetics and environment) case-control study. Am J Clin Nutr 2012;96:80–9.

[50] Mills JL, Hediger ML, Molloy CA, Chrousos GP, Manning-Courtney P, Yu KF, et al. Elevated levels of growth-related hormones in autism and autism spectrum disorder. Clin Endocrinol (Oxf) 2007;67:230–7.

[51] Kaminsky Z, Wang SC, Petronis A. Complex disease, gender and epigenetics. Ann Med 2006;38:530–44.

[52] Rodier PM, Ingram JL, Tisdale B, Croog VJ. Linking etiologies in humans and animal models: studies of autism. Reprod Toxicol 1997;11:417–22.

[53] Arndt TL, Stodgell CJ, Rodier PM. The teratology of autism. Int J Dev Neurosci 2005;23:189–99.

[54] Miller MT, Stromland K, Ventura L, Johansson M, Bandim JM, Gillberg C. Autism associated with conditions characterized by developmental errors in early embryogenesis: a mini review. Int J Dev Neurosci 2005;23:201–19.

[55] Christensen J, Gronborg TK, Sorensen MJ, Schendel D, Parner ET, Pedersen LH, et al. Prenatal valproate exposure and risk of autism spectrum disorders and childhood autism. JAMA 2013;309:1696–703.

[56] Croen LA, Connors SL, Matevia M, Qian Y, Newschaffer C, Zimmerman AW. Prenatal exposure to beta2-adrenergic receptor agonists and risk of autism spectrum disorders. J Neurodev Disord 2011;3:307–15.

[57] Malm H, Artama M, Brown AS, Gissler M, Gyllenberg D, Hinkka-Yli-Salomaki S, et al. Infant and childhood neurodevelopmental outcomes following prenatal exposure to selective serotonin reuptake inhibitors: overview and design of a Finnish Register-Based Study (FinESSI). BMC Psychiatry 2012;12:217.

[58] Rodier P, Miller RK, Brent RL. Does treatment of premature labor with terbutaline increase the risk of autism spectrum disorders? Am J Obstet Gynecol 2011;204:91–4.

[59] Grandjean P, Landrigan PJ. Developmental neurotoxicity of industrial chemicals. Lancet 2006;368:2167–78.

[60] Guilarte TR, Opler M, Pletnikov M. Is lead exposure in early life an environmental risk factor for schizophrenia? Neurobiological connections and testable hypotheses. Neurotoxicology 2012;33:560–74.

[61] Volk HE, Hertz-Picciotto I, Delwiche L, Lurmann F, McConnell R. Residential proximity to freeways and autism in the charge study. Environ Health Perspect 2011;119:873–7.

[62] Kalkbrenner AE, Schmidt RJ, Penlesky AC. Environmental chemical exposures and autism spectrum disorders: a review of the epidemiological evidence. Curr Probl Pediatr Adolesc Health Care 2014;44:277–318.

[63] Mendola P, Selevan SG, Gutter S, Rice D. Environmental factors associated with a spectrum of neurodevelopmental deficits. Ment Retard Dev Disabil Res Rev 2002;8:188–97.

[64] Eubig PA, Aguiar A, Schantz SL. Lead and PCBs as risk factors for attention deficit/hyperactivity disorder. Environ Health Perspect 2010;118:1654–67.

[65] Stamou M, Streifel KM, Goines PE, Lein PJ. Neuronal connectivity as a convergent target of gene×environment interactions that confer risk for autism spectrum disorders. Neurotoxicol Teratol 2013;36:3–16.

[66] Visser JC, Rommelse N, Vink L, Schrieken M, Oosterling IJ, van der Gaag RJ, et al. Narrowly versus broadly defined autism spectrum disorders: differences in pre- and perinatal risk factors. J Autism Dev Disord 2013;43:1505–16.

[67] Bill BR, Geschwind DH. Genetic advances in autism: heterogeneity and convergence on shared pathways. Curr Opin Genet Dev 2009;19:271–8.

[68] Hu VW. From genes to environment: using integrative genomics to build a "systems-level" understanding of autism spectrum disorders. Child Dev 2013;84:89–103.

[69] Bourgeron T. A synaptic trek to autism. Curr Opin Neurobiol 2009;19:231–4.

[70] Geschwind DH, Levitt P. Autism spectrum disorders: developmental disconnection syndromes. Curr Opin Neurobiol 2007;17:103–11.

[71] Rubenstein JL, Merzenich MM. Model of autism: increased ratio of excitation/inhibition in key neural systems. Genes Brain Behav 2003;2:255–67.

[72] Garey L. When cortical development goes wrong: schizophrenia as a neurodevelopmental disease of microcircuits. J Anat 2010;217:324–33.

[73] Penzes P, Cahill ME, Jones KA, VanLeeuwen JE, Woolfrey KM. Dendritic spine pathology in neuropsychiatric disorders. Nat Neurosci 2011;14:285–93.

[74] Svitkina T, Lin WH, Webb DJ, Yasuda R, Wayman GA, Van Aelst L, et al. Regulation of the postsynaptic cytoskeleton: roles in development, plasticity, and disorders. J Neurosci 2010;30:14937–42.

[75] Delorme R, Ey E, Toro R, Leboyer M, Gillberg C, Bourgeron T. Progress toward treatments for synaptic defects in autism. Nat Med 2013;19:685–94.

[76] Belmonte MK, Bourgeron T. Fragile X syndrome and autism at the intersection of genetic and neural networks. Nat Neurosci 2006;9:1221–5.

[77] Hutsler JJ, Zhang H. Increased dendritic spine densities on cortical projection neurons in autism spectrum disorders. Brain Res 2010;1309:83–94.

[78] Irwin SA, Patel B, Idupulapati M, Harris JB, Crisostomo RA, Larsen BP, et al. Abnormal dendritic spine characteristics in the temporal and visual cortices of patients with fragile-X syndrome: a quantitative examination. Am J Med Genet 2001;98:161–7.

[79] Glausier JR, Lewis DA. Dendritic spine pathology in schizophrenia. Neuroscience 2013;251:90–107.

[80] Watanabe Y, Khodosevich K, Monyer H. Dendrite development regulated by the schizophrenia-associated gene FEZ1 involves the ubiquitin proteasome system. Cell Rep 2014;7:552–64.

[81] Baribeau DA, Anagnostou E. A comparison of neuroimaging findings in childhood onset schizophrenia and autism spectrum disorder: a review of the literature. Front Psychiatry 2013;4:175.

[82] Zoghbi HY, Bear MF. Synaptic dysfunction in neurodevelopmental disorders associated with autism and intellectual disabilities. Cold Spring Harb Perspect Biol 2012;4: pii: a009886. http://dx.doi.org/10.1101/cshperspect.a009886.

[83] Gauthier J, Siddiqui TJ, Huashan P, Yokomaku D, Hamdan FF, Champagne N, et al. Truncating mutations in NRXN2 and NRXN1 in autism spectrum disorders and schizophrenia. Hum Genet 2011;130:563–73.

[84] Okerlund ND, Cheyette BN. Synaptic Wnt signaling-a contributor to major psychiatric disorders? J Neurodev Disord 2011;3:162–74.

[85] Olney JW, Farber NB, Wozniak DF, Jevtovic-Todorovic V, Ikonomidou C. Environmental agents that have the potential to trigger massive apoptotic neurodegeneration in the developing brain. Environ Health Perspect 2000;108(Suppl. 3):383–8. full.html 88olney/abstract.html.

[86] Lein P, Silbergeld E, Locke P, Goldberg AM. In vitro and other alternative approaches to developmental neurotoxicity testing (DNT). Environ Toxicol Pharmacol 2005;19:735–44.

[87] Baron-Cohen S, Lombardo MV, Auyeung B, Ashwin E, Chakrabarti B, Knickmeyer R. Why are autism spectrum conditions more prevalent in males? PLoS Biol 2011;9:e1001081.

[88] Williams GR. Neurodevelopmental and neurophysiological actions of thyroid hormone. J Neuroendocrinol 2008;20:784–94.

[89] Weiss B. The intersection of neurotoxicology and endocrine disruption. Neurotoxicology 2012;33:1410–9.

[90] Hornig M. The role of microbes and autoimmunity in the pathogenesis of neuropsychiatric illness. Curr Opin Rheumatol 2013;25:488–795.

[91] Mulle JG, Sharp WG, Cubells JF. The gut microbiome: a new frontier in autism research. Curr Psychiatry Rep 2013;15:337.

[92] Giulivi C, Zhang YF, Omanska-Klusek A, Ross-Inta C, Wong S, Hertz-Picciotto I, et al. Mitochondrial dysfunction in autism. JAMA 2010;304:2389–96.

[93] Rossignol DA, Frye RE. Evidence linking oxidative stress, mitochondrial dysfunction, and inflammation in the brain of individuals with autism. Front Physiol 2014;5:150.

[94] Colvin SM, Kwan KY. Dysregulated nitric oxide signaling as a candidate mechanism of fragile X syndrome and other neuropsychiatric disorders. Front Genet 2014;5:239.

[95] Kaur P, Armugam A, Jeyaseelan K. MicroRNAs in neurotoxicity. J Toxicol 2012;2012:870150.

[96] Tal TL, Tanguay RL. Non-coding RNAs–novel targets in neurotoxicity. Neurotoxicology 2012;33:530–44.

[97] Szyf M. How do environments talk to genes? Nat Neurosci 2013;16:2–4.

[98] Sun E, Shi Y. MicroRNAs: small molecules with big roles in neurodevelopment and diseases. Exp Neurol 2014: pii: S0014–4886(14)00257-X. http://dx.doi.org/10.1016/j.expneurol.2014.08.005.

[99] LaSalle JM, Powell WT, Yasui DH. Epigenetic layers and players underlying neurodevelopment. Trends Neurosci 2013;36:460–70.

[100] Nowak JS, Michlewski G. miRNAs in development and pathogenesis of the nervous system. Biochem Soc Trans 2013;41:815–20.

[101] Millan MJ. An epigenetic framework for neurodevelopmental disorders: from pathogenesis to potential therapy. Neuropharmacology 2013;68:2–82.

[102] Lesiak A, Zhu M, Chen H, Appleyard SM, Impey S, Lein PJ, et al. The environmental neurotoxicant PCB 95 promotes synaptogenesis via ryanodine receptor-dependent miR132 upregulation. J Neurosci 2014;34:717–25.

[103] Klein ME, Lioy DT, Ma L, Impey S, Mandel G, Goodman RH. Homeostatic regulation of MeCP2 expression by a CREB-induced microRNA. Nat Neurosci 2007;10:1513–4.

[104] Zhou Z, Hong EJ, Cohen S, Zhao WN, Ho HY, Schmidt L, et al. Brain-specific phosphorylation of MeCP2 regulates activity-dependent Bdnf transcription, dendritic growth, and spine maturation. Neuron 2006;52:255–69.

[105] LaSalle JM, Yasui DH. Evolving role of MeCP2 in Rett syndrome and autism. Epigenomics 2009;1:119–30.

[106] Percy AK. Rett syndrome: exploring the autism link. Arch Neurol 2011;68:985–9.

[107] Cukier HN, Lee JM, Ma D, Young JI, Mayo V, Butler BL, et al. The expanding role of MBD genes in autism: identification of a MECP2 duplication and novel alterations in MBD5, MBD6, and SETDB1. Autism Res 2012;5:385–97.

[108] Edbauer D, Neilson JR, Foster KA, Wang CF, Seeburg DP, Batterton MN, et al. Regulation of synaptic structure and function by FMRP-associated microRNAs miR-125b and miR-132. Neuron 2010;65:373–84.

[109] Kim AH, Reimers M, Maher B, Williamson V, McMichael O, McClay JL, et al. MicroRNA expression profiling in the prefrontal cortex of individuals affected with schizophrenia and bipolar disorders. Schizophr Res 2010;124:183–91.

[110] Miller BH, Zeier Z, Xi L, Lanz TA, Deng S, Strathmann J, et al. MicroRNA-132 dysregulation in schizophrenia has implications for both neurodevelopment and adult brain function. Proc Natl Acad Sci USA 2012;109:3125–30.

[111] Mitchell MM, Woods R, Chi LH, Schmidt RJ, Pessah IN, Kostyniak PJ, et al. Levels of select PCB and PBDE congeners in human postmortem brain reveal possible environmental involvement in 15q11-q13 duplication autism spectrum disorder. Environ Mol Mutagen 2012;53:589–98.

[112] Hu D, Hornbuckle KC. Inadvertent polychlorinated biphenyls in commercial paint pigments. Environ Sci Technol 2010;44:2822–7.

[113] Thomas K, Xue J, Williams R, Jones P, Whitaker D. Polychlorinated biphenyls (PCBs) in school buildings: sources, environmental levels, and exposures. U.S Environ Prot Agency 2012:150.

[114] Thompson MR, Boekelheide K. Multiple environmental chemical exposures to lead, mercury and polychlorinated biphenyls among childbearing-aged women (NHANES 1999–2004): body burden and risk factors. Environ Res 2013;121:23–30.

[115] Patterson PH. Maternal infection and immune involvement in autism. Trends Mol Med 2011;17:389–94.

[116] Careaga M, Van de Water J, Ashwood P. Immune dysfunction in autism: a pathway to treatment. Neurotherapeutics 2010;7:283–92.

[117] Deverman BE, Patterson PH. Cytokines and CNS development. Neuron 2009;64:61–78.

[118] Garay PA, McAllister AK. Novel roles for immune molecules in neural development: implications for neurodevelopmental disorders. Front Synaptic Neurosci 2010;2:136.

[119] Goines PE, Ashwood P. Cytokine dysregulation in autism spectrum disorders (ASD): possible role of the environment. Neurotoxicol Teratol 2013;36:67–81.

[120] Goines P, Van de Water J. The immune system's role in the biology of autism. Curr Opin Neurol 2010;23:111–7.

[121] Gadient RA, Lein P, Higgins D, Patterson PH. Effect of leukemia inhibitory factor (LIF) on the morphology and survival of cultured hippocampal neurons and glial cells. Brain Res 1998;798:140–6.

[122] Guo X, Metzler-Northrup J, Lein P, Rueger D, Higgins D. Leukemia inhibitory factor and ciliary neurotrophic factor regulate dendritic growth in cultures of rat sympathetic neurons. Brain Res Dev Brain Res 1997;104:101–10.

[123] Guo X, Chandrasekaran V, Lein P, Kaplan PL, Higgins D. Leukemia inhibitory factor and ciliary neurotrophic factor cause dendritic retraction in cultured rat sympathetic neurons. J Neurosci 1999;19:2113–21.

[124] Morikawa Y, Tohya K, Tamura S, Ichihara M, Miyajima A, Senba E. Expression of interleukin-6 receptor, leukemia inhibitory factor receptor and glycoprotein 130 in the murine cerebellum and neuropathological effect of leukemia inhibitory factor on cerebellar Purkinje cells. Neuroscience 2000;100:841–8.

[125] Vereyken EJ, Bajova H, Chow S, de Graan PN, Gruol DL. Chronic interleukin-6 alters the level of synaptic proteins in hippocampus in culture and in vivo. Eur J Neurosci 2007;25:3605–16.

[126] Kim IJ, Beck HN, Lein PJ, Higgins D. Interferon gamma induces retrograde dendritic retraction and inhibits synapse formation. J Neurosci 2002;22:4530–9.

[127] Andres DA, Shi GX, Bruun D, Barnhart C, Lein PJ. Rit signaling contributes to interferon-gamma-induced dendritic retraction via p38 mitogen-activated protein kinase activation. J Neurochem 2008;107:1436–47.

[128] Vikman KS, Owe-Larsson B, Brask J, Kristensson KS, Hill RH. Interferon-gamma-induced changes in synaptic activity and AMPA receptor clustering in hippocampal cultures. Brain Res 2001;896:18–29.

[129] Victorio SC, Havton LA, Oliveira AL. Absence of IFNgamma expression induces neuronal degeneration in the spinal cord of adult mice. J Neuroinflammation 2010;7:77.

[130] Ashwood P, Van de Water J. Is autism an autoimmune disease? Autoimmun Rev 2004;3:557–62.

[131] Cohly HH, Panja A. Immunological findings in autism. Int Rev Neurobiol 2005;71:317–41.

[132] Enstrom AM, Van de Water JA, Ashwood P. Autoimmunity in autism. Curr Opin Investig Drugs 2009;10:463–73.

[133] Jyonouchi H, Geng L, Ruby A, Zimmerman-Bier B. Dysregulated innate immune responses in young children with autism spectrum disorders: their relationship to gastrointestinal symptoms and dietary intervention. Neuropsychobiology 2005;51:77–85.

[134] Krey JF, Dolmetsch RE. Molecular mechanisms of autism: a possible role for Ca^{2+} signaling. Curr Opin Neurobiol 2007;17:112–9.

[135] Pardo CA, Eberhart CG. The neurobiology of autism. Brain Pathol 2007;17:434–47.

[136] Sudhof TC. Neuroligins and neurexins link synaptic function to cognitive disease. Nature 2008;455:903–11.

[137] Jan YN, Jan LY. The control of dendrite development. Neuron 2003;40:229–42.

[138] McAllister AK. Cellular and molecular mechanisms of dendrite growth. Cereb Cortex 2000;10:963–73.

[139] Scott EK, Luo L. How do dendrites take their shape? Nat Neurosci 2001;4:359–65.

[140] Chen JL, Nedivi E. Neuronal structural remodeling: is it all about access? Curr Opin Neurobiol 2010;20:557–62.

[141] Cline HT. Dendritic arbor development and synaptogenesis. Curr Opin Neurobiol 2001;11:118–26.

[142] Pittenger C, Kandel ER. In search of general mechanisms for long-lasting plasticity: aplysia and the hippocampus. Philos Trans R Soc Lond B Biol Sci 2003;358:757–63.

[143] Lohmann C, Myhr KL, Wong RO. Transmitter-evoked local calcium release stabilizes developing dendrites. Nature 2002;418:177–81.

[144] Wayman GA, Lee YS, Tokumitsu H, Silva AJ, Soderling TR. Calmodulin-kinases: modulators of neuronal development and plasticity. Neuron 2008;59:914–31.

[145] Wayman GA, Impey S, Marks D, Saneyoshi T, Grant WF, Derkach V, et al. Activity-dependent dendritic arborization mediated by CaM-kinase I activation and enhanced CREB-dependent transcription of Wnt-2. Neuron 2006;50:897–909.

[146] Wayman GA, Bose DD, Yang D, Lesiak A, Bruun D, Impey S, et al. PCB-95 modulates the calcium-dependent signaling pathway responsible for activity-dependent dendritic growth. Environ Health Perspect 2012;120:1003–9.

[147] Splawski I, Timothy KW, Sharpe LM, Decher N, Kumar P, Bloise R, et al. Ca(V)1.2 calcium channel dysfunction causes a multisystem disorder including arrhythmia and autism. Cell 2004;119:19–31.

[148] Pasca SP, Portmann T, Voineagu I, Yazawa M, Shcheglovitov A, Pasca AM, et al. Using iPSC-derived neurons to uncover cellular phenotypes associated with Timothy syndrome. Nat Med 2011;17:1657–62.

[149] Deidda G, Bozarth IF, Cancedda L. Modulation of GABAergic transmission in development and neurodevelopmental disorders: investigating physiology and pathology to gain therapeutic perspectives. Front Cell Neurosci 2014;8:119.

[150] Samaco RC, Hogart A, LaSalle JM. Epigenetic overlap in autism-spectrum neurodevelopmental disorders: MECP2 deficiency causes reduced expression of UBE3A and GABRB3. Hum Mol Genet 2005;14:483–92.

[151] LaSalle JM, Hogart A, Thatcher KN. Rett syndrome: a Rosetta stone for understanding the molecular pathogenesis of autism. Int Rev Neurobiol 2005;71:131–65.

[152] Vincent JB, Horike SI, Choufani S, Paterson AD, Roberts W, Szatmari P, et al. An inversion inv(4)(p12-p15.3) in autistic siblings implicates the 4p GABA receptor gene cluster. J Med Genet 2006;43:429–34.

[153] Ashley-Koch AE, Mei H, Jaworski J, Ma DQ, Ritchie MD, Menold MM, et al. An analysis paradigm for investigating multi-locus effects in complex disease: examination of three GABA receptor subunit genes on 15q11-q13 as risk factors for autistic disorder. Ann Hum Genet 2006;70:281–92.

[154] Ma DQ, Whitehead PL, Menold MM, Martin ER, Ashley-Koch AE, Mei H, et al. Identification of significant association and gene-gene interaction of GABA receptor subunit genes in autism. Am J Hum Genet 2005;77:377–88.

[155] Bloomquist JR. Chloride channels as tools for developing selective insecticides. Arch Insect Biochem Physiol 2003;54:145–56.

[156] Chen L, Durkin KA, Casida JE. Structural model for gamma-aminobutyric acid receptor noncompetitive antagonist binding: widely diverse structures fit the same site. Proc Natl Acad Sci USA 2006;103:5185–90.

[157] Sammelson RE, Caboni P, Durkin KA, Casida JE. GABA receptor antagonists and insecticides: common structural features of 4-alkyl-1-phenylpyrazoles and 4-alkyl-1-phenyltrioxabicyclooctanes. Bioorg Med Chem 2004;12:3345–55.

Genetic Factors in Environmentally Induced Disease

John T. Greenamyre, Paul Barrett

INTRODUCTION

During the past few decades, great strides have been made to treat life-threatening diseases such as cancer, diabetes, and cardiovascular disease. A main outcome of these advancements is prolonged lifespan. An unfortunate consequence of an aging population is the increased occurrence of age-related diseases, specifically neurodegenerative diseases. Neurodegeneration is described as the loss of neuronal function and structure. Not only do neurodegenerative diseases create severe physical and emotional problems for both patient and family, but the steady increase in patients developing neurodegenerative diseases generates an economic burden on society. These facts highlight the pressing need to better understand the mechanism of neurodegenerative disease action, for developing better diagnostic methods as well as therapeutic measures.

Although numerous diseases could be characterized as neurodegenerative, many are extremely rare and result from purely inherited genetic mutations. Of the more common diseases, such as Parkinson's disease (PD), Alzheimer's disease (AD), and amyotrophic lateral sclerosis (ALS), the mechanism for disease onset is still poorly understood. Many genetic mutations have been linked to these diseases, but most are rare or show a low-penetrance level, indicating that other factors may be necessary to promote disease pathogenesis. The purpose of this chapter is to elucidate potential environmental contributors and how they may interact with genetic factors to promote neurodegenerative disease progression.

Gene–Environment Interactions

The interaction between an organism's genetic code and its surrounding environment is a key relationship that is often overlooked, but nonetheless important: individuals carry genetic factors that may confer susceptibility (or possible resistance) to certain disorders, and environmental factors can influence this susceptibility. Genetic factors refer to any component of the genetic code of an organism, whereas environmental factors refer to any influence that originates from outside the genome. A common example of this can be seen in the case of skin cancer. People who have inherited fair skin (genetic factor) have a higher susceptibility to developing skin cancer, and this risk is enhanced by prolonged exposure to the sun (environmental factor). Another example of gene–environment interactions can be seen in agriculture, in which plant species are bred to have altered genetic properties that allow them to adapt to specific environmental factors, such as sunlight or water levels.

When relating to neurodegenerative diseases, the concept of gene–environment interactions normally refers to how the interaction between genetic susceptibility and environmental toxins, such as pesticides, herbicides, insecticides, and so on, leads to increased risk for developing neurodegenerative disorders. Although it is true that certain neurodegenerative diseases can be characterized as purely genetic in nature, these account for a small percentage of total disease cases. These purely genetic forms of disease often result in a decrease in age of onset and are more severe in nature. The other form of disease is referred to as sporadic, and is often a combination of genetic susceptibility and environmental factors. Although sporadic forms of neurodegenerative diseases may include genetic abnormalities, such as gene mutations or changes in gene expansion length, these genetic changes do not account for disease onset alone. This concept is known as penetrance, which is defined as the percentage of people with a specific genetic change who develop an associated affliction. In the case of neurodegenerative diseases, penetrance is often quite low, indicating that other environmental factors (or genetic modifiers) are required for disease etiology. In further support of the concept that genes do not solely contribute to disease onset is the fact that multiple studies have shown that identical twins (with identical genetic codes) do not show the same disease outcome, indicating that environmental factors can influence disease progression [1].

When investigating the impact of toxins on the nervous system, two types of toxicant exposure are examined: acute and chronic. Acute exposure refers to the presence of neurological changes after short (hours to days) exposure to a toxicant. Acute exposures may or may not be reversible and may or may not result in cell death. Chronic exposure is defined as when a neurological phenotype is present decades after small exposures to a neurotoxicant, with exposures occurring over weeks to years. Many of the cases discussed in this chapter will highlight the incidence of neurodegeneration after chronic exposures, though it is possible that acute exposures can result in neurodegeneration.

No matter the type of exposure, toxins can function in two distinct manners. The first is that the toxin can interact with cells directly, often initiating cell death from within the cell. The second is that toxins can have an indirect effect, often by changing the metabolism of other substances, altering the permeability of protective membranes, overstimulating the immune system, or altering hormone signaling. Correlating toxicant exposure to disease onset is a difficult task as studies often are forced to rely on an individual's recollection on the duration, amount, and type of toxicant exposure, often resulting in recall bias. In addition, professionals exposed to potential harmful toxins are frequently exposed to multiple toxins over time, and the exposure criteria are often variable [2]. Despite these hurdles, advancements have been made correlating toxicant exposure with increased risk of neurodegenerative diseases.

Biological Components Involved in Neurodegenerative Diseases

Studying the interplay between genetic information and environmental exposures in the context of disease is a difficult task; moreover, the undertaking becomes increasingly complex in the framework of the human nervous system. Unlike other organs within the body, the brain is unable to change shape and size due to size constraints placed by the skull. This means that any disorder or environmental factor (toxin, injury, etc.) that produces swelling can be fatal due to the lack of compliance of the skull. Secondly, the brain has the highest energy requirement of any organ within the body. The brain is 2% of total body weight, but receives 15% of cardiovascular output and consumes nearly 20% of total body oxygen and 25% of total body glucose [3]. These energy requirements stem from the necessity of neuronal populations to transmit nervous signals and conduct extensive antero- and retrograde transport along axons. The energy requirements for these processes are met by the functions of the mitochondria. One of the functions of the mitochondria is to create energy for the cell (in the form of adenosine triphosphate (ATP)) through a series of steps utilizing five complexes of proteins. Mitochondria are unique cellular organelles in that, whereas they contain their own genome, about 99% of the proteins necessary for mitochondrial function are nuclear coded and imported. Changes in mitochondrial import can have a severe impact on mitochondrial function. Mitochondrial dysfunction has been identified as an early, and sometimes primary, event in multiple neurodegenerative diseases. This chapter will highlight many of the interactions between environmental toxins, genetic abnormalities, and mitochondrial dysfunction in neurodegeneration.

Despite the high demand for energy, the brain actually has a small energy reserve capacity. Thus, toxins that disrupt energy production, for example by disrupting mitochondrial function, can prove fatal. A well-known example of this is the poison cyanide, which inhibits complex IV of the mitochondria. Neurological changes can be noted within seconds of administration, demonstrating the importance of regulating brain energy requirements [4,5]. Importantly, once a neuron dies it is gone forever, as neurons do not regenerate or undergo mitosis. This was first noted by Cajal, who stated "Once the development was ended, the founts of growth and regeneration of the axons and dendrites dried up irrevocably. In the adult centers, the nerve paths are something fixed, ended, and immutable. Everything may die, nothing may be regenerated" [6].

To help protect the brain, several lines of defense have evolved. The first is known as the blood–brain barrier (BBB), which is a highly selective permeable barrier between the fluids in the circulatory system and the brain. The BBB is formed by a layer of endothelial cells that line cerebral microvessels [7]. The BBB acts as a physical barrier due to the connections between the endothelial cells, known as tight junctions. These junctions only allow small, lipophilic molecules to pass,

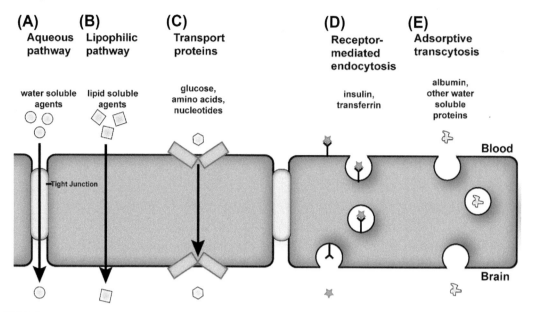

FIGURE 2.1 **Pathways across the blood–brain barrier.** A schematic diagram of the endothelial cells that form the blood–brain barrier (BBB). The main routes for molecular traffic across the BBB are shown. (A) Normally, the tight junctions severely restrict penetration of water-soluble compounds, including polar drugs. (B) Lipophilic molecules can cross through the BBB. (C) Transport proteins (carriers) move glucose, amino acids, purine bases, nucleosides, choline, and other substances across the BBB. Some transports are energy dependent (e.g., P-glycoprotein). (D) Certain proteins, such as insulin and transferrin, are taken up by specific receptor-mediated endocytosis and transcytosis. (E) Plasma proteins, such as albumin, are poorly transported, but can be transported by adsorptive-mediated endocytosis and transcytosis. Drug delivery across the brain endothelium depends on making use of pathways (B–E); most CNS drugs enter route (B). *Modified with permission from Ref. [7].*

such as O_2 or hydrophobic compounds, and all other molecules must be actively transported across. Movement in and out of the BBB occurs through diffusion, pinocytosis, carrier-mediated transport, and transcellular transport [8,9]. Methods of transport can be seen in Figure 2.1.

Although the BBB is a vital method of protecting the brain, it is not impervious to infiltration by toxins. Neurotoxicants that are lipid soluble, such as rotenone, can easily diffuse across the BBB. In addition, it is possible for genetic mutations to alter the permeability of the BBB. It has been shown that the protein P-glycoprotein, which regulates transport across the BBB, has altered expression levels in PD. In fact, the 3435T/T genotype showed decreased protein expression and function, and was present at the highest level in early-onset parkinsonian patients, and second highest in late-onset compared to age-matched controls [10]. Changes in P-glycoprotein expression can lead to increased BBB permeability, allowing potential toxins to enter the brain, which would normally be blocked. For example, it was shown that verapamil, which is normally excluded from the brain, was more readily absorbed into the brains of PD patients [11]. In addition, mice lacking the P-glycoprotein gene accumulate anticancer drugs, narcotics, and pesticides in brain tissue, all compounds that should be excluded by a properly intact BBB [12,13]. Figure 2.2 highlights methods of toxicant entry and potential interactions with various cell types found within the brain.

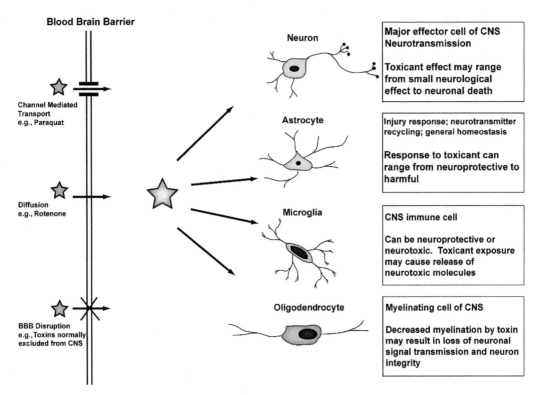

Blood Brain Barrier

Channel Mediated Transport
e.g., Paraquat

Diffusion
e.g., Rotenone

BBB Disruption
e.g., Toxins normally
excluded from CNS

Neuron

Major effector cell of CNS Neurotransmission

Toxicant effect may range from small neurological effect to neuronal death

Astrocyte

Injury response; neurotransmitter recycling; general homeostasis

Response to toxicant can range from neuroprotective to harmful

Microglia

CNS immune cell

Can be neuroprotective or neurotoxic. Toxicant exposure may cause release of neurotoxic molecules

Oligodendrocyte

Myelinating cell of CNS

Decreased myelination by toxin may result in loss of neuronal signal transmission and neuron integrity

FIGURE 2.2 Toxicant entry and cell-specific effects. To gain access to the CNS, a toxicant may enter through a specific transporter if it has structural similarities to endogenous molecules that are selectively transported across the BBB. Highly lipophilic molecules may pass directly through biological membranes gaining access to the CNS. A damaged or dysfunctional BBB could allow toxicants that would normally be excluded to enter the CNS. Once in the CNS, toxicants may interact and influence the physiology of a variety of very different cell types, including neurons, astrocytes, microglia, and oligodendrocytes. Toxicant action on each of these cell types may adversely affect neurological function. *Modified with permission from Ref. [1].*

PARKINSON'S DISEASE

Introduction

PD is the second most common neurodegenerative disease after Alzheimer's and is the most common neurodegenerative movement disorder, affecting greater than 1% of the worldwide population over age 65 [14]. As with many neurodegenerative diseases, the risk of developing PD dramatically increases with age, specifically after the fifth decade. Pathologically, PD is characterized by the progressive loss of dopaminergic neurons in the substantia nigra pars compacta (SNpc), resulting in striatal depletion of the neurotransmitter dopamine. Decreased dopamine levels in the striatum lead to some of the motor symptoms of PD, which often include bradykinesia (slowing of movement), resting tremor, rigidity, and gait disturbance. Evidence from neuroanatomical studies indicates that whereas nigral dopaminergic cell loss and striatal dopamine depletion are part of normal aging, the pattern in PD patients is strikingly different. Normal aged patients show nigral dopamine loss in the dorsal tier, whereas in PD patients, the loss is seen in the ventral tier [15]. This difference indicates that aging alone does not result in the development of PD and that other factors are involved. In addition to the degradation of dopaminergic neurons, PD pathology also includes the presence of ubiquitin and α-synuclein positive cytoplasmic inclusions known as Lewy bodies [14]. Lewy body formation is thought to be driven in part by the aggregation of α-synuclein, a protein that will be discussed in further detail. Agents that increase the accumulation or aggregation of this protein may be potential factors that promote PD progression.

Although bradykinesia, resting tremor, rigidity, and gait disturbance are the classical features of PD, if examined on a case-to-case basis, one will find great variability in the disease. A range of phenotypes is presented in PD, with some patients exhibiting pure asymmetric resting tremors, whereas others develop greater rigidity and bradykinesia [16]. In addition, within the PD phenotype, the pathology of the disease can be varied, with some patients showing classic Lewy body formation and others presenting with low or no Lewy body formation at all. These results suggest that multiple factors determine the age of onset and form of PD that patients will develop. This section will serve to elucidate both genetic and potential environmental factors that may contribute to PD etiology.

Genetic and Environmental Influences

Of all the neurodegenerative diseases discussed, PD research has the richest history of examining gene–environment interactions. Multiple genes of interest have been associated with the disease, and many are potentially linked to various environmental toxins. As will be shown, it appears that both genetic susceptibility and toxin exposure may act in similar/parallel mechanisms to promote PD onset.

Over the past few decades, great advances have been made correlating genetic changes to enhanced risk for PD. To date, 16 loci (Park 1–16) have been associated with PD, with mutations in five genes having been confirmed to cause parkinsonian syndromes that resemble PD [17,18]. These genes encode the following proteins: α-synuclein, leucine-rich repeat kinase 2 (LRRK2), parkin, phosphatase and tensin homolog-induced putative kinase 1 (Pink1), and DJ-1. Although a great depth of knowledge links genetic mutations to PD, it is also recognized that monogenic forms of PD account for only a small number of cases, roughly only 10–30% [19]. This infers that other factors may contribute to PD onset. In support of this, studies involving monozygotic and dizygotic twins demonstrate that the prevalence of PD is slightly lower in monozygotic twins, despite their identical genetic makeup, suggesting that environmental factors play a substantial role in conjunction with genetics [20]. Although the evidence points to an interplay between genetics and environmental exposures being crucial for PD etiology, the importance of family history and genetics cannot be understated, as these studies have given the most information in regard to disease progression and mechanism.

Of the proteins involved in PD development, the most heavily studied is α-synuclein, a key component of the Lewy body deposits often found in the brains of PD patients. The gene for α-synuclein encodes a protein of roughly 140 amino acids in length. The exact function of α-synuclein is not fully known, though it is hypothesized to play a role in the recycling of neurotransmitters through interactions with vesicular compartments. Thus, malfunctions in the protein's function could impact dopamine levels within the brain. To date, two different types of genetic alterations of α-synuclein can contribute to PD pathogenesis. Multiple point mutations in the α-synuclein amino acid sequence can lead to PD onset, with common mutations being A30P, A53T, and E46K. In addition to point mutations, gene duplication or triplication is also seen in some, but not all, patients with PD. α-Synculein gene triplication shows the highest penetrance as well as the most severe phenotype, hinting at a possible dose-dependent response between the levels of (wildtype) α-synuclein and PD. It is possible that mutations in the α-synuclein gene, and subsequent changes in the α-synuclein protein, could impact the turnover of the neurotransmitter dopamine, ultimately resulting in changes in dopaminergic neurons. It is also known that α-synuclein is a protein prone to aggregation, and is often a component of the Lewy body inclusions found within the brains of PD patients. Increasing evidence suggests that mutations, as well as increased protein levels through gene amplification, may contribute not to altered protein function, but rather to increased protein aggregation. New work suggests that these protein aggregates (either large fibrils or smaller soluble oligomers) may play a role in promoting neuron cell death by overwhelming vital cellular processes, such as proteosomal degradation of other proteins or ATP production by disrupting mitochondrial function.

Another protein genetically linked to PD is LRRK2. Like α-synuclein, the exact function of LRRK2 is not yet fully elucidated. It is known that LRRK2 is a large, multidomain protein with kinase activity. New phosphorylation targets of LRRK2 are still being discovered, and studies have shown that the kinase activity of LRRK2 might be a contributor to cellular toxicity [16]. Mutations in LRRK2 are thought to be one of the most common causes of inherited PD. To date, 43 mutations have been found in LRRK2, with 8 being linked to PD pathology. The most common mutation is G2019S (glycine to serine mutation), which increases the protein's kinase activity. Although the initial penetrance of the G2019S mutation is low (~28% at age 59), it steadily increases with age (51% at age 69 and 74% at age 79). Although information regarding the exact function of LRRK2 and how it contributes to PD are still being clarified, the importance of this protein and its role in PD etiology cannot be understated, as the G2019S mutation alone is associated with 15–30,000 PD cases in the United States, the same number of cases as ALS and Huntington's disease [2].

Of the three other proteins linked genetically to PD, all three are to be recessively inherited. These proteins include Parkin, Pink1, and DJ-1. The exact functions of all three are still being investigated, but a common thread between them relates to mitochondria function. Pink1 and Parkin play roles in maintaining mitochondria homeostasis whereas DJ-1 is important for regulating oxidative damage within the cell. Genetic data suggest that mutations in all three lead to loss of protein function and result in the onset of PD at an earlier age [1].

Although the link between genetic abnormalities and PD is well documented, the fact that all genetic mutations carry variable penetrance suggests that additional factors from outside the genome can contribute to PD etiology. In this regard, the possible link between environmental toxin exposure and incidences of PD has been extensively explored. Although no single agent has been conclusively associated with PD, analysis suggests that pesticide exposure could increase the risk of PD development by two- to three-fold [16]. We hypothesize that a combination of a person's genetic makeup and a lifetime of exposures to environmental toxins contributes to one's chances of developing PD.

One of the first insights into how toxins may contribute to PD (and neurodegenerative diseases in general) came in the early 1980s when it was noticed that a group of intravenous drug users in the San Francisco area developed parkinsonian symptoms. After further analysis, it was revealed the drug they were using contained a compound called 1-methyl-4-phenyl-1,2,3,6-tetrahydropyridine, commonly referred to as MPTP. MPTP is able to cross the BBB where it is taken up by astrocytes and converted to 1-methyl-4-phenylpyridinium (MPP+) by monoamine oxidase type B (MAOB). MPP+ is a toxic compound that is brought into neurons via the dopamine transporter, in which it initiates neuronal death by inhibiting complex I of the mitochondria. This mitochondrial dysfunction leads to decreased ATP levels, increased reactive oxygen species (ROS) production, and ultimately cell death. This improbable series of events was the first evidence that environmental products can lead to PD-like symptoms, and prompted researchers to investigate whether other compounds could have similar effects.

One of the first compounds investigated was paraquat, a bipyridyl contact pesticide, which has a structure very similar to MPP+. This led to speculation that paraquat might also be a potential dopamine-specific neurotoxin [21]. Several epidemiological studies have linked paraquat exposure to increased risk of PD [22,23], and this risk is elevated in farmers exposed to both paraquat and the fungicide maneb. In addition, in experimental animals, the paraquat/maneb combination was able to induce decreased motor activity, changes in dopamine levels, and decreased dopamine transporter expression [24–26]. However, these results are considered controversial, with some laboratories providing conflicting reports. Although paraquat is structurally similar to MPP+, it does not inhibit complex I of the mitochondria and must stimulate toxicity in a different manner, presumably redox cycling [27].

Another toxin linked to PD is rotenone, a membrane-permeable, naturally occurring insecticide and fish poison. Similar to MPP+, rotenone acts as a powerful complex I inhibitor and causes various forms of mitochondrial dysfunction, including ATP depletion and increased ROS production. Studies in rodents have shown that chromic and subcutaneous exposure to rotenone can result in parkinsonian symptoms with selective dopaminergic neuron loss, oxidative damage, and cytoplasmic inclusions similar to Lewy bodies [28]. Although the effects of rotenone are well studied, it falls under the same category of paraquat exposure, in which its links to PD are most likely limited to isolated incidences of exposure. Rotenone has a relatively short half-life, which reduces its bioavailability. Additionally, rotenone has limited commercial use, and exposure is most likely restricted to administrators of the toxin. Nevertheless, in this population, rotenone exposure increases the risk of PD [29]. These facts do not diminish the information gained from animal models, however, as these studies have given great insight into the mechanisms of PD, especially related to mitochondrial function. In addition, even acute exposures are quite possible, when coupled with genetic susceptibility, and could lead to the development of PD over time.

A new family of toxins being linked to PD are organochlorine pesticides. They have low volatility, are highly lipophilic (can cross the BBB), and have a strong tendency to bioaccumulate, resulting in chromic exposure most commonly through contaminated food. Of the organochlorines being investigated, dieldrin is the most common. Dieldrin ranked second among agricultural chemicals used in the US with a peak in the mid-1960s, with the use of approximately 20 million pounds per year [30]. Studies using dopaminergic cell lines have shown that dieldrin can generate ROS, induce proteasome dysfunction, promote α-synulcein aggregation (and fibrillization), and disrupt mitochondria membrane potential [31,32]. In addition, studies in animal models have shown that dieldrin exposure replicates many of the neuropathological changes associated with PD [33]. However, one of the key factors not replicated in vivo is the loss of dopaminergic neurons, raising questions about the effect dieldrin has on PD etiology.

One last environmental toxin associated with PD is chronic exposure to trichloroethylene (TCE), a highly volatile organic chemical commonly used as an industrial solvent, dry cleaning agent, grain fumigant, anesthetic, and caffeine extractant [34]. Recent animal model data show that inhalation of TCE creates motor impairment after 60 days [35]. In addition, chronic TCE administration causes specific nigral dopamine neuronal loss [36]. TCE is able to cross the BBB and induce oxidative stress. Though unknown, the proposed mechanism of toxicity is inhibition of complex I of the mitochondria [37], linking it to other known PD-causing agents.

Summary

Work from the past few decades has illuminated a distinct connection between genetic susceptibility, toxicant exposure, and PD onset. Although no one causative agent has been identified, the evidence presented suggests that environmental factors can contribute to disease progression. By gaining a better understanding of the roles that genetic mutations and toxicants have on PD progression, we can better comprehend the mechanisms that impact disease onset and progression, with the ultimate goal of developing therapeutic agents to halt disease development.

By taking a closer look at the majority of inherited genetic changes and the target of multiple toxins, a link between mitochondrial dysfunction and PD can be drawn. Many of the toxins discussed (and others not) inhibit complex I of the mitochondria, leading to decreased ATP and increased ROS production. In addition, the proteins genetically modified in PD all play a role in maintaining mitochondrial homeostasis. As mentioned earlier, maintaining proper mitochondrial function is paramount to preserving neuronal integrity, and even slight changes in mitochondrial function can lead to neuronal death. Thus, it is reasonable to hypothesize that mitochondrial dysfunction may be a key contributor in the pathogenesis of PD, and that gene–environment interaction studies may help elucidate potential therapeutics that help protect mitochondrial function under stressed conditions.

AMYOTROPHIC LATERAL SCLEROSIS

Introduction

ALS is an idiopathic fatal neurodegenerative disease arising from the loss of lower motor neurons in the brain stem and ventral horn of the spinal cord and loss of afferent upper motor neurons in the cortex [38]. Symptoms include initial muscle weakness localized to a single limb and/or bulbar region, which becomes progressively widespread, leading to quadriparesis, dysarthria, dysphagia, and ultimately death [39]. ALS is a rapidly progressive disease, normally resulting in death within three to five years of diagnosis. Roughly 10% of ALS cases are purely genetic, and variable penetrance is noted. In the remaining cases, mutations in the proteins superoxide dismutase 1 (SOD1) and thrombocytopenia-absent radius DNA-binding protein 43 kDa (TDP-43) may contribute to disease etiology, with new hypotheses linking accumulation and/or aggregation of these proteins with degenerating neuronal cells [40]. The role of environmental exposures in ALS pathogenesis needs to be further explored, but results linking another form of ALS (Guam's disease) to a known toxicant indicate that environmental factors may play a role in ALS development.

Genetic and Environmental Influences

Research during the past few decades has begun to investigate the correlation between genetic susceptibility and environmental exposure in the risk for developing ALS. It is known that approximately 10% of all ALS cases are purely genetic in nature, with 2% of these cases being caused by mutations in the SOD1 protein [41]. In the remaining 90% of cases, genetic penetrance is variable, indicating additional factors are required for disease onset to occur. This hypothesis is strengthened by the findings that a unique form of ALS, called ALS and PD and dementia (ALS–PDC, commonly referred to as Guam's disease), is potentially triggered by exposure to a unique compound found on Guam. In addition, incidences of ALS–PDC are also identified in individuals exposed to certain cyanobacteria, once again supporting the idea of environmental factors influencing disease progression. This has led to an increase in the study of potential environmental toxins and the role they play in ALS etiology.

Of the genetic factors known to be associated with ALS, the most widely studied is SOD1. SOD1 functions as a cytoplasmic antioxidant and is ubiquitously expressed. The normal function of SOD1 is to catalytically convert highly reactive superoxide to either hydrogen peroxide or oxygen, protecting cells from oxidative stress. Missense mutations at almost any amino acid in SOD1 can lead to ALS, independent of the effect the mutation has on the antioxidant activity [42,43]. This indicates that mutant forms of SOD1 can cause ALS by one or more acquired toxicities, rather than by reduced superoxide dismutase activity alone. Many forms of these toxicities have been identified and will be outlined below.

Of the multitude of mutations in SOD1, surprisingly 150 mutations do not impact protein folding [44]. This is interesting as misfolded SOD1 is prone to form ubiquitinated cytoplasmic inclusions (aggregates). These aggregates occur early in disease development and escalate as the disease progresses [45]. Aggregated mutant forms of SOD1 then accumulate at the endoplasmic reticulum (ER), in which they can bind to the ER luminal polypeptide chain binding protein (BiP), a key chaperone that regulates activation of ER stress transducers [46]. ER stress has been linked to glutamate excitotoxicity in animal models, in which excess calcium influx can create excessive firing of motor neurons derived from the failure to rapidly remove synaptic glutamate [47]. This increased signaling ultimately leads to death of the firing motor neurons.

Another potential toxic function related to ER interference relates to how mutant forms of SOD1 have been implicated with inhibition of the routine protein degradation pathway of the cell, known as the ER-associated degradation (ERAD) pathway. ERAD is a method utilized by cells to remove potentially harmful proteins when they fail to fold correctly. Inhibition of this pathway can lead to the accumulation of toxic proteins which can promote cell death. The disruption of ERAD is dismutase activity independent as both dismutase active mutants (G93A, A4V) and dismutase inactive mutants (G85R) can inhibit protein degradation. These mutants are able to interact with the protein derlin-1, a transmembrane protein in the ER that is instrumental for dislocation of misfolded proteins from the ER to the cytosol.

This interaction with derlin-1 inhibits the ERAD pathway and causes ER stress [48]. However, the role of mutant SOD1 ERAD inhibition in ALS etiology is still unknown, as inhibition has only been seen after disease onset has occurred.

In addition to disrupting ER function in motor neurons, mutant SOD1 has also been implicated with disturbing mitochondria function during ALS. Mutant forms of SOD1 have been shown to interact with the outer membrane of mitochondria, potentially associating with the voltage-dependent anion channel [47]. Multiple reports have corroborated this, showing that SOD1 accumulates in mitochondria [49,50]. Recent work from the Miller lab has also revealed that mutant SOD1 can inhibit protein import into the mitochondria [51]. Decreased mitochondrial import results in mitochondrial dysfunction and can lead to decreased ATP levels and increased ROS production. Although controversial, decreased ATP levels have been seen in presymptomatic [52] and symptomatic [53] ALS models using SOD1 mutants.

Toxicity associated with SOD1 mutations is not limited to motor neurons in ALS. Damage has also been seen in microglia, astrocytes, and the endothelial cells found in the BBB [47]. In fact, mutant SOD1 expressed by motor neurons alone is not sufficient to drive disease, it must be produced by other cells [54,55]. Mutant SOD1 produced by astrocytes is able to promote cell death by contributing to glutamate excitotoxicity. Wildtype SOD1 produced by astrocytes induces the upregulation of the glutamate receptor subunit GluR2 in neighboring motor neurons, which prevents Ca^{2+} excitotoxic damage [55]. Mutant SOD1 expressed by astrocytes leaves neurons vulnerable to excitotoxicity [55,56]. In addition, mutant SOD1 has been shown to damage the BBB by reducing levels of proteins required to maintain tight junction formation [57]. This damage to the BBB can increase protein levels in surrounding fluid above baseline values [58]. Damage to the BBB leads to early microhemorrhaging, as seen in animal models [59], which can release neurotoxic hemoglobin products and drive disease initiation by damaging motor neurons.

Other proteins have been implicated with ALS, though they are less well understood. One protein often studied is TDP-43, which is involved in RNA processing and can be dominantly inherited to cause ALS [60,61]. Mutant TDP-43 is able to form large aggregates in motor neurons and astrocytes [62,63]. Although the exact role these aggregates play in ALS etiology is still being investigated, their presence is similar to other neurodegenerative diseases that possess aggregated forms of proteins. In addition, information suggests that TDP-43 aggregates may function in a prion-like manner, and serve to propagate the disease [64].

As with the study of genetic factors influencing ALS, considerable research has been performed investigating potential environmental factors that contribute to ALS onset. Of these factors, one of the most commonly studied is the naturally occurring amino acid β-methylamino-L-alanine (BMAA). BMAA is incorporated into proteins, which leads to misfolding and potential cell death [65,66]. BMAA is found in cycad seeds (discussed later under Guam's disease) and is also produced by over 90% of all cyanobacteria [39]. Bioaccumulation of BMAA has been identified in various aquatic species, such as zooplankton, fish, mussels, and oysters. Tainted species have been found in the Baltic Sea [67], South Florida [68], Finland [69], and lakes in China, Michigan, and New Hampshire [70–72], as well as in desert dust from the Middle East [73,74]. In fact, epidemiological evidence suggests that veterans of the Gulf War exposed to BMAA have increased prevalence of ALS, though these reports are controversial [73,74]. To date, no known genetic susceptibility to neurotoxicity induced by BMAA exists.

Recent evidence suggests that apolipoprotein E (ApoE) may play a role in BMAA distribution and metabolism, as animal models have shown that BMAA toxicity is partially regulated by ApoE variants [75]. Intriguing future targets to investigate are genetic mutations in genes encoding the L-1 amino acid transporter for BMAA [76], which regulates its transport across the BBB, as well as the seyrl-tRNA synthetase [77], which modulates the incorporation of BMAA into neuronal proteins.

Other potential environmental factors influencing ALS progression are various heavy metals, such as mercury, zinc, and copper. Clinical symptoms of ALS are actually similar to those of long-term mercury exposure [78], leading to epidemiological studies linking mercury exposure to an increased risk for developing ALS [79,80]. These findings are supported by animal models linking mercury exposure to ALS etiology. These studies demonstrate that a single dose of mercuric chloride in mice is enough to cause the deposition of mercury in upper and lower motor neurons [81,82]. Interestingly, in animal models overexpressing a common SOD1 mutation (G93A), chronic exposure to low levels of mercury shows an earlier onset of ALS [83]. The impact of mercury on ALS etiology in humans is seen in a population case study from a Wisconsin community that is characterized by large consumption of freshwater fish rich in mercury content [71], further supporting the hypothesis that mercury exposure may play an influential role in ALS development. Other metals potentially involved in the etiology of ALS are copper and zinc, both essential cofactors for SOD1 function. ALS-associated mutations in SOD1 have been shown to inhibit the binding of copper and zinc by SOD1, leading to changes in protein stability and promoting SOD1 aggregation. In addition, it has been shown that abnormal levels of both copper and zinc are associated with ALS pathology in animal models [84,85], though the exact role of copper and zinc on ALS etiology is still being investigated.

A recent development in ALS etiology has found a further link between gene–environment interactions implicated with disease, linking changes in ALS patients' abilities to detoxify certain pesticides, metals, and other chemicals. This impaired ability to detoxify chemicals has been linked to polymorphisms in a metallothionein family of genes, specifically the paraoxonasse gene cluster (PON1) [86]. PON1 normally functions to hydrolyze pesticides, such as organophosphates [87], and mutations may potentially influence this activity. Although the role of PON1 mutations is still controversial in the impact they have on ALS, it is important to note that specific mutations in proteins may influence the body's ability to detoxify compounds, ultimately leading to disease onset.

Guam's Disease

As mentioned previously, ALS development and exposure to the amino acid BMAA are strongly associated. BMAA can be produced by cyanobacteria, but is also found in the colloid roots of cycad trees. At the end of World War II, ALS-like cases in Guam increased 50–100 times over the rest of the world [88]. In particular, this dramatic increase in ALS-like symptoms was seen specifically in the Chamorro population in Guam. It was also noted that in addition to ALS-like symptoms, patients also exhibited parkinsonian and dementia symptoms, leading to the disorder being termed ALS–parkinsonism–dementia complex (ALS–PDC) [89]. Initial research suggested that the cause of the symptoms resulted from ingestion of BMAA through cycad flour, which the Chamorro people used in tortillas, soups, and dumplings [90,91].

Indeed, it was noted that BMAA was neurotoxic and able to reproduce key features of the disease both in vitro and in vivo, being toxic to rats and primates [92–95]. In addition, it was noted that cycad treatment in rats also produced a PD-like phenotype [96].

Further inspection showed that BMAA caused disease-like symptoms in a multitude of ways. BMAA was able to produce changes in expression of histones, calcium- and calmodulin-binding proteins, and guanine nucleotide-binding proteins. In addition, BMAA provoked severe lesions in the hippocampus with neuronal degradation, cell loss, increased calcium deposits, and astrogliosis in animal models [95]. In the Chamorro population, increased BMAA was found in the brain tissue of patients with ALS–PDC [97,98], and it was subsequently found that BMAA was misincorporated into brain proteins, inducing misfolding and cell death [65,66].

Although the incorporation into brain tissue shows that BMAA at least partially contributes to ALS–PDC, when cycad seeds are washed, almost 100% of BMAA is removed, indicating that cycad flour is not responsible for the increase of ALS–PDC in the Chamorro population. It was then noted that BMAA can greatly bioaccumulate in various species, such as flying foxes, fruit bats, pigs, and deer that feed on the cycad seeds [97,99]. All species are a substantial part of the Chamorro diet, and due to the bioaccumulation effect, BMAA concentrations are approximately equivalent (per weight) up to 1014 kg of processed cycad flour [97], indicating a potential mechanism for the neurotoxic effect of BMAA in the Chamorro population. The overall effects of BMAA contributing to the dramatic increase of ALS–PDC seen in the Chamorro population in Guam are still controversial, but the link between an environmental toxin leading to a neurodegenerative disease is strong and indicates environmental exposure as a distinct risk factor for developing neurodegenerative diseases.

Summary

Evidence provided over the past decades has produced links between genetic susceptibility and environmental factors in ALS, highlighted by the increased occurrence of ALS–PDC in the Chamorro people in Guam. Many of these findings are highly controversial, and, although useful, many disease models fail to replicate all aspects of human disease. However, these models provide useful information regarding possible mechanisms for ALS etiology.

Gaining a better understanding of what mutations to SOD1 and TDP-43 do to produce ALS will also help identify key mechanisms that promote specific degeneration of motor neurons. Already, key findings linking mutations in SOD1 with altered copper and zinc activity, changes to the BBB, and mitochondrial impairment link possible genetic changes with altered resistance to environmental toxins. In addition, gaining a better understanding of mutations to PON1 will help identify which toxins are poorly cleared, hopefully leading to potential therapeutic design.

ALZHEIMER'S DISEASE

Introduction

AD is the most prevalent neurodegenerative disease, and it is anticipated to affect one out of 85 people worldwide by 2050 [100]—and one in two people over age 85. AD is characterized pathologically by the presence of neurofibrillary tangles containing the protein tau in a hyperphosphorylated state and extracellular plaques (aggregates) containing the protein Aβ [101]. Clinically, gross brain atrophy is seen in the temporal lobe, parietal lobe, cingulate gyrus, and portions of the frontal cortex [102]. In AD patients, approximately 50% greater loss of synapses occurs in the hippocampus, 20–30% decrease in the cortex, and 15–35% loss in cortical neurons when compared to age-matched controls [103].

Much debate continues regarding the initial cause of the disease, but it is generally agreed that both tau and Aβ play important and prominent roles. Tau is a microtubule-stabilizing protein that when hyperphosphorylated is prone to aggregation, leading to microtubule destabilization. The exact function of the Aβ protein is still poorly understood, as it has been shown to play roles in gene transcriptional regulation, cholesterol homeostasis, and cell signaling. What is well known is that Aβ is a cleavage product from a larger protein named the amyloid precursor protein (APP). APP is a transmembrane protein expressed in neuronal cells that is subject to two cleavage pathways. The nonamyloidogenic cleavage pathway is initiated when APP is cleaved by an enzyme named α-secretase. This cleavage event sheds a large, soluble ectodomain and produces a transmembrane segment called C83. C83 is subject to a transmembrane enzyme complex called γ-secretase, which consists of the proteins presenilin, nicastrin, anterior pharynx-defective 1 (APH-1), and presenilin enhancer 2 (PEN-2). This is referred to as the nonamyloidogenic pathway as none of the products contribute to AD. Conversely, the amyloidogenic cleavage pathway is initiated when APP is cleaved by another enzyme named β-secretase. This cleavage event also sheds a large, soluble ectodomain, but due to differences in cleavage sites, a transmembrane protein 99 amino acids in length is now generated (C99). C99 is subject to cleavage by γ-secretase, but, due to the change in protein length, the Aβ peptide is now generated, which over time can become toxic and cause AD onset. A variety of genetic and environmental factors can influence these events, either creating a greater risk of developing AD or accelerating the rate at which the disease progresses.

Genetic and Environmental Influences

Two distinct forms of AD occur, familial (purely genetic) and sporadic, each with unique risk factors. Familial AD is linked to mutations found in APP (and Aβ) and components of the γ-secretase complex. These mutations are highly penetrant and result in the development of AD at a premature age, and account for roughly only 2% of all AD cases [104]. Because these cases result in the development of AD at a much earlier age and are purely genetic in nature, they will not be discussed further in this section. Sporadic AD is less well understood, due in part to the low penetrance of mutations. Thus far, three proteins have been linked to increased risks of developing sporadic AD. These include ApoE, Tom40, and sortilin-related receptor 1 (SOR1). Interestingly, genome-wide association studies (GWAS) have mapped the ApoE and Tom40 genes to the same genetic region [105].

ApoE is a protein within the class of apolipoproteins the function of which is to bind lipids (and cholesterol) and transport them throughout the body. In the brain, ApoE serves to shuttle cholesterol from astrocytes to neurons. The ApoE gene can be spliced to create four isoforms, with ApoE ε4 being the biggest risk factor for developing AD [106]. In fact, 40–80% of all AD patients express at least one ApoE ε4 allele [107], and the risk factor is linked in a dose-dependent manner to the number of alleles expressed [106]. The exact mechanism of how the ApoE ε4 isoform increases the risk for AD is still not well understood, but it is known that ApoE protein helps maintain cholesterol homeostasis within the brain. In addition, it plays a role in facilitating the clearance of the Aβ polypeptide out of the brain; thus, dysregulation of either or both of these functions brought forth by the ApoE ε4 isoform may contribute to increased AD risk [108]. A protein linked to ApoE function that is also a risk factor for AD is sortilin-related receptor 1 (SOR1). SOR1 functions as a receptor for ApoE and can bind the protein with its cargo and internalize them into the cell. Lack of SOR1 expression is correlated with increased risk of developing AD, as a lack of SOR1 expression has been found in the brain tissue of AD patients [109]. In addition, SOR1 has been found to play a role in modulating APP sorting and cleavage events, implicating changes in SOR1 function with increased Aβ production [110].

A recent protein that has genetic mutations linked to AD is translocase of the outer membrane 40 (Tom40), an integral membrane protein found within the import complex of the outer membrane of mitochondria. Tom40 serves as the main channel functioning to import proteins from the cytoplasm into the mitochondria. Recent work has shown that polymorphisms present in the Tom40 gene are linked to an increased risk of developing AD [105]. In addition to this link, the study also showed that the length of the polymorphism had an impact on the age of disease onset, with longer polymorphism tracks resulting in earlier disease etiology. Interestingly, the Tom40 gene is in close proximity to the ApoE4 encoding region, indicating that changes in one gene may influence the other. In fact, it has been shown that polymorphisms in the Tom40 gene can affect levels of ApoE4 in the hippocampus of AD patients [111]. Although the exact impact of Tom40 polymorphisms is still being investigated, any changes to mitochondrial protein import have the potential to lead to drastic changes in mitochondria function, ultimately resulting in cell death. It is known that mitochondria dysfunction occurs in the early stages of AD, showing a potential link between Tom40 polymorphisms and AD onset [112,113]. It should be noted, however, that recent work has called into question the linkage between AD and Tom40 [114].

In terms of environmental risk factors associated with AD, scientific findings are highly controversial. Unlike with PD or ALS, the evidence for toxins or other environmental agents causing AD is limited. The best connections have been to metals, such as aluminum and lead, or lifestyle choices, such as exercise and diet, though the results are far from conclusive. A possible link between AD and aluminum exposure was first identified in the early 1970s [115], and was further supplemented by the fact that researchers found aluminum deposits in both Aβ plaques [116] and neurofibrillary tangles [117]. However, it was never clear whether aluminum was a pathogenic agent or was accumulating in plaques and tangles due to an increased affinity for the aggregates after they had already formed, or a laboratory artifact. In addition, large epidemiological studies have yet to identify high aluminum exposure or intake as a causative factor for AD [118,119]. Lastly, animal models have also shown conflicting results, as aluminum exposure to rodents caused some symptoms of AD, but failed to reproduce any classical pathology [1].

Another metal classically linked with AD etiology is lead. Reports have shown that developmental exposure of lead to rodents can alter APP expression and cleavage [120], with increased exposure resulting in the overexpression of APP. Additionally, in aged monkeys, it was shown that developmental exposure to lead increased the expression of AD-regulating genes and created an AD-like pathology in the prefrontal cortex [121]. Although these results show an interesting link between developmental exposure to lead and increased incidence of

AD in animal models, further work must be done to show a link for increased risk of AD in humans. Increased exposures to metals may play a role in AD and other neurodegenerative diseases, but given how tightly regulated metal ion concentrations are in the brain, increased exposure by a single source alone, such as dietary consumption (the main method of metal exposure), is highly unlikely.

Summary

The role of gene–environment interactions in AD is still being uncovered. Although a great wealth of information is available relating to purely inherited, early-onset AD, the combination of events leading to the development of sporadic AD is still quite unknown. The prevalence of the ApoE ε4 isoform in sporadic AD has been well identified, but how it contributes to AD etiology is unclear, though possible roles in cholesterol homeostasis or impacting Aβ clearance by interacting with the BBB are interesting concepts. In addition, recent work linking polymorphisms in the Tom40 gene to sporadic AD shows another connection between mitochondrial dysfunction and neurodegenerative diseases. When coupled with the fact that Aβ has been implicated with mitochondrial impairment, this shows how vital a role mitochondria may play in regulating neurodegeneration.

CONCLUSIONS AND FUTURE DIRECTIONS

Neurodegenerative diseases are currently creating a large socioeconomic burden on society worldwide, and based on current treatment and diagnostic methods the number of cases is estimated to dramatically increase. It was estimated that by 2020, neurodegenerative diseases will be the eighth leading cause of death in developed countries, and by mid-century neurodegenerative diseases will be the world's second leading cause of death, overtaking cancer [122,123]. However, it is estimated that by delaying the age of onset of neurodegenerative diseases, the disease burden could be greatly decreased. For example, if the age of onset of sporadic AD was delayed by one to two years, the decrease in disease burden by 2050 could be 9–23 million cases [100]. To gain better therapeutic and diagnostic methods, we must gain a better understanding of the mechanism of disease etiology. Based on recent evidence, research indicates an intimate relationship exists between genetic susceptibility, environmental exposure, and the risk for developing various neurodegenerative diseases.

As indicated previously, a variety of different proteins and toxins are involved in different neurodegenerative diseases. However, when looked at more closely, common links may be drawn between different neurodegenerative diseases and potential mechanisms. The first common thread that connects PD, ALS, and AD appears to be mitochondrial dysfunction. Mitochondrial dysfunction in neurodegenerative diseases makes sense due to the brain's high energy requirements. The fact that the brain is highly vulnerable to oxidative stress owing to its high lipid content, high oxygen metabolism, and low levels of antioxidants also hints that mitochondrial dysfunction may play an important role in neurodegeneration [124].

In PD, the most widely studied toxins linked to PD etiology are potent mitochondrial complex I inhibitors, such as rotenone and MPP+. In addition, of the five most common inherited proteins with disease-associated mutations all have been implicated with maintaining proper mitochondria function. In ALS, it has been seen that SOD1 mutations can cause the protein to accumulate at mitochondria, block mitochondrial import, and increase ROS production. In addition, mutant SOD1 has also been shown to decrease ATP levels in motor neurons, further indicating mitochondria dysfunction. AD patients show increased levels of lipid peroxidation, nitrotyrosine reactive carboynls, and nucleic acid oxidation, all signs of increased ROS production and mitochondrial stress [125]. The Aβ polypeptide itself has also been shown to play a role in mitochondrial dysfunction. Studies have shown that the Aβ peptide accumulates at mitochondria, and can decrease mitochondrial membrane potential, increase ROS production, change mitochondrial morphology, and potentially block mitochondrial import [126]. In addition, other studies have shown that mitochondrial dysfunction can push APP toward amyloidogenic cleavage [127–129], producing higher levels of Aβ. It is possible that, in AD, mitochondria are caught in a vicious cycle in which the Aβ peptide creates mitochondrial dysfunction, and this dysfunction in turn leads to increased Aβ production.

In addition to mitochondrial dysfunction, another potential common link between various neurodegenerative diseases is changes to the BBB, either allowing the entrance of potential toxins normally excluded from the brain or preventing the removal of harmful substances. In PD, genetic mutations have been linked to the P-glycoprotein, a key component in maintaining the integrity of the BBB. Various genetic alterations decrease the expression of P-glycoprotein, allowing for a more permeable BBB. In both animal models and human brain samples, these mutations allow compounds to cross the BBB that would be excluded with normal P-glycoprotein expression, indicating that mutations associated with PD could allow harmful toxins into the brain. In ALS, SOD1 mutations have also been shown to increase the permeability of the BBB. Mutant forms of SOD1 are able to alter the expression levels of certain proteins, such as ZO-1, Occludin, and Claudin-5, which are required for maintaining the tight junctions within the epithelial layer of the BBB. The reduced levels of these proteins allow compounds to pass through tight-junction interfaces, from which they would normally be repelled. A similar effect has been seen in AD, in which the Aβ peptide was able to decrease Claudin-5 and ZO-1 levels, resulting in a more permeable BBB.

In summary, abundant evidence suggests that both genetic variation and environmental factors contribute to most neurodegenerative diseases. By using various cellular and animal models, we are gaining a better understanding of how potential genetic mutations help promote disease onset. In addition, these same models allow us to better comprehend how environmental toxins may also contribute to disease etiology. Already the common threads of mitochondrial dysfunction and BBB disruption have been identified in a multitude of neurodegenerative diseases. Whether these are key events in all neurodegenerative diseases or simply a function of the methods used is yet to be determined, but for now the commonalities cannot be ignored, as these findings may help drive future neurotherapeutics.

References

[1] Cannon JR, Greenamyre JT. The role of environmental exposures in neurodegeneration and neurodegenerative diseases. Toxicol Sci 2011;124(2):225–50. http://dx.doi.org/10.1093/toxsci/kfr239.

[2] Horowitz MP, Greenamyre JT. Gene-environment interactions in Parkinson's disease: the importance of animal modeling. Clin Pharmacol Ther 2010;88(4):467–74.

[3] Clarke DD, Sokoloff L. Circulation and energy metabolism of the brain. In: Siegl GJ, Agranoff BW, Albers RW, Fisher SK, Uhler MD, editors. Basic Neurochemistry: Molecular, Cellular and Medical Aspects. Philadelphia: Lippincott-Raven; 1999, pp. 637–669.

[4] Dawson R, Beal MF, Bondy SC, DiMonte D, Isom GE. Excitotoxins, aging, and environmental neurotoxins: implications for understanding human neurodegenerative diseases. Toxicol Appl Pharmacol 1995;134(1):1–17.

[5] Way JL. Cyanide intoxication and its mechanism of antagonism. Annu Rev Pharmacol Toxicol 1984;24(1):451–81.

[6] Ramón S. Estudios sobre la degeneración y regeneración del sistema nerviosa. Imprenta de Hijos de Nicolás Moya; 1913.

[7] Abbott NJ, Ronnback L, Hansson E. Astrocyte-endothelial interactions at the blood–brain barrier. Nat Rev Neurosci 2006;7(1):41–53.

[8] Kotyk A, Janáček DK. Cell membrane transport: principles and techniques. New York: Plenum Press; 1975.

[9] Engelhardt B, Sorokin L, editors. The blood–brain and the blood–cerebrospinal fluid barriers: function and dysfunction. Seminars in immunopathology. Springer; 2009.

[10] Furuno T, Landi MT, Ceroni M, Caporaso N, Bernucci I, Nappi G, et al. Expression polymorphism of the blood–brain barrier component P-glycoprotein (MDR1) in relation to Parkinson's disease. Pharmacogenetics 2002;12(7):529–34.

[11] Kortekaas R, Leenders KL, van Oostrom JCH, Vaalburg W, Bart J, Willemsen ATM, et al. Blood–brain barrier dysfunction in parkinsonian midbrain in vivo. Ann Neurology 2005;57(2):176–9. http://dx.doi.org/10.1002/ana.20369.

[12] Schinkel AH, Smit JJM, Van Tellingen O, Beijnen JH, Wagenaar E, Van Deemter L, et al. Disruption of the mouse mdr1a P-glycoprotein gene leads to a deficiency in the blood–brain barrier and to increased sensitivity to drugs. Cell 1994;77(4):491–502.

[13] Schinkel AH, Wagenaar E, Mol CAAM, Van Deemter L. P-glycoprotein in the blood–brain barrier of mice influences the brain penetration and pharmacological activity of many drugs. J Clin Invest 1996;97(11):2517–24.

[14] Hatcher JM, Pennell KD, Miller GW. Parkinson's disease and pesticides: a toxicological perspective. Trends Pharmacol Sci 2008;29(6):322–9.

[15] Fearnley JM, Lees AJ. Ageing and Parkinson's disease: substantia nigra regional selectivity. Brain 1991;114(5): 2283–301.

[16] Ross CA, Smith WW. Gene–environment interactions in Parkinson's disease. Park Relat Disord 2007;13:S309–15.

[17] Lesage S, Brice A. Parkinson's disease: from monogenic forms to genetic susceptibility factors. Hum Mol Genet 2009;18(R1):R48–59.

[18] Bonifati V. Genetics of Parkinson's disease – state of the art, 2013. Parkinsonism Relat Disord 2014;20 (Suppl. 1(0)):S23–8. http://dx.doi.org/10.1016/S1353-8020(13)70009-9.

[19] Shulman JM, De Jager PL, Feany MB. Parkinson's disease: genetics and pathogenesis. Annu Rev Pathol Mech Dis 2011;6:193–222.

[20] Wirdefeldt K, Gatz M, Bakaysa SL, Fiske A, Flensburg M, Petzinger GM, et al. Complete ascertainment of Parkinson disease in the Swedish Twin Registry. Neurobiol Aging 2008;29(12):1765–73.

[21] Dawson TM, Dawson VL. Molecular pathways of neurodegeneration in Parkinson's disease. Science 2003;302(5646):819–22.

[22] Hertzman C, Wiens M, Bowering D, Snow B, Calne D. Parkinson's disease: a case–control study of occupational and environmental risk factors. Am J Ind Med 1990;17(3):349–55.

[23] Liou H, Tsai M, Chen C, Jeng J, Chang Y, Chen S, et al. Environmental risk factors and Parkinson's disease a case–control study in Taiwan. Neurology 1997;48(6):1583–8.

[24] Thiruchelvam M, Brockel B, Richfield E, Baggs R, Cory-Slechta D. Potentiated and preferential effects of combined paraquat and maneb on nigrostriatal dopamine systems: environmental risk factors for Parkinson's disease? Brain Res 2000;873(2):225–34.

[25] Barlow BK, Thiruchelvam MJ, Bennice L, Cory–Slechta DA, Ballatori N, Richfield EK. Increased synaptosomal dopamine content and brain concentration of paraquat produced by selective dithiocarbamates. J Neurochem 2003;85(4):1075–86.

[26] Cory-Slechta DA. Studying toxicants as single chemicals: does this strategy adequately identify neurotoxic risk? Neurotoxicology 2005;26(4):491–510.

[27] Richardson JR, Quan Y, Sherer TB, Greenamyre JT, Miller GW. Paraquat neurotoxicity is distinct from that of MPTP and rotenone. Toxicol Sci 2005;88(1):193–201. http://dx.doi.org/10.1093/toxsci/kfi304.

[28] Betarbet R, Sherer TB, MacKenzie G, Garcia-Osuna M, Panov AV, Greenamyre JT. Chronic systemic pesticide exposure reproduces features of Parkinson's disease. Nat Neurosci 2000;3(12):1301–6.

[29] Tanner CM, Kamel F, Ross G, Hoppin JA, Goldman SM, Korell M, et al. Rotenone, paraquat, and Parkinson's disease. Environ Health Perspect 2011;119(6):866–72.

[30] Jorgenson JL. Aldrin and dieldrin: a review of research on their production, environmental deposition and fate, bio-accumulation, toxicology, and epidemiology in the United States. Environ Health Perspect 2001;109(Suppl. 1):113.

[31] Kanthasamy AG, Kitazawa M, Kanthasamy A, Anantharam V. Dieldrin-induced neurotoxicity: relevance to Parkinson's disease pathogenesis. Neurotoxicology 2005;26(4):701–19.

[32] Kitazawa M, Anantharam V, Kanthasamy A. Dieldrin induces apoptosis by promoting caspase-3-dependent proteolytic cleavage of protein kinase Cδ in dopaminergic cells: relevance to oxidative stress and dopaminergic degeneration. Neuroscience 2003;119(4):945–64.

[33] Hatcher JM, Richardson JR, Guillot TS, McCormack AL, Di Monte DA, Jones DP, et al. Dieldrin exposure induces oxidative damage in the mouse nigrostriatal dopamine system. Exp Neurol 2007;204(2):619–30.

[34] Jollow DJ, Bruckner JV, McMillan DC, Fisher JW, Hoel DG, Mohr LC. Trichloroethylene risk assessment: a review and commentary. Crit Rev Toxicol 2009;39(9):782–97.

[35] Waseem M, Ali M, Dogra S, Dutta KK, Kaw JL. Toxicity of trichloroethylene following inhalation and drinking contaminated water. J Appl Toxicol 2001;21(6):441–4.

[36] Liu M, Choi DY, Hunter RL, Pandya JD, Cass WA, et al. Trichloroethylene induces dopaminergic neurodegeneration in Fisher 344 rats. J Neurochem 2010;112(3):773–83.

[37] Gash DM, Rutland K, Hudson NL, Sullivan PG, Bing G, Cass WA, et al. Trichloroethylene: parkinsonism and complex 1 mitochondrial neurotoxicity. Ann Neurol 2008;63(2):184–92.

[38] Bento-Abreu A, Van Damme P, Van Den Bosch L, Robberecht W. The neurobiology of amyotrophic lateral sclerosis. Eur J Neurosci 2010;31(12):2247–65.

[39] Trojsi F, Monsurrò MR, Tedeschi G. Exposure to environmental toxicants and pathogenesis of amyotrophic lateral sclerosis: state of the art and research perspectives. Int J Mol Sci 2013;14(8):15286–311.

[40] Soo KY, Farg M, Atkin JD. Molecular motor proteins and amyotrophic lateral sclerosis. Int J Mol Sci 2011;12(12):9057–82.

[41] Rosen DR, Bowling AC, Patterson D, Usdin TB, Sapp P, Mezey E, et al. A frequent ala 4 to val superoxide dismutase-1 mutation is associated with a rapidly progressive familial amyotrophic lateral sclerosis. Hum Mol Genet 1994;3(6):981–7.

[42] Borchelt DR, Lee MK, Slunt HS, Guarnieri M, Xu Z-S, Wong PC, et al. Superoxide dismutase 1 with mutations linked to familial amyotrophic lateral sclerosis possesses significant activity. Proc Natl Acad Sci USA 1994;91(17):8292–6.

[43] Robberecht W, Sapp P, Viaene MK, Rosen D, McKenna–Yasek D, Haines J, et al. Rapid communication: Cu/Zn superoxide dismutase activity in familial and sporadic amyotrophic lateral sclerosis. J Neurochem 1994;62(1):384–7.

[44] Turner BJ, Talbot K. Transgenics, toxicity and therapeutics in rodent models of mutant SOD1-mediated familial ALS. Prog Neurobiol 2008;85(1):94–134.

[45] Bruijn L, Becher M, Lee M, Anderson K, Jenkins N, Copeland N, et al. ALS-linked SOD1 mutant G85R mediates damage to astrocytes and promotes rapidly progressive disease with SOD1-containing inclusions. Neuron 1997;18(2):327–38.

[46] Kikuchi H, Almer G, Yamashita S, Guégan C, Nagai M, Xu Z, et al. Spinal cord endoplasmic reticulum stress associated with a microsomal accumulation of mutant superoxide dismutase-1 in an ALS model. Proc Natl Acad Sci U.S.A 2006;103(15):6025–30.

[47] Ilieva H, Polymenidou M, Cleveland DW. Non–cell autonomous toxicity in neurodegenerative disorders: ALS and beyond. J Cell Biol 2009;187(6):761–72.

[48] Nishitoh H, Kadowaki H, Nagai A, Maruyama T, Yokota T, Fukutomi H, et al. ALS-linked mutant SOD1 induces ER stress-and ASK1-dependent motor neuron death by targeting Derlin-1. Genes Dev 2008;22(11):1451–64.

[49] Higgins CM, Jung C, Ding H, Xu Z. Mutant Cu, Zn superoxide dismutase that causes motoneuron degeneration is present in mitochondria in the CNS. J Neurosci 2002;22(6):RC125.

[50] Velde CV, Miller TM, Cashman NR, Cleveland DW. Selective association of misfolded ALS-linked mutant SOD1 with the cytoplasmic face of mitochondria. Proc Natl Acad Sci USA 2008;105(10):4022–7.

[51] Li Q, Velde CV, Israelson A, Xie J, Bailey AO, Dong M-Q, et al. ALS-linked mutant superoxide dismutase 1 (SOD1) alters mitochondrial protein composition and decreases protein import. Proc Natl Acad Sci USA 2010;107(49):21146–51.

[52] Browne SE, Yang L, DiMauro J-P, Fuller SW, Licata SC, Beal MF. Bioenergetic abnormalities in discrete cerebral motor pathways presage spinal cord pathology in the G93A SOD1 mouse model of ALS. Neurobiol Dis 2006;22(3):599–610.

[53] Mattiazzi M, D'Aurelio M, Gajewski CD, Martushova K, Kiaei M, Beal MF, et al. Mutated human SOD1 causes dysfunction of oxidative phosphorylation in mitochondria of transgenic mice. J Biol Chem 2002;277(33):29626–33.

[54] Clement A, Nguyen M, Roberts E, Garcia M, Boillee S, Rule M, et al. Wild-type nonneuronal cells extend survival of SOD1 mutant motor neurons in ALS mice. Science 2003;302(5642):113–7.

[55] Yamanaka K, Chun SJ, Boillee S, Fujimori-Tonou N, Yamashita H, Gutmann DH, et al. Astrocytes as determinants of disease progression in inherited amyotrophic lateral sclerosis. Nat Neurosci 2008;11(3):251–3.

[56] Van Damme P, Bogaert E, Dewil M, Hersmus N, Kiraly D, Scheveneels W, et al. Astrocytes regulate GluR2 expression in motor neurons and their vulnerability to excitotoxicity. Proc Natl Acad Sci USA 2007;104(37):14825–30.

[57] Zhong Z, Deane R, Ali Z, Parisi M, Shapovalov Y, O'Banion MK, et al. ALS-causing SOD1 mutants generate vascular changes prior to motor neuron degeneration. Nat Neurosci 2008;11(4):420–2.

[58] Leonardi A, Abbruzzese G, Arata L, Cocito L, Vische M. Cerebrospinal fluid (CSF) findings in amyotrophic lateral sclerosis. J Neurology 1984;231(2):75–8.

[59] Garbuzova-Davis S, Haller E, Saporta S, Kolomey I, Nicosia SV, Sanberg PR. Ultrastructure of blood–brain barrier and blood–spinal cord barrier in SOD1 mice modeling ALS. Brain Res 2007;1157:126–37.

[60] Sreedharan J, Blair IP, Tripathi VB, Hu X, Vance C, Rogelj B, et al. TDP-43 mutations in familial and sporadic amyotrophic lateral sclerosis. Science 2008;319(5870):1668–72.

[61] Van Deerlin VM, Leverenz JB, Bekris LM, Bird TD, Yuan W, Elman LB, et al. TARDBP mutations in amyotrophic lateral sclerosis with TDP-43 neuropathology: a genetic and histopathological analysis. Lancet Neurol 2008;7(5):409–16.

[62] Kwiatkowski TJ, Bosco D, Leclerc A, Tamrazian E, Vanderburg C, Russ C, et al. Mutations in the FUS/TLS gene on chromosome 16 cause familial amyotrophic lateral sclerosis. Science 2009;323(5918):1205–8.

[63] Vance C, Rogelj B, Hortobágyi T, De Vos KJ, Nishimura AL, Sreedharan J, et al. Mutations in FUS, an RNA processing protein, cause familial amyotrophic lateral sclerosis type 6. Science 2009;323(5918):1208–11.

[64] Wegorzewska I, Bell S, Cairns NJ, Miller TM, Baloh RH. TDP-43 mutant transgenic mice develop features of ALS and frontotemporal lobar degeneration. Proc Natl Acad Sci USA 2009;106(44):18809–14.

[65] Xie X, Mondo K, Basile M, Bradley W, Mash D, editors. Tracking brain uptake and protein incorporation of cyanobacterial toxin BMAA. 22nd Annual Symposium on ALS/MND, Sydney, Australia; 2011.

[66] Rodgers K, Dunlop R, editors. The cyanobacteria-derived neurotoxin BMAA can be incorporated into cell proteins and could thus be an environmental trigger for ALS and other neurological diseases associated with protein misfolding. 22nd Annual Symposium on ALS/MND, Sydney, Australia; 2011.

[67] Jonasson S, Eriksson J, Berntzon L, Spáčil Z, Ilag LL, Ronnevi L-O, et al. Transfer of a cyanobacterial neurotoxin within a temperate aquatic ecosystem suggests pathways for human exposure. Proc Natl Acad Sci USA 2010;107(20):9252–7.

[68] Brand LE, Pablo J, Compton A, Hammerschlag N, Mash DC. Cyanobacterial blooms and the occurrence of the neurotoxin, beta-N-methylamino-L-alanine (BMAA), in South Florida aquatic food webs. Harmful Algae 2010;9(6):620–35.

[69] Sabel C, Boyle P, Löytönen M, Gatrell AC, Jokelainen M, Flowerdew R, et al. Spatial clustering of amyotrophic lateral sclerosis in Finland at place of birth and place of death. Am J Epidemiol 2003;157(10):898–905.

[70] Li A, Tian Z, Li J, Yu R, Banack SA, Wang Z. Detection of the neurotoxin BMAA within cyanobacteria isolated from freshwater in China. Toxicon 2010;55(5):947–53.

[71] Sienko DG, Davis JP, Taylor JA, Brooks BR. Amyotrophic lateral sclerosis: a case-control study following detection of a cluster in a small Wisconsin community. Arch Neurol 1990;47(1):38.

[72] Caller TA, Doolin JW, Haney JF, Murby AJ, West KG, Farrar HE, et al. A cluster of amyotrophic lateral sclerosis in New Hampshire: a possible role for toxic cyanobacteria blooms. Amyotroph Lateral Scler 2009;10(s2):101–8. http://dx.doi.org/10.3109/17482960903278485.

[73] Cox PA, Richer R, Metcalf JS, Banack SA, Codd GA, Bradley WG. Cyanobacteria and BMAA exposure from desert dust: a possible link to sporadic ALS among Gulf War veterans. Amyotroph Lateral Scler 2009;10(S2):109–17.

[74] Metcalf J, Richer R, Cox P, Codd G. Cyanotoxins in desert environments may present a risk to human health. Sci Total Environ 2012;421:118–23.

[75] Wilson JM, Petrik M, Moghadasian M, Shaw C. Examining the interaction of apo E and neurotoxicity on a murine model of ALS-PDC. Can J Physiol Pharmacol 2005;83(2):131–41.

[76] Smith QR, Nagura H, Takada Y, Duncan MW. Facilitated transport of the neurotoxin, β-N-methylamino-L-alanine, across the blood–brain barrier. J Neurochem 1992;58(4):1330–7.

[77] Lee JW, Beebe K, Nangle LA, Jang J, Longo-Guess CM, Cook SA, et al. Editing-defective tRNA synthetase causes protein misfolding and neurodegeneration. Nature 2006;443(7107):50–5.

[78] Schwarz S, Husstedt I, Bertram H, Kuchelmeister K. Amyotrophic lateral sclerosis after accidental injection of mercury. J Neurol Neurosurg Psychiatry 1996;60(6):698.

[79] Adams CR, Ziegler DK, Lin JT. Mercury intoxication simulating amyotrophic lateral sclerosis. JAMA 1983;250(5):642–3.

[80] Praline J, Guennoc A-M, Limousin N, Hallak H, de Toffol B, Corcia P. ALS and mercury intoxication: a relationship? Clin Neurol Neurosurg 2007;109(10):880–3.

[81] Arvidson B. Inorganic mercury is transported from muscular nerve terminals to spinal and brainstem motoneurons. Muscle Nerve 1992;15(10):1089–94.

[82] Chuu J-J, Liu S-H, Lin-Shiau S-Y. Differential neurotoxic effects of methylmercury and mercuric sulfide in rats. Toxicol Lett 2007;169(2):109–20.

[83] Postnatal exposure to methylmercury enhances development of paralytic phenotype. In: Johnson F, Atchison W, editors. Sod1–G93a female mice. The toxicologist, supplement to toxicological. 48th Annual Meeting and Expo. 2009.

[84] Groeneveld G, De Leeuw van Weenen J, Van Muiswinkel F, Veldman H, Veldink J, Wokke J, et al. Zinc amplifies mSOD1-mediated toxicity in a transgenic mouse model of amyotrophic lateral sclerosis. Neurosci Lett 2003;352(3):175–8.

[85] Tokuda E, Ono S-I, Ishige K, Naganuma A, Ito Y, Suzuki T. Metallothionein proteins expression, copper and zinc concentrations, and lipid peroxidation level in a rodent model for amyotrophic lateral sclerosis. Toxicology 2007;229(1):33–41.

[86] Ticozzi N, LeClerc AL, Keagle PJ, Glass JD, Wills AM, van Blitterswijk M, et al. Paraoxonase gene mutations in amyotrophic lateral sclerosis. Ann Neurol 2010;68(1):102–7.

[87] Costa LG, Richter RJ, Li W-F, Cole T, Guizzetti M, Furlong CE. Paraoxonase (PON 1) as a biomarker of susceptibility for organophosphate toxicity. Biomarkers 2003;8(1):1–12.

[88] Kurland LK, Mulder DW. Epidemiologic investigations of amyotrophic lateral sclerosis 1. Preliminary report on geographic distribution, with special reference to the Mariana Islands, including clinical and pathologic observations. Neurology 1954;4(5):355.

[89] Karamyan VT, Speth RC. Animal models of BMAA neurotoxicity: a critical review. Life Sci 2008;82(5):233–46.

[90] Whiting M, editor. Food practices in ALS foci in Japan, the marianas, and new guinea. Federation proceedings. 1964.

[91] Whiting M, Spatz M, Matsumoto H. Research progress on cycads. Econ Bot 1966;20(1):98–102.

[92] Nunn PB, Seelig M, Zagoren JC, Spencer PS. Stereospecific acute neuronotoxicity of 'uncommon' plant amino acids linked to human motor-system diseases. Brain Res 1987;410(2):375–9.

[93] Spencer PS, Nunn PB, Hugon J, Ludolph AC, Ross SM, Roy DN, et al. Guam amyotrophic lateral sclerosis-parkinsonism-dementia linked to a plant excitant neurotoxin. Science 1987;237(4814):517–22.

[94] Karlsson O, Roman E, Berg A-L, Brittebo EB. Early hippocampal cell death, and late learning and memory deficits in rats exposed to the environmental toxin BMAA (β-N-methylamino-L-alanine) during the neonatal period. Behav Brain Res 2011;219(2):310–20.

[95] Karlsson O, Berg A-L, Lindström A-K, Hanrieder J, Arnerup G, Roman E, et al. Neonatal exposure to the cyanobacterial toxin BMAA induces changes in protein expression and neurodegeneration in adult hippocampus. Toxicol Sci 2012;130(2):391–404.

[96] Shen WB, McDowell KA, Siebert AA, Clark SM, Dugger NV, Valentino KM, et al. Environmental neurotoxin–induced progressive model of parkinsonism in rats. Ann Neurol 2010;68(1):70–80.

[97] Banack SA, Cox PA. Biomagnification of cycad neurotoxins in flying foxes implications for ALS-PDC in Guam. Neurology 2003;61(3):387–9.

I. NEURODEVELOPMENTAL DISORDERS

[98] Cox PA, Banack SA, Murch SJ. Biomagnification of cyanobacterial neurotoxins and neurodegenerative disease among the Chamorro people of Guam. Proc Natl Acad Sci USA 2003;100(23):13380–3.

[99] Cox PA, Sacks OW. Cycad neurotoxins, consumption of flying foxes, and ALS-PDC disease in Guam. Neurology 2002;58(6):956–9.

[100] Brookmeyer R, Johnson E, Ziegler-Graham K, Arrighi HM. Forecasting the global burden of Alzheimer's disease. Alzheimer's Dement 2007;3(3):186–91.

[101] Tiraboschi P, Hansen L, Thal L, Corey-Bloom J. The importance of neuritic plaques and tangles to the development and evolution of AD. Neurology 2004;62(11):1984–9.

[102] Wenk GL. Neuropathologic changes in Alzheimer's disease. J Clin Psychiatry 2003;64:7–10.

[103] Reddy PH, Beal MF. Amyloid beta, mitochondrial dysfunction and synaptic damage: implications for cognitive decline in aging and Alzheimer's disease. Trends Mol Med 2008;14(2):45–53.

[104] Reddy PH, McWeeney S. Mapping cellular transcriptosomes in autopsied Alzheimer's disease subjects and relevant animal models. Neurobiol Aging 2006;27(8):1060–77. http://dx.doi.org/10.1016/j.neurobiolaging.2005.04.014.

[105] Roses A, Lutz M, Amrine-Madsen H, Saunders A, Crenshaw D, Sundseth S, et al. A TOMM40 variable-length polymorphism predicts the age of late-onset Alzheimer's disease. Pharmacogenomics J 2010;10(5):375–84.

[106] Corder E, Saunders A, Strittmatter W, Schmechel D, Gaskell P, Small G, et al. Gene dose of apolipoprotein E type 4 allele and the risk of Alzheimer's disease in late onset families. Science 1993;261(5123):921–3.

[107] Strittmatter WJ, Saunders AM, Schmechel D, Pericak-Vance M, Enghild J, Salvesen GS, et al. Apolipoprotein E: high-avidity binding to beta-amyloid and increased frequency of type 4 allele in late-onset familial Alzheimer disease. Proc Natl Acad Sci USA 1993;90(5):1977–81.

[108] Leduc V, Jasmin-Bélanger S, Poirier J. APOE and cholesterol homeostasis in Alzheimer's disease. Trends Mol Med 2010;16(10):469–77.

[109] Scherzer CR, Offe K, Gearing M, Rees HD, Fang G, Heilman CJ, et al. Loss of apolipoprotein E receptor LR11 in Alzheimer disease. Arch Neurol 2004;61(8):1200–5.

[110] Andersen OM, Reiche J, Schmidt V, Gotthardt M, Spoelgen R, Behlke J, et al. Neuronal sorting protein-related receptor sorLA/LR11 regulates processing of the amyloid precursor protein. Proc Natl Acad Sci USA 2005;102(38):13461–6.

[111] Bekris LM, Galloway NM, Montine TJ, Schellenberg GD, Yu CE. APOE mRNA and protein expression in postmortem brain are modulated by an extended haplotype structure. Am J Med Genet Part B Neuropsychiatr Genet 2010;153(2):409–17.

[112] Crouch P, Cimdins K, Duce J, Bush A, Trounce I. Mitochondria in aging and Alzheimer's disease. Rejuvenation Res 2007;10(3):349–58.

[113] Atamna H, Frey II WH. Mechanisms of mitochondrial dysfunction and energy deficiency in Alzheimer's disease. Mitochondrion 2007;7(5):297–310.

[114] Guerreiro RJ, Hardy J. TOMM40 association with Alzheimer disease: tales of APOE and linkage disequilibrium. Arch Neurol 2012;69(10):1243–4.

[115] Crapper D, Krishnan S, Dalton A. Brain aluminum distribution in Alzheimer's disease and experimental neurofibrillary degeneration. Science 1973;180(4085):511–3.

[116] Candy J, Klinowski J, Perry R, Perry E, Fairbairn A, Oakley A, et al. Aluminosilicates and senile plaque formation in Alzheimer's disease. Lancet 1986;327(8477):354–6.

[117] Perl DP, Brody AR. Alzheimer's disease: X-ray spectrometric evidence of aluminum accumulation in neurofibrillary tangle-bearing neurons. Science 1980;208(4441):297–9.

[118] Forster D, Newens A, Kay D, Edwardson J. Risk factors in clinically diagnosed presenile dementia of the Alzheimer type: a case-control study in northern England. J Epidemiol Community Health 1995;49(3):253–8.

[119] McLachlan D, Bergeron C, Smith J, Boomer D, Rifat S. Risk for neuropathologically confirmed Alzheimer's disease and residual aluminum in municipal drinking water employing weighted residential histories. Neurology 1996;46(2):401–5.

[120] Basha MR, Wei W, Bakheet SA, Benitez N, Siddiqi HK, Ge Y-W, et al. The fetal basis of amyloidogenesis: exposure to lead and latent overexpression of amyloid precursor protein and β-amyloid in the aging brain. J Neurosci 2005;25(4):823–9.

[121] Wu J, Basha MR, Brock B, Cox DP, Cardozo-Pelaez F, McPherson CA, et al. Alzheimer's disease (AD)-like pathology in aged monkeys after infantile exposure to environmental metal lead (Pb): evidence for a developmental origin and environmental link for AD. J Neurosci 2008;28(1):3–9.

[122] Lopez AD, Murray C. The global burden OF disease. Nat Med 1998;4(11):1241–3.

[123] Menken M, Munsat TL, Toole JF. The global burden of disease study: implications for neurology. Arch Neurol 2000;57(3):418–20.

[124] Reddy PH. Amyloid precursor protein–mediated free radicals and oxidative damage: implications for the development and progression of Alzheimer's disease. J Neurochem 2006;96(1):1–13.

[125] Castellani RJ, Harris PL, Sayre LM, Fujii J, Taniguchi N, Vitek MP, et al. Active glycation in neurofibrillary pathology of Alzheimer disease: N^ε-(carboxymethyl) lysine and hexitol-lysine. Free Radic Biol Med 2001;31(2):175–80.

[126] Sirk D, Zhu Z, Wadia JS, Shulyakova N, Phan N, Fong J, et al. Chronic exposure to sub–lethal beta–amyloid (Aβ) inhibits the import of nuclear–encoded proteins to mitochondria in differentiated PC12 cells. J Neurochem 2007;103(5):1989–2003.

[127] Webster M-T, Pearce B, Bowen D, Francis P. The effects of perturbed energy metabolism on the processing of amyloid precursor protein in PC12 cells. J Neural Transm 1998;105(8–9):839–53.

[128] Gabuzda D, Busciglio J, Chen LB, Matsudaira P, Yankner BA. Inhibition of energy metabolism alters the processing of amyloid precursor protein and induces a potentially amyloidogenic derivative. J Biol Chem 1994;269(18):13623–8.

[129] Gasparini L, Racchi M, Benussi L, Curti D, Binetti G, Bianchetti A, et al. Effect of energy shortage and oxidative stress on amyloid precursor protein metabolism in COS cells. Neurosci Lett 1997;231(2):113–7.

Fetal Alcohol Spectrum Disorders: Effects and Mechanisms of Ethanol on the Developing Brain

Marina Guizzetti

INTRODUCTION

Ethanol exposure during gestation is associated with a variety of negative outcomes including reduced growth, cognitive dysfunctions, and behavioral deficits [1]. The fact that women who engage in heavy drinking give birth to problematic children has been known since antiquity with references to it in the Bible, Aristotle's work, documents from the Royal College of Physicians of London in 1726, and in an 1899 report from an English prison physician [2].

However, it was not until 1968 that the effects of drinking during gestation were first documented in the scientific literature following a French study [3]. Independently, another group in Seattle reported a similar phenotype in 11 children born to alcoholic mothers and introduced the term fetal alcohol syndrome (FAS) to describe this condition [4–6]. The studies from Jones and colleagues triggered a new and intense line of research aimed at characterizing, treating, and preventing the effects of ethanol in the developing brain.

Alcohol exposure during gestation causes a continuum of effects in the offspring that are more severe in children with greater exposure [1]. However, even mild to moderate levels of in utero alcohol exposure have been associated with milder but similar central nervous system (CNS) dysfunction [7]. The United States Surgeon General (http://www.surgeongeneral.gov/news/2005/02/sg02222005.html) as well as other American health authorities advise pregnant women and women who may become pregnant to abstain from alcohol consumption.

This chapter describes the main clinical, animal, and in vitro research on FASD with the intent of providing a picture of the body of knowledge currently available on the manifestations, diagnosis, and mechanisms involved in this devastating condition affecting individuals throughout life.

FETAL ALCOHOL SPECTRUM DISORDERS (FASD)

Fetal alcohol spectrum disorders (FASD) are a group of conditions defined as the physical, behavioral, and learning sequelae that occur in the offspring of women who drank alcohol during pregnancy. However, the manifestations of the disorder display heterogeneity with the domains and severity of functioning or appearance differing greatly between individuals identified as having FASD [8,9].

Consequently, FASD is not a clinical diagnosis, rather, as defined by the Institute of Medicine (IOM) nomenclature, includes a variety of syndromes, such as fetal alcohol syndrome (FAS), partial FAS (pFAS), alcohol-related neurodevelopmental disorders (ARND), and alcohol-related birth defects (ARBD) [10,11].

Fetal alcohol syndrome (FAS)

FAS is the first documented and most widely recognized consequence of prenatal alcohol exposure [6]. To render an FAS diagnosis, anomalies in three distinct areas need to be present: (1) *growth retardation*, (2) *a distinct facial appearance*, and (3) *some evidence of a CNS dysfunction*.

Guidelines for FAS diagnosis have been published by the Centers for Disease Control and Prevention (CDC) [8] and the IOM [10,11]. In this section, we will refer to the CDC guidelines.

Growth retardation. Growth retardation can be pre- or postnatal and is defined as weight and/or height below the 10th percentile (after race, sex, and gestational age are considered).

Distinct facial appearance. The three typical facial characteristics required for a FAS diagnosis are smooth philtrum, thin vermilion (upper lip), and short palpebral fissures [12]. Other facial characteristics are often present in individuals affected by FAS, including epicanthal folds, strabismus, ptosis, low nasal bridge, and ear anomalies, but are not required for the diagnosis.

CNS dysfunction. Diagnostic criteria for FAS are either structural brain abnormalities or microcephaly (a head circumference at or below the 10th percentile) and neurological or functional deficits (e.g., global cognitive or intellectual deficits or functional deficits one standard deviation below the mean in at least three domains of functioning).

Partial (p)FAS

Partial FAS diagnosis is described in the revised IOM guidelines. A diagnosis of pFAS is rendered when a confirmed history of maternal alcohol exposure, two of the facial characteristics, and either growth retardation or CNS dysfunction are present [10].

Alcohol-Related Neurodevelopmental Disorders (ARND)

Shortly after the first characterization of FAS, it became apparent that FAS was not the only outcome of in utero alcohol exposure, but that cognitive and behavioral effects could be observed in individuals exposed to ethanol during gestation even in the absence of growth retardation and distinct facial appearance. To describe these effects of in utero alcohol exposure, the term ARND is currently used (previously called fetal alcohol effects). Because the characteristics of ARND (which include behavioral and cognitive effects similar to individuals with FAS, although in the absence of facial feature and growth retardation) are not unique to prenatal alcohol exposure, an ARND diagnosis is rendered only when prenatal alcohol exposure is confirmed. Furthermore, evidence of CNS damage defined as clinically significant structural, neurological, or functional impairment in three or more of the following domains, achievement, adaptive behavior, attention, cognition, executive functioning, language, memory, motor skills, multisensory integration, and social communication [13], need to be present for an ARND diagnosis [8,10].

Alcohol-Related Birth Defects (ARBD)

Alcohol-related birth defects (ARBD) are congenital anomalies to the heart, kidney, skeleton, ears, and eyes, which are linked to maternal alcohol use. These anomalies may or may not be present in conjunction with CNS dysfunction. Maternal alcohol use during gestation needs to be confirmed for an ARBD diagnosis [8,11].

FASD PREVALENCE

The CDC reported a prevalence of FAS ranging from 0.2 to two cases per 1000 births between 1993 and 2002 [8], whereas studies carried out in a similar time frame from others reported a prevalence between 0.5 and two cases in 1000 live births [14]. A more recent, mostly in-school, study found a prevalence of two to seven cases per 1000 live births in the USA [15]. In the same study, an FASD prevalence of 2–5% was found in younger schoolchildren in the USA and Western Europe [15].

The prevalence of FASD is not homogeneous across populations; for instance, an extremely high prevalence of FAS and pFAS, ranging from 68.0 to 89.2 per 1000 live births, was found in a region of South Africa characterized by low socioeconomical status, malnutrition, and a long history of wine production [16]. Among certain groups of Native Americans (Plains and Plateau culture tribes) an average FAS rate of nine per 1000 live births was reported [17].

NEUROBEHAVIORAL DEFICITS IN FASD

Neurobehavioral deficits associated with heavy prenatal alcohol exposure include reduced IQ and impairments in several neurodevelopmental domains such as attention, reaction time, visuospatial abilities, executive functions, fine and gross motor skills, memory, language, and social and adaptive functions [18]. Children afflicted with neurobehavioral impairments may not have FAS facial features; however, these deficits in neurobehavioral functions are usually more severe in individuals with FAS facial features [19].

FAS is the leading preventable cause of mental retardation in the general population [20], although only 25% of individuals with FAS are officially mentally retarded (i.e., with an IQ lower than 70) [21]. The mean IQ in individuals with an FAS diagnosis and full FAS characteristics is estimated between 65 and 72, whereas the average IQ of individuals prenatally exposed to high levels of alcohol but without all the facial characteristics of FAS is about 80 [22].

In utero alcohol exposure is also associated with decreased academic achievement and increased learning disabilities [23], with verbal and nonverbal learning and memory impairment [24–26], and with visuospatial processing deficits [24]. FASD individuals present attention deficits and are often diagnosed with attention deficit hyperactivity disorder (ADHD) [27,28]. Examples of executive function deficits in FASD include problems with response inhibition, set shifting, planning, and concept formation [29,30]. Motor dysfunction such as tremors, weak grasp, poor hand/eye coordination, and gait and balance difficulties, which in heavily exposed individuals persists into adulthood, has also been documented [29,30].

FASD AND BRAIN STRUCTURES

The facial dysmorphological characteristics of FAS are important because they are suggestive of brain abnormalities. Indeed, the symmetry of the forebrain and the symmetry of the face develop together during early embryogenesis between the fifth and the sixth week of gestation in humans. Insults affecting the development of the brain at this specific time result in alterations in the face as well as brain structures known as holoprosencephaly [31].

Although the facial characteristics are the most apparent signs of heavy prenatal alcohol exposure, the most devastating consequences are due to brain abnormalities leading to behavioral dysfunction, which enormously impact individuals affected by prenatal alcohol exposure throughout their lives.

Early studies based on autopsies reported general damage throughout most of the brain, microcephaly, migration abnormalities, agenesis of the corpus callosum and anterior commissure, heterotopias of neuroglia, and ventricle, brain stem, basal ganglia, and cerebellar anomalies [4,18]. It should be mentioned, however, that reports from autopsies represent only the most severe examples of in utero alcohol exposure, which are incompatible with life and therefore not necessarily representative of the brain abnormalities that occur in most individuals with FAS.

Magnetic resonance imaging (MRI) followed by quantitative structural analyses consistently reported reduction of the cranial vault and the concomitant reduction in brain size in alcohol-exposed individuals, as reviewed in [18].

Cortical regions have been consistently found to be affected by in utero alcohol exposure. In the frontal, parietal, and temporal lobes less white matter, gray matter, and total lobe volume, higher gray matter density, thicker cortices, displacement, and reduced temporal asymmetry were reported. On the other hand, the occipital lobes appear to be relatively insensitive to the effects of prenatal alcohol. The effects of prenatal alcohol exposure on cortical structures may be responsible for impairments in verbal learning and recall, executive functions, language, and visuospatial processing in individuals with FASD [18,32,33].

Although one would expect the hippocampus to be affected by prenatal alcohol exposure based on commonly observed learning and memory impairments and depression in individuals with FASD, and because animal studies found hippocampal anomalies following developmental ethanol exposure, hippocampal-imaging studies have not revealed a consistent structural abnormality. Indeed, some studies report significantly smaller hippocampi after accounting for total brain volume, and other studies report significant differences only before correction. On the other hand, recent study reported an altered hippocampus shape in FAS children [34]. It has also been reported that, in some FASD cases, hippocampal volume unilaterally decreases, whereas the contralateral hippocampus is normal [18,32,33].

The cerebellum appears to be a major target of ethanol, and its volume and surface area have been found reduced in fetal alcohol exposed children with and without facial dysmorphology. The anterior vermis was found to be the most severely affected area of the cerebellum; in some cases, it was actually displaced. Domains of functioning found in children with FASD and associated with cerebellar anomalies include verbal learning and memory, balance, coordination, attention, and classically conditioned eye-blink learning [18,32,33].

Anomalies in the corpus callosum have been consistently found in subjects prenatally exposed to alcohol including agenesis, complete absence, hypoplasia, volume reduction, displacement, and increased variability of its shape [18,32,33]. Neurobehavioral functions altered in FASD and associated with the corpus callosum include motor function, attention, verbal learning, and executive function [33].

The basal ganglia and, in particular, the caudate nucleus appear to be especially sensitive to the effects of prenatal alcohol exposure, as their volume is reduced in children with FASD compared with controls. Because of basal ganglia connections with the motor and frontal cortex, deficits in executive and motor functions and coordination found in individuals with FASD can be ascribed at least in part to these consequences of prenatal alcohol exposure [18,32,33].

ANIMAL MODELS FOR FASD

Animal studies have been instrumental to the understanding of the effects of ethanol on the developing brain. By far, the most used animal models in FASD research are rodents (mostly mice and rats). Interestingly, several neurobehavioral outcomes observed in individuals with FASD have been qualitatively reproduced in rodent models including learning, motor performance, inhibition, attention, and social deficits; behavioral effects in rodents were observed at blood alcohol concentrations that are achieved in humans [35-37].

Rodent studies also confirmed that ethanol exposure during brain development induced microcephaly [38] and affected the cerebral neocortex [39-43], the hippocampus [44], the cerebellum [45,46], the basal ganglia [47], and the corpus callosum [48-50].

To understand the timing of ethanol exposure used in rodent models of FASD, a comparison of the main stages of brain development in rodents and humans is provided. CNS development begins with the formation of the neural plate and continues through gestation and in the postnatal period in both humans and rodents. However, important events occurring during the third trimester of gestation in humans occur in rodents mostly postnatally (between gestational day (GD) 19 and postnatal day 10).

CNS development is characterized by vulnerable periods during which exposure to teratogens may result in abnormalities specific to the ontogenic events occurring at the time of exposure [51]. Ethanol appears to interfere with all CNS developmental stages [45].

A first critical period of development is when organogenesis occurs and the neural tube and crest are formed; this occurs in rats between GD 5 and GD 11 and in humans in the first trimester of gestation [51]. Beginning in the second week of gestation in rodents (GD 7 in mice and GD 9.5 in rats) and the first month of gestation in humans, specific areas of the CNS begin to form with neurogenesis and migration of cells in the forebrain, midbrain, and hindbrain [51]. Mice exposed to ethanol during GD 7 or 8 exhibit the craniofacial anomalies associated with FAS, as well as forebrain deficiencies including hypoplasia or aplasia of the corpus callosum, and deficiency in the hippocampus and the anterior cingulated cortex, consistent with the holoprosencephaly spectrum of malformations [162]. In addition, prenatal exposure of primates to ethanol during this period reduces the number of neurons in the somatosensory-motor cortex [40].

The second critical period of development occurs in rats from GD 11–12 to GD 18–21, corresponding to the second trimester of gestation in humans [51]. During this time, most of the areas of the nervous system are differentiating; this phase is characterized by intense neurogenesis and neuronal migration in the cerebral cortex and hippocampus, two areas greatly affected by in utero alcohol exposure [45]. In rats, neurogenesis occurs prenatally in all brain regions, except for granule cells in the cerebellum and the dentate gyrus of the hippocampus. After neuroblast proliferation stops, differentiation of these cells into neurons and glia begins, followed by the development and elongation of processes that will become axons or dendrites in neurons. At this time of development, neurons from the cerebral cortex leave the germinal zone and migrate using radial glia fibers as scaffolds [51]. Ethanol exposure during this stage alters radial glia and depresses the generation, survival, and migration of neurons from the neocortex, hippocampus, and the principal sensory nucleus [39,41,42,52].

The last critical period in brain development occurs in rats from GD 18 to postnatal day 9 and is considered the equivalent of the third trimester in human gestation. Major events during this period include a massive increase in brain size, proliferation of astrocytes and oligodendrocytes, myelination, synaptogenesis, and dendritic arborization [51]. Ethanol exposure during this developmental stage induces microcephaly [38], cerebellar [38] and hippocampal [53] abnormalities, severe apoptotic neuronal death in the hippocampus and cerebral cortex [54], reduced dendritic arborization [55-60], and behavioral dysfunctions [61].

Large animal models of FASD (including nonhuman primates and sheep) have also been developed. These models have the advantage of having longer gestations than rodents. In these models, the brain growth spurt occurs prenatally, similar to humans [62]. Other animal models used in FASD research include avian embryos [63], *Drosophila* [64], and zebrafish [65].

MECHANISMS INVOLVED IN FASD

Gene–Ethanol Interactions

Studies carried out in twins revealed that monozygotic twins exposed to ethanol during gestation have a 100% concordance for FAS diagnosis, whereas dizygotic twins displayed a 63% concordance [66,67]. This indicates that a certain degree of genetic susceptibility may underlie the FASD phenotype.

Several human studies have shown an association between polymorphisms in the alcohol dehydrogenase 1B (ADH1B) gene and FASD. The product of the ADH1B gene encodes for the enzyme that catalyzes the first step of alcohol catabolism leading to the formation of acetaldehyde. The alleles ADH1B*2 or ADH1B*3 encode for proteins that are faster in metabolizing ethanol than the allele ADH1B*1. Most studies found an increased risk for FASD in individuals homozygous for ADH1B*1 and a protective effect of ADH1B*2 and ADH1B*3 [68-70].

Different animal species and strains were reported differentially susceptible to the teratogenic effects of ethanol, further supporting the notion of a degree of genetic predisposition to FASD. Animal studies identified several genes, in addition to ADH1B, that can potentially be involved in the genetic susceptibility to FASD, in particular with regard to facial dysmorphology, although further research in this area is necessary to validate these findings [71].

Neuronal Apoptosis

It has been shown that the early postnatal period in rodents, corresponding to the third trimester of gestation in humans, is highly sensitive to the pro-apoptotic effects of ethanol on neurons [47]. In mice, exposure to high doses of ethanol on postnatal day 7 induces massive neuronal degeneration throughout the forebrain, midbrain, cerebellum, brain stem, spinal cord, and retina, which was demonstrated to be caused by apoptosis [47,54,72,73].

The effects of ethanol on neuronal apoptosis involve decreased expression of phosphorylated extracellular signal-regulated protein kinase (pERK), and the activation of the intrinsic mitochondrial pathway leading to mitochondrial damage [47]. Alcohol-induced activation of gamma-aminobutyric acid A (GABA$_A$) receptors and inhibition of N-methyl-D-aspartate (NMDA) receptors are involved in ethanol-induced neuronal apoptosis in the developing brain [47].

The rodent cerebellum is also highly vulnerable to the effects of ethanol during the first 10 postnatal days [46]. At this developmental stage, Purkinje cells, which are generated prenatally, undergo massive dendritic outgrowth and synaptogenesis. On the other hand, cerebellar granule neurons, which are generated postnatally, are at an immature stage. Ethanol exposure during this time period reduces the number of both Purkinje and granule cells [74-77]. Cerebellar granule neurons, however, appear to be more sensitive to ethanol than Purkinje cells, particularly in the very early stage of their differentiation [76,78].

Mechanisms involved in the apoptotic effects of ethanol in cerebellar granule cells include interference with the antiapoptotic effects of glutamate NMDA receptor stimulation, interference with neurotrophic factors signaling (in particular insulin-like growth factor (IGF) I and II and brain-derived neurotrophic factor [BDNF]), activation of the intrinsic apoptotic signaling cascade, and induction of oxidative stress [46].

Early embryogenesis, before the onset of neurogenenesis, is another developmental stage highly sensitive to the apoptotic effects of ethanol. Indeed, exposure of mouse embryos to alcohol on GD 6.5–11, corresponding to the three to five weeks of human gestation, causes a marked increase in apoptotic degeneration of precursor cell populations of the developing brain and craniofacial region at developmental stages at which alcohol exposure affects concurrently the developing face, brain, eyes, and ear [79,80].

Glial Cells

In the past several years, increasing attention has been devoted to the investigation of glial cell vulnerability to ethanol [81]. Several studies have shown that ethanol, both in vivo and in vitro, can significantly affect glial cells. Abnormalities in glial cell development are indeed suspected of contributing to the adverse effects of ethanol on the developing brain [82]. Abnormal glial migration in humans with FAS, as well as in primates and rats exposed to ethanol during development, has been reported [42,83]. A reduction in glial cell number has been reported in rat models of FAS [41,84]. In children with FAS, hypoplasia of the corpus callosum and anterior commissure, two areas originally formed by neuroglial cells, have been reported [85]. The finding that microencephaly is strongly associated with ethanol exposure during the brain growth spurt [86], a period characterized by rapid glial cell proliferation and maturation, also suggests a potential effect of ethanol on the proliferation, growth, and maturation of glia.

Several studies also investigated the effects of ethanol on astrocytes and glial cell lines in culture. Prenatal alcohol exposure delays the expression of the astrocytic marker glial acidic fibrillary protein [81], and the incubation of primary astrocytes in culture with ethanol inhibits astrocyte proliferation induced by serum, M3 muscarinic receptor stimulation, and IGF-1 [87-89].

Our studies revealed a selectivity in the action of ethanol; indeed, we observed that muscarinic receptor-mediated activation of phospholipase C and subsequent increase in intracellular calcium levels, activation of novel protein kinase Cε, and of mitogen-activated protein kinases are relatively unaffected by ethanol [90-92]. On the other hand, phospholipase D-mediated formation of phosphatidic acid and sequential activation of atypical protein kinase Cζ, p70S6 kinase and of NF-κB, are all strongly inhibited by ethanol at concentrations of 25–50 mM [93-97].

Phospholipase D-induced phosphatidic acid formation was identified to be the direct target of ethanol in the inhibition of muscarinic receptor- and serum-stimulated signaling in astrocytes. Indeed, ethanol is a competitive substrate for phospholipase D leading to the formation of phosphatidylethanol instead of the physiological second-messenger phosphatidic acid [88,96,98].

Oligodendrocytes are glial cells that form myelin in the CNS. White matter abnormalities were reported in children and adolescents with FASDs [99]. In vivo studies have shown that prenatal alcohol exposure delays the expression of myelin basic protein and the maturation of oligodendrocytes, causing ultrastructural damage of the myelin sheaths, and reduces the number of myelinated axons in the optic nerve [81]. Recently, widespread oligodendrocyte apoptosis has been reported in the white matter regions of the fetal brain of monkeys exposed to high alcohol levels during the equivalent of the third trimester of human gestation [100].

Upon activation, microglia express and release inflammatory cytokines (TNFα, IL1β) and reactive oxidative species (NO, superoxide) leading to neuroinflammation, which contributes to several neuropathological conditions. Evidence exists that microglial cells can be directly activated by alcohol [101], that microglia-released factors contribute to ethanol-induced apoptosis in developing hypothalamic neurons [102,103], and that ethanol causes microglia cell death [104].

Neuronal Plasticity

Several paradigms of neuronal plasticity are affected by alcohol exposure during brain development in animal models of FASD. Impaired long-term potentiation and long-term depression [105-108], learning and memory abilities [44,109-112], barrel cortex plasticity [113], ocular dominance plasticity [114,115], and eye-blink conditioning [116,117] were observed in animal models of FASD.

A large body of evidence suggests that structural plasticity is highly affected by in utero alcohol exposure. Indeed, some cortical maps are altered in FASD models [118-120]. Furthermore, prenatal and/or neonatal alcohol exposure reduced dendritic branching and dendritic spines density of pyramidal neurons in the hippocampus and neocortex. Dendritic spine morphology was also affected by alcohol exposure during brain development [55-60].

Furthermore, several abnormalities were identified in neurite outgrowth of neurons exposed to ethanol in vitro [121-125]. One important mechanism involved in ethanol-induced inhibition of neurite outgrowth is the inhibition of L1 cell adhesion molecules [121,122,125-127].

Recent studies have investigated the effect of ethanol on neuronal plasticity modulated by astrocytes in vitro. Neuritogenesis is inhibited in cortical neurons grown in the presence of astrocytes prepared from rats prenatally exposed to ethanol in comparison to neurons incubated with astrocytes from unexposed animals [128]. Ethanol also inhibits axonal growth of cerebellar neurons induced by the active fragment of the astrocyte-released activity-dependent neuroprotective protein [129].

In addition, we have recently reported that the stimulation of muscarinic receptors in astrocytes induces neuritogenesis in hippocampal neurons cocultured with prestimulated astrocytes [130], and that this effect is inhibited by physiologically relevant concentrations of ethanol [131]. Ethanol-treated astrocytes also displayed a reduced ability to foster neurite out-growth when neurons are plated on top of ethanol-treated astrocytes [132]. We also found that ethanol profoundly affects astrocyte secretion leading to the generation of an environment that inhibits neuronal development. Indeed, ethanol reduced the levels of neuritogenic extra-cellular matrix proteins laminin and fibronectin through the modulation of the plasminogen activator proteolytic system [131,133]. Ethanol also increased the levels of the chondroitin sulfate proteoglycan neurocan, an inhibitor of neurite outgrowth, through the inhibition of the activity of arylsulfatase B, an enzyme involved in the degradation of chondroitin-4-sulfate [132]. Figure 3.1 schematically summarizes our findings.

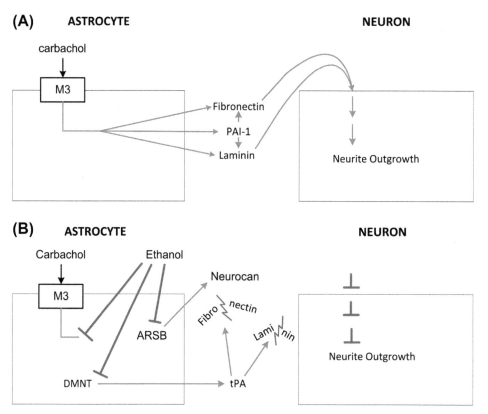

FIGURE 3.1 **Proposed mechanism of ethanol-treated astrocyte-induced inhibition of neurite outgrowth.** The stimulation of M3 muscarinic receptors in astrocytes increases the release of fibronectin and laminin and of the inhibitor of ECM degradation PAI-1, therefore inducing neuritogenesis (A). Ethanol, by inhibiting muscarinic signaling, inhibits the release of fibronectin, laminin, and PAI-1. Furthermore, ethanol in astrocytes increases the expression and release of tPA which promotes the degradation of neuritogenic ECM, through the inhibition of DNMT activity andof DNA methylation in the tPA promoter region. Ethanol also inhibits ARSB activity and increases the levels of inhibitorychondroitin-4-sulfate (C4S) and the chondroitin sulfate proteoglycan neurocan (B). All together, these mechanisms contribute to the inhibition of ethanol-treated astrocyte-induced neurite outgrowth.

Neuroendocrine System

In rodents, gestational and early postnatal ethanol exposures affect hypothalamic–pituitary–thyroid functions; indeed, decreased levels of circulating 3,5,3'-triiodothyronine (T3) and thyroxine (T4) and increased levels of thyroid-stimulating hormone were reported [134,135].

Prenatal alcohol exposure also increases the tone and dysregulates the hypothalamic–pituitary–adrenal axis throughout life, resulting in abnormally elevated responsiveness to stressors quantifiable biochemically and behaviorally; these effects of prenatal alcohol are often sexually dimorphic. These developmental effects of alcohol have been associated with increased depression and anxiety disorders observed in individuals affected by FASD [136,137].

Epigenetics

The term "epigenetics" describes mechanisms of gene expression regulation independent from coded DNA sequences that rely, instead, on modifications of histones, DNA, or RNA. Acetylation and methylation are two major epigenetic histone modifications; the best-characterized DNA modification is cytosine methylation; epigenetic changes of RNA are associated with the silencing of posttranscriptional RNA [138].

A delay in the DNA methylation program associated with a delay in growth, decreased promoter methylation, and increased expression of genes involved in development was reported in mouse embryo cultures exposed to ethanol during neurulation. Effects similar to the ones induced by ethanol were also observed after treating embryo cultures with an inhibitor of DNA methylation [139,140]. Alcohol also decreased DNA methylation, increased expression of genes involved in development, and inhibited migration and differentiation of neural stem cells in vitro [141]. It has been recently reported that prenatal ethanol modulates DNA methylation and histone acetylation and affects the expression of the proopiomelanocortin gene in hypothalamic neurons; gestational choline supplementation normalizes these effects of ethanol [142].

An increase in global DNA methylation in the prefrontal cortex and hippocampus of postnatal day 21 rats exposed to ethanol between postnatal days 4 and 9 was recently reported [143], whereas combined pre- and postnatal alcohol exposure resulted in a significant increase in DNA methyltransferase activity on postnatal day 21 [144].

In astrocyte cultures, ethanol decreases DNA methyltransferase (DNMT) activity and DNA methylation in the promoter region of tissue plasminogen activator (tPA), thereby increasing tPA expression [133]. Increases in tPA levels lead to increased activation of the proteolytic enzyme plasmin potentially contributing to the observed ethanol-induced decrease in neuritogenic extracellular matrix proteins [131].

A high dose of ethanol administrated on postnatal day 7 increases histone 3 lysine 9 and histone 3 lysine 27 dimethylation, which are associated with transcription silencing, and induce robust apoptotic neurodegeneration, whereas pharmacological inhibition of histone dimethylation prevents ethanol-induced neurodegeneration [145], indicating an involvement of epigenetic histone modifications in the developmental effects of alcohol.

MicroRNAs (miRNAs) are members of the larger family of nonprotein-coding RNA molecules regulating gene expression. MicroRNAs pair with complementary sequences in mRNAs silencing their translation and triggering their degradation. Ethanol suppressed the expression of four miRNAs (miR-9, miR-21, miR-153, and miR-335) in fetal neural stem cells (NSCs) and neural progenitor cells (NPCs) of the neuroepithelium, and increased the expression of proteins involved in the premature differentiation of NSC/NPC cells [146]. Suppressed expression of miR-9 and miR-153 miRNAs was confirmed in a zebrafish model during embryonic development, and these effects of ethanol were associated with craniofacial alterations and increased locomotor activity (hyperactivity), two hallmarks of FAS [147,148]. In an FAS mouse model, ethanol induced miR-10a and miR-10b expression and inhibited the expression of Hoxa1, a gene associated with craniofacial defects and mental retardation [149]. Changes in several miRNAs were also observed after chronic ethanol exposure and ethanol withdrawal in cortical neuron cultures [150]. Together, these published data suggest that miRNAs may play a major role in the facial feature and behavioral dysfunctions observed in FASD [151].

Oxidative Stress

Oxidative stress, which has been involved in the action of several teratogens, plays a role also in the effects of ethanol on the developing brain [152].

Gestational exposure to ethanol in rats and mice induced oxidative stress evidenced by increase lipid peroxidation, decreased levels of glutathione (GSH) and superoxide dismutase, and mitochondrial dysfunction. Oxidative stress after developmental alcohol exposure has been reported in the whole brain and in several brain regions, such as hippocampus, hypothalamus, and cerebellum. These effects can be long-lasting, and may lead to increased cell death and impaired neuronal plasticity [153-158].

Neonatal exposure of rodents to ethanol has also been shown to alter the GSH content and to increase the levels of lipid peroxides and protein carbonyls in several brain regions, effects that may cause cell death [159-163].

In summary, substantial evidence consistently indicates that reactive oxygen species (ROS) production, oxidative damage, and a dysregulation of the endogenous antioxidant system occur throughout gestation in different animal models of FASD and support the notion that oxidative stress is involved in the neuropathology of FASD.

EXPERIMENTAL TREATMENTS FOR FASD

Research on the mechanisms involved in brain dysfunction observed in FASD has been mostly directed to the identification of interventions, such as therapeutics, dietary supplements, or enhancement of the environment. The last section of this chapter describes FASD treatments that are currently under investigation in preclinical and clinical settings.

A substantial body of evidence derived from behavioral and neurochemical studies in rats supports the notion that choline deficiency during gestation negatively affects brain functions in the adult and aging brain. On the other hand, choline supplementation during gestation and neonatally improves memory performance and neuronal plasticity throughout life [164-166].

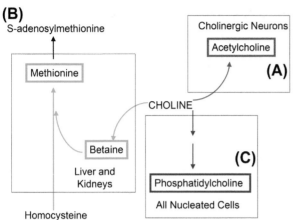

FIGURE 3.2   Choline is metabolized to: (A) acetylcholine; this reaction occurs only in cholinergic neurons expressing choline acyltransferase; (B) the universal methyl donor S-adenosylmethionine; the enzymes catalyzing the first 2 steps of this pathway, the conversion of choline to betaine and the transfer of a methy group from betaine to methionine are present only in the liver and kidneys; (C) choline-containing phospholipids of which by far the most abundant is PC; the enzymes necessary for this pathway are expressed by all mammalian nucleated cells [167].

Multiple mechanisms have been hypothesized for these effects of choline. Choline is the precursor of phosphatidylcholine and sphingomyelin, two important membrane phospholipids, which also can generate second messengers, such as phosphatidic acid, diacylglycerol, and ceramide. Choline is also the precursor of two signaling lipids, platelet-activating factor and sphingosyphosphorylcholine. Choline can be oxidized to form betaine, a methyl donor, which may play a role in protein synthesis and in epigenetic DNA methylation. Finally, choline is the precursor of the neurotransmitter acetylcholine [168]. Choline's main metabolic pathways are summarized in Figure 3.2.

Recent evidence suggests that choline supplementation may also be effective in improving cognitive functions in two neurodevelopmental disorders characterized by mental retardation, Down syndrome [169] and Rett syndrome [170].

A series of studies have shown that prenatal and neonatal choline supplementation reduces hyperactivity and improves hippocampal-associated spatial learning deficits and cerebellar-associated behavioral abnormalities in the development of reflexes and motor coordination in rats exposed to ethanol during brain development [112,171-175]. However, the exact mechanisms by which choline exerts its protective effects remain elusive. A clinical trial for testing the effectiveness of choline treatments in FASD children is currently in progress [176].

Behavioral deficits resulting from prenatal or neonatal ethanol exposure in rodents can be improved by enhancement of the postnatal environment by neonatal handling [177], environmental enrichment [178], or complex motor training [179,180]. However, whether these enhancements can revert the neuroanatomical changes induced by ethanol remains controversial. Indeed, although enrichment does not rescue the reduction in dendritic spine density in rats exposed prenatally to alcohol [181], nor the reduction in cortical thickness in mice [182], a combination of voluntary exercise followed by living in a complex environment partially restores adult neurogenesis in the dentate gyrus while improving hippocampus-dependent behavior in animals neonatally exposed to alcohol [183].

An area of investigation that deserves further examination is whether social interactions impact behavioral outcomes differently in control and ethanol-exposed animals [37].

Based on some of the mechanisms targeted by ethanol during brain development, therapeutics that may prevent the effects of prenatal alcohol exposure are under investigation in preclinical models. For instance, peptides that can antagonize ethanol-induced inhibition of L1 cell adhesion molecule have been tested in vitro and in vivo [184,185]; antioxidants, including superoxide dismutase [186], vitamin C [187], vitamin E [188], and green tea extract [189], have been shown to protect against ethanol-induced oxidative stress in FASD animal models. Serotonin agonists buspirone and ipsapirone were shown to protect against ethanol-induced serotoninergic neuron damage [190,191].

Acknowledgments

This work was supported by grants AA017180 and AA021876 from the National Institute of Alcoholism and Alcohol Abuse. The author is extremely grateful to Dr. David Gavin for critically reading this chapter and to Mr. Jeff Frkonja for the graphic design of Figure 3.1.

References

[1] Streissguth AP, Landesman-Dwyer S, Martin JC, Smith DW. Teratogenic effects of alcohol in humans and laboratory animals. Science 1980;209:353–61.

[2] Calhoun F, Warren K. Fetal alcohol syndrome: historical perspectives. Neurosci Biobehav Rev 2007;31:168–71.

[3] Lemoine P, Harousseau H, Borteyru JP, Menuet JC. Children of alcoholic parents: abnormalities observed in 127 cases. Ouest Med 1968;8:476–82.

[4] Jones KL, Smith DW. Recognition of the fetal alcohol syndrome in early infancy. Lancet 1973;302:999–1001.

[5] Jones KL, Smith DW. The fetal alcohol syndrome. Teratology 1975;12:1–10.

[6] Jones KL, Smith DW, Ulleland CN, Streissguth P. Pattern of malformation in offspring of chronic alcoholic mothers. Lancet 1973;1:1267–71.

[7] Flak AL, Su S, Bertrand J, Denny CH, Kesmodel US, Cogswell ME. The association of mild, moderate, and binge prenatal alcohol exposure and child neuropsychological outcomes: a meta-analysis. Alcohol Clin Exp Res 2014;38:214–26.

[8] Bertrand J, Floyd RL, Weber MK, O'Connor M, Riley EP, Johnson KA, et al. Fetal alcohol syndrome: guidelines for referral and diagnosis. Atlanta (GA): Centers for Disease Control and Prevention; 2004.

[9] Sokol RJ, Delaney-Black V, Nordstrom B. Fetal alcohol spectrum disorder. JAMA 2003;290:2996–9.

[10] Hoyme HE, MPA KW, Kodituwakku P, Gossage JP, Trujillo PM, Buckley D, et al. A practical clinical approach to diagnosis of fetal alcohol spectrum disorders: clarification of the 1996 Institute of Medicine criteria. Pediatrics 2005;115:37–49.

[11] Institute of Medicine (IOM), Howe CSK, Battaglia F, Committee to Study Fetal Alcohol Syndrome, Division of Biobehavioral Sciences and Mental Disorders. Fetal alcohol syndrome: diagnosis, epidemiology, prevention, and treatment. Washington (DC): National Academy Press; 1996.

[12] Astley SJ. Diagnostic guide for fetal alcohol spectrum disorders: the 4-digit diagnostic code. Seattle: University of Washington Publication Service; 2004.

[13] Lang J. Ten brain domains: a proposal for functional central nervous system parameters for fetal alcohol spectrum disorder diagnosis and follow-up. J FAS Inst 2006;4:1–11.

[14] May PA, Gossage JP. Estimating the prevalence of fetal alcohol syndrome. A summary. Alcohol Res Health 2001;25:159–67.

[15] May PA, Gossage JP, Kalberg WO, Robinson LK, Buckley D, Manning M, et al. Prevalence and epidemiologic characteristics of FASD from various research methods with an emphasis on recent in-school studies. Dev Disabil Res Rev 2009;15:176–92.

[16] May PA, Gossage JP, Marais AS, Adnams CM, Hoyme HE, Jones KL, et al. The epidemiology of fetal alcohol syndrome and partial FAS in a South African community. Drug Alcohol Depend 2007;88:259–71.

[17] May PA, Gossage JP. Epidemiology of alcohol consumption among American Indians living in four reservations and in nearby border towns. Drug Alcohol Depend 2001;63:S100.

[18] Riley EP, McGee CL. Fetal alcohol spectrum disorders: an overview with emphasis on changes in brain and behavior. Exp Biol Med (Maywood) 2005;230:357–65.

[19] Mattson SN, Riley EP. A review of the neurobehavioral deficits in children with fetal alcohol syndrome or prenatal exposure to alcohol. Alcohol Clin Exp Res 1998;22:279–94.

[20] Abel EL, Sokol RJ. Incidence of fetal alcohol syndrome and economic impact of FAS-related anomalies. Drug Alcohol Depend 1987;19:51–70.

[21] Streissguth AP, Barr HM, Kogan J, Bookstein FL. Primary and secondary disabilities. In: Streissguth AP, Kanter J, editors. Fetal alcohol syndrome in the challenge of fetal alcohol syndrome: overcoming secondary disabilities. Seattle (WA): University of Washington Press; 1997. p. 25–39.

[22] Mattson SN, Riley EP, Gramling L, Delis DC, Jones KL. Heavy prenatal alcohol exposure with or without physical features of fetal alcohol syndrome leads to IQ deficits. J Pediatr 1997;131:718–21.

[23] Howell KK, Lynch ME, Platzman KA, Smith GH, Coles CD. Prenatal alcohol exposure and ability, academic achievement, and school functioning in adolescence: a longitudinal follow-up. J Pediatr Psychol 2006;31:116–26.

[24] Mattson SN, Roebuck TM. Acquisition and retention of verbal and nonverbal information in children with heavy prenatal alcohol exposure. Alcohol Clin Exp Res 2002;26:875–82.

[25] Roebuck-Spencer TM, Mattson SN. Implicit strategy affects learning in children with heavy prenatal alcohol exposure. Alcohol Clin Exp Res 2004;28:1424–31.

[26] Mattson SN, Gramling L, Delis D, Jones KL, Riley EP. Global-local processing in children prenatally exposed to alcohol. Child Neuropsychol 1996;2:165–75.

[27] Coles CD, Platzman KA, Lynch ME, Freides D. Auditory and visual sustained attention in adolescents prenatally exposed to alcohol. Alcohol Clin Exp Res 2002;26:263–71.

[28] Fryer SL, McGee CL, Matt GE, Riley EP, Mattson SN. Evaluation of psychopathological conditions in children with heavy prenatal alcohol exposure. Pediatrics 2007;119:e733–41.

[29] Guerri C, Bazinet A, Riley EP. Foetal alcohol spectrum disorders and alterations in brain and behaviour. Alcohol Alcohol 2009;44:108–14.

[30] Mattson SN, Crocker N, Nguyen TT. Fetal alcohol spectrum disorders: neuropsychological and behavioral features. Neuropsychol Rev 2011;21:81–101.

[31] Cohen Jr MM, Shiota K. Teratogenesis of holoprosencephaly. Am J Med Genet 2002;109:1–15.

[32] Lebel C, Roussotte F, Sowell ER. Imaging the impact of prenatal alcohol exposure on the structure of the developing human brain. Neuropsychol Rev 2011;21:102–18.

[33] Norman AL, Crocker N, Mattson SN, Riley EP. Neuroimaging and fetal alcohol spectrum disorders. Dev Disabil Res Rev 2009;15:209–17.

[34] Joseph J, Warton C, Jacobson SW, Jacobson JL, Molteno CD, Eicher A, et al. Three-dimensional surface deformation-based shape analysis of hippocampus and caudate nucleus in children with fetal alcohol spectrum disorders. Hum Brain Mapp 2014;35:659–72.

[35] Driscoll CD, Streissguth AP, Riley EP. Prenatal alcohol exposure: comparability of effects in humans and animal models. Neurotoxicol Teratol 1990;12:231–7.

[36] Kelly SJ, Day N, Streissguth AP. Effects of prenatal alcohol exposure on social behavior in humans and other species. Neurotoxicol Teratol 2000;22:143–9.

[37] Kelly SJ, Goodlett CR, Hannigan JH. Animal models of fetal alcohol spectrum disorders: impact of the social environment. Dev Disabil Res Rev 2009;15:200–8.

[38] Bonthius DJ, West JR. Alcohol-induced neuronal loss in developing rats: increased brain damage with binge exposure. Alcohol Clin Exp Res 1990;14:107–18.

[39] Miller MW. Effects of alcohol on the generation and migration of cerebral cortical neurons. Science 1986;233:1308–11.

[40] Miller MW. Exposure to ethanol during gastrulation alters somatosensory-motor cortices and the underlying white matter in the macaque. Cereb Cortex 2007;17:2961–71.

[41] Miller MW, Potempa G. Numbers of neurons and glia in mature rat somatosensory cortex: effects of prenatal exposure to ethanol. J Comp Neurol 1990;293:92–102.

[42] Miller MW, Robertson S. Prenatal exposure to ethanol alters the postnatal development and transformation of radial glia to astrocytes in the cortex. J Comp Neurol 1993;337:253–66.

[43] Mooney SM, Miller MW. Nerve growth factor neuroprotection of ethanol-induced neuronal death in rat cerebral cortex is age dependent. Neuroscience 2007;149:372–81.

[44] Berman RF, Hannigan JH. Effects of prenatal alcohol exposure on the hippocampus: spatial behavior, electrophysiology, and neuroanatomy. Hippocampus 2000;10:94–110.

[45] Guerri C. Neuroanatomical and neurophysiological mechanisms involved in central nervous system dysfunctions induced by prenatal alcohol exposure. Alcohol Clin Exp Res 1998;22:304–12.

[46] Luo J. Mechanisms of ethanol-induced death of cerebellar granule cells. Cerebellum 2012;11:145–54.

[47] Creeley CE, Olney JW. Drug-induced apoptosis: mechanism by which alcohol and many other drugscan disrupt brain development. Brain Sci 2013;3:1153–81.

[48] Deng J, Elberger AJ. Corpus callosum and visual cortex of mice with deletion of the NMDA-NR1 receptor. II. Attenuation of prenatal alcohol exposure effects. Brain Res Dev Brain Res 2003;144:135–50.

[49] Livy DJ, Elberger AJ. Alcohol exposure during the first two trimesters-equivalent alters the development of corpus callosum projection neurons in the rat. Alcohol 2008;42:285–93.

[50] Qiang M, Wang MW, Elberger AJ. Second trimester prenatal alcohol exposure alters development of rat corpus callosum. Neurotoxicol Teratol 2002;24:719–32.

[51] Rice D, Barone Jr S. Critical periods of vulnerability for the developing nervous system: evidence from humans and animal models. Environ Health Perspect 2000;108(Suppl. 3):511–33.

[52] Sulik KK. Genesis of alcohol-induced craniofacial dysmorphism. Exp Biol Med (Maywood) 2005;230:366–75.

[53] Barnes DE, Walker DW. Prenatal ethanol exposure permanently reduces the number of pyramidal neurons in rat hippocampus. Brain Res 1981;227:333–40.

[54] West JR, Hamre KM, Cassell MD. Effects of ethanol exposure during the third trimester equivalent on neuron number in rat hippocampus and dentate gyrus. Alcohol Clin Exp Res 1986;10:190–7.

[55] Ikonomidou C, Bittigau P, Ishimaru MJ, Wozniak DF, Koch C, Genz K, et al. Ethanol-induced apoptotic neurodegeneration and fetal alcohol syndrome. Science 2000;287:1056–60.

[56] Cui ZJ, Zhao KB, Zhao HJ, Yu DM, Niu YL, Zhang JS, et al. Prenatal alcohol exposure induces long-term changes in dendritic spines and synapses in the mouse visual cortex. Alcohol Alcohol 2010;45:312–9.

[57] Davies DL, Smith DE. A Golgi study of mouse hippocampal CA1 pyramidal neurons following perinatal ethanol exposure. Neurosci Lett 1981;26:49–54.

[58] Granato A, Di Rocco F, Zumbo A, Toesca A, Giannetti S. Organization of cortico-cortical associative projections in rats exposed to ethanol during early postnatal life. Brain Res Bull 2003;60:339–44.

[59] Hamilton DA, Akers KG, Rice JP, Johnson TE, Candelaria-Cook FT, Maes LI, et al. Prenatal exposure to moderate levels of ethanol alters social behavior in adult rats: relationship to structural plasticity and immediate early gene expression in frontal cortex. Behav Brain Res 2010;207:290–304.

[60] Hamilton GF, Whitcher LT, Klintsova AY. Postnatal binge-like alcohol exposure decreases dendritic complexity while increasing the density of mature spines in mPFC layer II/III pyramidal neurons. Synapse 2010;64:127–35.

[61] Smith DE, Davies DL. Effect of perinatal administration of ethanol on the CA1 pyramidal cell of the hippocampus and Purkinje cell of the cerebellum: an ultrastructural survey. J Neurocytol 1990;19:708–17.

[62] Melcer T, Gonzalez D, Barron S, Riley EP. Hyperactivity in preweanling rats following postnatal alcohol exposure. Alcohol 1994;11:41–5.

[63] Cudd TA. Animal model systems for the study of alcohol teratology. Exp Biol Med (Maywood) 2005;230:389–93.

[64] Smith SM. The avian embryo in fetal alcohol research. Methods Mol Biol 2008;447:75–84.

[65] McClure KD, French RL, Heberlein UA. Drosophila model for fetal alcohol syndrome disorders: role for the insulin pathway. Dis Model Mech 2011;4:335–46.

[66] Cole GJ, Zhang C, Ojiaku P, Bell V, Devkota S, Mukhopadhyay S. Effects of ethanol exposure on nervous system development in zebrafish. Int Rev Cell Mol Biol 2012;299:255–315.

[67] Chasnoff IJ. Fetal alcohol syndrome in twin pregnancy. Acta Genet Med Gemellol (Roma) 1985;34:229–32.

[68] Streissguth AP, Dehaene P. Fetal alcohol syndrome in twins of alcoholic mothers: concordance of diagnosis and IQ. Am J Med Genet 1993;47:857–61.

[69] Das UG, Cronk CE, Martier SS, Simpson PM, McCarver DG. Alcohol dehydrogenase 2*3 affects alterations in offspring facial morphology associated with maternal ethanol intake in pregnancy. Alcohol Clin Exp Res 2004;28:1598–606.

[70] Green RF, Stoler JM. Alcohol dehydrogenase 1B genotype and fetal alcohol syndrome: a HuGE minireview. Am J Obstet Gynecol 2007;197:12–25.

[71] Jacobson SW, Carr LG, Croxford J, Sokol RJ, Li TK, Jacobson JL. Protective effects of the alcohol dehydrogenase-ADH1B allele in children exposed to alcohol during pregnancy. J Pediatr 2006;148:30–7.

[72] McCarthy N, Eberhart JK. Gene-ethanol interactions underlying fetal alcohol spectrum disorders. Cell Mol Life Sci 2014;71:2699–706.

[73] Dikranian K, Qin YQ, Labruyere J, Nemmers B, Olney JW. Ethanol-induced neuroapoptosis in the developing rodent cerebellum and related brain stem structures. Brain Res Dev Brain Res 2005;155:1–13.

[74] Tenkova T, Young C, Dikranian K, Labruyere J, Olney JW. Ethanol-induced apoptosis in the developing visual system during synaptogenesis. Invest Ophthalmol Vis Sci 2003;44:2809–17.

[75] Karacay B, Li S, Bonthius DJ. Maturation-dependent alcohol resistance in the developing mouse: cerebellar neuronal loss and gene expression during alcohol-vulnerable and -resistant periods. Alcohol Clin Exp Res 2008;32:1439–50.

[76] Maier SE, Miller JA, Blackwell JM, West JR. Fetal alcohol exposure and temporal vulnerability: regional differences in cell loss as a function of the timing of binge-like alcohol exposure during brain development. Alcohol Clin Exp Res 1999;23:726–34.

[77] Maier SE, West JR. Regional differences in cell loss associated with binge-like alcohol exposure during the first two trimesters equivalent in the rat. Alcohol 2001;23:49–57.

[78] Napper RM, West JR. Permanent neuronal cell loss in the cerebellum of rats exposed to continuous low blood alcohol levels during the brain growth spurt: a stereological investigation. J Comp Neurol 1995;362:283–92.

[79] Wozniak DF, Hartman RE, Boyle MP, Vogt SK, Brooks AR, Tenkova T, et al. Apoptotic neurodegeneration induced by ethanol in neonatal mice is associated with profound learning/memory deficits in juveniles followed by progressive functional recovery in adults. Neurobiol Dis 2004;17:403–14.

[80] Dunty Jr WC, Chen SY, Zucker RM, Dehart DB, Sulik KK. Selective vulnerability of embryonic cell populations to ethanol-induced apoptosis: implications for alcohol-related birth defects and neurodevelopmental disorder. Alcohol Clin Exp Res 2001;25:1523–35.

[81] Guerri C, Pascual M, Renau-Piqueras J. Glia and fetal alcohol syndrome. Neurotoxicology 2001;22:593–9.

[82] Phillips DE, Krueger SK. Effects of combined pre- and postnatal ethanol exposure (three trimester equivalency) on glial cell development in rat optic nerve. Int J Dev Neurosci 1992;10:197–206.

[83] Clarren SK, Alvord Jr EC, Sumi SM, Streissguth AP, Smith DW. Brain malformations related to prenatal exposure to ethanol. J Pediatr 1978;92:64–7.

[84] Perez-Torrero E, Duran P, Granados L, Gutierez-Ospina G, Cintra L, Diaz-Cintra S. Effects of acute prenatal ethanol exposure on Bergmann glia cells early postnatal development. Brain Res 1997;746:305–8.

[85] Riley EP, Mattson SN, Sowell ER, Jernigan TL, Sobel DF, Jones KL. Abnormalities of the corpus callosum in children prenatally exposed to alcohol. Alcohol Clin Exp Res 1995;19:1198–202.

[86] Samson H. Microenchephaly and fetal alcohol syndrome: human and animal studies. In: West J, editor. Alcohol and brain development. Oxford: Oxford Press; 1986. p. 167–83.

[87] Guizzetti M, Costa LG. Inhibition of muscarinic receptor-stimulated glial cell proliferation by ethanol. J Neurochem 1996;67:2236–45.

[88] Kotter K, Klein J. Ethanol inhibits astroglial cell proliferation by disruption of phospholipase D-mediated signaling. J Neurochem 1999;73:2517–23.

[89] Resnicoff M, Rubini M, Baserga R, Rubin R. Ethanol inhibits insulin-like growth factor-1-mediated signalling and proliferation of C6 rat glioblastoma cells. Lab Invest 1994;71:657–62.

[90] Catlin MC, Guizzetti M, Costa LG. Effect of ethanol on muscarinic receptor-induced calcium responses in astroglia. J Neurosci Res 2000;60:345–55.

[91] Guizzetti M, Costa LG. Muscarinic receptors, protein kinase C isozymes and proliferation of astroglial cells: effects of ethanol. Neurotoxicology 2000;21:1117–21.

[92] Yagle K, Lu H, Guizzetti M, Moller T, Costa LG. Activation of mitogen-activated protein kinase by muscarinic receptors in astroglial cells: role in DNA synthesis and effect of ethanol. Glia 2001;35:111–20.

[93] Guizzetti M, Bordi F, Dieguez-Acuna FJ, Vitalone A, Madia F, Woods JS, et al. Nuclear factor kappaB activation by muscarinic receptors in astroglial cells: effect of ethanol. Neuroscience 2003;120:941–50.

[94] Guizzetti M, Costa LG. Possible role of protein kinase C ζ in muscarinic receptor-induced proliferation of astrocytoma cells. Biochem Pharmacol 2000;60:1457–66.

[95] Guizzetti M, Thompson BD, Kim Y, VanDeMark K, Costa LG. Role of phospholipase D signaling in ethanol-induced inhibition of carbachol-stimulated DNA synthesis of 1321N1 astrocytoma cells. J Neurochem 2004;90:646–53.

[96] Tsuji R, Guizzetti M, Costa LG. In vivo ethanol decreases phosphorylated MAPK and p70S6 kinase in the developing rat brain. Neuroreport 2003;14:1395–9.

[97] Guizzetti M, Costa LG. Effect of ethanol on protein kinase C ζ and p70S6 kinase activation by carbachol: a possible mechanism for ethanol-induced inhibition of glial cell proliferation. J Neurochem 2002;82:38–46.

[98] Klein J. Functions and pathophysiological roles of phospholipase D in the brain. J Neurochem 2005;94:1473–87.

[99] Sowell ER, Johnson A, Kan E, Lu LH, Van Horn JD, Toga AW, et al. Mapping white matter integrity and neurobehavioral correlates in children with fetal alcohol spectrum disorders. J Neurosci 2008;28:1313–9.

[100] Creeley CE, Dikranian KT, Johnson SA, Farber NB, Olney JW. Alcohol-induced apoptosis of oligodendrocytes in the fetal macaque brain. Acta Neuropathol Commun 2013;1:23.

[101] Fernandez-Lizarbe S, Pascual M, Guerri C. Critical role of TLR4 response in the activation of microglia induced by ethanol. J Immunol 2009;183:4733–44.

[102] Boyadjieva NI, Sarkar DK. Cyclic adenosine monophosphate and brain-derived neurotrophic factor decreased oxidative stress and apoptosis in developing hypothalamic neuronal cells: role of microglia. Alcohol Clin Exp Res 2013;37:1370–9.

[103] Boyadjieva NI, Sarkar DK. Microglia play a role in ethanol-induced oxidative stress and apoptosis in developing hypothalamic neurons. Alcohol Clin Exp Res 2013;37:252–62.

[104] Kane CJ, Phelan KD, Han L, Smith RR, Xie J, Douglas JC, et al. Protection of neurons and microglia against ethanol in a mouse model of fetal alcohol spectrum disorders by peroxisome proliferator-activated receptor-gamma agonists. Brain Behav Immun 2011;25(Suppl. 1):S137–45.

[105] Izumi Y, Kitabayashi R, Funatsu M, Izumi M, Yuede C, Hartman RE, et al. A single day of ethanol exposure during development has persistent effects on bi-directional plasticity, N-methyl-D-aspartate receptor function and ethanol sensitivity. Neuroscience 2005;136:269–79.

[106] Richardson DP, Byrnes ML, Brien JF, Reynolds JN, Dringenberg HC. Impaired acquisition in the water maze and hippocampal long-term potentiation after chronic prenatal ethanol exposure in the guinea-pig. Eur J Neurosci 2002;16:1593–8.

[107] Servais L, Hourez R, Bearzatto B, Gall D, Schiffmann SN, Cheron G. Purkinje cell dysfunction and alteration of long-term synaptic plasticity in fetal alcohol syndrome. Proc Natl Acad Sci USA 2007;104:9858–63.

[108] Sutherland RJ, McDonald RJ, Savage DD. Prenatal exposure to moderate levels of ethanol can have long-lasting effects on hippocampal synaptic plasticity in adult offspring. Hippocampus 1997;7:232–8.

[109] Clements KM, Girard TA, Ellard CG, Wainwright PE. Short-term memory impairment and reduced hippocampal c-Fos expression in an animal model of fetal alcohol syndrome. Alcohol Clin Exp Res 2005;29:1049–59.

[110] Girard TA, Xing HC, Ward GR, Wainwright PE. Early postnatal ethanol exposure has long-term effects on the performance of male rats in a delayed matching-to-place task in the Morris water maze. Alcohol Clin Exp Res 2000;24:300–6.

[111] Goodlett CR, Johnson TB. Neonatal binge ethanol exposure using intubation: timing and dose effects on place learning. Neurotoxicol Teratol 1997;19:435–46.

[112] Thomas JD, Biane JS, O'Bryan KA, O'Neill TM, Dominguez HD. Choline supplementation following third-trimester-equivalent alcohol exposure attenuates behavioral alterations in rats. Behav Neurosci 2007;121:120–30.

[113] Rema V, Ebner FF. Effect of enriched environment rearing on impairments in cortical excitability and plasticity after prenatal alcohol exposure. J Neurosci 1999;19:10993–1006.

[114] Medina AE, Krahe TE, Coppola DM, Ramoa AS. Neonatal alcohol exposure induces long-lasting impairment of visual cortical plasticity in ferrets. J Neurosci 2003;23:10002–12.

[115] Medina AE, Ramoa AS. Early alcohol exposure impairs ocular dominance plasticity throughout the critical period. Brain Res Dev Brain Res 2005;157:107–11.

[116] Johnson TB, Stanton ME, Goodlett CR, Cudd TA. Eyeblink classical conditioning in the preweanling lamb. Behav Neurosci 2008;122:722–9.

[117] Stanton ME, Goodlett CR. Neonatal ethanol exposure impairs eyeblink conditioning in weanling rats. Alcohol Clin Exp Res 1998;22:270–5.

[118] Chappell TD, Margret CP, Li CX, Waters RS. Long-term effects of prenatal alcohol exposure on the size of the whisker representation in juvenile and adult rat barrel cortex. Alcohol 2007;41:239–51.

[119] Margret CP, Chappell TD, Li CX, Jan TA, Matta SG, Elberger AJ, et al. Prenatal alcohol exposure (PAE) reduces the size of the forepaw representation in forepaw barrel subfield (FBS) cortex in neonatal rats: relationship between periphery and central representation. Exp Brain Res 2006;172:387–96.

[120] Medina AE, Krahe TE, Ramoa AS. Early alcohol exposure induces persistent alteration of cortical columnar organization and reduced orientation selectivity in the visual cortex. J Neurophysiol 2005;93:1317–25.

[121] Bearer CF, Swick AR, O'Riordan MA, Cheng G. Ethanol inhibits L1-mediated neurite outgrowth in postnatal rat cerebellar granule cells. J Biol Chem 1999;274:13264–70.

[122] Littner Y, Tang N, He M, Bearer CF. L1 cell adhesion molecule signaling is inhibited by ethanol in vivo. Alcohol Clin Exp Res 2013;37:383–9.

[123] Yanni PA, Lindsley TA. Ethanol inhibits development of dendrites and synapses in rat hippocampal pyramidal neuron cultures. Brain Res Dev Brain Res 2000;120:233–43.

[124] Yanni PA, Rising LJ, Ingraham CA, Lindsley TA. Astrocyte-derived factors modulate the inhibitory effect of ethanol on dendritic development. Glia 2002;38:292–302.

[125] Yeaney NK, He M, Tang N, Malouf AT, O'Riordan MA, Lemmon V, et al. Ethanol inhibits L1 cell adhesion molecule tyrosine phosphorylation and dephosphorylation and activation of pp60(src). J Neurochem 2009;110:779–90.

[126] Chen S, Charness ME. Ethanol disrupts axon outgrowth stimulated by netrin-1, GDNF, and L1 by blocking their convergent activation of Src family kinase signaling. J Neurochem 2012;123:602–12.

[127] Dou X, Wilkemeyer MF, Menkari CE, Parnell SE, Sulik KK, Charness ME. Mitogen-activated protein kinase modulates ethanol inhibition of cell adhesion mediated by the L1 neural cell adhesion molecule. Proc Natl Acad Sci USA 2013;110:5683–8.

[128] Pascual M, Guerri C. The peptide NAP promotes neuronal growth and differentiation through extracellular signal-regulated protein kinase and Akt pathways, and protects neurons co-cultured with astrocytes damaged by ethanol. J Neurochem 2007;103:557–68.

[129] Chen S, Charness ME. Ethanol inhibits neuronal differentiation by disrupting activity-dependent neuroprotective protein signaling. Proc Natl Acad Sci USA 2008;105:19962–7.

[130] Guizzetti M, Moore NH, Giordano G, Costa LG. Modulation of neuritogenesis by astrocyte muscarinic receptors. J Biol Chem 2008;283:31884–97.

[131] Guizzetti M, Moore NH, Giordano G, VanDeMark KL, Costa LG. Ethanol inhibits neuritogenesis induced by astrocyte muscarinic receptors. Glia 2010;58:1395–406.

[132] Zhang X, Bhattacharyya S, Kusumo H, Goodlett CR, Tobacman JK, Guizzetti M. Arylsulfatase B modulates neurite outgrowth via astrocyte chondroitin-4-sulfate: dysregulation by ethanol. Glia 2014;62:259–71.

[133] Zhang X, Kusumo H, Sakharkar AJ, Pandey SC, Guizzetti M. Regulation of DNA methylation by ethanol induces tissue plasminogen activator expression in astrocytes. J Neurochem 2014;128:344–9.

[134] Portoles M, Sanchis R, Guerri C. Thyroid hormone levels in rats exposed to alcohol during development. Horm Metab Res 1988;20:267–70.

[135] Wilcoxon JS, Redei EE. Prenatal programming of adult thyroid function by alcohol and thyroid hormones. Am J Physiol Endocrinol Metab 2004;287:E318–26.

[136] Hellemans KG, Sliwowska JH, Verma P, Weinberg J. Prenatal alcohol exposure: fetal programming and later life vulnerability to stress, depression and anxiety disorders. Neurosci Biobehav Rev 2010;34:791–807.

[137] Weinberg J, Sliwowska JH, Lan N, Hellemans KG. Prenatal alcohol exposure: foetal programming, the hypothalamic-pituitary-adrenal axis and sex differences in outcome. J Neuroendocrinol 2008;20:470–88.

[138] Egger G, Liang G, Aparicio A, Jones PA. Epigenetics in human disease and prospects for epigenetic therapy. Nature 2004;429:457–63.

[139] Zhou FC, Chen Y, Love A. Cellular DNA methylation program during neurulation and its alteration by alcohol exposure. Birth Defects Res A Clin Mol Teratol 2011;91:703–15.

[140] Liu Y, Balaraman Y, Wang G, Nephew KP, Zhou FC. Alcohol exposure alters DNA methylation profiles in mouse embryos at early neurulation. Epigenetics 2009;4:500–11.

[141] Zhou FC, Balaraman Y, Teng M, Liu Y, Singh RP, Nephew KP. Alcohol alters DNA methylation patterns and inhibits neural stem cell differentiation. Alcohol Clin Exp Res 2011;35:735–46.

[142] Bekdash RA, Zhang C, Sarkar DK. Gestational choline supplementation normalized fetal alcohol-induced alterations in histone modifications, DNA methylation, and proopiomelanocortin (POMC) gene expression in beta-endorphin-producing POMC neurons of the hypothalamus. Alcohol Clin Exp Res 2013;37:1133–42.

[143] Otero NK, Thomas JD, Saski CA, Xia X, Kelly SJ. Choline supplementation and DNA methylation in the hippocampus and prefrontal cortex of rats exposed to alcohol during development. Alcohol Clin Exp Res 2012;36:1701–9.

[144] Perkins A, Lehmann C, Lawrence RC, Kelly SJ. Alcohol exposure during development: impact on the epigenome. Int J Dev Neurosci 2013;31:391–7.

[145] Subbanna S, Shivakumar M, Umapathy NS, Saito M, Mohan PS, Kumar A, et al. G9a-mediated histone methylation regulates ethanol-induced neurodegeneration in the neonatal mouse brain. Neurobiol Dis 2013;54:475–85.

[146] Sathyan P, Golden HB, Miranda RC. Competing interactions between micro-RNAs determine neural progenitor survival and proliferation after ethanol exposure: evidence from an ex vivo model of the fetal cerebral cortical neuroepithelium. J Neurosci 2007;27:8546–57.

[147] Pappalardo-Carter DL, Balaraman S, Sathyan P, Carter ES, Chen WJ, Miranda RC. Suppression and epigenetic regulation of MiR-9 contributes to ethanol teratology: evidence from zebrafish and murine fetal neural stem cell models. Alcohol Clin Exp Res 2013;37:1657–67.

I. NEURODEVELOPMENTAL DISORDERS

[148] Tal TL, Franzosa JA, Tilton SC, Philbrick KA, Iwaniec UT, Turner RT, et al. MicroRNAs control neurobehavioral development and function in zebrafish. FASEB J 2012;26:1452–61.

[149] Wang LL, Zhang Z, Li Q, Yang R, Pei X, Xu Y, et al. Ethanol exposure induces differential microRNA and target gene expression and teratogenic effects which can be suppressed by folic acid supplementation. Hum Reprod 2009;24:562–79.

[150] Guo Y, Chen Y, Carreon S, Qiang M. Chronic intermittent ethanol exposure and its removal induce a different miRNA expression pattern in primary cortical neuronal cultures. Alcohol Clin Exp Res 2012;36:1058–66.

[151] Balaraman S, Tingling JD, Tsai PC, Miranda RC. Dysregulation of microRNA expression and function contributes to the etiology of fetal alcohol spectrum disorders. Alcohol Res 2013;35:18–24.

[152] Brocardo PS, Gil-Mohapel J, Christie BR. The role of oxidative stress in fetal alcohol spectrum disorders. Brain Res Rev 2011;67:209–25.

[153] Chu J, Tong M, de la Monte SM. Chronic ethanol exposure causes mitochondrial dysfunction and oxidative stress in immature central nervous system neurons. Acta Neuropathol 2007;113:659–73.

[154] Dembele K, Yao XH, Chen L, Nyomba BL. Intrauterine ethanol exposure results in hypothalamic oxidative stress and neuroendocrine alterations in adult rat offspring. Am J Physiol Regul Integr Comp Physiol 2006;291:R796–802.

[155] Dong J, Sulik KK, Chen SY. The role of NOX enzymes in ethanol-induced oxidative stress and apoptosis in mouse embryos. Toxicol Lett 2010;193:94–100.

[156] Ramachandran V, Perez A, Chen J, Senthil D, Schenker S, Henderson GI. In utero ethanol exposure causes mitochondrial dysfunction, which can result in apoptotic cell death in fetal brain: a potential role for 4-hydroxynonenal. Alcohol Clin Exp Res 2001;25:862–71.

[157] Reyes E, Ott S, Robinson B. Effects of in utero administration of alcohol on glutathione levels in brain and liver. Alcohol Clin Exp Res 1993;17:877–81.

[158] Patten AR, Brocardo PS, Sakiyama C, Wortman RC, Noonan A, Gil-Mohapel J, et al. Impairments in hippocampal synaptic plasticity following prenatal ethanol exposure are dependent on glutathione levels. Hippocampus 2013;23:1463–75.

[159] Heaton MB, Paiva M, Madorsky I, Mayer J, Moore DB. Effects of ethanol on neurotrophic factors, apoptosis-related proteins, endogenous antioxidants, and reactive oxygen species in neonatal striatum: relationship to periods of vulnerability. Brain Res Dev Brain Res 2003;140:237–52.

[160] Heaton MB, Paiva M, Madorsky I, Shaw G. Ethanol effects on neonatal rat cortex: comparative analyses of neurotrophic factors, apoptosis-related proteins, and oxidative processes during vulnerable and resistant periods. Brain Res Dev Brain Res 2003;145:249–62.

[161] Kumral A, Tugyan K, Gonenc S, Genc K, Genc S, Sonmez U, et al. Protective effects of erythropoietin against ethanol-induced apoptotic neurodegeneration and oxidative stress in the developing C57BL/6 mouse brain. Brain Res Dev Brain Res 2005;160:146–56.

[162] Marino MD, Aksenov MY, Kelly SJ. Vitamin E protects against alcohol-induced cell loss and oxidative stress in the neonatal rat hippocampus. Int J Dev Neurosci 2004;22:363–77.

[163] Smith AM, Zeve DR, Grisel JJ, Chen WJ. Neonatal alcohol exposure increases malondialdehyde (MDA) and glutathione (GSH) levels in the developing cerebellum. Brain Res Dev Brain Res 2005;160:231–8.

[164] Li Q, Guo-Ross S, Lewis DV, Turner D, White AM, Wilson WA, et al. Dietary prenatal choline supplementation alters postnatal hippocampal structure and function. J Neurophysiol 2004;91:1545–55.

[165] Meck WH, Williams CL, Cermak JM, Blusztajn J. Developmental periods of choline sensitivity provide an ontogenic mechanism for regulating memory capacity and age-related dementia. Front Integr Neurosci 2008;1:1–11.

[166] Zeisel SH, Niculescu MD. Perinatal choline influences brain structure and function. Nutr Rev 2006;64:197–203.

[167] Li Z, Vance DE. Phosphatidylcholine and choline homeostasis. J Lipid Res 2008;49:1187–94.

[168] Blusztajn JK. Choline, a vital amine. Science 1998;281:794–5.

[169] Moon J, Chen M, Gandhy SU, Strawderman M, Levitsky DA, Maclean KN, et al. Perinatal choline supplementation improves cognitive functioning and emotion regulation in the Ts65Dn mouse model of Down syndrome. Behav Neurosci 2010;124:346–61.

[170] Nag N, Berger-Sweeney JE. Postnatal dietary choline supplementation alters behavior in a mouse model of Rett syndrome. Neurobiol Dis 2007;26:473–80.

[171] Thomas JD, Abou EJ, Dominguez HD. Prenatal choline supplementation mitigates the adverse effects of prenatal alcohol exposure on development in rats. Neurotoxicol Teratol 2009;31:303–11.

[172] Thomas JD, Garrison M, O'Neill TM. Perinatal choline supplementation attenuates behavioral alterations associated with neonatal alcohol exposure in rats. Neurotoxicol Teratol 2004;26:35–45.

[173] Thomas JD, Idrus NM, Monk BR, Dominguez HD. Prenatal choline supplementation mitigates behavioral alterations associated with prenatal alcohol exposure in rats. Birth Defects Res A Clin Mol Teratol 2010;88:827–37.

[174] Thomas JD, La Fiette MH, Quinn VR, Riley EP. Neonatal choline supplementation ameliorates the effects of prenatal alcohol exposure on a discrimination learning task in rats. Neurotoxicol Teratol 2000;22:703–11.

[175] Thomas JD, Tran TD. Choline supplementation mitigates trace, but not delay, eyeblink conditioning deficits in rats exposed to alcohol during development. Hippocampus 2012;22:619–30.

[176] Wozniak JR, Fuglestad AJ, Eckerle JK, Kroupina MG, Miller NC, Boys CJ, et al. Choline supplementation in children with fetal alcohol spectrum disorders has high feasibility and tolerability. Nutr Res 2013;33:897–904.

[177] Weinberg J, Kim CK, Yu W. Early handling can attenuate adverse effects of fetal ethanol exposure. Alcohol 1995;12:317–27.

[178] Hannigan JH, Berman RF, Zajac CS. Environmental enrichment and the behavioral effects of prenatal exposure to alcohol in rats. Neurotoxicol Teratol 1993;15:261–6.

[179] Klintsova AY, Cowell RM, Swain RA, Napper RM, Goodlett CR, Greenough WT. Therapeutic effects of complex motor training on motor performance deficits induced by neonatal binge-like alcohol exposure in rats. I. Behavioral results. Brain Res 1998;800:48–61.

[180] Klintsova AY, Scamra C, Hoffman M, Napper RM, Goodlett CR, Greenough WT. Therapeutic effects of complex motor training on motor performance deficits induced by neonatal binge-like alcohol exposure in rats: II. A quantitative stereological study of synaptic plasticity in female rat cerebellum. Brain Res 2002;937:83–93.

[181] Berman RF, Hannigan JH, Sperry MA, Zajac CS. Prenatal alcohol exposure and the effects of environmental enrichment on hippocampal dendritic spine density. Alcohol 1996;13:209–16.

[182] Wainwright PE, Levesque S, Krempulec L, Bulman-Fleming B, McCutcheon D. Effects of environmental enrichment on cortical depth and Morris-maze performance in B6D2F2 mice exposed prenatally to ethanol. Neurotoxicol Teratol 1993;15:11–20.

[183] Hamilton GF, Jablonski SA, Schiffino FL, St Cyr SA, Stanton ME, Klintsova AY. Exercise and environment as an intervention for neonatal alcohol effects on hippocampal adult neurogenesis and learning. Neuroscience 2014;265:274–90.

[184] Chen SY, Charness ME, Wilkemeyer MF, Sulik KK. Peptide-mediated protection from ethanol-induced neural tube defects. Dev Neurosci 2005;27:13–9.

[185] Wilkemeyer MF, Chen SY, Menkari CE, Brenneman DE, Sulik KK, Charness ME. Differential effects of ethanol antagonism and neuroprotection in peptide fragment NAPVSIPQ prevention of ethanol-induced developmental toxicity. Proc Natl Acad Sci USA 2003;100:8543–8.

[186] Kotch LE, Chen SY, Sulik KK. Ethanol-induced teratogenesis: free radical damage as a possible mechanism. Teratology 1995;52:128–36.

[187] Peng Y, Kwok KH, Yang PH, Ng SS, Liu J, Wong OG, et al. Ascorbic acid inhibits ROS production, NF-kappa B activation and prevents ethanol-induced growth retardation and microencephaly. Neuropharmacology 2005;48:426–34.

[188] Heaton MB, Mitchell JJ, Paiva M. Amelioration of ethanol-induced neurotoxicity in the neonatal rat central nervous system by antioxidant therapy. Alcohol Clin Exp Res 2000;24:512–8.

[189] Long L, Li Y, Wang YD, He QY, Li M, Cai XD, et al. The preventive effect of oral EGCG in a fetal alcohol spectrum disorder mouse model. Alcohol Clin Exp Res 2010;34:1929–36.

[190] Kim JA, Druse MJ. Protective effects of maternal buspirone treatment on serotonin reuptake sites in ethanol-exposed offspring. Brain Res Dev Brain Res 1996;92:190–8.

[191] Tajuddin NF, Druse MJ. In utero ethanol exposure decreased the density of serotonin neurons. Maternal ipsapirone treatment exerted a protective effect. Brain Res Dev Brain Res 1999;117:91–7.

Prenatal Infection: Setting the Course of Brain Aging and Alzheimer's Disease?

Irene Knuesel

INTRODUCTION

Epidemiological and experimental evidence suggests that the prenatal exposure to infections increases the risk for neuropsychiatric diseases including autism spectrum disorder and schizophrenia [1–3]. Numerous experimental studies involving prenatal infection and immune activation in rodents have directly tested the hypothesis of causality and explored crucial genetic, immunological, neurodevelopmental, as well as environmental

Environmental Factors in Neurodevelopmental and Neurodegenerative Disorders
http://dx.doi.org/10.1016/B978-0-12-800228-5.00004-2

67

factors implicated in determining the vulnerability to maternal infection-induced neurodevelopmental diseases in humans [4]. Notwithstanding the limitations of modeling complex neuropsychiatric traits in rodents, prenatally triggered abnormalities in neuroimmune mechanisms are increasingly accepted as crucial etiological factors in neurodevelopmental and potentially other neurological disorders as well [5]. Despite the recent appreciation that aging-associated neurodegenerative diseases are initiated decades before manifestation of dementia symptoms [6], the view that they may even have a neurodevelopmental origin is rather unexpected and, for many scientists, far-fetched. However, support for this hypothesis has recently been provided by two magnetic resonance imaging (MRI) studies in infants [7,8]. The two groups independently showed that infants with a high genetic predisposition for Alzheimer's disease (AD; ApoE4 allele carriers), had significant white-matter changes and reduced gray-matter volume in areas preferentially affected by AD as compared to noncarriers, highlighting that impairments in early brain development may indeed increase the risk for aging-associated neurodegenerative diseases. Interestingly, the discussion on this putative link is as old as AD. It has been initiated more than 100 years ago; concomitantly with the first descriptions of amyloid plaques and neurofibrillary tangles by Alois Alzheimer [9] and Oskar Fischer [10] and the definition of AD by the German psychiatrist Emil Kraeplin [11]. In contrast to presenile dementia described by Alzheimer, Fischer with his clinicopathological studies of almost 300 brains showed a strong interest in senile forms of dementia [12]. In his 1907 paper, Fischer described for the first time the neuritic plaques (Figure 4.1), characterized by club-shaped dystrophic neurites accompanying amyloid deposits [10]. He interpreted this as "proliferative changes of the neuronal skeleton" that reminded him of developing axonal growth cones, which were first described by Ramon Y. Cajal [13]. Although both Alzheimer and Fischer believed them to be degenerative, Cajal thought that the axonal component of the Fischer plaques were the "genuine expression of regeneration" [14]. Despite decades of extensive research in the field of neurodegeneration in general and AD in particular, the question of what triggers the formation of the neuropathological hallmarks and initiates a pathophysiological cascade that culminates in progressive neurodegeneration has remained unanswered ever since. Although the focus in the past has clearly been on degenerative processes and molecular mechanisms underlying plaque-associated neurotoxicity during aging, the latest developments in genetics, imaging, neuropsychology, and neuropathology have revived Cajal's view and indicated a considerable interest in delineating the biological mechanisms that facilitate and support successful brain aging [15]. In an attempt to bridge the early interpretation of regenerative events during brain aging

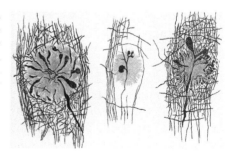

FIGURE 4.1 Oskar Fischer's drawing of neuritic plaques from patients with senile dementia. Fischer interpreted the abnormal, club-shaped neurites as "proliferative changes of the neuronal skeleton"; reminiscent of developing axonal growth cones, as first described by Ramon y Cajal. Cajal commented that the axonal component of the Fischer plaques were the "genuine expression of regeneration." See text for further discussion. *Figure adapted from [12].*

with the current status of AD research, the present article discusses the role of genetic and environmental factors in regulating fundamental neurodevelopmental programs that are essential for proper neuronal functions and deterministic for the brain's responsiveness to cumulative exposures to detrimental environmental elements during adulthood and aging.

WHAT CHARACTERIZES ALZHEIMER'S DISEASE?

The young age of Alois Alzheimer's first patient, Auguste Deter, as well as the neuropathological description of four additional cases of dementia with combined plaques and tangles, were the basis for Emil Kraeplin's decision to name and define AD as a rare form of presenile dementia in the eighth edition of his textbook of Psychiatry [11]. The recent identification of a presenilin 1 (PSEN1) mutation in the genome of Auguste Deter [16] retrospectively supported Kraeplin's view to separate this condition from senile dementia, a well-known neurological disorder at the time. Besides PSEN1, also PSEN2, and amyloid precursor protein (APP) mutations are causative for the rare early-onset or familial form of the disease [17–20]. The mutations have been shown to profoundly alter APP metabolism and favor the production of aggregation-prone amyloid-β (Aβ) species [21,22], as well as their accumulation in extracellular amyloid plaques [23]. The second neuropathological hallmark of AD, the intracellular neurofibrillary tangles (NFTs), which were shown to consist of hyperphosphorylated tau [24], are not linked to dominant mutations in the tau-encoding gene (MAPT) in AD. MAPT mutations are, however, causative for the development of frontotemporal dementia and other forms of tauopathies [25–27]. Based on the observation that tau-associated pathologies can exist without amyloid plaques but not vice versa, tau was discounted as the fundamental cause in AD [28]. However, the neuropathological staging of AD by Braak and Braak demonstrated that abnormal tau is present from the beginning until the end phase of the disorder [29]; initiated in catecholaminergic brain-stem neurons [30]. In contrast to amyloid plaque deposition, the tau-associated pathology correlates with cognitive performance [31], confirming its central role in progressive neurodegeneration. Most recently, Braak and colleagues revealed that pretangle tau inclusions in the proximal axon of locus coeruleus projection neurons develop already during childhood, however, not all of these inclusions convert into argyrophilic neurofibrillary lesions [32–34]. The biological significance of this phenomenon, particularly the mechanisms involved in the aging-related conversion from soluble to stable fibrillary inclusions, as well as the link to amyloid plaque deposition, remains to be elucidated. Although the data fit with the novel concept that genetic and/or environmental disruptions of neurodevelopmental processes enhance the risk for accelerated brain aging and neurodegeneration, it is apparent that the early initiation and temporal progression of the tau pathology is not in agreement with the prevalent "amyloid cascade hypothesis." It is based on the genetics and experimental discoveries related to the dominant-inherited AD mutations and places the generation of Aβ peptide as the most upstream event in the pathophysiology of AD [35,36], reviving the old question of whether the two subforms are distinct entities. The discussion on the molecular, cellular, as well as clinical differences between familial and sporadic AD are highly relevant for the present discussion of therapeutics, because they are almost exclusively based on pathogenic mutations, yet the target patient population is sporadic. The fusion of the two subtypes into one disease entity dates back to the 1970s when it became widely accepted that the neuropathology of patients with clinically diagnosed presenile (onset before 65 years) and senile (onset after 65) dementia were indistinguishable, with the consequence that the concept of AD widened significantly. Indeed, this redefinition ranked AD as one of the most common causes of death [37] and turned this progressive degenerative condition into a household name; synonymous with the most common form of progressive, aging-related memory loss.

TOWARD A NOVEL ANIMAL MODEL OF SPORADIC ALZHEIMER'S DISEASE

Besides the recent discovery of a rare coding mutation (A673T) in the APP gene, which protects against AD [38], the strong support for the amyloid cascade hypothesis is also tightly linked to the generation of transgenic mice overexpressing single or multiple pathogenic mutations that develop AD-like neuropathology. Although they have greatly enhanced our understanding of the biology of AD-causing mutations as drivers of the genetic variant of the disease, they do not model the etiological factors implicated in sporadic AD. Moreover, the transgenic mice are incomplete models because they do not develop NFTs and show no widespread, progressive neuronal cell death, nor significant cognitive decline [39,40]. Related to the mild to moderate cell loss, they do not reveal strong neuroinflammatory reactions; the typical accumulation of activated microglia and astrocytes seen in the vicinity of Aβ plaques and neuronal lesions in AD patients [41–47]. Through cross talk with neurons, other glia, and circulating immune cells, microglia function as primary central nervous system (CNS) guardians and contribute to tissue homeostasis and optimal neuronal functioning by constant surveillance of the brain parenchyma and elimination of potential mediators of infection and cellular damage [48]. Peripheral or local exposure to noxious stimuli induces potent inflammatory responses via toll-like receptor signaling and nuclear factor 'kappa-light-chain-enhancer' of activated B-cells (NF-κB) activation. However, this response depends on context-dependent gene activation patterns and is comparable to M1 ("proinflammatory"/classical) and M2 ("anti-inflammatory"/alternative) states of resident tissue macrophages [49]. Excessive inflammatory conditions favor M1-like states in microglia, a phenotype that is exacerbated by diminished blood–brain barrier (BBB) functions and enhanced recruitment of peripheral immune cells [50]. Thus, the functional and morphological changes in astrocytes and microglia are not only an invariant pathological feature of AD; they also likely reflect significant alterations in innate immune activation and failure to respond appropriately toward aging-associated accumulation of cellular debris (Figure 4.2). In line, the abundant presence of inflammatory mediators is being increasingly acknowledged as a process associated with the onset and progression of AD rather than representing a consequence and passive inflammatory reaction [51]. This hypothesis has recently received strong support by genome-wide association studies, which provided evidence of a significant overrepresentation of genetic risk loci linked to cellular immune functions in sporadic AD patients [52]. However, despite the accumulating genetic, neuropathological, and immunological findings, it has been difficult to monitor the temporal evolution of the neuroinflammatory changes and identify causal factors and downstream effects in AD. To elucidate the role of inflammatory modulators in the etiology of sporadic AD, we initiated a project to develop a murine model that should allow us to investigate the mechanistic link between inflammatory conditions and the development of AD-like neuropathology, neurodegeneration, and

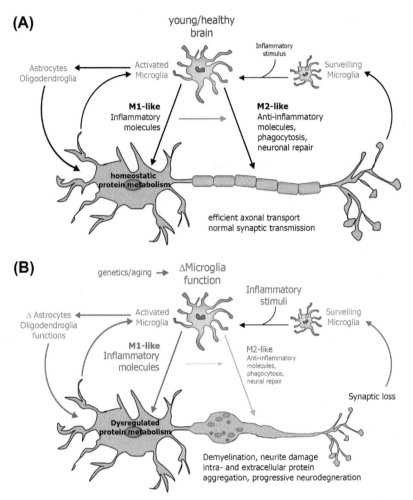

FIGURE 4.2 **Model of glial-mediated neuroprotection.** (A) In the young/healthy brain, inflammatory stimuli (e.g., injury, cellular debris, etc.) are perceived by surveiling microglia and transmitted to neighboring cells. This will provoke—in both glia and neurons—the induction of a "defense-program." The polarization of microglia to an M1-like state is transient and followed by the activation of signaling pathways to resolve inflammatory signaling and induce phagocytotic activity (M2-like state). (B) The persistence of inflammatory signals, together with an aging-associated decline in microglia functions, is expected to result in chronic cellular stress and dysregulated protein homeostasis in neurons. Accumulation of misfolded proteins further impairs fundamental cellular processes including axonal transport. The viscous cycle of pathophysiological events in neurons and glia ultimately results in degeneration and progressive atrophy characteristic of AD.

cognitive decline in the physiological context of aging. The approach was based on previous studies from our and other groups, demonstrating that systemic administration of the synthetic cytokine releaser and viral mimic polyinosinic:polycytidylic acid (PolyI:C) in mice results in long-term behavioral and neurochemical abnormalities in the adult offspring [3,53,54]. We had shown that the middle and late gestation periods corresponded to two windows of enhanced vulnerability to inflammatory cytokines, leading to an acute response in the fetal brain, delayed brain neuropathology, and significant behavioral changes [55–60]. Importantly, whereas mid-gestational immune activation did not result in persistent alterations in the levels of cytokines and chemokines, late-gestational PolyI:C exposure was characterized by long-term alterations of several cytokines, chemokines, growth factors, and other immune modulators in brain and plasma [61,62]. Moreover, we demonstrated for the first time that late-gestational viral-like infections in wildtype mice induce a distinct neuropathy in the hippocampus that is characterized by abnormal protein accumulations in axonal and dendritic varicosities [63]. In comparison to age-matched controls, our three-dimensional (3D) immunoelectron microscopical analysis revealed that prenatally immune-challenged mice showed a significantly higher density of the neuritic varicosities containing mitochondria, vacuoles, and cellular debris, confirming their intracellular origin and degenerative state [63]. The immunological and morphological changes in aged immune-challenged wildtype mice were paralleled by a significant elevation of murine APP and its proteolytic peptides, mislocalization of phosphorylated tau into somatodendritic compartments, as well as significant memory impairments [61], indicating that a systemic immune challenge during late gestation predisposes wildtype mice to develop AD-like neuropathology and cognitive decline during the course of aging. To test whether this phenotype could be aggravated further, we exposed the offspring to a second viral-like infection during aging. This second immune challenge strongly exacerbated the phenotype and indeed precipitated AD-like neuropathological changes, including a significant accumulation of amyloid precursor plaques, aggregation of phosphorylated tau in neuronal somata, as well as significant microglia activation, and reactive gliosis in the hippocampal formation [61]. Although Aβ peptides were not significantly enriched in extracellular deposits of double immune-challenged wildtype mice at 15 months, they dramatically increased in age-matched immune-challenged transgenic mice, precisely around the inflammation-induced accumulations of APP and its fragments, in striking similarity to the postmortem findings in human AD patients [61]. These findings suggest that amyloid plaques may originate in axonal varicosities that are filled with stuck APP-containing transport vesicles.

With this study, we also demonstrated that chronic inflammatory conditions induce an AD-like phenotype in wildtype mice during aging; representing a novel model of sporadic AD without the potential confounding factors of transgene overexpression. A major advantage of this model, however, was the possibility to investigate the temporal and spatial evolution of the pathophysiological changes during aging. Indeed, we were able to demonstrate that tau hyperphosphorylation precedes the accumulation of APP, in line with the postmortem findings in humans [64]. In addition, the abnormal accumulation of APP within axonal varicosities likely not only represents a significant precursor condition for the deposition of aggregation-prone peptides, but also for the mis-sorting of tau into somatodendritic compartments, in which it might transform into insoluble aggregates. Based on all these observations, we constructed a sequence of events and integrated the major findings identified in postmortem human studies accumulated in the past. The result was a proposition for a molecular mechanism that may underlie the earliest pathophysiological changes involved in Aβ plaque and NFT formations in sporadic AD [65]. The key points of our formulated inflammation hypothesis of AD presented in this paper include: (1) chronic inflammation coupled with neuronal aging induces cellular stress and concomitant impairments in basic cellular functions; (2) inflammation-induced hyperphosphorylation and mis-sorting of tau represents one of the earliest neuropathological changes in sporadic AD; (3) molecular changes underlying late-onset AD involve impairments in cytoskeletal stability and axonal transport, which trigger axonal degeneration, the formation of senile plaques and NFTs ultimately resulting in neuronal death; and (4) amyloid plaques form in the projection zone of tangle-bearing neurons; an observation that we recently discussed in depth [66] and received support by Braak and Del Tredici [67].

In the following, I will provide an expanded and updated view of the proposed molecular and cellular mechanisms that are affected by the pre- and postnatal immune challenges and implicated in the altered responsiveness of peripheral and brain immune cells during aging and AD.

CHRONIC NEUROINFLAMMATION—IMPACT ON BASIC CELLULAR FUNCTIONS

Several lines of research have provided convincing evidence for an involvement of inflammatory processes in AD. In line with microarray data involving aged human and murine mouse brain tissue pointing to the increased transcriptional activity in genes related to cellular stress and inflammation [68,69], a prominent view considers a shift from a neuroprotective to a neurodegenerative role of the innate immunity upon release of damage signals that are triggered by accumulating cellular distress in the aging brain [70]. It is based on the assumption that chronic exposure of microglia to Aβ leads to uncontrolled inflammation and the release of neurotoxic free radicals and reactive oxygen species (ROS) [71]. It is unclear though, how Aβ influences microglia activity in vivo, because knockout and genetic inactivation of the suggested Aβ-binding receptors and downstream effectors, Toll-like receptors (TLR)-4 [72,73], TLR2 [74], and MyD88 [75], TNFα receptors [76] in transgenic AD mouse models all aggravate amyloid deposition and cognitive function. Moreover, selective depletion of microglia in vivo was shown to have no effect on initiation or density of Aβ deposition in transgenic AD mice [77]. It is conceivable that this is due to the rapid repopulation by peripheral monocytes [78] or perivascular microglia precursor cells [79]. However, it remains to be investigated whether the new/naïve microglia would be more effective in clearing amyloid deposition than chronically Aβ-exposed microglia that may have lost their clearance capacity. The latter view is in line with experimental data pointing to a dysfunctional and senescent state of microglia in aging and AD [80,81], indicating that a hyperactive state and increased secretion of inflammatory mediators, combined with diminished neurotrophic support and downregulation of phagocytic functions by microglia, may lead to inefficient clearance of neurotoxic protein aggregates and progressive neuronal cell death [82].

The experimental data acquired with our mouse model support this view; however, the data include a complementary interpretation that involves a detrimental effect of chronically elevated inflammatory modulators on basic neuronal functions that culminate in progressive axonal and dendritic degeneration during aging, a process that appears to favor the focal accumulation of Aβ deposits and mislocalization and hyperphosphorylation of tau in somatodendritic compartments [61,63]. In line with their role in immune surveillance and phagocytosing apoptotic neurons [83], microglia might be primarily recruited to clear the fragmented neurons. Accordingly, their engagement in removal of

Aβ aggregates could be interpreted as an accompanying effect of the clearing process of degenerating neurites containing misfolded and damaged proteins. This hypothesis in line with the findings that undigested Aβ species can be released by microglia [84] and the above-mentioned study that microglia ablation does not affect amyloid deposition [77]. In the old human brain, however, the homeostatic balance is likely lost and replacement of degenerated glia hypothesized to decline due to replicative senescence [85]. Aimed at further supporting a proximal role of early neurodegenerative processes in the pathogenesis of sporadic AD, the following paragraphs highlight, based on our data and a selection of published data, the impact of chronic, systemically induced brain inflammation on synaptic and neuronal functions that are proposed to initiate a vicious cycle ultimately leading to progressive neurodegeneration and aging-associated dementia.

Systemic Infections and Their Impact on Neuronal Functions

The increase in proinflammatory cytokines in the serum accompanying systemic inflammations have been shown to play a fundamental role in immune to brain communication by activating the central innate immunity and initiation of a behavioral response, known as sickness behavior; an evolutionary conserved protective mechanism characterized by increased anxiety, depressed mood, and apathy (for review, [86]). A similar acute and transient infection-induced state that is frequently observed in the elderly is delirium, a well-described impairment in global cognitive function, which causes marked impairments in consciousness, attention, immediate recall, orientation (temporal and spatial), and perception [87,88]. Although increasingly accepted that peripheral infections with and without episodes of delirium significantly accelerate cognitive decline, specifically in AD patients [89,90], the molecular mechanisms underlying the infection-induced behavioral responses are poorly understood. Recent experimental data indicate that the perivascular macrophages play a crucial role in immune-brain communication by mediating prostaglandin-induced activation of the hypothalamic–pituitary–adrenal axis [91]. The brain-resident macrophages have been shown to exert an anti-inflammatory function that suppresses endothelial and CNS responses to inflammatory insults [92,93]. Moreover, astrocytes, which are in close apposition to the cerebral vasculature, help maintain BBB integrity and immune quiescence through contact-dependent mechanisms and through the release of soluble factors, such as transforming growth factor beta [94], neurotrophic factors [95,96], fibroblast growth factors [97], and retinoic acid [98] that contribute to optimal BBB functioning. In line, ablation of astrocyte reactivity has been shown to increase leukocyte trafficking into the CNS, BBB repair deficiency with vasogenic brain edema production, demyelination, and neuronal degeneration [99–101]. Astrocytes also express a number of receptors for molecules classically referred to as brain morphogens, including netrins and sonic hedgehog, which not only have crucial roles during development but recently have also been implicated in regulating BBB function, CNS homeostasis, and immune quiescence [102]. Accordingly, chronic inflammatory conditions and impairments of the BBB during aging are expected to profoundly increase central cytokine expression by microglia, astrocytes, and potentially also neurons [103,104]. In line, numerous preclinical studies have provided evidence for a highly consistent impairment in cognitive functions upon acute and chronic systemic infections, with aged rats and mice showing higher vulnerability as indicated by an exaggerated and long-lasting proinflammatory cytokine response that is not observed in young subjects [105]. The potentiated cytokine response in aged rodents has been attributed to a priming effect of microglia by aging-associated increase in antigen-presenting protein complexes, such as major histocompatibility complex class II molecules [106] or prior neurodegenerative pathology [107,108]. Our own data support these observations by demonstrating that a double viral-like immune challenge involving PolyI:C injection during late gestation and aging resulted in a pronounced increase in activated microglia in the hippocampus as compared to the single-exposed animals [61]. It is conceivable that these changes are linked to the detrimental effect of the prenatal immune challenge on the developing microglia, the yolk sac-derived immune cells of the brain, which start to colonize the brain during late-gestational stages [109,110]. This insult likely not only affects their proliferative and phagocytotic capacity during adulthood and aging, but maybe also induces their priming, thereby creating an innate immune memory that allows a faster and exaggerated response upon further immune challenges and exposures to adverse stimuli [5,111]. Moreover, the early exposure of microglia to excessive levels of inflammatory modulators may interfere with their recently discovered role in surveillance of synaptic maturation and pruning [112] and indicates that deficits in microglia function may contribute to synaptic abnormalities. They are not only characteristic features of neurodevelopmental and neurological disorders, but also constitute one of the earliest features of AD [113]. Given the recent findings of greater amounts of vesicle-associated membrane protein, complexins I and II, and the synaptosomal-associated protein (SNAP)-25/syntaxin interaction that were associated with lower odds of dementia [114], it is tempting to speculate that prenatal infection-induced synaptic abnormalities may constitute a crucial structural and functional component of the cognitive reserve that strongly influences the risk of dementia with aging [115].

Inflammatory Cytokines and Their Impact on Reelin-Mediated Neuronal Functions

In agreement with their proposed detrimental role on synaptic plasticity and learning and memory [116–118], proinflammatory cytokines released by activated microglia likely affect a multitude of molecules at synaptic sites, either directly or indirectly via their significant influence on astrocytic functions [119]. One of the molecules strongly affected by chronic inflammation is Reelin [55], a highly conserved extracellular protein secreted by the first neurons formed in the brain [120]. The protein and its downstream-signaling members control neuronal migration and cortical lamination during development (for recent review, [121]). The same pathway is used in the adult brain in which Reelin and its lipoprotein receptors are pivotal regulators of glutamatergic synapses, neuronal plasticity [122], and dendritic remodeling [123–126]. The Reelin signaling pathway is also involved in suppressing tau hyperphosphorylation [122,127,128], inhibiting APP shedding [129], decelerating the formation and reducing the density of Aβ plaque deposition [130], and modulating receptor recycling in competition with ApoEε4 [131], the major genetic risk factor of late-onset AD [132,133], thereby converging multiple disease-relevant signaling cascades in the adult brain.

Recent work from our group provided the first evidence that normal aging is accompanied by a reduction in Reelin protein levels and the number of Reelin-positive gamma-aminobutyric acid (GABA)ergic interneurons in the hippocampal formation in rodents and nonhuman primates [55], a finding, which has been recently reproduced in aged rats [134], as well as seen in Alzheimer's patients [135]. We discovered that Reelin accumulates in amyloid-like aggregates within the hippocampal formation [55,136–138], representing a highly conserved phenomenon also seen in postmortem human brain tissue [139], providing a putative cause for the reduction of functional Reelin protein at synaptic sites and the concomitant impairments in cognitive performance [134]. This aging-related process is dramatically accelerated in prenatally immune-challenged mice, resulting in much earlier and more aggravated accumulations of Reelin along the perforant path in the adult hippocampus [55,63]. By applying 2D- and 3D-immunoelectron microscopy, we have recently confirmed that Reelin accumulates in axonal and dendritic compartments in spheroid-like varicosities, a morphological feature that appears to involve budding-like extrusions in normal aged and saline-exposed mice, potentially reflecting a neuroprotective strategy of postmitotic neurons [63]. In line with Cajal's view, this phenomenon may also reflect axonal sprouting and the attempt of a potentially damaged neuron to form collaterals. The significant increase in Reelin-containing spheroids could indeed also reflect a reinitiation of a developmental program to promote neuritic outgrowth [140]. However, upon chronic inflammatory conditions and aging-associated impairments in basic cellular functions, these attempts may fail and even contribute to progressive failures in proper neuritic transport, as indicated by our findings of significantly more mitochondria, vacuoles, and silver-positive cellular debris present in Reelin-positive varicosities as compared to age-matched controls [63]. Interestingly, a similar phenotype has been described in *reeler orleans* (*Reln^{Orl}*) mice, a spontaneous mutant that harbors an L1 transposon in the *reelin* coding region, which results in the translation of a C-terminally truncated form of the protein that aggregates intracellularly and therefore cannot be secreted [141]. Characteristic morphological feature of these mice are the numerous axonal varicosities each containing spheroidal cisterns filled with Reelin-positive fibrillary material [142]. This distinct axonal phenotype in *Reln^{Orl}* mice, which also includes a significant increase in the numerical density of the axonal beads compared to wildtype mice, is not seen in full *reeler* knockout mice [142], indicating that the neuropathology is not due to a loss of Reelin function but directly related to the accumulation of the truncated protein.

Based on the striking morphological similarity between the axonal phenotype of immune-challenged wildtype mice [63] and the *Reln^{Orl}* mutants expressing a truncated form of Reelin [142], we hypothesized that inflammation-induced changes in proteolytic processing of Reelin may underlie the development of Reelin-positive neuritic varicosities. First biochemical evidence from our group indeed suggests that abnormal proteolytic cleavage of Reelin through the serine (tissue Plasminogen Activator, tPA) and matrix metalloproteinase (A Disintegrin And Metalloproteinase with thrombo-spondin motif, ADAMTS4/5) alters the ability of Reelin to form dimers and facilitates its oligomerization [143]. In line with the inflammatory cytokine-mediated induction of tPA [144] and ADAMTS4/5 [145,146] as well as the strong association of tPA with neuronal injuries [147], chronic inflammatory conditions are expected to profoundly impair Reelin-dependent signaling through alterations in its proteolytic processing. In agreement with the prediction that accumulation of Reelin and its fragment in axonal terminals is restricted to projection zones of Reelin-expressing neurons, the earliest aging-associated neuritic varicosities are indeed seen along the olfactory–limbic pathways in which Reelin is normally secreted from axonal terminals [66]. Moreover, these are also the brain areas with earliest amyloid pathology and that are innervated by noradrenergic and serotonergic brain-stem nuclei, which are—as indicated previously—also the first with abnormal accumulation of tau [30,67].

FROM AXONAL VARICOSITIES TO AXONAL DEGENERATION

Besides Reelin and its fragments, several proteolytic fragments of APP, as well as α-synuclein co-accumulate with Reelin in neuritic varicosities [148], indicating that both aging and chronic inflammatory states favor the aggregation of a multitude of abnormally processed and/or misfolded proteins in neuronal compartments. This hypothesis is in line with recent findings showing that IL1β, interferon-γ, and numerous other inflammatory stimuli induce through the mTor signaling pathway an increase in protein synthesis [149–151]. However, a large fraction of newly generated proteins are defective in folding, translation, and/or assembly, likely linked to the effect of oxygen radicals that are the result of the concomitant activation of the inducible nitric oxide synthase system through the inflammatory stimulus [152]. Seifert and colleagues recently provided the first evidence that these inflammation-induced pool of damaged proteins are selectively degraded through the immunoproteasome [151], a fast acting variant proteasome that—in addition to promoting an adaptive immune response through increased antigen presentation—also protects cells from the damaging side effects of an innate inflammatory response [152]. It is conceivable that chronic elevation of inflammatory cytokines as seen following a prenatal immune challenge [61] impairs the activity of the immunoproteasome in aging neurons. This in turn is expected to favor the intracellular accumulation of misfolded proteins, in agreement with previous findings demonstrating that several fundamental cellular processes including proteosomal degradation decelerates during aging [153]. The inflammatory stress-induced increase in protein synthesis may also reflect—again—an attempt to provide neuroprotection by selective induction of proteins involved in neuritic outgrowth and synapse formation/stabilization. In agreement, the pronounced induction of APP following systemic viral- [61] and bacterial-like infections [154], or direct application of IL1β to primary cultured neurons [155,156] as well as the intact brain [157] could be interpreted as a cellular stress-related response [158]. A similar attempt at neuroprotection could potentially underlie the significant increase in APP following traumatic head injury, in both rodents and humans [159,160]. In line with this view and the high turnover, rapid anterograde transport, and processing of APP in distal compartments [161–163]; genetically induced fiber tract degeneration also provokes distinct axonal accumulations of APP, as described in the gracile axonal dystrophy mouse [164], implying that focal blockage of axonal transport results in rapid and significant enrichments of APP in distal neurites. Disruption of axonal and dendritic transport, accompanied by increased levels of C-terminally cleaved APP fragments, has also been described following impaired lysosomal proteolysis [165], in agreement with previous reports showing that focal accumulation of autophagic vacuoles containing incompletely digested proteins and disintegrated organelles in focal axonal swellings are distinct neuropathological hallmarks of AD [166,167], reinforcing the crucial contribution of impaired autophagy-lysosomal pathways in aging-associated neurodegeneration (for recent review, [168]). Experimental evidence involving transgenic animals highlighted that the axonal enlargements with the accumulation of multiple axonal cargoes and cytoskeletal proteins precede the typical disease-related pathology, a feature that has been suggested to contribute to amyloid-β deposition [169,170]. Focal accumulation of lysosome-related vesicles might be sufficient to induce amyloidogenic processing of APP as hypothesized earlier by Bernstein et al. [171] and Cataldo and Nixon [172] based on their postmortem studies involving patients with AD and by us [65]. Based on the findings by Adalbert and colleagues demonstrating that dystrophic axons associated with amyloid plaques remain continuous and connected to neuronal somata in a transgenic mouse model of AD [173], early stages of the disease conceivably still allow certain transport processes but a lack of stabilizing presynaptic proteins, including APP [174], is expected to trigger synaptic disconnections, a well-described feature of early stages of AD pathophysiology [113]. Secondary impairments in mitochondrial transport and energy supply for the distal compartments are expected to promote axonal degeneration [175], potentially involving a "dying back" mechanism as suggested by Coleman [176].

Formation of Senile Plaques: Plaques and Neurofibrillary Tangles United

Dystrophic axons associated with amyloid-β plaques are ideally placed to link Aβ with the microtubule-stabilizing tau protein and NFT neuropathology, a view that is in line with the recently formulated hypothesis by Braak and Del Tredici [67]. So far, investigations on the molecular link between amyloid and NFTs have mainly centered on the amyloid cascade-derived upstream role of Aβ peptides on tau localization and phosphorylation. The concept included either Aβ as neurotoxic agent that interferes with proper synaptic functions, thereby inducing mis-sorting of tau from axonal to somatodendritic compartments and concomitant destabilization of microtubules [177,178] or suggests a dendritic function of tau that can interfere with proper NMDA receptor function at glutamatergic synapses and mediate Aβ-related toxicity [179,180]. Rarely, experimental investigations included APP as a putative contributing factor. In support of the latter, our findings of a significant increase in tau phosphorylation and aggregation, as well as mis-sorting from axonal to dendritic compartments following chronic inflammatory conditions [61] indicates that tau hyperphosphorylation impairs axonal transport, which in turn results in focal accumulations of APP; ultimately resulting in synaptic and axonal degeneration. This in turn may prevent the axonal targeting of tau and hence results in the accumulation of tau in soma and dendrites [32,65]. Inflammatory stimuli are expected to aggravate tau aggregation under chronic conditions [61,181], a phenomenon that is expected to further impair axonal transport mechanisms and ultimately result in degeneration of the whole cell. Our hypothesis challenges the current view that Aβ (in oligomeric form) is the initiating neurotoxic species, but rather confirms the importance of proper basic neuronal functions that are strongly affected by aging and chronic inflammatory conditions. In agreement with earlier reports, certain C-terminal APP species [182] could indeed exert secondary neurotoxic effects, but it is equally likely that other APP fragments including Aβ species may enhance and promote neuronal survival and repair, as proposed in the late 1980s [183] and reviewed recently [158].

CONCLUSIONS

The sequence of events proposed in this review is based on our recent experimental findings obtained from aged wildtype mice that received multiple challenges with a viral-like infection. The first was applied during late gestation, a time point of extensive synaptogenesis and initiation of microglia functions. This inflammatory insult is expected to strongly interfere with the recently discovered neurodevelopmental functions of microglia in surveillance of synaptic maturation that might result in long-term structural and functional alterations in neuronal circuitries. Moreover, the early exposure to a strong immune challenge likely results in microglia priming and hence altered responsiveness toward further inflammatory stimuli. In our model, we also applied a second immune challenge during aging; a vulnerable phase due to the decline in basic cellular functions and impairments in BBB integrity. Together, these manipulations were sufficient to induce long-term immunological, neurological, and behavioral changes that mimicked several features of early stages of sporadic AD. Our data suggest that the chronic inflammatory state observed in the double immune-challenged mice strongly contributes to protein misfolding and impairments in protein clearance. In addition to putative aging- and inflammation-induced impairments in mitochondrial functions, local accumulations of proteins destined to be transported to synaptic sites but stuck in axonal varicosities, which further impair axonal transport. It is conceivable that the concomitant loss of function of these proteins at synaptic sites will result in destabilization of neuronal networks and loss of synaptic contacts. Under chronic insulting conditions, progressive axonal dystrophy will result in complete degeneration of neurons. Importantly, the development of the neuropathology in the limbic–olfactory brain areas can be explained without the requirement of a "*trans*-synaptic spread" that recently received much attention [184,185]. Rather, we propose that the consistent progression follows interconnected brain areas and can be interpreted as a "dys-connection phenomenon" with the terminal NFT pathology seen in neuronal somata and amyloid plaque deposition in their axonal projection zone. This view is in line with the Braak staging [29] and the recent reports on pathological tau changes in neurons of catecholaminergic brain-stem nuclei [34], progressing from olfactory–limbic to isocortical association areas as the disease evolves. Interestingly, this stereotypic progression overlaps with the presence of axonal projections derived from Reelin-expressing cells located in the olfactory bulb, piriform and entorhinal cortex, as well as hippocampus and cerebellum [66]. We have therefore recently proposed that the reduction of Reelin, because of cumulative environmental injury/disease/infection-induced chronic inflammation, accelerates the age-dependent phosphorylation of tau and the instability of interneuronal connections [66]. The temporal sequence of the synaptic deficits would accordingly first predict impairments in olfactory information processing, followed by entorhinal cortex/hippocampus-dependent cognitive dysfunctions, clinical features that are indeed typical of early AD [186]. This view highlights the importance of Reelin-mediated signaling and other neurodevelopmental programs that are maintained in the adult brain and instrumental to modulate synaptic functions, as well as protect neuronal connectivity and integrity during aging. Inflammation/injury-induced dysfunctions of these crucial neurodevelopmental programs may indeed underlie the apparent "spread" of the neuropathology across brain networks in patients diagnosed with AD.

References

[1] Gilmore JH, Jarskog LF. Exposure to infection and brain development: cytokines in the pathogenesis of schizophrenia. Schizophr Res 1997;24:365–7.

[2] Patterson PH. Immune involvement in schizophrenia and autism: etiology, pathology and animal models. Behav Brain Res 2009;204:313–21.

[3] Shi L, Fatemi SH, Sidwell RW, Patterson PH. Maternal influenza infection causes marked behavioral and pharmacological changes in the offspring. J Neurosci 2003;23:297–302.

[4] Meyer U. Prenatal poly(i:C) exposure and other developmental immune activation models in rodent systems. Biol Psychiatry 2014;75:307–15.

[5] Knuesel I, Chicha L, Britschgi M, Schobel SA, Bodmer M, Hellings JA, et al. Maternal immune activation and abnormal brain development across CNS disorders. Nat Rev Neurol 2014;10:643–60.

[6] Musiek ES, Holtzman DM. Origins of Alzheimer's disease: reconciling cerebrospinal fluid biomarker and neuropathology data regarding the temporal sequence of amyloid-beta and tau involvement. Curr Opin Neurol 2012;25:715–20.

[7] Dean 3rd DC, Jerskey BA, Chen K, Protas H, Thiyyagura P, Roontiva A, et al. Brain differences in infants at differential genetic risk for late-onset Alzheimer disease: a cross-sectional imaging study. JAMA Neurol 2014;71:11–22.

[8] Knickmeyer RC, Wang J, Zhu H, Geng X, Woolson S, Hamer RM, et al. Common variants in psychiatric risk genes predict brain structure at birth. Cereb Cortex 2014;24:1230–46.

[9] Alzheimer A. Über eine eigenartige Erkrankung der Hirnrinde. Allg Z Psychiatr 1907;64:146–8.

[10] Fischer O. Miliare Nekrosen mit drusigen Wucherungen der Neurofibrillen, eine regelmässige Veränderung der Hirnrinde bei seniler Demenz. Monatsschr Psychiat Neurol 1907;22:361–72.

[11] Kraeplin E. 8. Auflage Psychiatrie. Ein Lehrbuch für Studierende und Aertze, Band II. Barth, Leipzig: Klinische Psychiatrie; 1910.

[12] Goedert M. Oskar Fischer and the study of dementia. Brain: J Neurol 2009;132:1102–11.

[13] de Castro F, Lopez-Mascaraque L, De Carlos JA. Cajal: lessons on brain development. Brain Res Rev 2007;55:481–9.

[14] May RM. Cajal's degeneration and regeneration of the nervous system. New York: Oxford University Press Inc.; 1991.

[15] Glahn DC, Kent Jr JW, Sprooten E, Diego VP, Winkler AM, Curran JE, et al. Genetic basis of neurocognitive decline and reduced white-matter integrity in normal human brain aging. Proc Natl Acad Sci USA 2013;110:19006–11.

[16] Muller U, Winter P, Graeber MB. A presenilin 1 mutation in the first case of Alzheimer's disease. Lancet Neurol 2013;12:129–30.

[17] Chartier-Harlin MC, Crawford F, Houlden H, Warren A, Hughes D, Fidani L, et al. Early-onset Alzheimer's disease caused by mutations at codon 717 of the beta-amyloid precursor protein gene. Nature 1991;353:844–6.

[18] Goate A, Chartier-Harlin MC, Mullan M, Brown J, Crawford F, Fidani L, et al. Segregation of a missense mutation in the amyloid precursor protein gene with familial Alzheimer's disease. Nature 1991;349:704–6.

[19] Levy E, Carman MD, Fernandez-Madrid IJ, Power MD, Lieberburg I, van Duinen SG, et al. Mutation of the Alzheimer's disease amyloid gene in hereditary cerebral hemorrhage, Dutch type. Science 1990;248:1124–6.

[20] Van Broeckhoven C, Haan J, Bakker E, Hardy JA, Van Hul W, Wehnert A, et al. Amyloid beta protein precursor gene and hereditary cerebral hemorrhage with amyloidosis (Dutch). Science 1990;248:1120–2.

[21] Haass C, Hung AY, Selkoe DJ, Teplow DB. Mutations associated with a locus for familial Alzheimer's disease result in alternative processing of amyloid beta-protein precursor. J Biol Chem 1994;269:17741–8.

[22] Suzuki N, Cheung TT, Cai XD, Odaka A, Otvos Jr L, Eckman C, et al. An increased percentage of long amyloid beta protein secreted by familial amyloid beta protein precursor (beta APP717) mutants. Science 1994;264:1336–40.

[23] Glenner GG, Wong CW. Alzheimer's disease: initial report of the purification and characterization of a novel cerebrovascular amyloid protein. Biochem Biophys Res Commun 1984;120:885–90.

[24] Grundke-Iqbal I, Iqbal K, Tung YC, Quinlan M, Wisniewski HM, Binder LI. Abnormal phosphorylation of the microtubule-associated protein tau (tau) in Alzheimer cytoskeletal pathology. Proc Natl Acad Sci USA 1986;83:4913–7.

[25] Hutton M, Lendon CL, Rizzu P, Baker M, Froelich S, Houlden H, et al. Association of missense and 5′-splice-site mutations in tau with the inherited dementia FTDP-17. Nature 1998;393:702–5.

[26] Spillantini MG, Murrell JR, Goedert M, Farlow MR, Klug A, Ghetti B. Mutation in the tau gene in familial multiple system tauopathy with presenile dementia. Proc Natl Acad Sci USA 1998;95:7737–41.

[27] van Swieten JC, Rosso SM, Heutink P. MAPT-related disorders. In: Pagon RA, Adam MP, Ardinger HH, Bird TD, Dolan CR, Fong CT, Smith RJH, Stephens K, editors. GeneReviews(R). 1993. [Seattle (WA)].

[28] Wisniewski HM, Iqbal K, Bancher C, Miller D, Currie J. Cytoskeletal protein pathology and the formation of beta-amyloid fibers in Alzheimer's disease. Neurobiol Aging 1989;10:409–12. Discussion 412–404.

[29] Braak H, Braak E. Neuropathological stageing of Alzheimer-related changes. Acta Neuropathol 1991;82:239–59.

[30] Braak H, Thal DR, Matschke J, Ghebremedhin E, Del Tredici K. Age-related appearance of dendritic inclusions in catecholaminergic brainstem neurons. Neurobiol Aging 2013a;34:286–97.

[31] Negash S, Bennett DA, Wilson RS, Schneider JA, Arnold SE. Cognition and neuropathology in aging: multidimensional perspectives from the rush religious orders study and rush memory and aging project. Curr Alzheimer Res 2011;8:336–40.

[32] Braak H, Del Tredici K. Where, when, and in what form does sporadic Alzheimer's disease begin? Curr Opin Neurol 2012;25:708–14.

[33] Braak H, Del Tredici K. Evolional aspects of Alzheimer's disease pathogenesis. J Alzheimers Dis 2013b;33(Suppl. 1):S155–61.

[34] Braak H, Thal DR, Ghebremedhin E, Del Tredici K. Stages of the pathologic process in Alzheimer disease: age categories from 1 to 100 years. J Neuropathol Exp Neurol 2011;70:960–9.

[35] Hardy J, Selkoe DJ. The amyloid hypothesis of Alzheimer's disease: progress and problems on the road to therapeutics. Science 2002;297:353–6.

[36] Hardy JA, Higgins GA. Alzheimer's disease: the amyloid cascade hypothesis. Science 1992;256:184–5.

[37] Katzman R. Editorial: the prevalence and malignancy of Alzheimer disease. A major killer. Arch Neurol 1976;33:217–8.

[38] Jonsson T, Atwal JK, Steinberg S, Snaedal J, Jonsson PV, Bjornsson S, et al. A mutation in APP protects against Alzheimer's disease and age-related cognitive decline. Nature 2012;488:96–9.

[39] Ashe KH, Zahs KR. Probing the biology of Alzheimer's disease in mice. Neuron 2010;66:631–45.

[40] Gotz J. Tau and transgenic animal models. Brain Res Brain Res Rev 2001;35:266–86.

[41] Eikelenboom P, Rozemuller AJ, Hoozemans JJ, Veerhuis R, van Gool WA. Neuroinflammation and Alzheimer disease: clinical and therapeutic implications. Alzheimer Dis Assoc Disord 2000;14(Suppl 1):S54–61.

[42] Eikelenboom P, van Exel E, Hoozemans JJ, Veerhuis R, Rozemuller AJ, van Gool WA. Neuroinflammation - an early event in both the history and pathogenesis of Alzheimer's disease. Neurodegener Dis 2010;7:38–41.

[43] Griffin WS, Nicoll JA, Grimaldi LM, Sheng JG, Mrak RE. The pervasiveness of interleukin-1 in Alzheimer pathogenesis: a role for specific polymorphisms in disease risk. Exp Gerontol 2000;35:481–7.

[44] Griffin WS, Stanley LC, Ling C, White L, MacLeod V, Perrot LJ, et al. Brain interleukin 1 and S-100 immunoreactivity are elevated in down syndrome and Alzheimer disease. Proc Natl Acad Sci USA 1989;86:7611–5.

[45] McGeer EG, McGeer PL. Inflammatory processes in Alzheimer's disease. Prog Neuro-psychopharmacol Biol psychiatry 2003;27:741–9.

[46] McGeer PL, Itagaki S, Tago H, McGeer EG. Occurrence of HLA-DR reactive microglia in Alzheimer's disease. Ann NY Acad Sci 1988;540:319–23.

[47] McGeer PL, McGeer EG. The inflammatory response system of brain: implications for therapy of Alzheimer and other neurodegenerative diseases. Brain Res Brain Res Rev 1995;21:195–218.

[48] Nimmerjahn A, Kirchhoff F, Helmchen F. Resting microglial cells are highly dynamic surveillants of brain parenchyma in vivo. Science 2005;308:1314–8.

[49] Colton C, Wilcock DM. Assessing activation states in microglia. CNS Neurol Disord Drug Targets 2010;9:174–91.

[50] Ransohoff RM, Brown MA. Innate immunity in the central nervous system. J Clin Invest 2012;122:1164–71.

[51] Mosher KI, Wyss-Coray T. Microglial dysfunction in brain aging and Alzheimer's disease. Biochem Pharmacol 2014;88:594–604.

[52] Lambert JC, Ibrahim-Verbaas CA, Harold D, Naj AC, Sims R, Bellenguez C, et al. Meta-analysis of 74,046 individuals identifies 11 new susceptibility loci for Alzheimer's disease. Nat Genet 2013;45:1452–8.

[53] Meyer U, Feldon J, Schedlowski M, Yee BK. Towards an immuno-precipitated neurodevelopmental animal model of schizophrenia. Neurosci Biobehav Rev 2005;29:913–47.

[54] Zuckerman L, Rehavi M, Nachman R, Weiner I. Immune activation during pregnancy in rats leads to a post-pubertal emergence of disrupted latent inhibition, dopaminergic hyperfunction, and altered limbic morphology in the offspring: a novel neurodevelopmental model of schizophrenia. Neuropsychopharmacology 2003;28:1778–89.

[55] Knuesel I, Nyffeler M, Mormede C, Muhia M, Meyer U, Pietropaolo S, et al. Age-related accumulation of Reelin in amyloid-like deposits. Neurobiol Aging 2009;30:697–716.

[56] Meyer U, Knuesel I, Nyffeler M, Feldon J. Chronic clozapine treatment improves prenatal infection-induced working memory deficits without influencing adult hippocampal neurogenesis. Psychopharmacol Berl 2010;208:531–43.

[57] Meyer U, Nyffeler M, Engler A, Urwyler A, Schedlowski M, Knuesel I, et al. The time of prenatal immune challenge determines the specificity of inflammation-mediated brain and behavioral pathology. J Neurosci 2006a;26:4752–62.

[58] Meyer U, Nyffeler M, Schwendener S, Knuesel I, Yee BK, Feldon J. Relative prenatal and postnatal maternal contributions to schizophrenia-related neurochemical dysfunction after in utero immune challenge. Neuropsychopharmacology 2008;33:441–56.

[59] Meyer U, Schwendener S, Feldon J, Yee BK. Prenatal and postnatal maternal contributions in the infection model of schizophrenia. Exp Brain Res 2006b;173:243–57.

[60] Nyffeler M, Meyer U, Yee BK, Feldon J, Knuesel I. Maternal immune activation during pregnancy increases limbic GABAA receptor immunoreactivity in the adult offspring: implications for schizophrenia. Neuroscience 2006;143:51–62.

[61] Krstic D, Madhusudan A, Doehner J, Vogel P, Notter T, Imhof C, et al. Systemic immune challenges trigger and drive Alzheimer-like neuropathology in mice. J Neuroinflammation 2012a;9:151.

[62] Knuesel I, Wyss-Coray T. Maternal immune activation during late gestation results in long-term changes in plasma cytokine and chemokine levels in the adult murine offspring. Unpublished data.

[63] Doehner J, Genoud C, Imhof C, Krstic D, Knuesel I. Extrusion of misfolded and aggregated proteins - a protective strategy of aging neurons? Eur J Neurosci 2012;35:1938–50.

[64] Braak H, Zetterberg H, Del Tredici K, Blennow K. Intraneuronal tau aggregation precedes diffuse plaque deposition, but amyloid-beta changes occur before increases of tau in cerebrospinal fluid. Acta Neuropathol 2013b;126:631–41.

[65] Krstic D, Knuesel I. Deciphering the mechanism underlying late-onset Alzheimer disease. Nat Rev Neurol 2013b;9:25–34.

[66] Krstic D, Pfister S, Notter T, Knuesel I. Decisive role of Reelin signaling during early stages of Alzheimer's disease. Neuroscience 2013;246:108–16.

[67] Braak H, Del Tredici K. Amyloid-beta may be released from non-junctional varicosities of axons generated from abnormal tau-containing brainstem nuclei in sporadic Alzheimer's disease: a hypothesis. Acta Neuropathol 2013a;126:303–6.

[68] Lee CK, Weindruch R, Prolla TA. Gene-expression profile of the ageing brain in mice. Nat Genet 2000;25:294–7.

[69] Lu T, Pan Y, Kao SY, Li C, Kohane I, Chan J, et al. Gene regulation and DNA damage in the ageing human brain. Nature 2004;429:883–91.

[70] Maccioni RB, Rojo LE, Fernandez JA, Kuljis RO. The role of neuroimmunomodulation in Alzheimer's disease. Ann NY Acad Sci 2009;1153:240–6.

[71] Cagnin A, Brooks DJ, Kennedy AM, Gunn RN, Myers R, Turkheimer FE, et al. In-vivo measurement of activated microglia in dementia. Lancet 2001;358:461–7.

[72] Song M, Jin J, Lim JE, Kou J, Pattanayak A, Rehman JA, et al. TLR4 mutation reduces microglial activation, increases Abeta deposits and exacerbates cognitive deficits in a mouse model of Alzheimer's disease. J Neuroinflammation 2011;8:92.

[73] Tahara K, Kim HD, Jin JJ, Maxwell JA, Li L, Fukuchi K. Role of toll-like receptor signalling in Abeta uptake and clearance. Brain: J Neurol 2006;129:3006–19.

[74] Richard KL, Filali M, Prefontaine P, Rivest S. Toll-like receptor 2 acts as a natural innate immune receptor to clear amyloid beta 1-42 and delay the cognitive decline in a mouse model of Alzheimer's disease. J Neurosci 2008;28:5784–93.

[75] Michaud JP, Richard KL, Rivest S. Hematopoietic MyD88-adaptor protein acts as a natural defense mechanism for cognitive deficits in Alzheimer's disease. Stem Cell Rev 2012;8:898–904.

[76] Montgomery SL, Mastrangelo MA, Habib D, Narrow WC, Knowlden SA, Wright TW, et al. Ablation of TNF-RI/RII expression in Alzheimer's disease mice leads to an unexpected enhancement of pathology: implications for chronic pan-TNF-alpha suppressive therapeutic strategies in the brain. Am J Pathol 2011;179:2053–70.

[77] Grathwohl SA, Kalin RE, Bolmont T, Prokop S, Winkelmann G, Kaeser SA, et al. Formation and maintenance of Alzheimer's disease beta-amyloid plaques in the absence of microglia. Nat Neurosci 2009;12:1361–3.

[78] Varvel NH, Grathwohl SA, Baumann F, Liebig C, Bosch A, Brawek B, et al. Microglial repopulation model reveals a robust homeostatic process for replacing CNS myeloid cells. Proc Natl Acad Sci USA 2012;109:18150–5.

[79] Elmore MR, Najafi AR, Koike MA, Dagher NN, Spangenberg EE, Rice RA, et al. Colony-stimulating factor 1 receptor signaling is necessary for microglia viability, unmasking a microglia progenitor cell in the adult brain. Neuron 2014;82:380–97.

[80] Streit WJ, Sammons NW, Kuhns AJ, Sparks DL. Dystrophic microglia in the aging human brain. Glia 2004;45:208–12.

[81] Streit WJ, Xue QS. Microglial senescence. CNS Neurol Disord Drug Targets 2013;12:763–7.

[82] Lucin KM, Wyss-Coray T. Immune activation in brain aging and neurodegeneration: too much or too little? Neuron 2009;64:110–22.

[83] Peri F, Nusslein-Volhard C. Live imaging of neuronal degradation by microglia reveals a role for v0-ATPase a1 in phagosomal fusion in vivo. Cell 2008;133:916–27.

[84] Chung H, Brazil MI, Soe TT, Maxfield FR. Uptake, degradation, and release of fibrillar and soluble forms of Alzheimer's amyloid beta-peptide by microglial cells. J Biol Chem 1999;274:32301–8.

[85] Streit WJ. Microglial senescence: does the brain's immune system have an expiration date? Trends Neurosci 2006;29:506–10.

[86] Dantzer R, Kelley KW. Twenty years of research on cytokine-induced sickness behavior. Brain Behav Immun 2007;21:153–60.

[87] Burns A, Gallagley A, Byrne J. Delirium. J Neurol Neurosurg Psychiatry 2004;75:362–7.

[88] Meagher D. Motor subtypes of delirium: past, present and future. Int Rev Psychiatry 2009;21:59–73.

[89] Fong TG, Jones RN, Shi P, Marcantonio ER, Yap L, Rudolph JL, et al. Delirium accelerates cognitive decline in Alzheimer disease. Neurology 2009;72:1570–5.

[90] Holmes C, Cunningham C, Zotova E, Woolford J, Dean C, Kerr S, et al. Systemic inflammation and disease progression in Alzheimer disease. Neurology 2009;73:768–74.

[91] Schiltz JC, Sawchenko PE. Signaling the brain in systemic inflammation: the role of perivascular cells. Front Biosci: J Virtual Libr 2003;8:S1321–9.

[92] Serrats J, Schiltz JC, Garcia-Bueno B, van Rooijen N, Reyes TM, Sawchenko PE. Dual roles for perivascular macrophages in immune-to-brain signaling. Neuron 2010;65:94–106.

[93] Teijaro JR, Walsh KB, Cahalan S, Fremgen DM, Roberts E, Scott F, et al. Endothelial cells are central orchestrators of cytokine amplification during influenza virus infection. Cell 2011;146:980–91.

[94] Fabry Z, Topham DJ, Fee D, Herlein J, Carlino JA, Hart MN, et al. TGF-beta 2 decreases migration of lymphocytes in vitro and homing of cells into the central nervous system in vivo. J Immunol 1995;155:325–32.

[95] Koyama Y, Baba A, Matsuda T. Endothelins stimulate the expression of neurotrophin-3 in rat brain and rat cultured astrocytes. Neuroscience 2005;136:425–33.

[96] Koyama Y, Tsujikawa K, Matsuda T, Baba A. Endothelin-1 stimulates glial cell line-derived neurotrophic factor expression in cultured rat astrocytes. Biochem Biophys Res Commun 2003;303:1101–5.

[97] Reuss B, Dono R, Unsicker K. Functions of fibroblast growth factor (FGF)-2 and FGF-5 in astroglial differentiation and blood–brain barrier permeability: evidence from mouse mutants. J Neurosci 2003;23:6404–12.

[98] Mizee MR, Wooldrik D, Lakeman KA, van het Hof B, Drexhage JA, Geerts D, et al. Retinoic acid induces blood–brain barrier development. J Neurosci 2013;33:1660–71.

[99] Bush TG, Puvanachandra N, Horner CH, Polito A, Ostenfeld T, Svendsen CN, et al. Leukocyte infiltration, neuronal degeneration, and neurite outgrowth after ablation of scar-forming, reactive astrocytes in adult transgenic mice. Neuron 1999;23:297–308.

[100] Faulkner JR, Herrmann JE, Woo MJ, Tansey KE, Doan NB, Sofroniew MV. Reactive astrocytes protect tissue and preserve function after spinal cord injury. J Neurosci 2004;24:2143–55.

[101] Skripuletz T, Hackstette D, Bauer K, Gudi V, Pul R, Voss E, et al. Astrocytes regulate myelin clearance through recruitment of microglia during cuprizone-induced demyelination. Brain: J Neurol 2013;136:147–67.

[102] Schwartz M, Kipnis J, Rivest S, Prat A. How do immune cells support and shape the brain in health, disease, and aging? J Neurosci 2013;33:17587–96.

[103] Anisman H, Gibb J, Hayley S. Influence of continuous infusion of interleukin-1beta on depression-related processes in mice: corticosterone, circulating cytokines, brain monoamines, and cytokine mRNA expression. Psychopharmacol Berl 2008;199:231–44.

[104] Obuchowicz E, Marcinowska A, Drzyzga L, Wojcikowski J, Daniel WA, Herman ZS. Effect of chronic treatment with perazine on lipopolysaccharide-induced interleukin-1 beta levels in the rat brain. Naunyn Schmiedeb Arch Pharmacol 2006;373:79–84.

I. NEURODEVELOPMENTAL DISORDERS

[105] Barrientos RM, Higgins EA, Biedenkapp JC, Sprunger DB, Wright-Hardesty KJ, Watkins LR, et al. Peripheral infection and aging interact to impair hippocampal memory consolidation. Neurobiol Aging 2006;27:723–32.

[106] Frank MG, Barrientos RM, Biedenkapp JC, Rudy JW, Watkins LR, Maier SF. mRNA up-regulation of MHC II and pivotal pro-inflammatory genes in normal brain aging. Neurobiol Aging 2006;27:717–22.

[107] Cunningham C, Wilcockson DC, Campion S, Lunnon K, Perry VH. Central and systemic endotoxin challenges exacerbate the local inflammatory response and increase neuronal death during chronic neurodegeneration. J Neurosci 2005;25:9275–84.

[108] Field RH, Gossen A, Cunningham C. Prior pathology in the basal forebrain cholinergic system predisposes to inflammation-induced working memory deficits: reconciling inflammatory and cholinergic hypotheses of delirium. J Neurosci 2012;32:6288–94.

[109] Ginhoux F, Greter M, Leboeuf M, Nandi S, See P, Gokhan S, et al. Fate mapping analysis reveals that adult microglia derive from primitive macrophages. Science 2010;330:841–5.

[110] Herbomel P, Thisse B, Thisse C. Zebrafish early macrophages colonize cephalic mesenchyme and developing brain, retina, and epidermis through a M-CSF receptor-dependent invasive process. Dev Biol 2001;238:274–88.

[111] Perry VH, Nicoll JA, Holmes C. Microglia in neurodegenerative disease. Nat Rev Neurol 2010;6:193–201.

[112] Paolicelli RC, Bolasco G, Pagani F, Maggi L, Scianni M, Panzanelli P, et al. Synaptic pruning by microglia is necessary for normal brain development. Science 2011;333:1456–8.

[113] Selkoe DJ. Alzheimer's disease is a synaptic failure. Science 2002;298:789–91.

[114] Honer WG, Barr AM, Sawada K, Thornton AE, Morris MC, Leurgans SE, et al. Cognitive reserve, presynaptic proteins and dementia in the elderly. Transl Psychiatry 2012;2:e114.

[115] Cummings JL, Vinters HV, Cole GM, Khachaturian ZS. Alzheimer's disease: etiologies, pathophysiology, cognitive reserve, and treatment opportunities. Neurology 1998;51:S2–17. Discussion S65–S67.

[116] Barrientos RM, Frank MG, Hein AM, Higgins EA, Watkins LR, Rudy JW, et al. Time course of hippocampal IL-1 beta and memory consolidation impairments in aging rats following peripheral infection. Brain Behav Immun 2009;23:46–54.

[117] Chapman TR, Barrientos RM, Ahrendsen JT, Maier SF, Patterson SL. Synaptic correlates of increased cognitive vulnerability with aging: peripheral immune challenge and aging interact to disrupt theta-burst late-phase long-term potentiation in hippocampal area CA1. J Neurosci 2010;30:7598–603.

[118] Lynch MA. Age-related neuroinflammatory changes negatively impact on neuronal function. Front Aging Neurosci 2010;1:6.

[119] Zhang D, Hu X, Qian L, O'Callaghan JP, Hong JS. Astrogliosis in CNS pathologies: is there a role for microglia? Mol Neurobiol 2010;41:232–41.

[120] D'Arcangelo G, Miao GG, Chen SC, Soares HD, Morgan JI, Curran T. A protein related to extracellular matrix proteins deleted in the mouse mutant reeler. Nature 1995;374:719–23.

[121] Knuesel I. Reelin-mediated signaling in neuropsychiatric and neurodegenerative diseases. Prog Neurobiol 2010;91:257–74.

[122] Weeber EJ, Beffert U, Jones C, Christian JM, Forster E, Sweatt JD, et al. Reelin and ApoE receptors cooperate to enhance hippocampal synaptic plasticity and learning. J Biol Chem 2002;277:39944–52.

[123] Chameau P, Inta D, Vitalis T, Monyer H, Wadman WJ, van Hooft JA. The N-terminal region of reelin regulates postnatal dendritic maturation of cortical pyramidal neurons. Proc Natl Acad Sci USA 2009;106:7227–32.

[124] Jossin Y, Goffinet AM. Reelin signals through phosphatidylinositol 3-kinase and Akt to control cortical development and through mTor to regulate dendritic growth. Mol Cell Biol 2007;27:7113–24.

[125] Niu S, Renfro A, Quattrocchi CC, Sheldon M, D'Arcangelo G. Reelin promotes hippocampal dendrite development through the VLDLR/ApoER2-Dab1 pathway. Neuron 2004;41:71–84.

[126] Niu S, Yabut O, D'Arcangelo G. The reelin signaling pathway promotes dendritic spine development in hippocampal neurons. J Neurosci 2008;28:10339–48.

[127] Beffert U, Weeber EJ, Morfini G, Ko J, Brady ST, Tsai LH, et al. Reelin and cyclin-dependent kinase 5-dependent signals cooperate in regulating neuronal migration and synaptic transmission. J Neurosci 2004;24:1897–906.

[128] Trommsdorff M, Gotthardt M, Hiesberger T, Shelton J, Stockinger W, Nimpf J, et al. Reeler/disabled-like disruption of neuronal migration in knockout mice lacking the VLDL receptor and ApoE receptor 2. Cell 1999;97:689–701.

[129] Rice HC, Young-Pearse TL, Selkoe DJ. Systematic evaluation of candidate ligands regulating ectodomain shedding of amyloid precursor protein. Biochemistry 2013;52:3264–77.

[130] Kocherhans S, Madhusudan A, Doehner J, Breu KS, Nitsch RM, Fritschy JM, et al. Reduced Reelin expression accelerates amyloid-beta plaque formation and tau pathology in transgenic Alzheimer's disease mice. J Neurosci 2010a;30:9228–40.

[131] Chen Y, Durakoglugil MS, Xian X, Herz J. ApoE4 reduces glutamate receptor function and synaptic plasticity by selectively impairing ApoE receptor recycling. Proc Natl Acad Sci USA 2010;107:12011–6.

[132] Corder EH, Saunders AM, Strittmatter WJ, Schmechel DE, Gaskell PC, Small GW, et al. Gene dose of apolipoprotein E type 4 allele and the risk of Alzheimer's disease in late onset families. Science 1993;261:921–3.

[133] Strittmatter WJ, Saunders AM, Schmechel D, Pericak-Vance M, Enghild J, Salvesen GS, et al. Apolipoprotein E: high-avidity binding to beta-amyloid and increased frequency of type 4 allele in late-onset familial Alzheimer disease. Proc Natl Acad Sci USA 1993;90:1977–81.

[134] Stranahan AM, Haberman RP, Gallagher M. Cognitive decline is associated with reduced reelin expression in the entorhinal cortex of aged rats. Cereb Cortex 2011;21:392–400.

[135] Herring A, Donath A, Steiner KM, Widera MP, Hamzehian S, Kanakis D, et al. Reelin depletion is an early phenomenon of Alzheimer's pathology. J Alzheimers Dis 2012;30:963–79.

[136] Doehner J, Knuesel I. Reelin-mediated signaling during normal and pathological forms of aging. Aging Dis 2010;1:12–29.

[137] Kocherhans S, Madhusudan A, Doehner J, Breu KS, Nitsch RM, Fritschy JM, et al. Reelin expression accelerates amyloid-β plaque formation and Tau pathology in transgenic AD mice. J Neurosci 2010;30:9228–40.

[138] Madhusudan A, Sidler C, Knuesel I. Accumulation of reelin-positive plaques is accompanied by a decline in basal forebrain projection neurons during normal aging. Eur J Neurosci 2009;30:1064–76.

[139] Notter T, Knuesel I. Reelin immunoreactivity in neuritic varicosities in the human hippocampal formation of non-demented subjects and Alzheimer's disease patients. Acta Neuropathol Commun 2013;1:27.

[140] Leemhuis J, Bouche E, Frotscher M, Henle F, Hein L, Herz J, et al. Reelin signals through apolipoprotein E receptor 2 and Cdc42 to increase growth cone motility and filopodia formation. J Neurosci 2010;30:14759–72.

[141] de Bergeyck V, Nakajima K, Lambert de Rouvroit C, Naerhuyzen B, Goffinet AM, Miyata T, et al. A truncated Reelin protein is produced but not secreted in the 'Orleans' reeler mutation (Reln[rl-Orl]). Brain Res Mol Brain Res 1997;50:85–90.

[142] Derer P, Derer M, Goffinet A. Axonal secretion of Reelin by Cajal-Retzius cells: evidence from comparison of normal and Reln(Orl) mutant mice. J Comp Neurol 2001;440:136–43.

[143] Krstic D, Rodriguez M, Knuesel I. Regulated proteolytic processing of Reelin through interplay of tissue plasminogen activator (tPA), ADAMTS-4, ADAMTS-5, and their modulators. PLoS One 2012b;7:e47793.

[144] Eberhardt W, Beck KF, Pfeilschifter J. Cytokine-induced expression of tPA is differentially modulated by NO and ROS in rat mesangial cells. Kidney Int 2002;61:20–30.

[145] Liacini A, Sylvester J, Zafarullah M. Triptolide suppresses proinflammatory cytokine-induced matrix metalloproteinase and aggrecanase-1 gene expression in chondrocytes. Biochem Biophys Res Commun 2005;327:320–7.

[146] Stradner MH, Angerer H, Ortner T, Fuerst FC, Setznagl D, Kremser ML, et al. The immunosuppressant FTY720 (fingolimod) enhances glycosaminoglycan depletion in articular cartilage. BMC Musculoskelet Disord 2011;12:279.

[147] Lemarchant S, Docagne F, Emery E, Vivien D, Ali C, Rubio M. tPA in the injured central nervous system: different scenarios starring the same actor? Neuropharmacology 2012;62:749–56.

[148] Doehner J, Madhusudan A, Konietzko U, Fritschy JM, Knuesel I. Co-localization of reelin and proteolytic AbetaPP fragments in hippocampal plaques in aged wild-type mice. J Alzheimers Dis 2010a;19:1339–57.

[149] Ma Z, Gibson SL, Byrne MA, Zhang J, White MF, Shaw LM. Suppression of insulin receptor substrate 1 (IRS-1) promotes mammary tumor metastasis. Mol Cell Biol 2006;26:9338–51.

[150] Reits EA, Hodge JW, Herberts CA, Groothuis TA, Chakraborty M, Wansley EK, et al. Radiation modulates the peptide repertoire, enhances MHC class I expression, and induces successful antitumor immunotherapy. J Exp Med 2006;203:1259–71.

[151] Seifert U, Bialy LP, Ebstein F, Bech-Otschir D, Voigt A, Schroter F, et al. Immunoproteasomes preserve protein homeostasis upon interferon-induced oxidative stress. Cell 2010;142:613–24.

[152] van Deventer S, Neefjes J. The immunoproteasome cleans up after inflammation. Cell 2010;142:517–8.

[153] Mattson MP, Magnus T. Ageing and neuronal vulnerability. Nat Rev Neurosci 2006;7:278–94.

[154] Lee JW, Lee YK, Yuk DY, Choi DY, Ban SB, Oh KW, et al. Neuro-inflammation induced by lipopolysaccharide causes cognitive impairment through enhancement of beta-amyloid generation. J Neuroinflammation 2008;5:37.

[155] Forloni G, Demicheli F, Giorgi S, Bendotti C, Angeretti N. Expression of amyloid precursor protein mRNAs in endothelial, neuronal and glial cells: modulation by interleukin-1. Brain Res Mol Brain Res 1992;16: 128–34.

[156] Grilli M, Goffi F, Memo M, Spano P. Interleukin-1beta and glutamate activate the NF-kappaB/Rel binding site from the regulatory region of the amyloid precursor protein gene in primary neuronal cultures. J Biol Chem 1996;271:15002–7.

[157] Sheng JG, Ito K, Skinner RD, Mrak RE, Rovnaghi CR, Van Eldik LJ, et al. In vivo and in vitro evidence supporting a role for the inflammatory cytokine interleukin-1 as a driving force in Alzheimer pathogenesis. Neurobiol Aging 1996;17:761–6.

[158] Krstic D, Knuesel I. The airbag problem-a potential culprit for bench-to-bedside translational efforts: relevance for Alzheimer's disease. Acta Neuropathol Commun 2013a;1:62.

[159] Griffin WS, Sheng JG, Gentleman SM, Graham DI, Mrak RE, Roberts GW. Microglial interleukin-1 alpha expression in human head injury: correlations with neuronal and neuritic beta-amyloid precursor protein expression. Neurosci Lett 1994;176:133–6.

[160] Masumura M, Hata R, Uramoto H, Murayama N, Ohno T, Sawada T. Altered expression of amyloid precursors proteins after traumatic brain injury in rats: in situ hybridization and immunohistochemical study. J Neurotrauma 2000;17:123–34.

[161] Cirrito JR, Kang JE, Lee J, Stewart FR, Verges DK, Silverio LM, et al. Endocytosis is required for synaptic activity-dependent release of amyloid-beta in vivo. Neuron 2008;58:42–51.

[162] Koo EH, Sisodia SS, Archer DR, Martin LJ, Weidemann A, Beyreuther K, et al. Precursor of amyloid protein in Alzheimer disease undergoes fast anterograde axonal transport. Proc Natl Acad Sci USA 1990;87:1561–5.

[163] Morales-Corraliza J, Mazzella MJ, Berger JD, Diaz NS, Choi JH, Levy E, et al. In vivo turnover of tau and APP metabolites in the brains of wild-type and Tg2576 mice: greater stability of sAPP in the beta-amyloid depositing mice. PLoS One 2009;4:e7134.

[164] Ichihara N, Wu J, Chui DH, Yamazaki K, Wakabayashi T, Kikuchi T. Axonal degeneration promotes abnormal accumulation of amyloid beta-protein in ascending gracile tract of gracile axonal dystrophy (GAD) mouse. Brain Res 1995;695:173–8.

[165] Lee S, Sato Y, Nixon RA. Lysosomal proteolysis inhibition selectively disrupts axonal transport of degradative organelles and causes an Alzheimer's-like axonal dystrophy. J Neurosci 2011;31:7817–30.

[166] Morfini GA, Burns M, Binder LI, Kanaan NM, LaPointe N, Bosco DA, et al. Axonal transport defects in neurodegenerative diseases. J Neurosci 2009;29:12776–86.

[167] Nixon RA, Wegiel J, Kumar A, Yu WH, Peterhoff C, Cataldo A, et al. Extensive involvement of autophagy in Alzheimer disease: an immuno-electron microscopy study. J Neuropathol Exp Neurol 2005;64:113–22.

[168] Nixon RA, Yang DS. Autophagy failure in Alzheimer's disease–locating the primary defect. Neurobiol Dis 2011;43:38–45.

[169] Stokin GB, Lillo C, Falzone TL, Brusch RG, Rockenstein E, Mount SL, et al. Axonopathy and transport deficits early in the pathogenesis of Alzheimer's disease. Science 2005;307:1282–8.

[170] Wirths O, Weis J, Szczygielski J, Multhaup G, Bayer TA. Axonopathy in an APP/PS1 transgenic mouse model of Alzheimer's disease. Acta Neuropathol 2006;111:312–9.

[171] Bernstein HG, Bruszis S, Schmidt D, Wiederanders B, Dorn A. Immunodetection of cathepsin D in neuritic plaques found in brains of patients with dementia of Alzheimer type. J Hirnforsch 1989;30:613–8.

[172] Cataldo AM, Nixon RA. Enzymatically active lysosomal proteases are associated with amyloid deposits in Alzheimer brain. Proc Natl Acad Sci USA 1990;87:3861–5.

[173] Adalbert R, Nogradi A, Babetto E, Janeckova L, Walker SA, Kerschensteiner M, et al. Severely dystrophic axons at amyloid plaques remain continuous and connected to viable cell bodies. Brain: J Neurol 2009;132:402–16.

[174] Weyer SW, Klevanski M, Delekate A, Voikar V, Aydin D, Hick M, et al. APP and APLP2 are essential at PNS and CNS synapses for transmission, spatial learning and LTP. Embo J 2011;30:2266–80.

[175] Misko AL, Sasaki Y, Tuck E, Milbrandt J, Baloh RH. Mitofusin2 mutations disrupt axonal mitochondrial positioning and promote axon degeneration. J Neurosci 2012;32:4145–55.

[176] Coleman M. Axon degeneration mechanisms: commonality amid diversity. Nat Rev Neurosci 2005;6:889–98.

[177] Zempel H, Mandelkow EM. Linking amyloid-beta and tau: amyloid-beta induced synaptic dysfunction via local wreckage of the neuronal cytoskeleton. Neurodegener Dis 2012;10:64–72.

[178] Zempel H, Thies E, Mandelkow E, Mandelkow EM. Abeta oligomers cause localized Ca(2+) elevation, missorting of endogenous Tau into dendrites, Tau phosphorylation, and destruction of microtubules and spines. J Neurosci 2010;30:11938–50.

[179] Ittner LM, Gotz J. Amyloid-beta and tau–a toxic pas de deux in Alzheimer's disease. Nat Rev Neurosci 2011;12:65–72.

[180] Ittner LM, Ke YD, Delerue F, Bi M, Gladbach A, van Eersel J, et al. Dendritic function of tau mediates amyloid-beta toxicity in Alzheimer's disease mouse models. Cell 2010;142:387–97.

[181] Li Y, Liu L, Barger SW, Griffin WS. Interleukin-1 mediates pathological effects of microglia on tau phosphorylation and on synaptophysin synthesis in cortical neurons through a p38-MAPK pathway. J Neurosci 2003;23:1605–11.

[182] Yankner BA, Dawes LR, Fisher S, Villa-Komaroff L, Oster-Granite ML, Neve RL. Neurotoxicity of afragment of the amyloid precursor associated with Alzheimer's disease. Science 1989;245:417–20.

[183] Whitson JS, Selkoe DJ, Cotman CW. Amyloid beta protein enhances the survival of hippocampal neurons in vitro. Science 1989;243:1488–90.

[184] Frost B, Diamond MI. Prion-like mechanisms in neurodegenerative diseases. Nat Rev Neurosci 2010;11:155–9.

[185] Raj A, Kuceyeski A, Weiner M. A network diffusion model of disease progression in dementia. Neuron 2012;73:1204–15.

[186] Devanand DP, Liu X, Tabert MH, Pradhaban G, Cuasay K, Bell K, et al. Combining early markers strongly predicts conversion from mild cognitive impairment to Alzheimer's disease. Biol Psychiatry 2008;64:871–9.

Neurobehavioral Effects of Air Pollution in Children

Michal Kicinski, Tim S. Nawrot

INTRODUCTION

Air pollutants are solid particles, gases, and liquid droplets in the air that can adversely affect ecosystems and the health of humans. Major ambient air pollutants include toxic metals, polycyclic aromatic hydrocarbons, benzene, particulate matter (PM), nitrogen oxides, sulfur oxides, carbon monoxide, and ozone. Power plants, factories, and vehicle and air traffic are the most important anthropogenic sources of air pollution. In addition, natural sources such as wildfires and volcanic activity contribute to ambient air pollution.

Air pollution is widespread. Therefore, even if it leads to a small increase in health risks for an individual, it has considerable implications for public health [1]. It has been estimated that ambient air pollution was responsible for 3.7 million premature deaths in 2012, representing 6.7% of the total deaths [2]. The cardiovascular and respiratory effects of this major environmental health risk factor have been extensively investigated [3–5]. Recently, a considerable amount of research has also studied whether air pollution is harmful to the brain.

Environmental Factors in Neurodevelopmental and Neurodegenerative Disorders
http://dx.doi.org/10.1016/B978-0-12-800228-5.00005-4

Exposure to air pollution is particularly high in children because they spend much time outdoors and have a high breathing rate. Moreover, children may be more vulnerable than adults to neurotoxic insults. A disturbance of rapid and extensive developmental processes undergone by children may have a large impact on the nervous system. An infant's blood–brain barrier is more susceptible to disruption, allowing toxicants to enter the brain [6]. The ability of newborns to metabolize toxicants is more limited than of adults [7]. Because of a high level of exposure and vulnerability to neurotoxicants, the neurobehavioral effects of ambient air pollution in children may be particularly serious.

In this chapter, we review the body of research on the neurobehavioral effects of ambient air pollution in children.

EVIDENCE FROM EPIDEMIOLOGICAL RESEARCH

Recently, a number of epidemiological studies have investigated the neurobehavioral effects of environmental exposure to air pollution in children. In this section, we present a review of these studies.

Perinatal Air Pollution Exposure and Neurodevelopment

Several studies have reported an inverse association between a perinatal exposure to air pollution and neurobehavioral outcomes during childhood (Tables 5.1 and 5.2). An American study assessed the exposure to airborne polycyclic aromatic hydrocarbons (PAHs) during the third trimester of pregnancy in a cohort of nonsmoking African-American and Dominican women residing in New York [8]. The personal PAHs assessment involved wearing a small backpack containing a monitor for two consecutive days and keeping it near the bed at night [9]. At ages one, two, and three, the Bayley Scales of Infant Development assessed mental and psychomotor development. Prenatal PAHs exposure above the third quartile was associated with a 5.7 points decrease in the Mental Development Index at age three (95% CI: −9.1 to −2.3) (Table 5.1) [9]. However, the exposure was not associated with the two neurodevelopmental outcomes at age one or two. In the next follow-up, the Wechsler Preschool and Primary Scale of Intelligence-Revised measured intelligence at five years of age [10]. Children with a prenatal PAHs exposure above the median had a 4.7 points (95% CI: −7.7 to −1.6) lower verbal IQ and 4.3 points (95% CI: −7.4 to −1.2) lower full-scale IQ than those with exposure below the median, adjusting for many potential confounders including maternal intelligence. Additionally, the American study administered the Child Behavior Checklist to the mothers in order to assess the behavior of the children at five and seven years of age [11]. A higher PAHs exposure indicated by personal PAHs monitoring and PAH–DNA adducts in cord blood collected at delivery were associated with an increase in anxiety and attention problems [11,12].

A Polish study supported the results observed in the birth cohort from New York [13]. A PAHs exposure above the median, assessed by personal air monitoring of the mother during the pregnancy, was associated with a 1.4 point decrease (95% CI: −2.5 to −0.2) in the Raven Progressive Matrices Test in a group of 214 five-year-old children. However, a large study on the neurobehavioral effects of perinatal air pollution exposure ($N = 1889$) reported no evidence of association between indicators of traffic-related air pollution and neurodevelopment [14]. Neither residential NO_2 nor benzene concentrations during pregnancy were associated

TABLE 5.1 Perinatal Air Pollution Exposure and Neurodevelopment: Overview of Epidemiological Studies

Reference	Study group	Age	Study design	Results summary
Edwards et al. [13]	A birth cohort from Krakow, Poland (N = 214)	5 years	Prospective cohort	• Prenatal PAHs exposure assessed by personal air monitoring in the mother associated with impaired intelligence (assessed using the Raven Progressive Matrices Test), adjusting for prenatal environmental tobacco smoke, gender, and maternal education.
Guxens et al. [14]	A birth cohort from Spain (N = 1889)	14 months	Prospective cohort	• Mean residential NO_2 and benzene concentrations during the pregnancy not associated with mental development assessed by the Bayley Scales of Infant Development.
Perera et al. [9]	Children of African-American and Dominican mothers from New York (N = 183)	1, 2, and 3 years	Prospective cohort	• Prenatal PAHs exposure above the third quartile measured by personal air monitoring of the mother associated with a lower Mental Development Index of the Bayley Scale of Infant Development at age 3, adjusting for ethnicity, gender, gestational age, and home environment. • No association with Psychomotor Development Index or behavioral problems assessed by the Child Behavior Checklist. • No associations at age 1 or 2.
Perera et al. [10]	Children of African-American and Dominican mothers from New York (N = 249)	5 years	Prospective cohort	• Prenatal PAHs exposure above the median measured by personal air monitoring in the mother associated with full-scale IQ and verbal IQ of the Wechsler Preschool and Primary Scale of Intelligence-Revised, adjusting for ethnicity, gender, prenatal environmental tobacco smoke, maternal education, maternal IQ, and quality of the home caretaking environment. • No associations with performance IQ.
Perera et al. [11]	Children of African-American and Dominican mothers from New York (N = 215)	5 years and 7 years	Retrospective cohort	• Higher PAH–DNA adducts in cord blood collected at delivery associated with higher symptom scores of Anxious/Depressed and anxiety problems at 5 years and attention problems at 5 and 7 years (assessed with the Child Behavior Checklist), adjusting for prenatal environmental tobacco smoke, gender, gestational age, maternal IQ, quality of the home caretaking environment, maternal education, ethnicity, maternal prenatal demoralization, age at assessment, and heating season. • Associations between the exposure and the evolution of scores over time not reported.
Perera et al. [12]	Children of African-American and Dominican mothers from New York (N = 253)	7 years	Prospective cohort	• Prenatal PAHs exposure associated with higher symptom scores of Anxious/Depressed, attention problems, and anxiety problems (the Child Behavior Checklist), adjusting for prenatal environmental tobacco smoke, gender, gestational age, maternal IQ, quality of the home caretaking environment, maternal education, ethnicity, maternal prenatal demoralization, age at assessment, and heating season.

TABLE 5.2   Perinatal Air Pollution Exposure and Neurodevelopmental Conditions: Overview of Epidemiological Studies

Reference	Study group	Age	Study design	Results summary
Newman et al. [15]	A subsample from the Cincinnati Childhood Allergy and Air Pollution Study (N = 576)	7 years	Retrospective cohort	• The average residential concentrations during the first year of life of elemental carbon attributed to traffic associated with hyperactivity measured by the Behavioral Assessment System for Children, Parent Rating Scale, adjusting for gender, environmental tobacco smoke exposure in the first year of life, and maternal education. • No associations with other ADHD-related outcomes (attention problems, aggression, conduct problems, or atypicality).
Roberts et al. [18]	Children of participants from the Nurses' Health Study II (325 cases of autism and 22,101 controls)	3–18 years	Case–control	• Residential annual average concentrations of diesel PM at the year of birth associated with the risk of autism, adjusting for maternal age, year of birth, maternal parents' education, census tract median income, census tract percent college educated, and hazardous air pollutant model year.
Volk et al. [17]	Cases (N = 304) and controls (N = 259) matched for gender, age, and area from the Childhood Autism Risks from Genetics and the Environment (CHARGE) study in California	24–60 months	Case–control	• Living <309 m (10th percentile) from a freeway at the time of delivery associated with a higher risk of autism (reference: >1.419 m), OR = 1.86, 95% CI: 1.04–3.45, adjusting for gender, ethnicity, maximum education of parents, maternal age, and maternal smoking during pregnancy. • Distance to major road not associated with autism.
Volk et al. [16]	Cases and controls matched for gender, age, and area from the Childhood Autism Risks from Genetics and the Environment (CHARGE) study in California (N = 524)	24–60 months	Case–control	• Higher risk of autism for the highest quartile of the average residential traffic-related air pollution concentrations (estimated using nitrogen oxides), PM_{10}, $PM_{2.5}$, and NO_2 during the first year of life, pregnancy, the first trimester, second trimester, and third trimester (reference: the lowest quartile), adjusting for gender, ethnicity, maximum education of parents, maternal age, and maternal smoking during pregnancy. • Residential ozone concentrations not associated with autism.

with mental development assessed by the Bayley Scales of Infant Development in 14-month-old children from Spain.

An American study estimated residential traffic exposure during the first year of life and administered the Behavioral Assessment System for Children, Parent Rating Scale when the children were seven years old (Table 5.2) [15]. This scale assesses five dimensions related to attention-deficit/hyperactivity disorder (ADHD): hyperactivity, attention problems, aggression, conduct problems, and atypicality. Residential concentrations of elemental carbon attributed to traffic were associated with hyperactivity (odds ratio (OR) = 1.7, 95% CI: 1.0–2.7), but not with the four remaining dimensions (N = 576) in a model adjusting for gender, environmental tobacco smoke exposure in the first year of life, and maternal education.

Autism is another clinical outcome that has received attention as a possible health effect of air pollution exposure. The American study Childhood Autism Risks from Genetics and the Environment (CHARGE) determined residential traffic exposure in a group of children with autism diagnosed between 24 and 60 months of age (N = 304) and normally developing controls matched for gender, age, and area (N = 259). Children in the highest quartile of the average concentrations of several pollutants (nitrogen oxides, PM_{10}, $PM_{2.5}$, and NO_2) during the first year of life, whole pregnancy, first trimester, second trimester, and third trimester had a higher risk of autism compared to those in the lowest quartile [16]. In the analysis adjusting for gender, ethnicity, maximum education of parents, maternal age, and maternal smoking during pregnancy the ORs ranged from 1.4 to 3.1. Additionally, distance from a freeway was more likely to be smaller than or equal to 309 m (10th percentile) for cases than for controls (OR = 1.86, 95% CI: 1.04–3.45) [17]. An association between a perinatal traffic exposure and the risk of autism was also observed in a case–control study among children of participants of the Nurses' Health Study II (325 cases, 22,101 controls) [18]. In the analysis adjusting for maternal age, year of birth, parents' education, census tract median income, census tract percent college educated, and hazardous air pollutant model year, residential annual average concentrations of diesel PM at the year of birth in the highest quintile were associated with a higher risk of autism compared to the lowest quintile.

Long-Term Air Pollution Exposure in Childhood and Neurobehavioral Performance

Several longitudinal and cross-sectional studies have investigated the association between long-term air pollution exposure and neurobehavioral performance in children (Table 5.3). In a Boston study, 202 children with a mean age of 9.7 years completed a cognitive battery of tests covering a broad range of cognitive domains [19]. An interquartile-range increase in the lifetime residential black carbon concentrations was associated with a 3.4 point decrease in IQ (95% CI: −6.6 to −0.3) measured by the Kaufman Brief Intelligence Test and a 3.9 point decrease (95% CI: −7.5 to −0.3) in a general index of the Wide Range Assessment of Memory and Learning, adjusting for age, gender, primary language spoken at home, mother's education, in utero tobacco smoke, secondhand smoke, birth weight, and blood lead level. In the same group, a negative association between lifetime black carbon exposure and sustained attention assessed by the Continuous Performance Test has been reported [20]. To our best knowledge, the results of the analyses of the associations between black carbon and performance in other cognitive tests administered in the Boston study have not been reported to date.

TABLE 5.3 Long-Term Air Pollution Exposure in Childhood and Neurobehavioral Performance: Overview of Epidemiological Studies

Reference	Study group	Age	Study design	Results summary
Calderón-Garcidueñas et al. [25]	Children from Mexico City (N = 55) and controls (N = 18) matched by socioeconomic status	Mean: 10.7 years	Cross-sectional	• Children from Mexico City but not controls showed cognitive deficits as indicated by the total score and for 7 of the 12 subscales (Information, Similarities, Vocabulary, Comprehension, Digit Span, Object Assembly, and Coding) of the Wechsler Intelligence Scale for Children-Revised.
Calderón-Garcidueñas et al. [24]	Children from Mexico City (N = 20) and controls (N = 10) matched by age and socioeconomic status	Mean age at baseline: 7 years	Prospective cohort	• Weaker performance of children from Mexico City in 5 of the 12 subscales of the Wechsler Intelligence Scale for Children-Revised (Arithmetic, Vocabulary, Digit Span, Picture Completion, and Coding) at baseline and/or after a 1-year follow-up. • The comparison of the evolution of cognitive performance over time not reported.
Chiu et al. [20]	A subsample from a birth cohort from Boston (N = 174)	7–14 years	Retrospective cohort	• Lifetime residential black carbon levels negatively associated with sustained attention as indicated by the hit reaction time and the number of commission errors in the Continuous Performance Test, adjusting for age, gender, child IQ, blood lead level, maternal education, pre- and postnatal tobacco smoke exposure, and community-level social stress.
Clark et al. [22]	Children from the RANCH project attending schools around London Heathrow Airport (N = 719)	9–10 years	Cross-sectional	• Mean annual NO_2 concentrations at schools not associated with reading comprehension, recognition memory, information recall, conceptual recall, or working memory.
Freire et al. [23]	A subsample from a birth cohort from the south of Spain (N = 210)	4–5 years	Cross-sectional	• Residential mean annual NO_2 concentrations not associated with any of the 11 cognitive domains from the McCarthy Scales of Children's Abilities.
Siddique et al. [26]	Children from Delhi (N = 969) and controls from a rural area (N = 850)	9–17 years	Cross-sectional	• An increase in PM_{10} exposure associated with a higher ADHD prevalence, controlling for gender, age, socioeconomic status, and BMI.

Reference	Sample	Age	Design	Findings
Suglia et al. [19]	A sample from a birth cohort from Boston ($N = 202$)	Mean age: 9.7 years	Retrospective cohort	• Lifetime residential black carbon levels associated with the Visual and General (but not with Verbal or Learning) scales of the Wide Range Assessment of Memory and Learning and with the Matrices and Composite (but not Vocabulary) scales of the Kaufman Brief Intelligence Test, adjusting for age, gender, primary language spoken at home, mother's education, in utero tobacco smoke, secondhand smoke, birth weight, and blood lead level.
Van Kampen et al. [21]	Children from the RANCH project attending schools around Schiphol Amsterdam Airport ($N = 553$)	9–11 years	Cross-sectional	• Mean annual school concentrations of NO_2 negatively associated with memory (measured by the Digit Memory Span Test from the Neurobehavioral Evaluation System), adjusting for gender, age, employment status, crowding, homeownership, mother's education, long-standing illness, main language spoken at home, parental support, double glazing at home, road- and air traffic noise. • No associations with performance in the Simple Reaction Time, Switching Attention, Hand–Eye Coordination, or Symbol Digit Substitution tests. • No associations of residential NO_2 with any neurobehavioral outcomes.
Wang et al. [27]	Children from Quanzhou, China, attending two schools: one located in a polluted and another one in a clean area ($N = 928$)	8–10 years	Cross-sectional	• Children from the school located in a clean area performed better in the Continuous Performance Test (from the Neurobehavioral Evaluation System) and in the Digit Symbol, Pursuit Aiming, and Sign Register tests (from the Jinyi Psychomotor Test Battery). • No differences found in performance in the Line Discrimination, Visual Retention, or Simple Reaction Time tests (Neurobehavioral Evaluation System) or in performance in the Digit Erase Test (Jinyi Psychomotor Test Battery).

Two cross-sectional studies from the Road Traffic and Aircraft Noise Exposure and Children's Cognition and Health project measured annual NO_2 concentrations at primary schools located near a major airport and administered cognitive tests [21,22]. Air pollution levels were not associated with reading comprehension, recognition memory, information recall, conceptual recall, or working memory in 719 children attending schools around London's Heathrow airport [22]. Exposure to NO_2 was associated with memory span in 485 children from 24 schools located near the Schiphol Amsterdam Airport [21]. However, the associations with the remaining nine neurobehavioral parameters assessed in this study (involving attention, locomotion, and perceptual coding) were not statistically significant. Additionally, residential NO_2 concentrations were not associated with any of the neurobehavioral outcomes assessed. Residential exposure to traffic-related air pollution indicated by NO_2 concentrations was not associated with cognition in a cross-sectional Spanish study either [23]. This study administered the McCarthy Scales of Children's Abilities (perceptual performance, verbal, quantitative, memory, motor, executive function, memory span, verbal memory, working memory, gross motor, and fine motor) in a group of 210 four-year-old children.

Several studies compared cognitive performance of children living in clean areas and areas with much air pollution [24–27]. Children from Mexico City scored poorer in several subtests of the Wechsler Intelligence Scale for Children-Revised than controls matched for socioeconomic status [24,25]. A Chinese study administered a battery of neurobehavioral tests in 928 nine-year-old children attending two schools with different concentrations of ambient PM_{10} and NO_2 [27]. Participants living in a polluted area performed weaker in most of the tasks, including tests of attention, visual perception, psychomotor stability, and coordination. A study from India observed a higher prevalence of ADHD in 969 urban children than in 850 controls, who lived in rural areas (OR = 4.47, 95% CI: 2.77–7.29) [26]. Controlling for gender, age, socioeconomic status, and body-mass index (BMI), residential PM_{10} concentrations above $200\,\mu g/m^3$ were associated with a 2.77 times higher odds of ADHD (95% CI: 1.38–5.56) compared with concentrations below $120\,\mu g/m^3$.

EVIDENCE FROM ANIMAL RESEARCH

Besides the epidemiological studies discussed in the previous section, a number of experiments using mouse and rat models have been recently conducted (Table 5.4). In most of these studies, animals were exposed continually to polluted air in special chambers. The exposure commonly lasted for at least a few hours per day. Most experiments investigated the effects of diesel exhausts or concentrated ambient particles. The commonly studied neurobehavioral outcomes included learning, memory, spontaneous motor activity, and anxiety.

Several experiments have investigated the effects of air pollution on higher-order cognitive functions [28–32]. Mice exposed 6 hours a day of 5 days a week to $PM_{2.5}$ at environmentally relevant concentrations ($94.38\,\mu g/m^3$) over a period of nine months starting at four weeks of age showed poorer spatial memory in a maze than control animals [28]. Another study observed a negative effect of a long-term $PM_{2.5}$ exposure at levels typical of many cities and areas located near highways ($16.2\,\mu g/m^3$) on discriminative memory [32]. Rats exposed to air pollution from gestation until 171 days of age showed a deteriorated ability to discriminate between a familiar and unfamiliar object compared to control animals at the day following

TABLE 5.4 Overview of Animal Studies

Reference	Animals	Exposure groups	Neurobehavioral outcomes
Allen et al. [36]	Mice (N=7 or N=8 per group)	1. Ultrafine particles (150 μg/m³) over postnatal days 4-7 and 10-13 for 4h/day + adult exposure over days 56-60 2. Only postnatal exposure 3. Only adult exposure 4. Control (filtered air)	• No differences in spontaneous motor activity at 6 months of age.
Bolton et al. [33]	Mice (N=6 per group)	Diesel exhaust (2000 μg/m³) 4 h/day for 9 consecutive days from embryonic days 9-17 and control (filtered air)	• Lower spontaneous motor activity of the exposed mice at 5 months of age. • No differences in the level of anxiety between the exposed and control groups.
Davis et al. [34]	Mice (N=4 per group)	Nanoscale PM (350 μg/m³) from 7 weeks before conception up to 2 days before birth and control (filtered air)	• No differences in spontaneous motor activity or anxiety as assessed by the Open Field Test[a] or the Elevated-Plus Maze Test[b] at 8 months of age.
Fonken et al. [28]	Mice	$PM_{2.5}$ (94.38 μg/m³) 6 h/day, 5 days/week for 9 months, starting at 4 weeks of age and control (filtered air)	• Impaired spatial learning in exposed mice in adulthood as indicated by the number of errors and latency in the Barnes maze[c] during the second day of the training, but the analysis involved multiple comparisons and didn't account for it. • Impaired spatial memory as evaluated by the Barnes maze[c]. • Increased anxiety as indicated by the amount of time spent outside the center of the field in the Open Field Test[a], but no differences in anxiety assessed in the Elevated-Plus maze. • Decreased motor activity in the Forced Swim Test[d]. • No differences in spontaneous motor activity in the Open Field Test[a]. • No differences in motor coordination in a task involving not falling off a rotating rod.
Hougaard et al. [30]	Mice (N=20 per group)	Diesel exhaust (19,000 μg/m³) 1 h/day on gestational days 9-19 and control (filtered air)	• No differences in learning or memory in the Morris water maze[e] at 3-4 months of age. • No differences in spontaneous motor activity.
Hougaard et al. [29]	Mice (exposed: N=22; control: N=23)	Nanoparticle UV-Titan (40,000 μg/m³) 1 h/day on gestational days 8-18 and control (filtered air)	• No differences in learning or memory in the Morris water maze[e] at 3-4 months of age. • No differences in spontaneous motor activity. • Higher level of anxiety in exposed mice.

(Continued)

TABLE 5.4   Overview of Animal Studies—cont'd

Reference	Animals	Exposure groups	Neurobehavioral outcomes
Suzuki et al. [35]	Mice (N = 10 per group)	Diesel exhaust (171 µg/m³) 8 h/day 5 days/week on gestational days 2–16 and control	• A lower level of spontaneous motor activity at 4 weeks of age in exposed mice.
Yokota et al. [42]	Mice (N = 15 per group)	Diesel exhaust (1000 µg/m³) 8 h/day 5 days/week on gestational days 2–17 and control	• Poorer motor coordination as indicated by the retention time in tasks involving not falling off a rotating rod at 5 weeks of age. • No differences in motor coordination in a hanging test. • No statistically significant differences in motor coordination as indicated by not jumping from a small platform.
Yokota et al. [31]	Rats (N = 10 per group)	1. Intranasal instillation of diesel exhaust (50 µg) once a week from 14 days of age for 4 weeks 2. 10 µg 3. Liquid containing diesel exhaust 4. Sham treatment	• Some differences in spontaneous motor activity at 6 weeks of age in analysis stratified by time of the day. • Global differences in spontaneous motor activity not reported. • No differences in learning ability at 9 weeks of age evaluated by the Shuttle Box Avoidance Test[f].
Yokota et al. [41]	Mice (N = 27 per group)	Diesel exhaust (1000 µg/m³) 8 h/day 5 days/week on gestational days 2–17 and control	• Increased level of spontaneous motor activity during the light phase. • Global differences in spontaneous motor activity not reported.
Zanchi et al. [32]	Rats (N = 12 per group)	1. Perinatal (until postnatal day 21) and childhood/adulthood (from postnatal day 21 for 150 days) exposure to non-filtered air in a chamber located in an area with much traffic-related air pollution (PM₂.₅ = 16.2 µg/m³) 2. Perinatal non-filtered air 3. Non-filtered air at childhood/adulthood 4. Control (filtered air)	• Poorer discriminative memory and decreased habituation at 172 days of age in rats exposed both perinatally and later in life, as indicated by the Spontaneous Nonmatching-to-Sample Recognition Test[g].

[a] *Open Field Test: a test tracking the movements of an animal in a small chamber (motor activity indicator) and the time spent close to the walls in the dark region of the field (anxiety indicator).*
[b] *Elevated-Plus Maze Test: a test measuring the time spent in open arms and the number of entries to open arms in a chamber consisting of two open arms bisected by two open arms enclosed by walls.*
[c] *Barnes maze: a brightly lit arena with a hole leading to a dark box.*
[d] *Forced Swim Test: a test assessing swimming behavior in a cylindrical tank.*
[e] *Morris water maze: a plastic pool with a small platform situated 1 cm below the water surface and invisible from the water.*
[f] *Shuttle Box Avoidance Test: a test involving avoiding electric currents preceded by a buzzer and light.*
[g] *Spontaneous Nonmatching-To-Sample Recognition Test: a test measuring the relationship between the time spent exploring a familiar and unfamiliar object (discrimination indicator) and the difference between the time spent exploring objects in the first and second session (habituation index).*

the exposure period. However, gestational high-level exposure to air pollution did not affect spatial learning or memory in a water-maze task [29,30].

Spontaneous motor activity has been the most commonly investigated outcome in experimental studies of neurobehavioral effects of air pollution [28–30,33–36]. In the Open Field Test, an animal is placed in a small chamber and an automatic system records characteristics of its behavior. A high-level (2000 µg/m³, 4 h/day) exposure to diesel exhausts during gestation caused a decrease of the distance traveled in an open field at six months of age [33]. A similar result was observed for lower air pollution levels and a longer daily exposure. Mice exposed during gestation 8 h/day of five days a week to 1000 µg/m³ of diesel exhausts showed a decreased motor activity at four weeks of age compared to control mice [35]. However, studies using lower air pollution concentrations, which were more typical of the ambient levels, did not find evidence of an effect of air pollution on spontaneous motor activity [28,36]. A high-level exposure to air pollution 1 h/day during gestation did not affect spontaneous motor activity in adulthood either [29,30].

The Open Field Test has been also used to evaluate anxiety. In this paradigm, the amount of time spent close to the walls in the dark region of the field indicates anxiety. Mice exposed to ambient concentrated $PM_{2.5}$ (94.38 µg/m³) 6 h/day of five days a week for nine months starting at four weeks of age spent less time in the center of the field than the control mice in an experiment at 11 months of age [28]. A high nanoparticle exposure (40,000 µg/m³) 1 h/day during gestation caused an increase in the level of anxiety at three months of age [29]. However, two other experiments did not find evidence of effects of gestational air pollution exposure on anxiety [33,34].

BIOLOGICAL PATHWAYS

A substantial amount of evidence shows that exposure to diesel exhausts, ultrafine particles, and concentrated ambient particles has a potential to induce neurotoxicity in the brain, which may be responsible for the neurobehavioral effects of air pollution. The effects observed in experimental studies in rodents include changes in gene expression [37–39], the level and turnover of neurotransmitters [35,40–44], oxidative stress [32,44,45], proinflammatory cytokine response [28,33,39,40,46–50], glial activation [49,51], and neuron apoptosis [52,53]. Dogs living in Mexico City showed increased neuroinflammation [54,55], compared to dogs living in clean areas. Additionally, studies investigating the brains of humans who died suddenly revealed that residential air pollution was associated with an increased level of inflammation in the brain [56,57].

Several pathways exist by which air pollution may exert the neurotoxic effects briefly summarized in the previous paragraph. The direct ones involve detrimental changes caused by nanoparticles that translocate to the brain via the olfactory bulb or the blood–brain barrier. The indirect pathways involve a proinflammatory response outside the central nervous system (CNS) that leads to increased systemic inflammation, which, in turn, may adversely affect the brain.

In rats, inhaled ultrafine particles translocate from the nose along the olfactory nerve to the olfactory bulb and other regions of the brain [58–61]. Additionally, studies in rodents have demonstrated that ultrafine particles translocate from the lungs into the blood, from which they can reach extrapulmonary organs including the brain [62,63]. Although we do not know whether a translocation via the olfactory pathway or via the blood–brain barrier plays the most important role in humans, it is clear that exposure to air pollution is associated with an accumulation of ultrafine particles in the human brain [57,64]. The presence of ultrafine particles in the brain may induce microglia activation [65], oxidative stress [66–68], proinflammatory cytokine response [68], neuronal death [65,67], and changes in neurotransmission [66], as revealed by in vitro studies.

Besides the direct pathways described in the preceding paragraph, air pollution may introduce changes in the respiratory and cardiovascular systems and the liver leading to a generalized systemic response, which, in turn, may adversely affect the brain. Lung macrophages, which play a key role in processing inhaled substances, produce a range of proinflammatory cytokines when exposed to air pollution [69–73]. Inflammatory mediators produced in the lung translocate into the circulation [74]. A release of cytokines from the lungs or translocation of ultrafine particles into the circulation may trigger a sequel of proinflammatory events including a production of leukocytes and platelets in the bone marrow, a release of acutephase proteins from the liver, and an activation of vascular endothelium [70,75–77]. Indeed, air pollution exposure increased the level of systemic inflammation in experimental rodent models [74,78,79]. A positive association between air pollution exposure and markers of systemic inflammation has also been observed in some [70,80–83], but not all [84–86], epidemiological studies.

The relevance of systemic inflammation for the central nervous system (CNS) has been demonstrated by a large body of recent evidence [87,88]. In rodents, exposure to systemic inflammatory stimuli caused neuroinflammation, neurodegeneration, cognitive decline, and motor symptoms [89–95]. An association between the level of systemic inflammation and neurodegeneration has also been observed in epidemiological studies [96–98].

EVALUATION OF THE EVIDENCE AND FUTURE RESEARCH DIRECTIONS

Multiple recent epidemiological studies have reported inverse associations between neurobehavioral performance of children and air pollution. The affected neurobehavioral domains included intelligence and specific cognitive skills such as memory and attention. Additionally, some preliminary evidence suggests that air pollution exposure during gestation and early childhood increases the risk of neurodevelopmental conditions such as autism and ADHD. In two American studies, a perinatal traffic exposure was associated with an increased risk of autism [16–18]. Residential traffic exposure during the first year of life was associated with hyperactivity in seven-year-old children [15]. Additionally, residential PM_{10} concentrations were associated with a higher risk of ADHD in a study from India including children from Delhi and a rural area [26].

Poor cognitive capabilities of parents adversely affect cognitive capabilities of their child and may increase the chance of living in an area with poor air quality, for example due to a low income. Observational human studies do not exclude that this phenomenon of reverse causation [99,100] affected the estimates of the association between air pollution and neurobehavioral outcomes. Additionally, some determinants of the neurobehavioral function may be correlated with exposure to air pollution. For example, living in an area with much air pollution may be associated with a poor quality of schools or other facilities stimulating the development of children. However, all reviewed studies reported estimates corrected for socioeconomic status and other important confounders. Moreover, in some regions including Europe, socioeconomic indicators are weakly correlated with air pollution exposure [101]. Nevertheless, the possibility of residual confounding caused by factors not included in the models or measured inaccurately cannot be excluded.

The inverse association between neurobehavioral outcomes and air pollution exposure has also been observed in experimental studies in rodents, which are completely free of the problems of reverse causation and confounding. However, most of these experiments used air pollution levels much higher than the ambient concentrations in highly polluted areas inhabited by humans. Three experiments [28,32,35] that applied environmentally relevant exposure levels supported the hypothesis of detrimental effects of air pollution on the brain. The neurobehavioral domains affected by air pollution at environmentally relevant exposure levels included spatial memory [28], discriminative memory [32], habituation [32], anxiety [28], and motor activity [28,35]. Future animal studies should concentrate on the effects of environmentally relevant exposure levels.

To date, all human studies investigated the effects of a long-term air pollution exposure. Although investigating the short-term effects of air pollution in children is less relevant for public health than studying the long-term effects, it can be accomplished using study designs that exclude the risk of reverse causation and largely limit the risk of confounding. A panel study, in which each subject serves as his own control, is an example of such a design.

CONCLUSIONS

Numerous recent epidemiological studies have observed an inverse association between neurobehavioral outcomes including intelligence, specific cognitive skills, and clinical conditions ADHD and autism, and exposure to air pollution. Although we do not know exactly by what mechanisms air pollution affects the brain, several plausible biological pathways have been identified. The epidemiological evidence is supported by experimental studies in rodents.

References

[1] Nawrot TS, Perez L, Kunzli N, Munters E, Nemery B. Public health importance of triggers of myocardial infarction: a comparative risk assessment. Lancet 2011;377(9767):732–40.
[2] Ambient (outdoor) air pollution and health. Fact sheet N 313. World Health Organization; 2014.
[3] Sun Q, Hong X, Wold LE. Cardiovascular effects of ambient particulate air pollution exposure. Circulation 2010;121(25):2755–65.
[4] Brook RD, Rajagopalan S, Pope CA 3rd, Brook JR, Bhatnagar A, Diez-Roux AV, et al. Particulate matter air pollution and cardiovascular disease: an update to the scientific statement from the American Heart Association. Circulation 2010;121(21):2331–78.
[5] Wong GW. Air pollution and health. Lancet Respir Med 2014;2(1):8–9.
[6] Saunders NR, Liddelow SA, Dziegielewska KM. Barrier mechanisms in the developing brain. Front Pharmacol 2012;3:46.
[7] Ginsberg G, Hattis D, Miller R, Sonawane B. Pediatric pharmacokinetic data: implications for environmental risk assessment for children. Pediatrics 2004;113(Suppl. 4):973–83.

[8] Perera FP, Rauh V, Tsai WY, Kinney P, Camann D, Barr D, et al. Effects of transplacental exposure to environmental pollutants on birth outcomes in a multiethnic population. Environ Health Perspect 2003;111(2):201–5.

[9] Perera FP, Rauh V, Whyatt RM, Tsai WY, Tang D, Diaz D, et al. Effect of prenatal exposure to airborne polycyclic aromatic hydrocarbons on neurodevelopment in the first 3 years of life among inner-city children. Environ Health Perspect 2006;114(8):1287–92.

[10] Perera FP, Li Z, Whyatt R, Hoepner L, Wang S, Camann D, et al. Prenatal airborne polycyclic aromatic hydrocarbon exposure and child IQ at age 5 years. Pediatrics 2009;124(2):e195–202.

[11] Perera FP, Wang S, Vishnevetsky J, Zhang B, Cole KJ, Tang D, et al. Polycyclic aromatic hydrocarbons-aromatic DNA adducts in cord blood and behavior scores in New York city children. Environ Health Perspect 2011;119(8):1176–81.

[12] Perera FP, Tang D, Wang S, Vishnevetsky J, Zhang B, Diaz D, et al. Prenatal polycyclic aromatic hydrocarbon (PAH) exposure and child behavior at age 6-7 years. Environ Health Perspect 2012;120(6):921–6.

[13] Edwards SC, Jedrychowski W, Butscher M, Camann D, Kieltyka A, Mroz E, et al. Prenatal exposure to airborne polycyclic aromatic hydrocarbons and children's intelligence at 5 years of age in a prospective cohort study in Poland. Environ Health Perspect 2010;118(9):1326–31.

[14] Guxens M, Aguilera I, Ballester F, Estarlich M, Fernandez-Somoano A, Lertxundi A, et al. Prenatal exposure to residential air pollution and infant mental development: modulation by antioxidants and detoxification factors. Environ Health Perspect 2012;120(1):144–9.

[15] Newman NC, Ryan P, Lemasters G, Levin L, Bernstein D, Hershey GK, et al. Traffic-related air pollution exposure in the first year of life and behavioral scores at 7 years of age. Environ Health Perspect 2013;121(6):731–6.

[16] Volk HE, Lurmann F, Penfold B, Hertz-Picciotto I, McConnell R. Traffic-related air pollution, particulate matter, and autism. JAMA Psychiatry 2013;70(1):71–7.

[17] Volk HE, Hertz-Picciotto I, Delwiche L, Lurmann F, McConnell R. Residential proximity to freeways and autism in the CHARGE study. Environ Health Perspect 2011;119(6):873–7.

[18] Roberts AL, Lyall K, Hart JE, Laden F, Just AC, Bobb JF, et al. Perinatal air pollutant exposures and autism spectrum disorder in the children of Nurses' Health Study II participants. Environ Health Perspect 2013;121(8):978–84.

[19] Suglia SF, Gryparis A, Wright RO, Schwartz J, Wright RJ. Association of black carbon with cognition among children in a prospective birth cohort study. Am J Epidemiol 2008;167(3):280–6.

[20] Chiu YH, Bellinger DC, Coull BA, Anderson S, Barber R, Wright RO, et al. Associations between traffic-related black carbon exposure and attention in a prospective birth cohort of urban children. Environ Health Perspect 2013;121(7):859–64.

[21] Van Kempen E, Fischer P, Janssen N, Houthuijs D, van Kamp I, Stansfeld S, et al. Neurobehavioral effects of exposure to traffic-related air pollution and transportation noise in primary schoolchildren. Environ Res 2012;115:18–25.

[22] Clark C, Sorqvist PA. 3 year update on the influence of noise on performance and behavior. NoiseHealth 2012;14(61):292–6.

[23] Freire C, Ramos R, Puertas R, Lopez-Espinosa MJ, Julvez J, Aguilera I, et al. Association of traffic-related air pollution with cognitive development in children. J Epidemiol Community Health 2010;64(3):223–8.

[24] Calderon-Garciduenas L, Engle R, Mora-Tiscareno A, Styner M, Gomez-Garza G, Zhu H, et al. Exposure to severe urban air pollution influences cognitive outcomes, brain volume and systemic inflammation in clinically healthy children. Brain Cogn 2011;77(3):345–55.

[25] Calderon-Garciduenas L, Mora-Tiscareno A, Ontiveros E, Gomez-Garza G, Barragan-Mejia G, Broadway J, et al. Air pollution, cognitive deficits and brain abnormalities: a pilot study with children and dogs. Brain Cogn 2008;68(2):117–27.

[26] Siddique S, Banerjee M, Ray MR, Lahiri T. Attention-deficit hyperactivity disorder in children chronically exposed to high level of vehicular pollution. Eur J Pediatr 2011;170(7):923–9.

[27] Wang S, Zhang J, Zeng X, Zeng Y, Wang S, Chen S. Association of traffic-related air pollution with children's neurobehavioral functions in Quanzhou, China. Environ Health Perspect 2009;117(10):1612–8.

[28] Fonken LK, Xu X, Weil ZM, Chen G, Sun Q, Rajagopalan S, et al. Air pollution impairs cognition, provokes depressive-like behaviors and alters hippocampal cytokine expression and morphology. Mol Psychiatry 2011;16(10):987–95, 973.

[29] Hougaard KS, Jackson P, Jensen KA, Sloth JJ, Loschner K, Larsen EH, et al. Effects of prenatal exposure to surface-coated nanosized titanium dioxide (UV-Titan). A study in mice. Part Fibre Toxicol 2010;7:16.

[30] Hougaard KS, Jensen KA, Nordly P, Taxvig C, Vogel U, Saber AT, et al. Effects of prenatal exposure to diesel exhaust particles on postnatal development, behavior, genotoxicity and inflammation in mice. Part Fibre Toxicol 2008;5:3.

[31] Yokota S, Takashima H, Ohta R, Saito Y, Miyahara T, Yoshida Y, et al. Nasal instillation of nanoparticle-rich diesel exhaust particles slightly affects emotional behavior and learning capability in rats. J Toxicol Sci 2011;36(3):267–76.

[32] Zanchi AC, Fagundes LS, Barbosa Jr F, Bernardi R, Rhoden CR, Saldiva PH, et al. Pre and post-natal exposure to ambient level of air pollution impairs memory of rats: the role of oxidative stress. Inhal Toxicol 2010;22(11):910–8.

[33] Bolton JL, Smith SH, Huff NC, Gilmour MI, Foster WM, Auten RL, et al. Prenatal air pollution exposure induces neuroinflammation and predisposes offspring to weight gain in adulthood in a sex-specific manner. FASEB J 2012;26(11):4743–54.

[34] Davis DA, Bortolato M, Godar SC, Sander TK, Iwata N, Pakbin P, et al. Prenatal exposure to urban air nanoparticles in mice causes altered neuronal differentiation and depression-like responses. PLoS One 2013;8(5):e64128.

[35] Suzuki T, Oshio S, Iwata M, Saburi H, Odagiri T, Udagawa T, et al. In utero exposure to a low concentration of diesel exhaust affects spontaneous locomotor activity and monoaminergic system in male mice. Part Fibre Toxicol 2010;7:7.

[36] Allen JL, Conrad K, Oberdorster G, Johnston CJ, Sleezer B, Cory-Slechta DA. Developmental exposure to concentrated ambient particles and preference for immediate reward in mice. Environ Health Perspect 2013;121(1):32–8.

[37] Yokota S, Hori H, Umezawa M, Kubota N, Niki R, Yanagita S, et al. Gene expression changes in the olfactory bulb of mice induced by exposure to diesel exhaust are dependent on animal rearing environment. PLoS One 2013;8(8):e70145.

[38] Tsukue N, Watanabe M, Kumamoto T, Takano H, Takeda K. Perinatal exposure to diesel exhaust affects gene expression in mouse cerebrum. Arch Toxicol 2009;83(11):985–1000.

[39] Win-Shwe TT, Yamamoto S, Fujitani Y, Hirano S, Fujimaki H. Nanoparticle-rich diesel exhaust affects hippocampal-dependent spatial learning and NMDA receptor subunit expression in female mice. Nanotoxicology 2012;6(5):543–53.

[40] Tin Tin Win S, Mitsushima D, Yamamoto S, Fukushima A, Funabashi T, Kobayashi T, et al. Changes in neurotransmitter levels and proinflammatory cytokine mRNA expressions in the mice olfactory bulb following nanoparticle exposure. Toxicol Appl Pharmacol 2008;226(2):192–8.

[41] Yokota S, Mizuo K, Moriya N, Oshio S, Sugawara I, Takeda K. Effect of prenatal exposure to diesel exhaust on dopaminergic system in mice. Neurosci Lett 2009;449(1):38–41.

[42] Yokota S, Moriya N, Iwata M, Umezawa M, Oshio S, Takeda K. Exposure to diesel exhaust during fetal period affects behavior and neurotransmitters in male offspring mice. J Toxicol Sci 2013;38(1):13–23.

[43] Balasubramanian P, Sirivelu MP, Weiss KA, Wagner JG, Harkema JR, Morishita M, et al. Differential effects of inhalation exposure to $PM_{2.5}$ on hypothalamic monoamines and corticotrophin releasing hormone in lean and obese rats. Neurotoxicology 2013;36:106–11.

[44] MohanKumar SM, Campbell A, Block M, Veronesi B. Particulate matter, oxidative stress and neurotoxicity. Neurotoxicology 2008;29(3):479–88.

[45] Van Berlo D, Albrecht C, Knaapen AM, Cassee FR, Gerlofs-Nijland ME, Kooter IM, et al. Comparative evaluation of the effects of short-term inhalation exposure to diesel engine exhaust on rat lung and brain. Arch Toxicol 2010;84(7):553–62.

[46] Tin Tin Win S, Yamamoto S, Ahmed S, Kakeyama M, Kobayashi T, Fujimaki H. Brain cytokine and chemokine mRNA expression in mice induced by intranasal instillation with ultrafine carbon black. Toxicol Lett 2006;163(2):153–60.

[47] Campbell A, Oldham M, Becaria A, Bondy SC, Meacher D, Sioutas C, et al. Particulate matter in polluted air may increase biomarkers of inflammation in mouse brain. Neurotoxicology 2005;26(1):133–40.

[48] Gerlofs-Nijland ME, Van Berlo D, Cassee FR, Schins RP, Wang K, Campbell A. Effect of prolonged exposure to diesel engine exhaust on proinflammatory markers in different regions of the rat brain. Part Fibre Toxicol 2010;7:12.

[49] Levesque S, Taetzsch T, Lull ME, Kodavanti U, Stadler K, Wagner A, et al. Diesel exhaust activates and primes microglia: air pollution, neuroinflammation, and regulation of dopaminergic neurotoxicity. Environ Health Perspect 2011;119(8):1149–55.

[50] Levesque S, Surace MJ, McDonald J, Block ML. Air pollution & the brain: subchronic diesel exhaust exposure causes neuroinflammation and elevates early markers of neurodegenerative disease. J Neuroinflammation 2011;8:105.

[51] Kleinman MT, Araujo JA, Nel A, Sioutas C, Campbell A, Cong PQ, et al. Inhaled ultrafine particulate matter affects CNS inflammatory processes and may act via MAP kinase signaling pathways. Toxicol Lett 2008;178(2):127–30.

[52] Sugamata M, Ihara T, Takano H, Oshio S, Takeda K. Maternal diesel exhaust exposure damages newborn murine brains. J Health Sci 2006;52(1):82–4.

[53] Sugamata M, Ihara T, Sugamata M, Takeda K. Maternal exposure to diesel exhaust leads to pathological similarity to autism in newborns. J Health Sci 2006;52(4):486–8.

[54] Calderon-Garciduenas L, Azzarelli B, Acuna H, Garcia R, Gambling TM, Osnaya N, et al. Air pollution and brain damage. Toxicol Pathol 2002;30(3):373–89.

[55] Calderon-Garciduenas L, Maronpot RR, Torres-Jardon R, Henriquez-Roldan C, Schoonhoven R, Acuna-Ayala H, et al. DNA damage in nasal and brain tissues of canines exposed to air pollutants is associated with evidence of chronic brain inflammation and neurodegeneration. Toxicol Pathol 2003;31(5):524–38.

[56] Calderon-Garciduenas L, Reed W, Maronpot RR, Henriquez-Roldan C, Delgado-Chavez R, Calderon-Garciduenas A, et al. Brain inflammation and Alzheimer's-like pathology in individuals exposed to severe air pollution. Toxicol Pathol 2004;32(6):650–8.

[57] Calderon-Garciduenas L, Solt AC, Henriquez-Roldan C, Torres-Jardon R, Nuse B, Herritt L, et al. Long-term air pollution exposure is associated with neuroinflammation, an altered innate immune response, disruption of the blood–brain barrier, ultrafine particulate deposition, and accumulation of amyloid beta-42 and alpha-synuclein in children and young adults. Toxicol Pathol 2008;36(2):289–310.

[58] Oberdorster G, Sharp Z, Atudorei V, Elder A, Gelein R, Kreyling W, et al. Translocation of inhaled ultrafine particles to the brain. Inhal Toxicol 2004;16(6–7):437–45.

[59] Elder A, Gelein R, Silva V, Feikert T, Opanashuk L, Carter J, et al. Translocation of inhaled ultrafine manganese oxide particles to the central nervous system. Environ Health Perspect 2006;114(8):1172–8.

[60] Dorman DC, McManus BE, Parkinson CU, Manuel CA, McElveen AM, Everitt JI. Nasal toxicity of manganese sulfate and manganese phosphate in young male rats following subchronic (13-week) inhalation exposure. Inhal Toxicol 2004;16(6–7):481–8.

[61] Tjalve H, Henriksson J, Tallkvist J, Larsson BS, Lindquist NG. Uptake of manganese and cadmium from the nasal mucosa into the central nervous system via olfactory pathways in rats. Pharmacol Toxicol 1996;79(6):347–56.

[62] Oberdorster G, Sharp Z, Atudorei V, Elder A, Gelein R, Lunts A, et al. Extrapulmonary translocation of ultrafine carbon particles following whole-body inhalation exposure of rats. J Toxicol Environ Health A 2002;65(20):1531–43.

[63] Furuyama A, Kanno S, Kobayashi T, Hirano S. Extrapulmonary translocation of intratracheally instilled fine and ultrafine particles via direct and alveolar macrophage-associated routes. Arch Toxicol 2009;83(5):429–37.

[64] Calderon-Garciduenas L, Franco-Lira M, Henriquez-Roldan C, Osnaya N, Gonzalez-Maciel A, Reynoso-Robles R, et al. Urban air pollution: influences on olfactory function and pathology in exposed children and young adults. Exp Toxicol Pathol 2010;62(1):91–102.

[65] Block ML, Wu X, Pei Z, Li G, Wang T, Qin L, et al. Nanometer size diesel exhaust particles are selectively toxic to dopaminergic neurons: the role of microglia, phagocytosis, and NADPH oxidase. FASEB J 2004;18(13):1618–20.

[66] Davis DA, Akopian G, Walsh JP, Sioutas C, Morgan TE, Finch CE. Urban air pollutants reduce synaptic function of CA1 neurons via an NMDA/NO pathway in vitro. J Neurochem 2013;127(4):509–19.

[67] Gillespie P, Tajuba J, Lippmann M, Chen LC, Veronesi B. Particulate matter neurotoxicity in culture is size-dependent. Neurotoxicology 2013;36:112–7.

[68] Hartz AM, Bauer B, Block ML, Hong JS, Miller DS. Diesel exhaust particles induce oxidative stress, proinflammatory signaling, and P-glycoprotein up-regulation at the blood–brain barrier. FASEB J 2008;22(8):2723–33.

[69] Refsnes M, Hetland RB, Ovrevik J, Sundfor I, Schwarze PE, Lag M. Different particle determinants induce apoptosis and cytokine release in primary alveolar macrophage cultures. Part Fibre Toxicol 2006;3:10.

[70] Van Eeden SF, Tan WC, Suwa T, Mukae H, Terashima T, Fujii T, et al. Cytokines involved in the systemic inflammatory response induced by exposure to particulate matter air pollutants (PM(10)). Am J Respir Crit Care Med 2001;164(5):826–30.

[71] Sawyer K, Mundandhara S, Ghio AJ, Madden MC. The effects of ambient particulate matter on human alveolar macrophage oxidative and inflammatory responses. J Toxicol Environ Health A 2010;73(1):41–57.

[72] Huang YC, Li Z, Harder SD, Soukup JM. Apoptotic and inflammatory effects induced by different particles in human alveolar macrophages. Inhal Toxicol 2004;16(14):863–78.

[73] Devlin RB, McKinnon KP, Noah T, Becker S, Koren HS. Ozone-induced release of cytokines and fibronectin by alveolar macrophages and airway epithelial cells. Am J Physiol 1994;266(6 Pt 1):L612–9.

[74] Kido T, Tamagawa E, Bai N, Suda K, Yang HH, Li Y, et al. Particulate matter induces translocation of IL-6 from the lung to the systemic circulation. Am J Respir Cell Mol Biol 2011;44(2):197–204.

[75] Tan WC, Qiu D, Liam BL, Ng TP, Lee SH, Van Eeden SF, et al. The human bone marrow response to acute air pollution caused by forest fires. Am J Respir Crit Care Med 2000;161(4 Pt 1):1213–7.

[76] Hogg JC, Van Eeden S. Pulmonary and systemic response to atmospheric pollution. Respirology 2009;14(3):336–46.

[77] Mukae H, Hogg JC, English D, Vincent R, Van Eeden SF. Phagocytosis of particulate air pollutants by human alveolar macrophages stimulates the bone marrow. Am J Physiol Lung Cell Mol Physiol 2000;279(5):L924–31.

[78] Chen T, Jia G, Wei Y, Li J. Beijing ambient particle exposure accelerates atherosclerosis in ApoE knockout mice. Toxicol Lett 2013;223(2):146–53.

[79] Chen TL, Liao JW, Chan WH, Hsu CY, Yang JD, Ueng TH. Induction of cardiac fibrosis and transforming growth factor-beta1 by motorcycle exhaust in rats. Inhal Toxicol 2013;25(9):525–35.

[80] Wittkopp S, Staimer N, Tjoa T, Gillen D, Daher N, Shafer M, et al. Mitochondrial genetic background modifies the relationship between traffic-related air pollution exposure and systemic biomarkers of inflammation. PLoS One 2013;8(5):e64444.

[81] Neophytou AM, Hart JE, Cavallari JM, Smith TJ, Dockery DW, Coull BA, et al. Traffic-related exposures and biomarkers of systemic inflammation, endothelial activation and oxidative stress: a panel study in the US trucking industry. Environ Health 2013;12:105.

[82] Bind MA, Baccarelli A, Zanobetti A, Tarantini L, Suh H, Vokonas P, et al. Air pollution and markers of coagulation, inflammation, and endothelial function: associations and epigene-environment interactions in an elderly cohort. Epidemiology 2012;23(2):332–40.

[83] Chuang KJ, Chan CC, Su TC, Lee CT, Tang CS. The effect of urban air pollution on inflammation, oxidative stress, coagulation, and autonomic dysfunction in young adults. Am J Respir Crit Care Med 2007;176(4):370–6.

[84] Johannesson S, Andersson EM, Stockfelt L, Barregard L, Sallsten G. Urban air pollution and effects on biomarkers of systemic inflammation and coagulation: a panel study in healthy adults. Inhal Toxicol 2014;26(2):84–94.

[85] Ruckerl R, Greven S, Ljungman P, Aalto P, Antoniades C, Bellander T, et al. Air pollution and inflammation (interleukin-6, C-reactive protein, fibrinogen) in myocardial infarction survivors. Environ Health Perspect 2007;115(7):1072–80.

[86] Forbes LJ, Patel MD, Rudnicka AR, Cook DG, Bush T, Stedman JR, et al. Chronic exposure to outdoor air pollution and markers of systemic inflammation. Epidemiology 2009;20(2):245–53.

[87] Clark IA, Alleva LM, Vissel B. The roles of TNF in brain dysfunction and disease. Pharmacol Ther 2010;128(3):519–48.

[88] Cunningham C. Microglia and neurodegeneration: the role of systemic inflammation. Glia 2013;61(1):71–90.

[89] Semmler A, Frisch C, Debeir T, Ramanathan M, Okulla T, Klockgether T, et al. Long-term cognitive impairment, neuronal loss and reduced cortical cholinergic innervation after recovery from sepsis in a rodent model. Exp Neurol 2007;204(2):733–40.

[90] Semmler A, Okulla T, Sastre M, Dumitrescu-Ozimek L, Heneka MT. Systemic inflammation induces apoptosis with variable vulnerability of different brain regions. J Chem Neuroanat 2005;30(2–3):144–57.

[91] Weberpals M, Hermes M, Hermann S, Kummer MP, Terwel D, Semmler A, et al. NOS2 gene deficiency protects from sepsis-induced long-term cognitive deficits. J Neurosci 2009;29(45):14177–84.

[92] Lee JW, Lee YK, Yuk DY, Choi DY, Ban SB, Oh KW, et al. Neuro-inflammation induced by lipopolysaccharide causes cognitive impairment through enhancement of beta-amyloid generation. J Neuroinflammation 2008;5:37.

[93] Pott Godoy MC, Tarelli R, Ferrari CC, Sarchi MI, Pitossi FJ. Central and systemic IL-1 exacerbates neurodegeneration and motor symptoms in a model of Parkinson's disease. Brain 2008;131(Pt 7):1880–94.

[94] Richwine AF, Parkin AO, Buchanan JB, Chen J, Markham JA, Juraska JM, et al. Architectural changes to CA1 pyramidal neurons in adult and aged mice after peripheral immune stimulation. Psychoneuroendocrinology 2008;33(10):1369–77.

[95] Qin L, Wu X, Block ML, Liu Y, Breese GR, Hong JS, et al. Systemic LPS causes chronic neuroinflammation and progressive neurodegeneration. Glia 2007;55(5):453–62.

[96] Schuitemaker A, Dik MG, Veerhuis R, Scheltens P, Schoonenboom NS, Hack CE, et al. Inflammatory markers in AD and MCI patients with different biomarker profiles. Neurobiol Aging 2009;30(11):1885–9.

[97] Holmes C, Cunningham C, Zotova E, Woolford J, Dean C, Kerr S, et al. Systemic inflammation and disease progression in Alzheimer disease. Neurology 2009;73(10):768–74.

[98] Laurin D, Curb JD, Masaki KH, White LR, Launer LJ. Midlife C-reactive protein and risk of cognitive decline: a 31-year follow-up. Neurobiol Aging 2009;30(11):1724–7.

[99] Rothman K, Greenland S, Lash T. Modern epidemiology. Philadelphia, PA: Wolters Kluwer/Lippincott Williams & Wilkins; 2008.

[100] Bowling A, Ebrahim S. Handbook of health research methods: investigation, measurement, and analysis. Maidenhead, Berkshire, England: McGraw-Hill, Open University Press; 2005.

[101] Temam S, Kuenzli N, Anto JM, Bousquet J, Jarvis D, Le Moual N, et al. Association between socioeconomic status and air pollution exposure in Europe. Basel: Environment and Health; 2013.

6

The Role of Methylmercury Exposure in Neurodevelopmental and Neurodegenerative Disorders

*Thomas M. Burbacher, Rafael Ponce,
Kimberly S. Grant*

Environmental Factors in Neurodevelopmental and Neurodegenerative Disorders
http://dx.doi.org/10.1016/B978-0-12-800228-5.00006-6

This chapter is dedicated to the memory of Dr Patricia Rodier, friend, colleague, and pioneer in science.

INTRODUCTION

For many young children living in developed nations, infectious disease is no longer the primary threat to good health. A more subtle danger has emerged over the last 60 years [1]. Scientific consensus is growing that early exposure to environmental chemicals may be contributing to the increased number of children with certain chronic illnesses or neurodevelopmental disorders [2–4]. Methylmercury (MeHg) is one of the most widely studied environmental chemicals in modern and historical scientific records. Present in many of the world's oceans, lakes, and rivers, MeHg persists in aquatic environments and bioaccumulates in predatory fish, such as tuna, swordfish, and shark. The consumption of MeHg-contaminated fish constitutes the principal means of exposure for both human and animal populations [5,6]. The action of MeHg on biological and psychological functions over a lifespan is complex, and the risk of neurotoxicity is not equally distributed across age groups [7,8]. The developing brain is highly sensitive to the consequences of MeHg exposure, and chemically induced injuries sustained early in life can have enduring consequences for health and behavior over the lifespan [9,10].

Catastrophic episodes of high-dose human exposure to MeHg focused global attention on the danger of MeHg exposure to pregnant women and their children. Over time, increasingly sophisticated environmental epidemiology studies have been initiated to improve the precision of our understanding of MeHg effects on child cognition and health. Some of the most important and instructive lessons in contemporary teratology and toxicology have been gleaned from studies of MeHg on mammalian biological function. This chapter will provide an interdisciplinary look at the form and function of MeHg-related injuries of the nervous system, including neurodevelopmental disorders early in life and neurodegenerative diseases associated with aging. The cellular and anatomical pathways of injury that are linked to MeHg exposure will be explored and, whenever possible, results across species will be examined to better understand the neurotoxic signature of MeHg for humans and animal models.

FETAL MINAMATA DISEASE AND ASSOCIATED DISORDERS

In the last century, episodes of high-dose human exposure to MeHg have occurred in Japan and Iraq [11,12]. In the mid-1950s, a mysterious disease syndrome in newborn infants was observed by Japanese physicians. Infants born to mothers residing near Minamata

Bay developed a constellation of neurological symptoms, later called fetal Minamata disease (MD), which included intellectual impairment, loss of coordinated muscle movement, primitive reflexes, and cerebral palsy [13,14]. Research into the origin of this medical phenomenon revealed that placental transfer of MeHg from the mother to the fetus resulted in high levels of in utero exposure to this metal. The source of maternal–fetal exposure was linked to the consumption of MeHg-contaminated fish and shellfish caught in Minamata Bay. The source of the pollution in Minamata Bay was ultimately traced back to Hg-contaminated effluent being released into the bay by an acetaldehyde plant manufacturing plastic products. Observations by medical staff during the outbreak clearly showed that mothers who displayed mild or no signs of MeHg toxicity gave birth to infants with clear evidence of MeHg poisoning [15]. This clinical lesson was one of the first-documented incidents of enhanced fetal sensitivity to a neurotoxicant, a groundbreaking principle in contemporary teratology. The expression of clinical symptoms in prenatally exposed infants varied widely from mild affliction to severe intellectual and physical disabilities [16]. Delays in motor development, inability to visually track objects, and poorly developed survival reflexes, such as sucking and swallowing, were the first exposure-related symptoms observed. As the affected children matured, clinical observations revealed persisting primitive reflexes, dysarthria (a motor-speech disorder), hyperkinesias (hyperactivity), and a severe cerebral palsy-like syndrome. Many fetal MD patients could not walk by age seven, most displayed strabismus (misalignment of the eyes), and all sustained severe intellectual disability (Figure 6.1). Autopsy data from MeHg-exposed children demonstrated that developmental

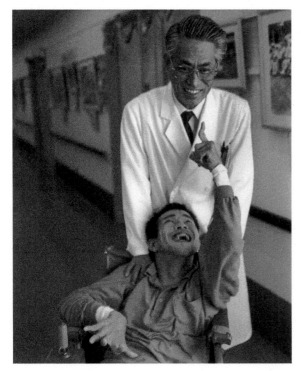

FIGURE 6.1 Kazumitsu Hannaga, a fetal Minamata disease patient, and his physician, Dr Hiroyuki Moriyama, at the Meisui-en Hospital in 1991 *(permission pending from EHP)*.

exposure results in widespread damage to the brain, primarily through hypoplasia and cell loss (see further discussion below) [17,18].

A report by Yorifuji et al. [19] provided exposure estimates associated with Hg-related disorders using Hg data from umbilical cord samples collected from infants born between 1925 and 1980 from coastal-dwelling families living near Minamata Bay or the Shiranui Sea area. The authors calculated estimates of maternal hair Hg concentrations and maternal daily intake of Hg from these samples. The estimated median maternal hair Hg concentration was 41 ppm, corresponding to a daily maternal intake of 225 μg (or 3.75 μg/kg/day for a 60 kg woman), for mothers delivering infants with fetal MD. A median maternal hair Hg concentration of ~17 ppm, corresponding to a daily maternal intake of 1.54 μg/kg/day, was reported for mothers delivering infants with less severe impairments (intellectual disabilities), whereas mothers delivering infants with no Hg-related symptoms had a median maternal hair Hg concentration of ~6 ppm, corresponding to a daily maternal intake of 0.55 μg/kg/day.

In the early 1970s, another outbreak of high-dose MeHg intoxication occurred in Iraq when grain treated with a MeHg-based fungicide was mistakenly used to make bread [11]. Across the country, 6530 cases of MeHg poisoning required hospital admittance, and of these patients, 459 died [20]. Pregnant women were among those who consumed contaminated bread, and neurobehavioral consequences in exposed infants ranged from mild to severe. Some asymptomatic women in Iraq gave birth to infants with clear evidence of MeHg-related neurological injuries including MD, corroborating clinical findings from Japan [21]. Iraqi children exposed to MeHg (Figure 6.2) during fetal development were evaluated during infancy and childhood, and highly exposed children showed severe sensory impairments, paralysis, pathological reflexes, intellectual disabilities, and, in some cases, cerebral palsy [22–24]. Exposure-related effects included marked delays in the attainment of important developmental milestones, such as walking and talking. Consistent with the concept of progressive injury, some neurological symptoms increased in severity over time.

A report by Cox et al. [25] provided an "Estimated Lowest Effect Level (ELEL)" for retarded walking using data from 83 Iraqi mother–infant pairs. The authors concluded

FIGURE 6.2 American scientist evaluating neurological function in a young Iraqi boy prenatally exposed to MeHg.

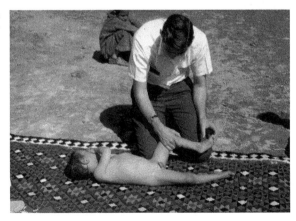

that "motor retardation should occur in children prenatally exposed to maternal hair concentrations of less than 50 ppm and may be expected in the range of 10–20 ppm." The statistical best estimate for a population threshold was estimated to be 10 ppm, or an estimated intake of 0.9 µg/kg/day.

ENVIRONMENTAL EPIDEMIOLOGY STUDIES OF LOW-LEVEL MeHg EXPOSURE: IMPROVING MEASUREMENTS AND METHODS

Prospective cohort epidemiologic studies of high fish-eating populations have provided additional insights into the dose–response relationship between MeHg intake during pregnancy and offspring neurodevelopmental effects. A New Zealand cohort of 73 mothers who consumed fish more than three times a week during pregnancy and had hair mercury levels over 6 ppm was enrolled in a study aimed at determining the effects of chronic low-level Hg exposure on child mental development [26]. Children of 31 women from the above group were matched with children whose mother's hair Hg level was below 6 ppm and tested on the Denver Developmental Screening Exam (DDSE) at four years of age. Results indicated that approximately 52% of children with maternal hair mercury values greater than 6 ppm showed marked delays in attaining important developmental milestones compared to 17% of controls [26]. At six years of age, a second cohort of children ($n = 61$) was examined with standardized neurobehavioral test measures commonly found in New Zealand schools [27]. Domains of intelligence, language, fine and gross motor development, academic achievement, and social competence were targeted. The endpoints most sensitive to high maternal MeHg body burden in exposed children were language development (Test of Language Development), intelligence (WISC-R), and psychomotor abilities, such as block building, puzzle solutions, finger tapping, and drawing (McCarthy Scales of Children's Abilities–Perceptual Scale). A benchmark dose analysis conducted on the data from the six-year cohort found that when maternal hair Hg was expressed as a continuous variable, the associations between maternal hair mercury and childhood performance were not significant [28]. These results were, however, clearly influenced by one child with extremely high maternal hair Hg values but normal psychometric scores. When this child was removed from the analysis, scores on language development and perceptual performance were inversely correlated with maternal hair Hg concentrations. The authors of the report conclude that decrements in neurobehavioral performance may be most highly associated with maternal hair Hg levels above 10 ppm. The New Zealand longitudinal cohort study is noteworthy for two reasons; confounding variables that can impact neurobehavioral performance, such as infant birth weight, breastfeeding, maternal age, maternal smoking and alcohol habits, and social class, were statistically controlled, and a broad battery of tests were given to the children, increasing the likelihood of identifying more subtle deficits in neurocognitive function.

The New Zealand prospective cohort study provided evidence that maternal intake of MeHg as low as 0.5–1 µg/kg/day may be related to the disruption of normal developmental processes in exposed children. These results were provocative, but systematic studies with greater numbers of children were needed to fully characterize the risks of exposure. To further investigate the relationship between maternal fish consumption, prenatal MeHg exposure, and neurodevelopmental effects, a large prospective study was launched in the Faroe

Islands in 1986–1987 [29]. The Faroe Islands is a rugged country that lies halfway between Norway and Iceland, and inhabitants rely heavily on the native harvesting of local fish and pilot whale for sustenance. Approximately 1000 children were enrolled at birth from three Faroese hospitals and samples of maternal hair and cord blood were collected at delivery for mercury analysis. Children were tested at six to seven years of age with a series of cognitive and motor assessments ($n = 917$), and hair samples were collected for analysis of current mercury exposure levels ($n = 672$) [30]. Tests of finger tapping, hand–eye coordination, reaction time, visuospatial ability, language, and three subtests from the Wechsler Intelligence Scale for Children (Digit Span, Similarities, and Block Design) were administered. Physical exams showed that children were generally in good health, and visual acuity and hearing were not affected by early MeHg exposure. Multiple regression analyses on neuropsychological data revealed exposure-related changes in language, attention, and verbal memory with smaller effects on motor and visuospatial function. Adverse effects on cognition remained in place after controlling for confounds, such as employment, alcohol and cigarette use, current place of residence, as well as obstetrical risk factors.

Additional follow-up testing of the Faroese cohort occurred at age 14 ($n = 815$) using an expanded version of the neuropsychological battery given at six to seven years [31]. After controlling for confounders that can also affect neurobehavioral performance (e.g., sex, age, maternal IQ, parental employment, smoking and alcohol, dietary habits, etc.), Faroese adolescents demonstrated adverse effects of prenatal MeHg exposure on motor speed, attention, and memory. The breadth of the neurobehavioral impairment suggests that widely divergent brain systems were affected. Although adverse effects at 14 years were weaker than at seven years, study results continued to suggest long-term, enduring effects of early MeHg exposure on brain function in exposed children. Separate reports by Murata et al. (2002, 2004) [32,33] described additional long-term effects in Faroese children at seven and 14 years of age using a neurophysiological approach, brain-stem auditory-evoked potentials (BAEPs). Children born to mothers with maternal hair concentrations of ~10 ppm displayed delayed BAEP latencies at both test ages.

The results of the initial Faroe Island study were used by the Environmental Protection Agency to calculate a reference dose (RfD) for MeHg. Mercury levels in cord blood associated with multiple childhood outcomes resulted in similar estimates of ingested dose, ~1 µg/kg/day. An uncertainty factor of 10 was used due to uncertainty in estimating an ingested mercury dose from cord-blood Hg concentrations and pharmacodynamic variability for a calculated RfD of 0.1 µg/kg/day (see the following discussion).

To further investigate the relationship between maternal hair mercury concentration and neurodevelopmental effects in exposed children, a third prospective birth cohort study was assembled in the Seychelles Islands in 1989 [34]. The Seychelles Islands is an archipelago in the Indian Ocean, northeast of Madagascar, where fishing figures prominently in the local economy (Figure 6.3). Finfish caught off the Seychelles Islands have mercury concentrations, which are similar to ocean fish caught for consumption in larger, more industrialized nations [35].

Investigators successfully recruited 779 mother–infant pairs to participate in a long-term study aimed at examining the effects of MeHg on child development. Over 80% of the women in this study reported eating fish on a daily basis, and the median number of fish meals consumed per week during pregnancy was 12 [36]. Hair samples were collected from pregnant women and their children for mercury analysis, and a battery of newborn, childhood, and adolescent assessments were used to examine the immediate and long-term

FIGURE 6.3 Fish sold at the Victoria market in Gerard Larose, Seychelles *(photograph used with permission from the Seychelles News Agency—Creative Commons License).*

neurodevelopmental effects of early MeHg exposure [37–43]. Results from the pediatric neurobehavioral exam given at 6.5 months of age showed that maternal Hg concentrations were not related to alterations in infant neurologic responses, muscle tone, and deep tendon reflexes. Tests of visual recognition memory (Fagan Test of Infant Intelligence, FTII) were unrelated to prenatal MeHg exposure as was performance on a screening test of cognitive and motor abilities. The achievement of important developmental milestones, such as independent walking, an index of developmental maturation, was also documented in this group, and maternal hair Hg was not associated with either endpoint. The children in the Seychelles cohort, now approximately 24–25 years old, have been carefully studied over time with a broad range of psychometric tools to examine memory, learning, language, perceptual skills, and motor development. To date, the results of this longitudinal investigation do not support an adverse relationship between fetal MeHg exposure, as measured by elevated maternal hair Hg levels, and key neurodevelopmental and life outcomes, such as health, intellectual ability, social competence, and school success. On some measures (Preschool Language Scale, Bender–Gestalt Test, and the Connor's teacher rating scale of attention-deficit hyperactivity), results indicated enhanced performance due to prenatal

MeHg exposure [40,42]. The authors indicated that these results were most likely due to the known benefits on development associated with seafood consumption, which provides an excellent source of protein and essential omega-3 polyunsaturated fatty acids (PUFAs) (particularly docosahexaenoic acid (DHA) and eicosapentaenoic acid (EPA)) [35]. These results indicated that future studies of the potential detrimental effects of MeHg exposure from seafood consumption during pregnancy would also need to address the possible benefits of a diet rich in seafood.

THE DELICATE BALANCE BETWEEN THE RISKS AND BENEFITS OF MATERNAL FISH CONSUMPTION FOR CHILDHOOD OUTCOMES

The lack of adverse neurobehavioral changes in the Seychellois cohort, coupled with improved performance on some cognitive tests by the more highly exposed children, suggested that maternal diet can change the dynamic nature of the developing nervous system in both positive and negative directions. To better characterize the risks and benefits of concurrent PUFA and MeHg exposure during pregnancy, a second cohort study was launched in the Seychelles Islands in 2001 [44,45]. Three hundred pregnant women were enrolled in this longitudinal study, and maternal nutritional status was characterized during pregnancy with a food survey and a diet diary. To biologically measure key fatty acids and MeHg exposure, a maternal blood sample was collected during pregnancy, and both maternal blood and hair were collected at delivery. Women in this study consumed approximately nine fish meals per week during pregnancy and had a mean hair Hg level of 5.9 ppm. The Bayley Scales of Infant Development (BSID-II) were used to measure mental and psychomotor performance during infancy and early childhood, along with a variety of other secondary cognitive measures. Results indicated that prenatal MeHg exposure alone was not related to any developmental endpoint at any of the ages studied (up to 30 months of postnatal age). However, when both prenatal MeHg exposure and nutritional variables were included in the statistical model, Bayley psychomotor scores (e.g., rolling, crawling, grasping) at 30 months were inversely related to maternal mercury hair concentrations. No beneficial associations with nutritional factors were found. These findings suggest that nutritional benefits derived from the maternal diet are potential confounders and may obscure the true risk of early MeHg exposure on the neurobehavioral development. A more recent paper from this investigative team showed that a new analysis of the Bayley data continued to support the finding of impaired psychomotor performance at 30 months in MeHg-exposed children [46]. Maternal PUFA levels, however, exerted an effect in the opposite direction and were associated with improved scores on the exam.

A second smaller cohort consisting of 182 births was also assembled in the Faroe Islands (1994–1995) to study the complex relationship between nutritional factors, such as infant PUFA status and elevated MeHg exposure. The average cord-blood Hg level in study infants was 22.8 µg/L. In this group, neurological fitness (reflexes, muscle tone, behavioral state, and responsivity) was measured using the Prechtl exam at approximately two weeks, and infant performance was interpreted in relation to maternal seafood diet during pregnancy and cord-blood Hg [47]. Using regression analysis and appropriate statistical controls for dietary influences, investigators found that a 10-fold increase in cord-blood Hg was associated with a decreased neurologic optimality score. Essential nutrients such as PUFAs did not exert protective effects on neonatal performance. At seven years of age, visual-evoked potentials were measured in this cohort to evaluate electrophysiological changes in relation to maternal hair Hg levels [48]. Higher mercury concentrations in maternal hair and infant cord blood were significantly associated with decrements in visual-evoked potential latencies only after adjusting for cord-serum PUFA levels [48]. Tests of neurobehavioral function were also administered at seven years of age in these children with a broad-based cognitive battery [49]. Results from this study revealed a significant relationship between prenatal MeHg exposure and memory performance in childhood that was strengthened when the data were adjusted for cord-serum PUFA concentrations. These results showcase the need to include measurements of essential fatty acids in birth cohort studies to avoid underestimating the adverse effects of MeHg exposure on cognitive performance.

One of the first reports of MeHg-related effects in a US cohort came from a project in the state of Massachusetts, Project Viva, examining the effects of diet and environmental exposures on pregnancy outcome and offspring development. As noted above, fish and other seafood contain essential nutrients such as PUFAs and serve as an important source of protein for pregnant women. In this prospective birth cohort study, women were recruited at their initial obstetrical visit and completed multiple questionnaires (including health history, behavioral habits, and food frequency—including seafood) [50]. Hair and blood samples were collected on a subset of the women to quantify mercury exposure. Infants were evaluated on a test of visual recognition memory at six months, and study results demonstrated that higher fish intake during pregnancy was associated with better infant memory scores. However, higher maternal hair mercury values were linked to lower memory scores. Children in this cohort were also evaluated on a number of standardized cognitive tests at three years of age [51]. Using maternal blood mercury values from the second trimester, higher levels of prenatal mercury exposure were related to poorer performance on tests of language and visual motor skills. Higher fish intake (>2 servings per week) was associated with better developmental test scores. During elementary school, data from children in this cohort suggested that in utero MeHg exposure was linked to the expression of behaviors consistent with attention-deficit/hyperactivity disorder (e.g., inattention and hyperactivity/impulsivity) [52]. However, maternal fish consumption during pregnancy was also highly protective against this behavioral symptomology and was associated with higher scores on standardized measures of intelligence (WISC-III). The results of the six month and three year evaluation indicate that offspring of women with the highest intake of low-mercury fish exhibited the best performance on early developmental assessments. For the visual recognition memory test, offspring of women who consumed more than two fish meals per week and had hair Hg levels below 1.2 ppm exhibited the highest memory scores, whereas offspring of women who consumed less than two fish meals per week and had hair mercury levels above 1.2 ppm showed reduced memory scores [50]. These results indicate that the maternal consumption of MeHg at 0.1 $\mu g/kg/day$ is associated with neurodevelopmental effects on early cognition.

A collective look at the Japanese, Iraqi, New Zealand, Faroese, and Seychellois cohorts illustrates how retrospective high-dose studies gave way to prospective studies of more environmentally relevant levels of MeHg exposure, each building upon the last. The rough-hewn but critically important medical observations made in Japan were augmented by the measurement of developmental milestones, such as walking and talking in Iraq. Studies in New Zealand and the Faroe Islands made further psychometric improvements by moving beyond clinical observations and milestone documentation to include detailed measurements of mental and psychomotor functioning. Prospective cohort studies such as those taking place in the republic of Seychelles and Project Viva have helped to unravel the complex relationship between maternal diet, MeHg exposure, and childhood development. The current EPA/FDA joint guidance on fish consumption for women of childbearing age attempts to provide some balance for this complex relationship by advising women to eat seafood for their health and the health of their baby, but to limit their seafood intake to two meals per week and to choose wisely by avoiding fish high in MeHg. The joint guidance is based on the current RfD for MeHg developed by the Environmental Protection Agency (EPA). As mentioned above, the RfD is based largely on the results of the initial Faroe Island study. Using the reported MeHg effects on a number of different childhood outcome measures, the EPA calculated an RfD dose of 0.1 µg/kg/day, using an uncertainty factor of 10. A review of the RfD was conducted by a National Academy of Sciences panel in 2000 who concluded that the RfD was based on sound science and was valid [53]. No updated review of the RfD has taken place since the publication of reports indicating MeHg-exposure-related effects at the RfD [50–52,54], essentially negating the uncertainty factor.

INSIGHTS ON NEUROBEHAVIORAL EFFECTS OF MeHg FROM ANIMAL MODEL STUDIES

Throughout the last 40 years, research with animal models has contributed significantly to our understanding of the adverse effects of early MeHg exposure on offspring health and development [55]. Studies using nonhuman primates have examined both prenatal and postnatal MeHg exposure effects. Nonhuman primates share a similar pattern of MeHg toxicokinetics [56–60]. Monkeys also share similar patterns in brain structure and neurobehavioral outcomes with humans (e.g., memory, complex problem solving, highly evolved social behavior) [61]. Three cohorts of nonhuman primates have been examined in longitudinal studies of early MeHg exposure and neurodevelopmental effects, each with a different dosing schedule (postnatal only, pre- plus postnatal, and prenatal only). Monkeys postnatally exposed to daily doses of 50 µg/kg MeHg from birth to seven years of age exhibited impaired high-frequency hearing and deficits on tests of spatial vision [62,63]. Losses in vision did not normalize over time, suggesting irreversible damage to the parvocellular region of the brain [10]. At 18 years of age, animals in this group showed increased vibration sensitivity thresholds, suggesting MeHg-related changes in somatosensory function [64]. Persisting somatosensory disorders in the form of increased touch thresholds have been reported by adult residents of Minamata Bay, more than 30 years after the cessation of MeHg exposure [65]. Psychomotor testing at 13 years of age revealed that exposed monkeys had slowed responses on a fruit-retrieval task when compared to controls [10]. This finding is consistent with the reduced finger-tapping speed that has been observed in children exposed to MeHg [30,31] and highlights the long-term sensitivity of integrated motor behavior to the effects of early MeHg exposure [10]. Nonhuman primates exposed to MeHg (10, 25, or 50 µg/kg/day) during both the pre- and postnatal periods (up to 4.5 years after birth) exhibited decreased visual contrast sensitivity thresholds on a test of spatial vision administered at five years of age [66]. Auditory testing at 19 years indicated that exposure was associated with elevated pure-tone thresholds across a range of frequencies [67]. Deficits in sensory acuity across multiple systems were clearly linked to early MeHg exposure and, in the case of hearing, decrements in performance became more evident as the monkeys aged. Significant losses in visual and auditory functioning, including blindness and deafness in extreme cases, have been documented in highly exposed human infants and children [22,24,68]. MeHg-exposed monkeys performed well on most tests of learning and memory [69]. During infancy and the juvenile periods, learning was evaluated with measures of visual discrimination reversal learning and/or fixed interval performance. MeHg-exposed subjects performed as well as controls on the discrimination reversal tasks, but results of the fixed-interval task suggested that MeHg exposure may have interfered with temporal discrimination. Studies of in utero MeHg exposure alone (50 µg/kg/day) have reported significant delays in the development of coordinated-reaching behavior [9]. These effects are consistent with the finding that delays in early psychomotor development (fine and gross motor skills) are associated with higher maternal and newborn cord-blood Hg levels in humans [70]. Prenatal MeHg effects were also observed on early cognitive assessments, such as object permanence [71] and visual recognition memory [72,73]. Decrements in visual recognition memory scores have also been found in human infants with a fetal history of MeHg exposure through maternal consumption of contaminated fish [50]. Other effects observed following prenatal MeHg exposure included reduced play and increased passive and nonsocial behaviors in infants throughout the first nine months of life [74,75] as well as a reduction in physical growth (weight gain) in male offspring during adolescence (2.5–4 years) [76]. Long-term effects on spatial vision (reduction in contrast-sensitivity thresholds at higher spatial frequencies) were also reported following in utero MeHg exposure alone [77], whereas adult learning, memory, and attentional skills appeared normal [78].

Most experimental studies of MeHg neurodevelopmental effects have been conducted in rodent animal models [55]. One of the initial MeHg studies conducted in the mouse model [79,80] reported both immediate and delayed effects (associated with aging) following prenatal MeHg exposure. Increased pup mortality and delays in physical growth were early effects of MeHg exposure, whereas abnormalities in psychomotor function (swimming) were observed as animals aged. Numerous reports on MeHg neurodevelopmental effects using the rat model come from the Collaborative Behavioral Teratology Study (CBTS) [81–83]. Pregnant rats were exposed to MeHg (2 or 6 mg/kg) on gestation days six to nine and exposed offspring were examined on a battery of behavioral assessments throughout the first four months of life. MeHg-related effects included increased auditory startle-habituation amplitudes, primarily at the high dose, increased activity, and decreased performance on a visual discrimination task. Associated studies using an expanded test battery or lower and longer exposures reported detrimental effects on early reflex responses, such as surface righting, pivoting, and negative geotaxis, as well as maze learning [84,85]. Studies also reported effects on the physical development of the offspring at the high dose [83,85]. Effects on postnatal physical development in exposed pups similar to the ones reported in the CBTS have been reported in other studies. These include delays in the appearance of physical landmarks (e.g., tooth eruption, eye opening) and reductions in growth and weight gain [86–88]. Reduced litter size and offspring mortality have also been reported at relatively high-dose prenatal MeHg exposure [87,89]. MeHg-induced effects on early reflexes and motor responses (e.g., auditory startle, surface-righting response, rotorod performance) have also been reported in numerous studies [86,88–96], as have effects on neonatal activity [80,94,97,98]. Two early studies of visual-evoked potentials have reported reductions in peak latencies from the visual cortex in MeHg-exposed rats [99,100]. Sensory changes associated with developmental MeHg exposure are not limited to the visual system. Recordings of the auditory brain-stem response in MeHg-exposed mice have shown changes in waveforms and latencies that reflect irreversible damage to hearing thresholds in exposed offspring [88]. Several studies using the rodent model have focused on examining the effects of MeHg exposure on learning and memory tasks. Deficits in avoidance learning have been reported in both mice and rats [97,101,102] as well as deficits in maze learning [85,91,92,103–106]. Studies using an operant conditioning Differential Reinforcement of High Rates paradigm [107,108] have reported effects on response rates at doses as low as 0.01 and 0.05 mg/kg in rats exposed during gestation days six to nine. Attempts to replicate these results have not been successful thus far [109]. A more recent report [92] examined mice exposed to 0.01 mg/kg MeHg during gestation days 8–18. The results of this study indicated MeHg-induced deficits in motor coordination, activity, and reference memory. In addition to the array of neurobehavioral outcomes outlined above, perinatal exposure to MeHg is associated with depression-like signs in young and adult male mice [105]. These adverse changes in behavior (immobility/floating during a forced-swim task) can be reversed through antidepressant treatment with fluoxetine and are linked to epigenetic changes in brain-derived neurotrophic factor gene expression [110]. These studies demonstrate the power of perinatal MeHg exposure to predispose treated offspring to the onset of disease during adulthood and aging.

Compelling evidence from animal work with nonhuman primates and rodents suggests that developmental MeHg exposure has adverse impacts across a number of important behavioral domains in postnatal life. In monkeys, clear developmental delays occur in cognition and abnormal patterns of social behavior in MeHg-exposed infants, but adult learning and memory skills appear to be intact. Treatment-related changes in psychomotor performance (uncoordinated-reaching behavior) have been observed across cohorts. Long-term neurotoxicity after perinatal exposure is primarily expressed as reductions in the acuity of the visual, auditory, and somatosensory systems. In rodents, disorganized or slowed motor coordination and delays in physical growth/maturation characterize the primary effects of early MeHg exposure. Treatment-related changes in cognitive abilities provide evidence that MeHg can negatively impact learning and memory in rats and mice as well. Like macaque monkeys, rodents exposed to MeHg in either the pre- or postnatal period (or both) express disturbances in adult sensory system functioning, which underscore the possibility of irreversible brain damage.

CROSS-SPECIES COMPARISONS OF MeHg NEURODEVELOPMENTAL EFFECTS

As mentioned above, the current MeHg RfD of 0.1 µg/kg/day is based on results from the Faroe Island studies. A critical dose of 1.1 µg/kg/day was used to develop the RfD based on effects associated with a maternal hair Hg level of 11 ppm. An uncertainty factor of 10 was used to arrive at 0.1 µg/kg/day. Cernichiari and colleagues calculated a maternal hair to infant brain Hg concentration ratio of 0.015 using values obtained from 27 infant brains from the Seychelles Island study [34]. Using this value, the critical dose for MeHg effects (1.1 μg/kg/day) would be associated with an infant brain Hg level of ~0.165 ppm. Thus far, the lowest in utero dose studied for nonhuman primates is 50 μg/kg/day MeHg [71]. Reported brain Hg levels for offspring from this study varied across different brain regions (2.10–7.92 ppm) with an overall average brain Hg concentration of ~4 ppm [111]. A recent study in mice reported MeHg exposure-related effects at a maternal intake of 10 μg/kg/day MeHg, which produced measured fetal brain Hg levels of ~0.05 ppm, nearly three times below that reported for humans [92]. Results from the nonhuman primate and mouse studies mentioned above did not produce a no-effect level for in utero MeHg exposure. An RfD based on these data would need to be adjusted by a factor of 1000 to account for cross-species comparisons, differences in toxicokinetics and toxicodynamics, and using a lowest-observed-adverse-effect level (LOAEL) instead of a no-observed-adverse-effect level (NOAEL). Thus, RfD values ranging from 0.01 μg/kg/day to 0.05 μg/kg/day would be calculated from studies using animal models.

MECHANISMS UNDERLYING MeHg-MEDIATED NEUROTOXICITY

Given the neurobehavioral effects identified above, what is known regarding the molecular, cellular, and organ-based mechanisms of MeHg neurotoxicity? As discussed above, systemic MeHg exposure leads to accumulation in the adult and developing brain. Although MeHg is lipophilic, MeHg appears to enter the brain as a MeHg–cysteine complex, which mimics methionine. This complex is transported via the L-type large neutral amino acid transporter, LAT1 [112]. Once inside the brain, MeHg can be demethylated, resulting in accumulation of inorganic Hg [113,114]. The extent to which organic and inorganic Hg contribute to neurotoxicity remains a matter of debate [115].

The histopathological lesions following MeHg exposure in the adult and developing central nervous system (CNS) are distinct [116–118]. Whereas histopathological effects are localized to specific regions in adult MeHg toxicity (e.g., neuronal loss in the cortical sulci, including the calcarine fissure of the occipital cortex, and the external and internal granule layers of the cerebellum), diffuse and more severe lesions are observed following developmental exposure (reviewed in [74,75,119]). Characteristic pathology following developmental exposure in humans and animal models includes decreased cell abundance and evidence of cell-cycle arrest, and ectopic and disoriented neuronal distribution [17,120–125]. Differences in the patterns of CNS lesions following adult vs. developmental exposure have been attributed to the unique developmental processes underlying CNS development. The developing CNS undergoes a highly synchronized progression of cell proliferation, migration, differentiation, and progressive cell loss to insure the correct patterning and localization of discrete nuclei in the developing brain. Disruption of these discrete stages may not be overcome [126,127]. The observations of neuronal disorganization and hypocellularity following in utero exposure have led to extensive research into the effects of MeHg on cell proliferation (and cell-cycle progression), cell death, and cell migration, and efforts to identify key molecular targets of MeHg toxicity.

MeHg has an extremely high affinity for protein sulfhydryl groups (log K range of 15–23), and its affinity for selenium is as much as a million-fold greater [128,129]. The high-affinity binding of MeHg to selenium groups underlies various observations indicating a protective role for dietary selenium in MeHg toxicity by various mechanisms including sequestration, glutathione (GSH) synthesis, increased GSH peroxidase activity, increased selenoprotein levels, and increased MeHg demethylation (see [115]). The high affinity for protein sulfhydryl groups also means that MeHg can act as a general enzyme poison. Various cellular processes are altered by MeHg exposure in vivo and in vitro, including RNA and protein synthesis [130–134], mitochondrial function (including loss of mitochondrial membrane potential, adenosine triphosphate (ATP) production, and calcium homeostasis, protein phosphorylation) [133,135], and production of reactive oxygen species (ROS) [136–139]. Although such broad-ranging effects on fundamental cellular processes challenge identification of principal drivers of MeHg toxicity, a review of the mechanistic data suggests several key targets.

EFFECTS ON MITOCHONDRIA, INCLUDING REDUCED CELLULAR ENERGETICS, ALTERED Ca^{2+} HOMEOSTASIS, AND PRODUCTION OF REACTIVE OXYGEN SPECIES

MeHg can accumulate in mitochondria [140,141], and mitochondrial function appears to be among the most sensitive targets of MeHg toxicity (see Table 6.1; Figure 6.4) (reviewed by [142–144]). MeHg exposure can cause mitochondrial K^+ accumulation, and H^+ efflux, mitochondrial swelling, and dissipation of inner membrane potential [145,146]. Potassium is a key regulator of mitochondrial volume, mitochondrial energy transfer to the cytosol, inner membrane potential (via the K^+/H^+ antiporter), and the production of ROS. Under conditions of decreased inner membrane potential, the mitochondrial volume will contract and activity of the K^+/H^+ antiport will decrease, thereby restoring membrane potential. Conversely, conditions that lead to mitochondrial K^+ uptake (along with water and other ions) will increase mitochondrial volume and dissipate the membrane potential [147]. Because MeHg also directly binds and inhibits electron-transport complexes, it can both directly and indirectly interfere with oxidative respiration and decrease ATP production in vivo and in vitro [119,141,148–151].

In addition to effects on cellular energetics, MeHg's effects on mitochondria include the increased production of ROS, loss of mitochondrial calcium sequestration, and cytochrome c release, which characterize the mitochondrial permeability transition [138–140,150–152]. Production of ROS is a by-product of mitochondrial electron transport. Disruption of electron transfer along the electron-transport chain will drive excessive ROS production and cause cellular injury. In vivo and in vitro MeHg exposure increases production of ROS, including superoxide anion (O_2^-) and hydrogen peroxide (H_2O_2). Consumption of GSH (reviewed in [140,152]) and MeHg-mediated ROS production and cellular injury can be reduced by GSH and various other antioxidants, including vitamin E and selenium, indicating a role for oxidative stress in MeHg neurotoxicity (reviewed in [139,152–158]).

Numerous reports document MeHg-mediated disruption in intracellular Ca^{2+} homeostasis (and potentially other divalent cations, see Table 6.1) (reviewed in [150,159–163]). This increase in intracellular Ca^{2+} appears to arise first through release of intracellular stores maintained in smooth endoplasmic reticulum and mitochondria, followed by influx of extracellular Ca^{2+} [164]. Intracellular Ca^{2+} is tightly regulated, particularly in neurons, which maintain a very high Ca^{2+} gradient across the cellular membrane to respond to small elevations in

TABLE 6.1   Mechanisms of MeHg-Mediated Neurotoxicity—Concentration- and Time-Dependent Effects In Vitro

Ni et al. [165]	0.1–1 μM, primary rat astrocytes and microglia	Decreased GSH/GSSG (≥0 μM) and increased ROS production (≥1 μM) within 1 min in microglia; minimal cytotoxicity at 6 h (≥1 μM only). Lower sensitivity of astrocytes attributed to lower MeHg uptake and higher basal GSH
Limke and Atchison [138]	0.2–0.5 μM, rat cerebellar granule cells	Increased $[Ca]_i$ at 18 min; precedes increased mitochondrial membrane permeability (evaluated only at 0.5 μM) at 25 min
Heidmann et al. [166]	0.25–0.5, chick forebrain neurons	After 2 h exposure, increased microtubule depolymerization observed immediately after end of treatment; protection with Ca^{2+} chelation
Marty and Atchison [167]	0.5–1 μM, rat cerebellar granule cells	Increased $[Ca]_i$ at ≤45 min precedes increased plasma membrane permeability/cell death at 3.5 h; survival improved with Ca^{2+} chelation
Castoldi et al. [168]	≥1 μM, rat cerebellar granule cells	Impaired mitochondrial function at 1 h. Increased microtubule depolymerization, neuronal network fragmentation, and apoptosis at 1.5 h
Faustman et al. [160]	1 μM, primary rat midbrain cells	Increased $[Ca]_i$ at 5 min, reduced cell division at 24 h
Allen et al. [169]	≥1 μM, primary rat cerebral astrocyte and neuronal co-cultures	Decreased uptake of the GSH precursor cystine in astrocytes, but not neurons, at 60 min
Polunas et al. [139]	1.5 μM, undifferentiated and neuronally differentiated murine embryonal carcinoma cells	Increased ROS and decreased mitochondrial membrane potential at ≥20 min in differentiated neurons, reduced by inhibitor of mitochondrial membrane transition pore opening
Yuan and Atchison [170]	≥4 μM, hippocampal CA1 neurons	Altered evoked action potentials at ≤150 min
Kunimoto and Suzuki [171]	≥3 μM, rat cerebellar slices	Impaired granule neuron migration at 20 h
Sarafian et al. [119]	5 μM, rat cerebellar granule cells	Decreased ATP levels and protein synthesis, no cell death at 15 min

(Continued)

TABLE 6.1   Mechanisms of MeHg-Mediated Neurotoxicity—Concentration- and Time-Dependent Effects In Vitro—cont'd

Aschner et al. [172]	10 μM, primary rat astrocyte cultures	Decreased Na+-dependent glutamate uptake at 1 min, loss of osmotic regulation (i.e., Na+ and K+) at ≥5 min
Aschner et al. [173]	10 μM, primary rat astrocyte cultures	Astrocytic swelling, loss of osmotic regulation (i.e., Na+ and K+) at ≥5 min
Sarafian and Verity [158]	10 μM, primary rat astrocyte cultures	Increased ROS at 1 h
Gasso et al. [174]	10 μM, rat cerebellar granule cells	Increased $[Ca]_i$ at 1 h
Sarafian and Verity [152]	10 μM, rat cerebellar granule cells	Increased membrane lipid peroxidation, decreased GSH, minimal cell death at 0.5 h
Yuan and Atchison [175,176]	≥10 μM, cerebellar granule cells and Purkinje cells	Altered spontaneous synaptic currents at ≤40 min; granule cells more sensitive than Purkinje cells
Toimela and Tahti [177]	22 μM, neuroblastoma, glioblastoma, and retinal pigment epithelial cell lines	Cytotoxicity (EC_{50}) at 15 min among most sensitive cell line (neuroblastoma)
Cheung and Verity [148]	25 μM, rat brain synaptosomes	Decreased protein synthesis, 20 min
Kauppinen et al. [149]	30 μM, guinea pig cerebral cortical synaptosomes	Decreased ATP, ADP, ATP/ADP and glycolysis rate at 5 min. Increased $[Ca]_i$ at 10 min

Ca^{2+} signaling [178,179]. Intracellular Ca^{2+} is localized in subcellular compartments, including mitochondria and smooth endoplasmic reticulum, to regulate cellular metabolism and communication [180]. Under physiological conditions, mitochondrial Ca^{2+} accumulation following increased intracytosolic Ca^{2+} occurs through recently identified mitochondrial calcium transporters, and is facilitated by the intracellular localization of mitochondria to calcium channel microclusters in the plasma membrane or endoplasmic reticulum [181,182]. Our emerging understanding of mitochondrial Ca^{2+} physiology has evolved from simply considering mitochondria as simple buffers of cytosolic Ca^{2+}, particularly under supraphysiological conditions [183], to recognizing the central role Ca^{2+} plays in directly regulating mitochondrial energetics and cellular metabolic homeostasis, neocortical activity, and thus, indirectly, ROS production—all of which are affected by MeHg [184,185].

At its extreme, elevations in intracellular Ca^{2+} and increased ROS can act synergistically to drive mitochondrial permeability transition, release of cytochrome c and other small molecules that stimulate apoptosis and necrosis [138,139,186,187]. Thus, many of the identified mechanisms underlying MeHg can be explained because of mitochondrial dysfunction. The high-energy demand and need for efficient mitochondrial function in nervous tissues may underlie the sensitivity of the brain to MeHg toxicity. In addition, despite the higher oxygen demand and ROS production in the CNS, glutathione peroxidase, superoxide dismutase, and glutathione levels are relatively low in specific brain regions (e.g., cerebellum), which may increase their susceptibility toward MeHg toxicity [140].

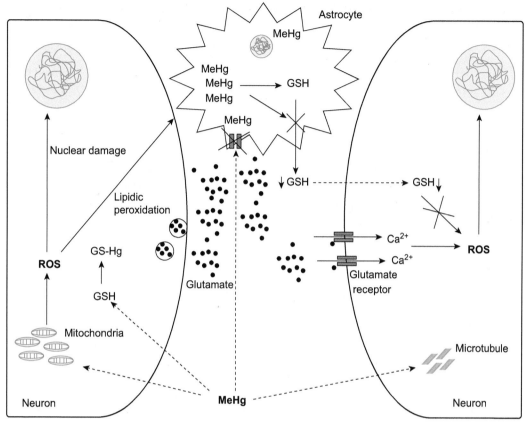

FIGURE 6.4   A schematic model of some of the currently proposed mechanisms for cellular damage induced by MeHg in the CNS. In the extracellular environment, MeHg inhibits glutamate uptake as well as a number of the amino acids that are associated with the synthesis of astrocytic glutathione (GSH). Accumulation of glutamate in the extracellular space and the resulting excessive activation of N-methyl-D-aspartate (NMDA) receptors can result in excitotoxicity and, ultimately, cell death. Other proposed mechanisms are related with mitochondrial MeHg-associated dysfunction, including impaired cytoplasmatic Ca^{2+} homeostasis and release of ROS, metabolic inhibition that leads to impaired ATP production, lipidic peroxidation, and nuclear damage. MeHg also can provoke microtubule chain disruption decreasing vesicular migration or genotoxicity. *From [143] with permission.*

EFFECTS OF MeHg ON MICROTUBULES AS A MECHANISM UNDERLYING DEVELOPMENTAL NEUROPATHOLOGY

Functional alterations in microtubule stability, which involve formation of disulfide bonds for their assembly, may be disrupted directly by MeHg. Immature microtubules may be more sensitive to the destabilizing effects of MeHg over mature microtubules [188,189]. Additional indirect effects on microtubules may occur via alterations in intracellular redox status [190], ATP/GTP, intracellular calcium, or protein phosphorylation [191,192]. The effects on microtubule dynamics likely underlie the observed mitotic inhibition observed in the developing CNS at lower doses [124,193]. Total cell-cycle arrest observed with higher MeHg doses likely reflect increasing dysregulation in cell-cycle regulatory proteins [194,195] and effects to cellular energetics/mitochondrial function, intracellular redox status, and protein and RNA synthesis as discussed previously. In addition to its critical role in mitotic division [196], normal microtubule function is required for neuronal migration and polarization during brain development [197]. Thus, the effects of MeHg on microtubules explain both the reduced brain size and neuronal heterotopias observed following in utero MeHg exposure [122,123,125].

EFFECTS OF MeHg ON ASTROCYTES AND MICROGLIA

The unique interdependence between astrocytes and neurons in the CNS makes neurons susceptible to excitotoxic stimulation secondary to MeHg-mediated effects on astrocytes [142,198,199]. Of particular importance is the key role astrocytes play in conditioning the extracellular microenvironment around neurons by removing neurotransmitters, particularly glutamate, thus protecting neurons from neurotoxicity [200]. In vivo and in vitro studies indicate that MeHg can preferentially accumulate in astrocytes, alter astrocyte intracellular redox status, alter intracellular osmolytes (i.e., Na^+, K^+), inhibit synaptic glutamate (and aspartate) uptake, and stimulate glutamate release, leading to neuronal excitotoxicity and dysfunction [142,172,173,199,201] (see Table 6.1). In addition, astrocytic mitochondria are susceptible to the same effects of MeHg as are neuronal mitochondria, including dissipation of inner membrane potential, loss of ATP synthesis and mitochondrial Ca^{2+} homeostasis, and generation of ROS/decreased GSH [142,202]. Thus, MeHg-mediated effects on astrocytes, including glutamate dysregulation, can potentiate neurotoxicity independently of its direct effects on neurons [203]. In addition, neurons are dependent on astrocytes for supply of GSH precursor molecules, and MeHg can impair uptake of the critical GSH precursor, cystine, by astrocytes, thus ultimately reducing both astrocytic and neuronal GSH levels, and sensitizing neurons toward glutamate toxicity. This situation is compounded by the production of hydrogen peroxide in neurons, which can interfere with astrocytic glutamate transport and thus exacerbate neuronal excitotoxicity (see [202]). In addition to effects on astrocytes, microglia appear to be highly susceptible to MeHg-mediated neurotoxicity, and demonstrate greater sensitivity to MeHg than do astrocytes, attributed to greater MeHg uptake and lower basal GSH levels [165]. Thus, the susceptibility of discrete cell types or brain regions to the impacts of MeHg appears to derive from differential cellular uptake, susceptibility to ROS, and the nature of intercellular interactions that drive either mutual protection or harm.

MECHANISMS OF MeHg AT LOW DOSES: EFFECTS ON NEURONAL SIGNALING

As documented above, concern over MeHg's neurotoxicity has evolved over time from characterizing the gross, histologically apparent effects on brain development that arose following mass poisoning episodes in Japan, Iraq, and elsewhere, toward research evaluating more subtle neurobehavioral effects in populations with exposure through high consumption of fish and seafood. To understand the neurobehavioral effects of MeHg at such low doses, mechanistic studies have focused on neuronal signaling and synaptic function.

MeHg alters both central and peripheral synaptic function, and affects both excitatory and inhibitory transmission. Using neuromuscular preparations, high concentrations of MeHg ($\geq 20\,\mu$M) initially stimulated, then suppressed miniature end-plate potential frequency (attributable to spontaneous acetylcholine release), whereas lower MeHg concentrations (i.e., $4\,\mu$M) decreased spontaneous acetylcholine release [204,205]. Subsequent studies indicated that this effect could be modulated by specific inhibition of mitochondrial Ca^{2+} uptake/release, indicating an important role for nerve terminal mitochondrial Ca^{2+} regulation in MeHg-mediated neuromuscular synaptic transmission alterations [206].

Investigations of the cause of motor deficits associated with MeHg exposure have evaluated the effects of MeHg on the cerebellum (see [207]). In an evaluation of synaptic transmission between parallel and climbing fibers and Purkinje cells using cerebellar slices, acute exposure to $\geq 10\,\mu$M MeHg initially stimulated, then suppressed synaptic transmission, as was seen with neuromuscular preparations. MeHg appeared to mediate these effects by blocking glutamate-mediated postsynaptic responses (e.g., altered postsynaptic glutamate receptors or transmitter release from parallel or climbing fibers; altered Purkinje cell membrane polarization; altered Na^+-dependent, parallel-fiber synaptic currents, and Ca^{2+}-dependent, climbing-fiber synaptic currents; and decreased threshold for Ca^{2+} channel activation) [175,176,207]. Subsequent studies demonstrated that the alterations in Purkinje cell activity coincided with MeHg-induced elevations in presynaptic intracellular Ca^{2+} concentrations in the molecular and granule layers of the cerebellum [208].

Using hippocampal slices to evaluate the effects of MeHg on CNS synaptic function, MeHg ($\geq 4\,\mu$M) initially hyperpolarized then depolarized the resting potential of CA1 neuronal membrane. This effect appeared associated with both pre- and postsynaptic effects, although a block in gamma-aminobutyric acid ionotropic receptor ($GABA_A$) receptor-mediated inhibitory synaptic transmission has been identified as a primary driver of excitatory hippocampal transmission [170,209]. More recent in vivo studies evaluating the effects of low-level MeHg exposure (0.75 or 1.5 mg/kg/day) on the developing visual system of the rat support a role for both dysregulation of $GABA_A$ receptor-mediated synaptic responses and cellular Ca^{2+} homeostasis [210].

Taken as a whole, MeHg can cause a range of effects on central and peripheral synaptic transmission, including effects on specific membrane channels (particularly inhibitory GABAergic receptors), membrane polarity, and ionic regulation (particularly intracellular Ca^{2+}). As indicated in Table 6.1, these effects occur at low levels of exposure and differentially across cell types, thus providing a mechanistic rationale for the observed clinical effects of MeHg toxicity.

DELAYED NEUROTOXICITY, NEURODEGENERATIVE DISEASE, AND INTERGENERATIONAL EFFECTS

The issue of whether exposure to a neurotoxic agent early in life can result in accelerated age-related decline or play a role in neurodegenerative disorders is a significant public health concern [211–213]. Longitudinal data clearly suggest that early life exposures to certain chemicals or viruses can exert adverse and debilitating effects years after cessation of exposure [214,215]. A well-known example of this is the link between childhood exposure to the poliovirus and increased risk for the onset of motor neuron disease during middle age [216].

Delayed or latent toxicity due to developmental MeHg exposure was identified over 40 years ago by Spyker, who observed kyphosis, neuromuscular deficits, and other severe behavioral abnormalities in MeHg-exposed mice as they aged [79]. Since these groundbreaking observations in mice, delayed neurotoxicity has also been documented in humans, monkeys, and rats [10,59,67,109,217–219]. Survey data from patients over 40 years of age with diagnosed MD revealed that these individuals had greater difficulty performing daily activities, such as eating, dressing, and bathing when compared to age-matched controls [218]. The relative loss in everyday living skills was accelerated in the MD patients, suggesting an interaction between previous MeHg exposure (20–30 years before assessment) and functional decline during adulthood. The role of childhood MeHg exposure on mental health and neurological disorders was recently investigated in several Japanese coastal communities that differed in degree of historical MeHg contamination [220]. Archived records of neuropsychiatric evaluations showed that the prevalence of symptoms, such as intellectual impairment and affective/behavioral disorders, was highest in developmentally exposed adults residing near Minamata Bay. This association remained significant when individuals with fetal MD were removed from the analysis, suggesting long-term impairment in neuropsychiatric function in individuals with moderate levels of MeHg exposure early in life.

Delayed neurotoxicity has also been observed in macaque monkeys developmentally exposed to MeHg [10]. When postnatally exposed animals reached middle age (approximately 13 years of age), they began exhibiting signs of clumsiness when climbing in their home cages. A loss in physical coordination was not observed in unexposed controls. Monkeys who received both prenatal and postnatal MeHg exposure were evaluated for high-frequency hearing at 11 and 19 years of age. The rate of deterioration in auditory function between 11 and 19 years was accelerated in the monkeys with a history of early MeHg exposure [67]. Both monkey cohorts were also evaluated on tests of spatial vision in later adulthood and 40% of MeHg-exposed animals showed a constriction of visual fields that was not present when they were clinically evaluated at younger ages [221]. These findings may be explained by an acceleration in neuronal cell loss, which normally occurs during aging, and the ultimate failure of remaining cells to compensate for this injury [215], although other mechanisms have been postulated [212]. In the case of MeHg exposure during brain development, impaired neuronal production early in life could result in the premature expression of functional losses during adulthood [222,223].

The role and relative risk of MeHg exposure in the development of neurodegenerative disorders has been investigated by several international teams. A records research study in the Faroe Islands, where the prevalence of Parkinson's disease (PD) is higher than most other Northern European countries, showed that early exposure to MeHg was not associated with an increased risk of PD [224]. In a separate study of a small group of patients with PD in the United States, 13/14 (93%) had measurable blood Hg values, whereas only 2 of the 14 (14%) control subjects showed a similar Hg body burden [225]. These findings led the authors to postulate that mercury may play a role in the etiology of PD. The role of Hg in the onset of multiple sclerosis (MS) has been investigated in a small Iranian cohort [226]. Patients affected with MS and age-matched, healthy controls provided serum samples for the biochemical analysis. Results showed that blood Hg concentrations were significantly higher in MS patients than in controls. Additionally, the highest blood Hg values tended to come from MS patients, suggesting that high Hg could play a role in susceptibility to MS. Environmental neurotoxicant exposure is actually known to play a role in the development of amyotrophic lateral sclerosis (ALS). In the 1950s, the Chamorro people of Guam presented with extremely high rates of ALS that were later linked to the consumption of flying fox (native fruit bats) and ground cycad seeds [227]. The role of Hg in the onset of ALS has been studied in epidemiologic and laboratory studies, but the findings do not provide solid evidence of a causal relationship (see [228,229] for literature reviews). Despite the plausible relationship between chronic MeHg exposure and neurodegeneration, the role of MeHg in progressive brain disease remains inconclusive.

Relatively little is known regarding the mechanisms by which either acute or chronic MeHg exposure may underlie neurodegeneration. Mechanistically, intracellular Ca^{2+} dysregulation is associated with a range of neurological disorders, including neurodegenerative diseases (reviewed in [230]). Alterations in mitochondrial function, including chronic oxidative stress, are also implicated in a variety of neurodegenerative disorders [231]. Because dysregulation of intracellular Ca^{2+} homeostasis and mitochondrial function are among the most sensitive effects of MeHg, it is plausible that chronic MeHg exposure may contribute to the increased burden of neurodegeneration with aging. In rats, prior dietary administration of selenium (0.06 or 0.6 ppm) protected against somatosensory deficits associated with subsequent chronic exposure to low levels of MeHg (0.5, 5, or 15 ppm) in drinking water for 16 months [232]. Such data support a role for supplemental selenium in either chelating MeHg and/or replacing selenium lost to MeHg binding in critical selenoproteins involved in antioxidant defenses [232,233].

Lastly, in a series of recent papers from investigators at the Karolinska Institute in Sweden, embryonic neural stem cells exposed to nanomolar levels of MeHg resulted in long-term, intergenerational effects on neural stem cell (NSC) proliferation and modified the expression of genes related to regulation of cell cycle and senescence [234,235]. Although MeHg exposure reduced total cell count across multiple generations of NSCs (daughter cells were unexposed), these effects were not due to cell death but rather to decreased rates of neuronal proliferation. These compelling data suggest that exposure to low-level MeHg may predispose the brain to neurodevelopmental and/or degenerative disorders through adverse and long-lasting heritable changes in NSC proliferation.

THE SOCIETAL AND HUMAN COSTS OF MeHg EXPOSURE

The impact of MeHg exposure on childhood health and development can be viewed from a variety of perspectives. Modern exposure scenarios, apart from isolated cases of poisoning, reflect a pattern of chronic, low-dose environmental exposure to MeHg. Adverse neurobehavioral effects are generally subtle in nature and represent significant but small shifts in cognitive, psychomotor, and sensory performance. Bellinger et al. [2] have provided a convincing argument that the long-term implications of neurobehavioral changes in performance can only be properly understood from a public health perspective, not a patient perspective. Small decrements in many individuals, when amortized over a generation, hold the power to have serious public health consequences. For MeHg, models have been developed that estimate a yearly loss of 284, 580 full-scale IQ points in young American children (0–5 years) due to environmental MeHg exposure. Reductions in intellectual resources have significant implications for economic productivity in the workplace and demonstrate the importance of applying a population-based perspective to policies designed to protect children's health. To assess the cost of MeHg-related health effects in real economic dollars, the environmentally attributable statistical model (developed by the Institute of Medicine) was employed with extant data, and mercury-related losses in cognitive function and work productivity were estimated at a staggering $8.7 billion [236]. The authors also found that prenatal MeHg exposure was associated with 1566 excess cases of intellectual disability (IQ<70) per year, and the proper care of these children required an additional $28 million to $3.3 billion from the American health care budget.

Collective lessons learned from epidemiological and laboratory studies showcase the lifelong and irreversible nature of MeHg-induced injury on brain and behavior. The adverse effects of early MeHg exposure may be delayed, or silent neurotoxicity may be expressed years after cessation of exposure. Currently, no direct evidence links perinatal or adult Hg exposure with the onset of neurodegenerative disease, but studies such as these should be undertaken if we are to better understand the role of environmental risk in the onset of progressive brain diseases.

Acknowledgments

We extend our sincere appreciation to Colin Eckhoff for his dedicated assistance with the preparation of the manuscript. We would also like to thank the Washington National Primate Research Center (P51OD010425-NIH) and the Center on Human Development and Disability (P51HD02274-NIH) for their support over the years.

References

[1] Landrigan PJ, Kimmel CA, Correa A, Eskenazi B. Children's health and the environment: public health issues and challenges for risk assessment. Environ Health Perspect 2004;122(2):257–65.

[2] Bellinger D. A strategy for comparing the contributions of environmental chemicals and other risk factors to neurodevelopment of children. Environ Health Perspect 2012;120(4):501–7.

[3] Landrigan PJ, Goldman LR. Children's vulnerability to toxic chemicals: a challenge and opportunity to strengthen health and environmental policy. Health 2011;30(5):842–50.

[4] Woodruff TJ, Axelrad DA, Kyle AD, Nweke O, Miller GG, Hurley BJ. Trends in environmentally related childhood illnesses. Pediatrics 2004;113(Suppl. 4):1133–40.

[5] Díez S. Human health effects of methylmercury exposure. Rev Environ Contam Toxicol 2009;198:111–32.

[6] National Research Council (U.S.). Toxicological effects of methylmercury. National Academies Press (U.S.); 2000.

[7] Karagas MR, Choi AL, Oken E, Horvat M, Schoeny R, Kamai E, et al. Evidence on the human health effects of low-level methylmercury exposure. Environ Health Perspect 2012;120(6):799–806.

[8] Mergler D, Anderson HA, Chan LH, Mahaffey KR, Murray M, Sakamoto M, et al. Panel on health risks and toxicological effects of methylmercury. Methylmercury exposure and health effects in humans: a worldwide concern. Ambio 2007;36(1):3–11.

[9] Burbacher TM, Monnett C, Grant KS, Mottet NK. Methylmercury exposure and reproductive dysfunction in the monkey. Toxicol Appl Pharmacol 1984;75:18–24.

[10] Rice DC. Evidence for delayed neurotoxicity produced by methylmercury. Neurotoxicology 1996;17(3–4):583–96.

[11] Bakir F, Damluji SF, Amin-Zaki L, Murtadha M, Khalidi A, al-Rawi NY, et al. Methylmercury poisoning in Iraq. Science 1973;181:230–41.

[12] Tokuomi H, Okajima T, Kanai J, Tsunoda M, Ichiyasu Y, Misumi H, et al. Minamata disease. World Neurol 1961;2:536–45.

[13] Harada M. Mental deficiance due to methylmercury poisoning. Brain Dev 1974;6:378–87.

[14] Kitamura S, Kakita T, Kojoo J. A supplement to the results of the epidemiological survey on Minamata disease. J Kumamoto Med Sci 1960;38:29–34.

[15] Harada M. Clinical studies on prolonged Minamata Disease. Psychiat Neurol Jap 1972;74:668–78.

[16] Harada Y. Fetal methylmercury poisoning. In: Tsubak K, Inekayama K, editors. Minamata disease. Amsterdam: Elsevier Science Publ., NY; 1977. p. 38–52.

[17] Eto K, Oyanagi S, Itai Y, Tokunaga H, Takizawa Y, Suda I. A fetal type of Minamata disease. An autopsy case report with special reference to the nervous system. Mol Chem Neuropathol 1992;16(1–2):171–86.

[18] Harada M. Congenital Minamata disease: intrauterine methylmercury poisoning. Teratology 1978;18(2):285–8.

[19] Yorifuji T, Kashima S, Tsuda T, Harada M. What has methylmercury in umbilical cords told us? - minamata disease. Sci Total Environ 2009;408(2):272–6.

[20] Clarkson TW, Amin-Zaki L, Al-Tikriti SK. An outbreak of methylmercury poisoning due to consumption of contaminated grain. Fed Proc 1976;35(12):2395–9.

[21] Marsh DO, Myers GJ, Clarkson TW, Amin-Zaki L, Tikriti S, Majeed MA. Fetal methylmercury poisoning: clinical and toxicological data on 29 cases. Ann Neurol 1980;7(4):348–53.

[22] Amin-Zaki L, Elhassani S, Majeed MA, Clarkson TW, Doherty RA, Greenwood MR. Studies of infants postnatally exposed to methylmercury. J Peds 1974;85:81–4.

[23] Amin-Zaki L, Majeed MA, Elhassani SB, Clarkson TW, Greenwood MR, Doherty RA. Prenatal methylmercury poisoning. Clinical observations over five years. Am J Dis Child 1979;133:172–7.

[24] Amin-Zaki L, Majeed MA, Greenwood MR, Elhassani SB, Clarkson TW, Doherty RA. Methylmercury poisoning in the Iraqi suckling infant: a longitudinal study over five years. J Appl Toxicol 1981;1:210–4.

[25] Cox C, Clarkson TW, Marsh DO, Amin-Zaki L, Tikriti S, Myers GG. Dose-response analysis of infants prenatally exposed to methyl mercury: an application of a single compartment model to single-strand hair analysis. Environ Res 1989;49(2):318–32.

[26] Kjellstrom T, Kennedy P, Wallis S, Mantell C. Physical and mental development of children with prenatal exposure to mercury from fish. Stage 1: preliminary tests at age 4. Report: 3080. National Swedish Environmental Protection Board; 1986.

[27] Kjellstrom T, Kennedy P, Wallis S, Stewart A, Friberg L, Lind B, et al. Physical and mental development of children with prenatal exposure to mercury from fish. Report: 3642. National Swedish Environmental Protection Board; 1989.

[28] Crump KS, Kjellström T, Shipp AM, Silvers A, Stewart A. Influence of prenatal mercury exposure upon scholastic and psychological test performance: benchmark analysis of a New Zealand cohort. Risk Anal 1998;18(6):701–13.

[29] Grandjean P, Weihe P. Neurobehavioral effects of intrauterine mercury exposure: potential sources of bias. Environ Res 1993;61(1):176–83.

[30] Grandjean P, Weihe P, White RF, Debes F, Araki S, Yokoyama K, et al. Cognitive deficit in 7-year-old children with prenatal exposure to methylmercury. Neurotoxicol Teratol 1997;19:417–28.

[31] Debes F, Budtz-Jorgenson E, Weihle P, White RF, Grandiean P. Impact of prenatal methylmercury exposure on neurobehavioral function at age 14 years. Neurotoxicol Teratol 2006;28:363–75.

[32] Murata K, Budtz-Jørgensen E, Grandjean P. Benchmark dose calculations for methylmercury-associated delays on evoked potential latencies in two cohorts of children. Risk Anal 2002;22(3):465–74.

[33] Murata K, Weihe P, Budtz-Jorgenson E, Jorgenson PJ, Grandjean P. Delayed brainstem auditory evoked potential latencies in 14-year-old children exposed to methylmercury. J Pediatr 2004;144:177–83.

[34] Marsh DO, Myers GJ, Clarkson TW, Davidson PW, Cox C, Cernichiari E, et al. The Seychelles study of fetal methylmercury exposure and child development: introduction. Neurotoxicology 1995;16(4):583–96.

[35] Myers GJ, Davidson PW, Strain JJ. Nutrient and methyl mercury exposure from consuming fish. J Nutr 2007;137(12):2805–8.

[36] Shamlaye CF, Marsh DO, Myers GJ, Cox C, Davidson PW, Choisy O, et al. The Seychelles child development study on neurodevelopmental outcomes in children following in utero exposure to methylmercury from a maternal fish diet: background and demographics. Neurotoxicology 1995;16(4):597–612.

[37] Davidson PW, Cory-Slechta DA, Thurston SW, Huang LW, Shamlaye CF, Gunzler D, et al. Fish consumption and prenatal methylmercury exposure: cognitive and behavioral outcomes in the main cohort at 17 years from the Seychelles child development study. Neurotoxicology 2011;32(6):711–7.

[38] Davidson PW, Myer GJ, Shamlaye CF, Cox C. Prenatal methyl mercury exposure from fish consumption and child development: a review of evidence and perspectives from the Seychelles child development study. Neurotoxicity 2006;27(6):1106–9.

[39] Davidson PW, Myers GJ, Cox C, Shamlaye C, Choisy O, Sloane-Reeves J, et al. Neurodevelopmental test selection, administration, and performance in the main Seychelles child development study. Neurotoxicology 1995;16(4):665–76.

[40] Davidson PW, Myers GJ, Cox C, Shamlaye C, Sloane-Reeves J, Cernichiari E, et al. Effects of prenatal and post-natal methylmercury exposure from fish consumption on neurodevelopment: outcomes at 66 months of age in the Seychelles child development study. JAMA 1998;280(8):701–7.

[41] Myers GJ, Davidson PW, Shamlaye CF, Axtell CD, Cernichiarim E, Choisy O, et al. Effects of prenatal methylmercury exposure from a high fish diet on developmental milestones in the Seychelles child development study. Neurotoxicology 1997;18(3):819–29.

[42] Myers GJ, Davidson PW, Shamlaye CF, Palumbo D, Cernichiarim E, Choisy O, et al. Prenatal methylmercury exposure from ocean fish consumption in the Seychelles child development study. Lancet 2003;361:1686–92.

[43] Wijngaarden Ev, Thurston SW, Myers GJ, Strain JJ, Weiss B, Zarcone T, et al. Prenatal methyl mercury exposure in relation to neurodevelopment and behavior at 19 years of age in the Seychelles child development study. Neurotoxicol Teratol 2013;39:19–25.

[44] Davidson PW, Strain JJ, Myers GJ, Thurston SW, Bonham MP, Shamlaye CF, et al. Neurodevelopmental effects of maternal nutritional status and exposure to methylmercury from eating fish during pregnancy. Neurotoxicology 2008;29(5):767–75.

[45] Strain JJ, Davidson PW, Bonham MP, Duffy EM, Stokes-Riner A, Thurston SW, et al. Associations of maternal long-chain polyunsaturated fatty acids, methyl mercury, and infant development in the Seychelles child development nutrition study. Neurotoxicology 2008;29(5):776–82.

[46] Stokes-Rine A, Thurston SW, Myers GJ, Duffy EM, Wallace J, Bonham M, et al. A longitudinal analysis of prenatal exposure to methylmercury and fatty acids in the Seychelles. Neurotoxicol Teratol 2011;33(2):325–8.

[47] Steuerwald U, Weihem P, Jørgensen PJ, Bjerve K, Brock J, Heinzow B, et al. Maternal seafood diet, methylmercury exposure, and neonatal neurologic function. J Pediatr 2000;136(5):599–605.

[48] Yorifuji T, Murata K, Bjerve KS, Choi AL, Weihe P, Grandjean P. Visual evoked potentials in children prenatally exposed to methylmercury. Neurotoxicity 2013;37:15–8.

[49] Choi AL, Mogense UB, Bjerve KS, Debes F, Weihe P, Grandjean P, et al. Negative confounding by essential fatty acids in methylmercury neurotoxicity associations. Neurotoxicol Teratol 2014;42:85–92.

[50] Oken E, Wright RO, Kleinman KP, Bellinger D, Amarasiriwardena CJ, Hu H, et al. Maternal fish consumption, hair mercury, and infant cognition in a U.S. Cohort. Environ Health Perspect 2005;113:1376–80.

[51] Oken E, Wright RO, Kleinman KP, Bellinger D, Amarasiriwardena CJ, Hu H, et al. Maternal fish intake during pregnancy, blood mercury levels, and child cognition at age 3 years in a US cohort. Am J Epidemiol 2008;167(10):1171–81.

[52] Sagiv SK, Thurston SW, Bellinger DC, Amarasiriwardena C, Korrick SA. Prenatal exposure to mercury and fish consumption during pregnancy and attention-deficit/hyperactivity disorder-related behavior in children. Pediatr Adolesc Med 2012;166(12):1123–31.

[53] Committee on Toxicological Effects of Methylmercury, National Research Council of the United States, National Academies of Science. Toxicological effects of methylmercury. Washington: National Academies Press; 2000.

[54] Xue F, Holzman C, Rahbar MH, Trosko K, Fischer L. Maternal fish consumption, mercury levels, and risk of preterm delivery. Envrion Health Perspect 2007;115(1):42–7.

[55] Castoldi AF, Onishchenko N, Johansson C, Coccini T, Roda E, Vahter M, et al. Neurodevelopmental toxicity of methylmercury: laboratory animal data and their contribution to human risk assessment. Regul Toxicol Pharmacol 2008;51(2):215–29.

[56] Cernichiari E, Brewer R, Myers GJ, Marsh DO, Lapham LW, Cox C, et al. Monitoring methylmercury during pregnancy: maternal hair predicts fetal brain exposure. Neurotoxicology 1995;16(4):705–10.

[57] Magos L. The absorption, distribution and excretion of methylmercury. In: Eccles CU, Annau Z, editors. The toxicity of methylmercury. Baltimore: John Hopkins; 1987. pp. 24–44.

[58] Mottet NK, Vahter ME, Charleston JS, Friberg LT. Metabolism of methylmercury in the brain and its toxicological significance. Met Ions Biol Syst 1997;34:371–403.

[59] Rice DC. Blood mercury concentrations following methylmercury exposure in adult and infant monkeys. Environ Res 1989;49:115–26.

[60] Vahter M, Mottet NK, Friberg L, Lind B, Shen DD, Burbacher T. Speciation of mercury in the primate blood and brain following long-term exposure to methyl mercury. Toxicol Appl Pharmacol 1994;124(2):221–9.

[61] Grant KS, Rice DC. Exposure to environmental chemicals and developmental risk: contributions from studies with monkeys. In: Burbacher T, Sackett G, Grant K, editors. Nonhuman primate models of Children's health and developmental disabilities. Elsevier Academic Press; 2008. pp. 377–420.

[62] Rice DC, Gilbert SG. Early chronic low-level methylmercury poisoning in monkeys impairs spatial vision. Sci 1982;216:759–61.

[63] Rice DC, Gilbert SG. Exposure to methyl mercury from birth to adulthood impairs high-frequency hearing in monkeys. Toxicol Appl Pharmacol 1992;115:6–10.

[64] Rice DC, Gilbert SG. Effects of developmental methylmercury exposure or lifetime lead exposure on vibration sensitivity function in monkeys. Toxicol Appl Pharmacol 1995;134:161–9.

[65] Ninomiya T, Imamura K, Kuwahata M, Kindaichi M, Susa M, Ekino S. Reappraisal of somatosensory disorders in methylmercury poisoning. Neurotoxicol Teratol 2005;27(4):643–53.

[66] Rice DC, Gilbert SG. Effects of developmental exposure to methylmercury on spatial and temporal visual function in monkeys. Toxicol Appl Pharmacol 1990;102:151–63.

[67] Rice DC. Age-related increase in auditory impairment in monkeys exposed in-utero plus postnatally to methylmercury. Toxicol Sci 1998;44:191–6.

[68] Harada M. Minamata disease: methylmercury poisoning in Japan caused by environmental pollution. Crit Rev Toxicol 1995;25:1–24.

[69] Rice DC. Effects of pre- plus postnatal exposure to methylmercury in the monkey on fixed interval and discrimination reversal performance. Neurotoxicol 1992;13:443–52.

[70] Jedrychowski W, Jankowski J, Flak E, Skarupa A, Mroz E, Sochacka-Tatara E, et al. Effects of prenatal exposure to mercury on cognitive and psychomotor function in one-year-old infants: epidemiologic cohort study in Poland. Ann Epidemiol 2006;16:439–47.

[71] Burbacher TM, Grant KS, Mottet NK. Retarded object permanence development in methylmercury exposed *Macaca fascicularis* infants. Dev Psychol 1986;22:771–6.

[72] Gunderson VM, Grant-Webster KS, Burbacher TM, Mottet NK. Visual recognition memory deficits in methylmercury-exposed *Macaca fascicularis* infants. Neurotoxicol Teratol 1988;10:373–9.

[73] Gunderson VM, Grant KS, Burbacher TM, Fagan 3rd JF, Mottet NK. The effect of low-level prenatal methylmercury exposure on visual recognition memory in infant crab-eating macaques. Child Dev 1986;57:1076–83.

[74] Burbacher TM, Rodier PM, Weiss B. Methylmercury developmental neurotoxicity: a comparison of effects in humans and animals. Neurotoxicol Teratol 1990;12(3):191–202.

[75] Burbacher TM, Sackett GP, Mottet NK. Methylmercury effects on the social behavior of *Macaca fascicularis* infants. Neurotoxicol Teratol 1990;12:65–71.

[76] Grant-Webster KS, Burbacher TM, Motter NK. Puberal growth retardation in primates: a latent effect of in utero exposure to methylmercury. Toxicol 1992;12:310.

[77] Burbacher TM, Grant KS, Mayfield DB, Gilbert SG, Rice DC. Prenatal methylmercury exposure affects spatial vision in adult monkeys. Toxicol Appl Pharmacol 2005;208:21–8.

[78] Gilbert SG, Burbacher TM, Rice DC. Effects of in utero methylmercury exposure on aspatial delayed alternation task in monkeys. Toxicol Appl Pharmacol 1993;123:130–6.

[79] Spyker JM. Assessing the impact of low level chemicals on development: behavioral and latent effects. Fed Proc 1975;34:1835–44.

[80] Spyker JM, Sparber SB, Goldberg AM. Subtle consequences of methylmercury exposure: behavioral deviations in offspring of treated mothers. Science 1972;177:621–3.

[81] Adams J, Buelke-Sam J, Kimmel CA, Nelson CJ, Miller DR. Collaborative behavioral teratology study: preliminary research. Neurobehav Toxicol Teratol 1985;7(6):555–78.

[82] Adams J, Buelke-Sam J, Kimmel CA, Nelson CJ, Reiter LW, Sobotka TJ, et al. Collaborative behavioral teratology study: protocol design and testing procedures. Neurobehav Toxicol Teratol 1985;7(6):579–86.

[83] Buelke-Sam J, Kimmel CA, Adams J, Nelson CJ, Vorhees CV, Wright DC, et al. Collaborative behavioral teratology study: results. Neurobehav Toxicol Teratol 1985;7:591–624.

[84] Geyer MA, Butcher RE, Fite K. A study of startle and locomotor activity in rats exposed prenatally to methylmercury. Neurobehav Toxicol Teratol 1985;7:759–65.

[85] Vorhees C. Behavioral effects of prenatal methylmercury in rats: a parallel trial to the collaborative behavioral teratology study. Neurobehav Toxicol Teratol 1985;7:128–35.

[86] Elsner J, Hodel B, Suter KE, Oelke D, Ulbrich B, Schreiner G, et al. Detection limits of different approaches in behavioral technology, and correlation of effects with neurochemical parameters. Neurotoxicol Teratol 1988;10:155–67.

[87] Goldey ES, O'Callaghan JP, Stanton ME, Barone S, Krofton KM. Developmental neurotoxicity: evaluation of testing procedures with methylazoxymethanol and methylmercury. Fundam Appl Toxicol 1994;23:447–64.

[88] Huang CF, Liu SH, Hsu CJ, Lin-Shiau SY. Neurotoxicological effects of low-dose methylmercury and mercuric chloride in developing offspring mice. Toxicol Lett 2011;201(3):196–204.

[89] Inouye M, Murao K, Kajiwara Y. Behavioral and neuropathological effects of prenatal methylmercury exposure in mice. Neurobehav Toxicol Teratol 1985;7:227–32.

[90] Fujimura M, Cheng J, Zhao W. Perinatal exposure to low-dose methylmercury induces dysfunction of motor coordination with decreases in synaptophysin expression in the cerebellar granule cells of rats. Brain Res 2012;1464:1–7.

[91] Goulet S, Doré FY, Mirault ME. Neurobehavioral changes in mice chronically exposed to methylmercury during fetal and early postnatal development. Neurotoxicol Teratol 2003;25(3):335–47.

[92] Montgomery KS, Mackey J, Thuett K, Ginestra S, Bizon JL, Abbott LC. Chronic, low-dose prenatal exposure to methylmercury impairs motor and mnemonic function in adult C57/B6 mice. Behav Brain Res 2008;191(1):55–61.

[93] Sakamoto M, Kakita A, Wakabayashi K, Takahashi H, Nakano A, Agaki H. Evaluation of changes in methylmercury accumulation in the developing rat brain and its effects: a study with consecutive and moderate dose exposure throughout gestation and lactation periods. Brain Res 2002;949(1–2):51–9.

[94] Su MQ, Okita GT. Behavioral effects on the progeny of mice treated with methylmercury. Toxicol Appl Pharmacol 1976;38:195–205.

[95] Suter KE, Schoen H. Testing strategies in behavioral teratology: I. Testing battery approach. Neurobehav Toxicol Teratol 1986;8:561–6.

[96] Weiss B, Stern S, Cox C, Balys M. Perinatal and lifetime exposure to methylmercury in the mouse: behavioral effects. Neurotoxicology 2005;26(4):675–90.

[97] Eccles CU, Annau Z. Prenatal methylmercury exposure: I. Alterations in neonatal activity. Neurobehav Toxicol Teratol 1982;4:371–6.

[98] Eccles CU, Annau Z. Prenatal methylmercury exposure: II. Alterations in learning and psychotropic drug sensitivity in adult offspring. Neurobehav Toxicol Teratol 1982;4:377–82.

[99] Dyer RS, Eccles CU, Annau Z. Evoked potential alterations following prenatal methylmercury exposure. Pharmacol Biochem Behav 1978;8:137–41.

[100] Zenick H. Evoked potential alterations in methylmercury chloride toxicity. Pharmacol Biochem Behav 1976;5:253–5.

[101] Hughes JA, Annou Z. Postnatal behavioral effects in mice after prenatal exposure to methylmercury. Pharmacol Biochem Behav 1976;4:385–91.

[102] Schalock RL, Brown WJ, Kark RA, Menon NK. Perinatal methylmercury intoxication: behavioral effects in rats. Dev Psychobiol 1981;14(3):213–9.

[103] Falluel-Morel A, Sokolowski K, Sisti HM, Zhou X, Shors TJ, Dicicco-Bloom E. Developmental mercury exposure elicits acute hippocampal cell death, reductions in neurogenesis, and severe learning deficits during puberty. J Neurochem 2007;103(5):1968–81.

[104] Liang J, Inskip M, Newhook D, Messier C. Neurobehavioral effect of chronic and bolus doses of methylmercury following prenatal exposure in C57BL/6 weanling mice. Neurotoxicol Teratol 2009;31(6):372–81.

[105] Onishchenko N, Tamm C, Vahter M, Hökfelt T, Johnson JA, Johnson DA, et al. Developmental exposure to methylmercury alters learning and induces depression-like behavior in male mice. Toxicol Sci 2007;97(2):428–37.

[106] Zenick H. Behavioral and biochemical consequences in methylmercury chloride toxicity. Pharmacol Biochem Behav 1974;2(6):709–13.

[107] Bornhausen M, Musch HR, Greim H. Operant behavior changes in rats after prenatal methylmercury exposure. Toxicol Appl Pharmacol 1980;56:305–48.

[108] Musch HR, Bornhausen M, Kriegel H, Greim H. Methylmercury chloride induces learning deficits in prenatally treated rats. Arch Toxicol 1978;126:103–8.

[109] Newland MC, Reile PA, Langston JL. Gestational exposure to methylmercury retards choice in transition in aging rats. Neurotoxicol Teratol 2004;26:179–94.

[110] Onishchenko N, Karpova N, Sabri F, Castrén E, Ceccatelli S. Long-lasting depression-like behavior and epigenetic changes of BDNF gene expression induced by perinatal exposure to methylmercury. J Neurochem 2008;106(3):1378–87.

[111] Stinson CH, Shen DM, Burbacher TM, Mohamed MK, Mottet NK. Kinetics of methylmercury in blood and brain during chronic exposure in the monkey *Macaca fascicularis*. Pharmacol Toxicol 1989;65(3):223–30.

[112] Yin Z, Jiang H, Syversen T, Rocha JB, Farina M, Aschner M. The methylmercury-L-cysteine conjugate is a substrate for the L-type large neutral amino acid transporter. J Neurochem 2008;107(4):1083–90.

[113] Friberg L, Mottet NK. Accumulation of methylmercury and inorganic mercury in the brain. Biol Trace Elem Res 1989;21:201–6.

[114] Vahter ME, Mottet NK, Friberg LT, Lind SB, Charleston JS, Burbacher TM. Demethylation of methyl mercury in different brain sites of *Macaca fascicularis* monkeys during long-term subclinical methyl mercury exposure. Toxicol Appl Pharmacol 1995;134(2):273–84.

[115] Syversen T, Kaur P. The toxicology of mercury and its compounds. J Trace Elem Med Biol 2012;26(4):215–26.

[116] Bakir F, Rustam H, Tikriti S, Al-Damluji SF, Shihristani H. Clinical and epidemiological aspects of methylmercury poisoning. Postgrad Med J 1980;56(651):1–10.

[117] Clarkson TW, Nordberg GF, Sager PR. Reproductive and developmental toxicity of metals. Scand J Work Environ Health 1985;11(3 Spec No):145–54.

[118] Takeuchi T. Pathology of Minamata disease. With special reference to its pathogenesis. Acta Pathol Jpn 1982;32(Suppl. 1):73–99.

[119] Sarafian T, Hagler J, Vartavarian L, Verity MA. Rapid cell death induced by methyl mercury in suspension of cerebellar granule neurons. J Neuropathol Exp Neurol 1989;48(1):1–10.

[120] Choi BH. Methylmercury poisoning of the developing nervous system: I. Pattern of neuronal migration in the cerebral cortex. Neurotoxicology 1986;7(2):591–600.

[121] Choi BH. Effects of methylmercury on neuroepithelial germinal cells in the developing telencephalic vesicles of mice. Acta Neuropathol 1991;81(4):359–65.

[122] Choi BH, Lapham LW, Amin-Zaki L, Saleem T. Abnormal neuronal migration, deranged cerebral cortical organization, and diffuse white matter astrocytosis of human fetal brain: a major effect of methylmercury poisoning in utero. J Neuropathol Exp Neurol 1978;37(6):719–33.

[123] Matsumoto H, Koya G, Takeuchi T. Fetal Minamata disease. A neuropathological study of two cases of intra-uterine intoxication by a methyl mercury compound. J Neuropathol Exp Neurol 1965;24(4):563–74.

[124] Sager PR, Aschner M, Rodier PM. Persistent, differential alterations in developing cerebellar cortex of male and female mice after methylmercury exposure. Brain Res 1984;314(1):1–11.

[125] Takeuchi T. Pathology of fetal Minamata disease. Pediatrician 1977;6:69–87.

[126] Herschkowitz N. Brain development in the fetus, neonate and infant. Biol Neonate 1988;54(1):1–19.

[127] Rodier PM. Vulnerable periods and processes during central nervous system development. Environ Health Perspect 1994;102(Suppl. 2):121–4.

[128] Dyrssen D, Wedborg M. The sulphur-mercury(II) system in natural waters. Water Air Soil Pollut 1991;56:507–19.

[129] Hughes WL. A physicochemical rationale for the biological activity of mercury and its compounds. Ann N Y Acad Sci 1957;65(5):454–60.

[130] Cheung MK, Verity MA. Experimental methyl mercury neurotoxicity: locus of mercurial inhibition of brain protein synthesis in vivo and in vitro. J Neurochem 1985;44(6):1799–808.

[131] Kuznetsov DA, Zavijalov NV, Govorkov AV, Sibileva TM. Methyl mercury-induced nonselective blocking of phosphorylation processes as a possible cause of protein synthesis inhibition in vitro and in vivo. Toxicol Lett 1987;36(2):153–60.

[132] Sarafian T, Verity MA. Inhibition of RNA and protein synthesis in isolated cerebellar cells by in vitro and in vivo methyl mercury. Neurochem Pathol 1985;3(1):27–39.

[133] Sarafian T, Verity MA. Altered patterns of protein phosphorylation and synthesis caused by methyl mercury in cerebellar granule cell culture. J Neurochem 1990;55(3):922–9.

[134] Sarafian TA, Verity MA. Changes in protein phosphorylation in cultured neurons after exposure to methyl mercury. Ann N Y Acad Sci 1993;679:65–77.

[135] Sarafian TA. Methyl mercury increases intracellular Ca2+ and inositol phosphate levels in cultured cerebellar granule neurons. J Neurochem 1993;61(2):648–57.

[136] Atchison WD, Hare MF. Mechanisms of methylmercury-induced neurotoxicity. FASEB J 1994;8(9):622–9.

[137] Bondy SC, McKee M. Disruption of the potential across the synaptosomal plasma membrane and mitochondria by neurotoxic agents. Toxicol Lett 1991;58(1):13–21.

[138] Limke TL, Atchison WD. Acute exposure to methylmercury opens the mitochondrial permeability transition pore in rat cerebellar granule cells. Toxicol Appl Pharmacol 2002;178(1):52–61.

[139] Polunas M, Halladay A, Tjalkens RB, Philbert MA, Lowndes H, Reuhl K. Role of oxidative stress and the mitochondrial permeability transition in methylmercury cytotoxicity. Neurotoxicology 2011;32(5):526–34.

[140] Mori N, Yasutake A, Hirayama K. Comparative study of activities in reactive oxygen species production/defense system in mitochondria of rat brain and liver, and their susceptibility to methylmercury toxicity. Arch Toxicol 2007;81(11):769–76.

[141] Yoshino Y, Mozai T, Nakao K. Biochemical changes in the brain in rats poisoned with an alkymercury compound, with special reference to the inhibition of protein synthesis in brain cortex slices. J Neurochem 1966;13(11):1223–30.

[142] Aschner M, Syversen T, Souza DO, Rocha JB, Farina M. Involvement of glutamate and reactive oxygen species in methylmercury neurotoxicity. Braz J Med Biol Res 2007;40(3):285–91.

[143] do Nascimento JL, Oliveira KR, Crespo-Lopez ME, Macchi BM, Maues LA, Pinheiro Mda C, et al. Methylmercury neurotoxicity & antioxidant defenses. Indian J Med Res 2008;128(4):373–82.

[144] Roos D, Seeger R, Puntel R, Vargas Barbosa N. Role of calcium and mitochondria in MeHg-mediated cytotoxicity. J Biomed Biotechnol 2012;2012:248764.

[145] Sone N, Larsstuvold MK, Kagawa Y. Effect of methyl mercury on phosphorylation, transport, and oxidation in mammalian mitochondria. J Biochem 1977;82(3):859–68.

[146] Verity MA, Brown WJ, Cheung M. Organic mercurial encephalopathy: in vivo and in vitro effects of methyl mercury on synaptosomal respiration. J Neurochem 1975;25(6):759–66.

[147] Garlid KD, Paucek P. Mitochondrial potassium transport: the K(+) cycle. Biochim Biophys Acta 2003;1606(1–3):23–41.

[148] Cheung M, Verity MA. Methylmercury inhibition of synaptosome protein synthesis: the role of mitochondrial dysfunction. Environ Res 1981;24:286–98.

[149] Kauppinen RA, Komulainen H, Taipale H. Cellular mechanisms underlying the increase in cytosolic free calcium concentration induced by methylmercury in cerebrocortical synaptosomes from guinea pig. J Pharmacol Exp Ther 1989;248(3):1248–54.

[150] Levesque PC, Atchison WD. Disruption of brain mitochondrial calcium sequestration by methylmercury. J Pharmacol Exp Ther 1991;256(1):236–42.

[151] Mori N, Yasutake A, Marumoto M, Hirayama K. Methylmercury inhibits electron transport chain activity and induces cytochrome c release in cerebellum mitochondria. J Toxicol Sci 2011;36(3):253–9.

[152] Sarafian T, Verity MA. Oxidative mechanisms underlying methyl mercury neurotoxicity. Int J Dev Neurosci 1991;9(2):147–53.

[153] Choi BH, Yee S, Robles M. The effects of glutathione glycoside in methyl mercury poisoning. Toxicol Appl Pharmacol 1996;141(2):357–64.

[154] Farina M, Rocha JB, Aschner M. Mechanisms of methylmercury-induced neurotoxicity: evidence from experimental studies. Life Sci 2011;89(15–16):555–63.

[155] Kaur P, Aschner M, Syversen T. Glutathione modulation influences methyl mercury induced neurotoxicity in primary cell cultures of neurons and astrocytes. Neurotoxicology 2006;27(4):492–500.

[156] Kaur P, Evje L, Aschner M, Syversen T. The in vitro effects of selenomethionine on methylmercury-induced neurotoxicity. Toxicol In Vitro 2009;23(3):378–85.

[157] Li S, Thompson SA, Woods JS. Localization of gamma-glutamylcysteine synthetase mRNA expression in mouse brain following methylmercury treatment using reverse transcription in situ PCR amplification. Toxicol Appl Pharmacol 1996;140(1):180–7.

[158] Shanker G, Aschner M. Methylmercury-induced reactive oxygen species formation in neonatal cerebral astrocytic cultures is attenuated by antioxidants. Brain Res Mol Brain Res 2003;110(1):85–91.

[159] Denny MF, Atchison WD. Mercurial-induced alterations in neuronal divalent cation homeostasis. Neurotoxicology 1996;17(1):47–61.

[160] Faustman EM, Ponce RA, Ou YC, Mendoza MA, Lewandowski T, Kavanagh T. Investigations of methylmercury-induced alterations in neurogenesis. Environ Health Perspect 2002;110(Suppl 5):859–64.

[161] Hare MF, Atchison WD. Methylmercury mobilizes Ca++ from intracellular stores sensitive to inositol 1,4,5-trisphosphate in NG108-15 cells. J Pharmacol Exp Ther 1995;272(3):1016–23.

[162] Komulainen H, Bondy SC. Increased free intrasynaptosomal Ca2+ by neurotoxic organometals: distinctive mechanisms. Toxicol Appl Pharmacol 1987;88(1):77–86.

[163] Limke TL, Heidemann SR, Atchison WD. Disruption of intraneuronal divalent cation regulation by methylmercury: are specific targets involved in altered neuronal development and cytotoxicity in methylmercury poisoning? Neurotoxicology 2004;25(5):741–60.

[164] Limke TL, Otero-Montanez JK, Atchison WD. Evidence for interactions between intracellular calcium stores during methylmercury-induced intracellular calcium dysregulation in rat cerebellar granule neurons. J Pharmacol Exp Ther 2003;304(3):949–58.

[165] Ni M, Li X, Yin Z, Sidoryk-Wegrzynowicz M, Jiang H, Farina M, et al. Comparative study on the response of rat primary astrocytes and microglia to methylmercury toxicity. Glia 2011;59(5):810–20.

[166] Heidemann SR, Lamoureux P, Atchison WD. Inhibition of axonal morphogenesis by nonlethal, submicromolar concentrations of methylmercury. Toxicol Appl Pharmacol 2001;174(1):49–59.

[167] Marty MS, Atchison WD. Elevations of intracellular Ca²⁺ as a probable contributor to decreased viability in cerebellar granule cells following acute exposure to methylmercury. Toxicol Appl Pharmacol 1998;150(1):98–105.

[168] Castoldi AF, Barni S, Turin I, Gandini C, Manzo L. Early acute necrosis, delayed apoptosis and cytoskeletal breakdown in cultured cerebellar granule neurons exposed to methylmercury. J Neurosci Res 2000;59(6):775–87.

[169] Allen JW, Shanker G, Aschner M. Methylmercury inhibits the in vitro uptake of the glutathione precursor, cystine, in astrocytes, but not in neurons. Brain Res 2001;894(1):131–40.

[170] Yuan Y, Atchison WD. Methylmercury acts at multiple sites to block hippocampal synaptic transmission. J Pharmacol Exp Ther 1995;275(3):1308–16.

[171] Kunimoto M, Suzuki T. Migration of granule neurons in cerebellar organotypic cultures is impaired by methylmercury. Neurosci Lett 1997;226(3):183–6.

[172] Aschner M, Eberle NB, Miller K, Kimelberg HK. Interactions of methylmercury with rat primary astrocyte cultures: inhibition of rubidium and glutamate uptake and induction of swelling. Brain Res 1990;530(2):245–50.

[173] Aschner M, Vitarella D, Allen JW, Conklin DR, Cowan KS. Methylmercury-induced astrocytic swelling is associated with activation of the Na+/H+ antiporter, and is fully reversed by amiloride. Brain Res 1998;799(2):207–14.

[174] Gasso S, Cristofol RM, Selema G, Rosa R, Rodriguez-Farre E, Sanfeliu C. Antioxidant compounds and Ca(2+) pathway blockers differentially protect against methylmercury and mercuric chloride neurotoxicity. J Neurosci Res 2001;66(1):135–45.

[175] Yuan Y, Atchison WD. Methylmercury differentially affects GABA(A) receptor-mediated spontaneous IPSCs in Purkinje and granule cells of rat cerebellar slices. J Physiol 2003;550(Pt 1):191–204.

[176] Yuan Y, Atchison WD. Methylmercury induces a spontaneous, transient slow inward chloride current in Purkinje cells of rat cerebellar slices. J Pharmacol Exp Ther 2005;313(2):751–64.

[177] Toimela T, Tahti H. Mitochondrial viability and apoptosis induced by aluminum, mercuric mercury and methylmercury in cell lines of neural origin. Arch Toxicol 2004;78(10):565–74.

[178] Berridge MJ. Neuronal calcium signaling. Neuron 1998;21(1):13–26.

[179] Bootman MD, Berridge MJ. The elemental principles of calcium signaling. Cell 1995;83(5):675–8.

[180] Simpson PB. The local control of cytosolic Ca2+ as a propagator of CNS communication - integration of mitochondrial transport mechanisms and cellular responses. J Bioenerg Biomembr 2000;32:5–13.

[181] Drago I, Pizzo P, Pozzan T. After half a century mitochondrial calcium in- and efflux machineries reveal themselves. EMBO J 2011;30(20):4119–25.

[182] Pozzan T, Rizzuto R. The renaissance of mitochondrial calcium transport. Eur J Biochem 2000;267(17):5269–73.

[183] Williams GS, Boyman L, Chikando AC, Khairallah RJ, Lederer WJ. Mitochondrial calcium uptake. Proc Natl Acad Sci USA 2013;110(26):10479–86.

[184] Glancy B, Balaban RS. Role of mitochondrial Ca2+ in the regulation of cellular energetics. Biochemistry 2012;51(14):2959–73.

[185] Sanganahalli BG, Herman P, Hyder F, Kannurpatti SS. Mitochondrial calcium uptake capacity modulates neocortical excitability. J Cereb Blood Flow Metab 2013;33(7):1115–26.

[186] Lemasters JJ, Theruvath TP, Zhong Z, Nieminen AL. Mitochondrial calcium and the permeability transition in cell death. Biochim Biophys Acta 2009;1787(11):1395–401.

[187] Orrenius S, Zhivotovsky B, Nicotera P. Regulation of cell death: the calcium-apoptosis link. Nat Rev Mol Cell Biol 2003;4(7):552–65.

[188] Graff RD, Philbert MA, Lowndes HE, Reuhl KR. The effect of glutathione depletion on methyl mercury-induced microtubule disassembly in cultured embryonal carcinoma cells. Toxicol Appl Pharmacol 1993;120(1):20–8.

[189] Vogel DG, Margolis RL, Mottet NK. The effects of methyl mercury binding to microtubules. Toxicol Appl Pharmacol 1985;80(3):473–86.

[190] Luduena RF, Roach MC. Tubulin sulfhydryl groups as probes and targets for antimitotic and antimicrotubule agents. Pharmacol Ther 1991;49(1–2):133–52.

[191] Norman SG, Johnson GV. Compromised mitochondrial function results in dephosphorylation of tau through a calcium-dependent process in rat brain cerebral cortical slices. Neurochem Res 1994;19(9):1151–8.

[192] Prokop A. The intricate relationship between microtubules and their associated motor proteins during axon growth and maintenance. Neural Dev 2013;8:17.

[193] Ponce RA, Kavanagh TJ, Mottet NK, Whittaker SG, Faustman EM. Effects of methyl mercury on the cell cycle of primary rat CNS cells in vitro. Toxicol Appl Pharmacol 1994;127(1):83–90.

[194] Mendoza MA, Ponce RA, Ou YC, Faustman EM. p21(WAF1/CIP1) inhibits cell cycle progression but not G2/M-phase transition following methylmercury exposure. Toxicol Appl Pharmacol 2002;178(2):117–25.

[195] Ou YC, Thompson SA, Ponce RA, Schroeder J, Kavanagh TJ, Faustman EM. Induction of the cell cycle regulatory gene p21 (Waf1, Cip1) following methylmercury exposure in vitro and in vivo. Toxicol Appl Pharmacol 1999;157(3):203–12.

[196] Kline-Smith SL, Walczak CE. Mitotic spindle assembly and chromosome segregation: refocusing on microtubule dynamics. Mol Cell 2004;15(3):317–27.

[197] Sakakibara A, Ando R, Sapir T, Tanaka T. Microtubule dynamics in neuronal morphogenesis. Open Biol 2013;3(7):130061.

[198] Aschner M, Du YL, Gannon M, Kimelberg HK. Methylmercury-induced alterations in excitatory amino acid transport in rat primary astrocyte cultures. Brain Res 1993;602(2):181–6.

[199] Aschner M, Yao CP, Allen JW, Tan KH. Methylmercury alters glutamate transport in astrocytes. Neurochem Int 2000;37(2–3):199–206.

[200] De Keyser J, Mostert JP, Koch MW. Dysfunctional astrocytes as key players in the pathogenesis of central nervous system disorders. J Neurol Sci 2008;267(1–2):3–16.

[201] Aschner M, Chen R, Kimelberg HK. Effects of mercury and lead on rubidium uptake and efflux in cultured rat astrocytes. Brain Res Bull 1991;26(4):639–42.

[202] Allen JW, Shanker G, Tan KH, Aschner M. The consequences of methylmercury exposure on interactive functions between astrocytes and neurons. Neurotoxicology 2002;23(6):755–9.

[203] Park ST, Lim KT, Chung YT, Kim SU. Methylmercury-induced neurotoxicity in cerebral neuron culture is blocked by antioxidants and NMDA receptor antagonists. Neurotoxicology 1996;17(1):37–45.

[204] Atchison WD, Narahashi T. Methylmercury-induced depression of neuromuscular transmission in the rat. Neurotoxicology 1982;3(3):37–50.

[205] Traxinger DL, Atchison WD. Reversal of methylmercury-induced block of nerve-evoked release of acetylcholine at the neuromuscular junction. Toxicol Appl Pharmacol 1987;90(1):23–33.

[206] Levesque PC, Atchison WD. Interactions of mitochondrial inhibitors with methylmercury on spontaneous quantal release of acetylcholine. Toxicol Appl Pharmacol 1987;87(2):315–24.

[207] Yuan Y, Atchison WD. Comparative effects of methylmercury on parallel-fiber and climbing-fiber responses of rat cerebellar slices. J Pharmacol Exp Ther 1999;288(3):1015–25.

[208] Yuan Y, Atchison WD. Methylmercury-induced increase of intracellular Ca2+ increases spontaneous synaptic current frequency in rat cerebellar slices. Mol Pharmacol 2007;71(4):1109–21.

[209] Yuan Y, Atchison WD. Action of methylmercury on GABA(A) receptor-mediated inhibitory synaptic transmission is primarily responsible for its early stimulatory effects on hippocampal CA1 excitatory synaptic transmission. J Pharmacol Exp Ther 1997;282(1):64–73.

[210] Dasari S, Yuan U. Low level postnatal methymercury exposure in vivo alters developmental forms of short-term synaptic plasticity in the visual cortex of the rat. Toxicol Appl Pharmacol 2009;240(3):412–22.

[211] Giordano G, Costa LG. Developmental neurotoxicity: some old and new issues. ISRN Toxicol 2012; Article ID 814795.

[212] Weiss B, Clarkson TW, Simon W. Silent latency periods in methylmercury poisoning and in neurodegenerative disease. Environ Health Perspect 2002;110(Suppl. 5):851–4.

[213] Weiss B. Risk assessment: the insidious nature of neurotoxicity and the aging brain. Neurotoxicology 1990;11(2):305–13.

[214] Reuhl KR. Delayed expression of neurotoxicity: the problem of silent damage. Neurotoxicol 1991;12:341–6.

[215] Rice D, Barone Jr S. Critical periods of vulnerability for the developing nervous system: evidence from humans and animal models. Environ Health Perspect 2000;108:511–33.

[216] Martyn CN, Barker DJ, Osmond C. Motoneuron disease and past poliomyelitis in England and Wales. Lancet June 11, 1988;1(8598):1319–22.

[217] Ekino S, Susa M, Imamura T, Kitamura T. Minamata disease revisited: an update on the acute and chronic manifestations of methyl mercury poisoning. J Neurol Sci 2007;262(1–2):131–44.

[218] Kinjo Y, Higashi H, Nakano A, Sakamoto M, Sakai R. Profile of subjective complaints and activities of daily living among current patients with Minamata disease after 3 decades. Environ Res 1993;63:241–51.

[219] Newland MC, Rasmussen EB. Aging unmasks adverse effects of gestational exposure to methylmercury in rats. Neurotoxicol Teratol 2000;22:819–28.

[220] Yorifuji T, Tsuda T, Inoue S, Takao S, Harada M. Long-term exposure to methylmercury and psychiatric symptoms in residents of Minamata, Japan. Envrion Int 2011;37(5):907–13.

[221] Rice DC, Hayward S. Comparison of visual function at adulthood and during aging in monkeys exposed to lead or methylmercury. Neurotoxicology 1999;20(5):767–84.

[222] Landrigan PJ, Sonawane B, Butler RN, Trasande L, Callan R, Droller D. Early environmental origins of neurodegenerative disease in later life. Environ Health Perspect 2005;113(9):1230–3.

[223] Langston JW, Forno LS, Tetrud J, Reeves AG, Kaplan JA, Karluk D. Evidence of active nerve cell degeneration in the substantia nigra of humans years after 1-methyl-4-phenyl-1,2,3,6-tetrahydropyridine exposure. Ann Neurol 1999;46(4):598–605.

[224] Petersen MS, Weihe P, Choi A, Grandjean P. Increased prenatal exposure to methylmercury does not affect the risk of Parkinson's disease. Neurotoxicology 2008;29(4):591–5.

[225] Dantzig PI. Parkinson's disease, macular degeneration and cutaneous signs of mercury toxicity. J Occup Envrion Med 2006;48(7):656.

[226] Attar AM, Kharkhaneh A, Etemadifar M, Keyhanian K, Davoudi V, Saadatnia M. Serum mercury level and multiple sclerosis. Biol Trace Elem Res 2012;146(2):150–3.

[227] Cox P, Sacks O. Cycad neurotoxins, consumption of flying foxes and ALS/PDC disease in Guam. Neurology 2002;58:956–9.

[228] Callaghan B, Feldman D, Gruis K, Feldman E. The association of exposure to lead, mercury, and selenium and the development of amyotrophic lateral sclerosis and the epigenetic implications. Neurodegener Dis 2011;8(1–2):1–8.

[229] Johnson FO, Atchison WD. The role of environmental mercury, lead and pesticide exposure in development of amyotrophic lateral sclerosis. Neurotoxicology 2009;30(5):761–5.

[230] Brini M, Cali T, Ottolini D, Carafoli E. Neuronal calcium signaling: function and dysfunction. Cell Mol Life Sci 2014;71(15):2787–814.

[231] Federico A, Cardaioli E, Da Pozzo P, Formichi P, Gallus GN, Radi E. Mitochondria, oxidative stress and neurodegeneration. J Neurol Sci 2012;322(1–2):254–62.

[232] Heath JC, Banna KM, Reed MN, Pesek EF, Cole N, Li J, et al. Dietary selenium protects against selected signs of aging and methylmercury exposure. Neurotoxicology 2010;31(2):169–79.

[233] Ralston NV, Raymond LJ. Dietary selenium's protective effects against methylmercury toxicity. Toxicology 2010;278(1):112–23.

[234] Bose R, Onishchenko N, Edoff K, Janson Lang AM, Ceccatelli S. Inherited effects of low-dose exposure to methylmercury in neural stem cells. Toxicol Sci 2012;130(2):383–90.

[235] Ceccatelli S, Bose R, Edoff K, Onishchenko N, Spulber S. Long-lasting neurotoxic effects of exposure to methylmercury during development. J Intern Med 2013;273(5):490–7.

[236] Trasande L, Schechter C, Haynes KA, Landrigan PJ. Applying cost analyses to drive policy that protects children: mercury as a case study. Ann N Y Acad 2006;1076:911–23.

Developmental Exposure to Lead: Overview and Integration of Neurobehavioral Consequences and Mediation

Deborah A. Cory-Slechta

INTRODUCTION AND FOCUS

The scientific literature documenting the neurotoxicity arising from exposure to the inorganic metal lead is extensive and well beyond the potential for coverage in any single book chapter. Much of this literature, in both human studies and animal models, has

Environmental Factors in Neurodevelopmental and Neurodegenerative Disorders
http://dx.doi.org/10.1016/B978-0-12-800228-5.00007-8

focused on the consequences of developmental exposures to lead on cognition and associated central nervous system (CNS) mediators, and, as is becoming increasingly clear, such consequences extend well into adult stages of life. Although early development appears to be a period of particular vulnerability, and thus also of particular relevance to public health screening and intervention programs, it is important to appreciate that because of residual contamination, exposure to lead is virtually lifelong. Much remains to be understood about the impact of such cumulative exposures during adult and senior periods of life. Also pressing is the question of whether developmental exposures alone can produce programming effects with lifelong and perhaps even transgenerational consequences. The current chapter focuses on behavioral and CNS consequences of developmental exposures that to date have been reported even out to young adulthood. It does not attempt to be fully inclusive of all published studies but to provide an overall characterization and hypothesized interrelationships of these consequences that may be of utility both for further advancing the understanding of the lifetime trajectory of developmental lead exposure and for the purpose of developing effective behavioral and public health intervention strategies.

HISTORY OF LEAD AS AN ENVIRONMENTAL AND OCCUPATIONAL CONTAMINANT

Awareness of the potential for lead to damage the nervous system dates back to antiquity, as a consequence of the very early recognition of the utility of lead ores for such items as coins, jewelry, etc. Lead mines were found in Turkey as early as 6500 BC. The use of lead was markedly expanded by the ancient Romans, with extensive mining and smelting of lead for use in coinage and jewelry, aqueducts for water movement, vessels for food and wine storage, and even as a sweetener. These multiple uses, as would be expected, also resulted in extensive and significant exposures of the population, such as has been documented in bone lead measurements from skeletal remains at Pompeii. An even broader expansion of the use of lead was triggered by the Industrial Revolution, with corresponding increases both in occupational exposures and in recognition of lead toxicity. Such recognition included descriptions by Charles Dickens of neurotoxicity in women who worked in the white lead mills of London, and by Benjamin Franklin, who described peripheral neuropathy-associated signs in painters and typesetters. But implementation of effective industrial hygiene measures was still years away. Numerous articles have documented this historical saga [1,2].

It was, however, the addition of lead to gasoline and to paint that ultimately produced widespread environmental contamination and human environmental exposures. The addition of lead into gasoline as an antiknock agent in the United States began in the 1920s, a story that has been previously detailed [3]. While white lead paint has existed at least since the time of the ancient Greeks, US paint companies began broad marketing of this product at approximately the same time as gasoline [4]; it was already in use in other countries. These practices resulted in the eventual incorporation of lead into dust and soil, air, water, and food supplies. The consequences of such use were found as early as

1892 in Brisbane Australia, with children exhibiting lead intoxication and encephalopathy, toxicity ultimately attributed to ingestion of peeling chips of lead-based house paint. By the 1970s, cases of lead intoxication and mortality in children were increasingly being reported in the United States [5–8]. Underscoring the fact that the scope of the problem was of far greater magnitude than was even being conveyed by these reports of mortality were the findings of Byers and Lord [9] demonstrating that even children who survived acute lead encephalopathy were highly likely to experience long-term residual behavioral and CNS damage.

Lead poisoning of children eventually forced the decision to phase lead out of gasoline in the United States. With that decision came a gradual reduction in lead levels in air and other media and corresponding gradual reductions in population blood lead levels [1]. This trajectory of reductions in population blood lead levels made possible a much broader characterization of the concentration–effect relationship between blood lead (the most widely used biomarker of lead exposure) and CNS toxicity. These studies at ever lower blood lead levels have now established that it is not possible to define a level of lead, in terms of measurable blood lead values in field instruments, that are without adverse effects in children. As these data unfolded, the Centers for Disease Control (CDC) implemented a sequential series of reductions in the blood lead level defined as a level of concern (see below) to be used in screening and remediation programs. Based on this evolving science, in 2012, the CDC finally eliminated the use of the phrase "level of concern" altogether in relation to blood lead (http://www.cdc.gov/nceh/lead/ACCLPP/blood_lead_levels.htm), noting that no level without adverse effects had been scientifically identified. Instead, it adopted a National Health and Nutrition Examination Survey (NHANES)-based reference value for blood lead to be used by state and other governmental programs for lead cleanup and renovation of housing. Over this same period of declining blood lead values, reported blood lead effect levels in adults also significantly declined, particularly for CNS effects (http://ntp.niehs.nih.gov/?objectid=4F04B8EA-B187-9EF2-9F9413C68E76458E). These findings are likely to further influence occupational exposure limit guidelines for lead as set by the National Institute of Occupational Safety and Health (NIOSH).

Unfortunately, warnings from public health and safety officials of the potential for an environmental "epidemic" of lead poisoning that could have prevented this historic tragedy, such as those of Dr Alice Hamilton, went unheeded. Her testimony at the hearings regarding the addition of tetraethyl lead to gasoline warned of the insidious onset of lead toxicity and its cumulative impact [10]. As a result, millions of US children were, and some continue to be, exposed to elevated levels of lead [11].

Regrettably, the adage that history repeats itself is unfortunately true for lead exposure and its neurotoxic consequences. Indeed, significantly elevated blood lead levels are the norm for many children in countries that have not phased lead out of gasoline [12]. Furthermore, occupational exposures are a particular problem for children in many developing countries through activities such as lead smelting of electronic waste and destruction of old ships [13–17]. A notable tragedy has unfolded over the past several years in Nigeria, where activities related to gold mining have resulted in widespread lead exposures and what has been described as the worst outbreak of lead poisoning in modern history, killing hundreds of children [18–21].

With its long history and tragic consequences, lead exposure is undoubtedly the most widely studied of all known neurotoxicants, with the current understanding that even the lowest measurable blood lead levels today are not without adverse consequences. This chapter will focus largely on the understanding of the impact of lead exposure on cognitive and behavioral functions, specifically on the impact of the neurodevelopmental consequences of lead exposures. This understanding has been facilitated not only by the ability to study ever lower levels of lead exposure in human cohorts, but by the fact that the toxicokinetics, effect levels, and profile of adverse effects in animal models are similar to those in humans and, thus, have provided a rich base of information on behavioral mechanisms and neurobiological underpinnings of lead neurotoxicity [22].

THE PERSISTENT QUESTION OF WHETHER ELEVATED LEAD EXPOSURE REDUCES IQ

Byers and Lord's discovery [9] that children who survived acute lead encephalopathy nevertheless suffered pronounced and enduring residual neurological consequences of lead poisoning raised significant concerns as to the concentration–effect relationship between lead exposure and CNS impairments, that is, to the question of whether even lower levels of exposure might produce neurotoxicity. Efforts to address this issue were pioneered by Needleman and colleagues. In a landmark 1979 paper, Needleman and colleagues demonstrated significant reductions of IQ test scores, auditory speech processing, and measures of attention with increasing levels of lead in teeth, a marker of cumulative body burden of lead [23]. Importantly, the results were shown not to be due to 39 other variables that were measured in this population of 3329 children. Additionally, the number of behavioral problems, labeled as distractible, nonpersistent, dependent, unorganized, hyperactive, impulsive, frustrated, daydreamer, unable to follow simple directions or sequences and low overall findings, increased with increasing tooth lead concentrations, based on ratings by teachers and parents, who were blind to the tooth lead levels of the children.

These findings provoked numerous additional cross-sectional studies, but also, importantly, the implementation of a series of prospective longitudinal cohort studies over the next 15 years [24–31]. The pervasive question of many of these studies continued to be whether lead exposure reduced IQ and did so independently of other known influences such as parental education, maternal IQ, socioeconomic status, etc. These studies were carried out in populations of children from multiple different countries, a key feature given that various environmental conditions differed in different countries, allowing a determination of the generality of elevated blood lead on IQ. Later studies continued to improve on earlier studies, with greater sample sizes and often better validated and more sensitive measures of outcome. Furthermore, with the elimination of lead from gasoline and paint in the late 1970s and the consequent decline in blood lead levels in the United States over the next 20 years, studies of dose–effect relationships between lead exposure and IQ and other behaviors were able to examine the consequences of increasingly lower blood lead levels. Considered collectively and in several meta-analyses (see below), these studies have consistently shown (three to four point) IQ reductions associated with lead exposure even after adjustment for relevant covariates. With IQ reductions reported at increasingly lower blood lead levels, the CDC continued to reduce the designated "blood lead level of concern" from 60 μg/dL in 1960, to 40 μg/dL in 1973, to 30 μg/dL in 1975, to 25 μg/dL in 1985, and to 10 μg/dL in 1991.

A prospective study in 2003 directly addressed the safety of the CDC's 1991 established blood lead level of concern of 10 μg/dL, by examining IQ changes in children whose blood lead level never exceeded 10 μg/dL [32], assessing IQ at three and five years of age and using covariates that included the child's sex, birth weight, iron status, maternal IQ, years of education, race, tobacco use during pregnancy, household income, and HOME score. It found that IQ reductions were significantly associated with blood lead levels below the then defined level of concern of 10 μg/dL and, even further, demonstrated that the magnitude of IQ reductions (based on increases in blood lead from 1 to 10 μg/dL) at 7.4 points was actually significantly greater than the 4.6 point reduction that occurred across the full range of blood lead values (i.e., including those exceeding 10 μg/dL), findings subsequently confirmed in other longitudinal studies [33].

A subsequent pooled analysis to further evaluate blood lead levels ≤10 μg/dL included seven international population-based longitudinal studies and confirmed both associations between blood lead levels and reduced IQ below the 10 μg/dL level of concern and the prior reported associations of greater reductions in IQ at lower blood lead concentrations [34].

The most recent studies have now encompassed the lowest levels of measurable blood lead. For example, in a Taiwanese cohort of children followed from 2–3 to 8–9 years of age, geometric mean blood lead values of 2.48, 2.49, and 1.97 μg/dL were associated with significant reductions in IQ scores adjusted for maternal education, maternal blood lead level, HOME scores, and child sex [35]. Lucchini et al. reported that a benchmark blood lead level associated with a loss of one IQ point was 0.19 μg/dL in a population of 299 Italian adolescents aged 11–14 years [36]. Another pooled analysis of the seven longitudinal studies yielded a lower confidence limit for a benchmark dose of approximately 0.1–1.0 μg/dL for a concentration leading to a one-point reduction in IQ [37]. In sum, the available evidence now suggests that it is not possible to determine a safe level of blood lead in children with respect to reductions in IQ, notably a full 20 years after an original report of the inability to find a threshold in the relationship between blood lead and IQ [38].

Challenges to such findings that have been raised include difficulty in disentangling the impact of low socioeconomic status as a cause of IQ reductions, or even of reverse causality. However, several papers argue against this assertion. As reported by Bellinger et al. [39], scores on the Bayley Scales of Infant Development declined with increasing blood lead level to a greater extent in children of lower socioeconomic status, but they also nevertheless declined in higher socioeconomic status children. In the Yugoslavia prospective study that reported reductions in Wechsler Intelligence Scale IQ scores with increasing blood lead concentrations [40], the town with the elevated blood lead exposures in which a smelter was located also had the higher HOME scores and maternal IQ scores, that is, it was the higher socioeconomic status community. Hansen et al. [41], in a type of "case–control" study, matched children with high versus low lead levels by sex and socioeconomic status, and they still found that high lead-exposed children scored lower on IQ tests than did low lead-exposed children. A meta-analysis of studies of low-level lead exposure and children's IQ explicitly indicated that the effect was not limited to disadvantaged children [38].

An additional argument that has been proffered in response to these findings is that as blood leads have declined, there should be a corresponding increase in achievement test scores, IQ etc.

if lead exposure was the causative factor. Such a question is particularly difficult to address in human studies given that in the 30+ years since the phase out of lead from gasoline and paint, so many other pertinent factors have also changed, including in conditions that would likely be detrimental to cognitive achievement that could mask any effect of lead abatement. For example, according to the US Census Bureau, the number of people living in poverty has risen steadily since approximately 1980 (http://www.census.gov/prod/2013pubs/p60-245.pdf). Rates of childhood obesity (http://www.cdc.gov/obesity/data/trends.html#State) and diabetes (http://www.diabetesandenvironment.org/home/incidence/historical) have also increased dramatically, changes that have been associated with diminished cognitive achievement [42,43]. Furthermore, the fact that bone lead stores will be mobilized back to blood if lead exposure is terminated complicates the ability to study this issue directly in humans or in animal models. Nevertheless, to this point, Liu et al. [44] noted an improvement in cognitive test scores of 4.0 points per each $10\,\mu g/dL$ decline in blood lead in children over a period of follow-up from baseline to 36 months. This effect was seen in only children receiving placebo and not in those who had been randomized to receive succimer chelation to lower lead levels, leaving residual issues as to efficacy of such interventions. A privately funded effort (http://www.bos.frb.org/econoomic/wp/index.htm) directly addressing this issue reported small but significant improvements in scores on standardized English and mathematics tests with falling blood lead levels.

Several studies have also attempted to define specific critical periods for lead-induced developmental neurotoxicity including IQ reduction, for example, prenatal vs. postnatal periods of exposure and specific times during pregnancy. Such questions are important as they relate to public health screening and intervention efforts. Studies in animal models have reported behavioral and neurobiological consequences of lead at virtually all stages of development ranging from gestational to adult exposures, suggesting that susceptibility is not restricted to the earlier periods of development (e.g., [45–48]). In human populations, some early studies suggested that prenatal, but not postnatal, exposures were critical to early cognitive development based on the Bayley Scales of Infant Development [49], with another study even suggesting that the first trimester was most critical [50]. However, subsequent studies confirmed the importance of postnatal exposures [51–53], likely reflecting the somewhat older age of children at the time of analyses and, thus, the ability to utilize more sensitive outcome measures. In addition, the significance of concurrent blood lead was also demonstrated in a study across cohorts reporting that the associations of blood lead with IQ scores increased in strength as children aged, while those between prenatal blood lead and IQ were attenuated [54], and of significant associations of blood lead levels at seven years of age, but not at two years of age on externalizing behavior [55].

The apparent nonlinearity of lead effects on IQ, that is, greater reductions at lower than higher blood lead concentrations [32], has also caused some to suggest reverse causality. However, a long history of reports in the literature in both human and animal studies documents nonlinear effects of lead exposure. As noted as early as 1990 [56], nonlinear curves related lead exposure to multiple physiological systems and function, including locomotor activity in rats [57], cerebral cortical growth in rats [58], rates of operant behavior in rats [59,60], alterations in maze performance in rats [61], age at vaginal opening in rats [62], discrimination learning in monkeys [63], serum urea and urate in men [64], body weight and height of children [65], brainstem auditory-evoked responses of children [66], nutrient intakes of vitamin D, vitamin C, and iron [67], blood pressure in rodents [68], and alterations in glucocorticoid negative feedback in rats have been reported [69]. While there may be multiple reasons for such nonlinearity, it is notable that nonlinearity also characterizes the effects of lead on calcium-related reactions, with the substitution of lead for calcium considered to be a key mechanism for multiple effects of lead. Audesirk and colleagues [70,71] demonstrated that lower concentrations of lead increased (whereas higher concentrations decreased) activation of calcineurin, a protein with activity dependent upon Ca^{2+}/calmodulin binding.

WHAT DOES AN IQ REDUCTION MEAN BEHAVIORALLY: CHANGES IN LEARNING

Tests of intelligence encompass a broad array of cognitive/executive functions to yield integrated composite scores, raising the question as to which such specific cognitive function(s) are impaired by elevated lead exposure. Moving beyond assessment of IQ score to a more precise understanding of the types of CNS behavioral deficits associated with elevated lead exposure is critical to any type of behavioral interventions that might be implemented to assist affected children. A number of human studies have addressed the question, and this has been a prominent component of the experimental animal literature. As pointed out by Bellinger [72], a broad array of cognitive deficits have been associated with elevated lead exposure, that is, there is no one specific behavioral biomarker or signature. Deficits have included impaired executive functions, motor function, and language. This breadth of effects and variability across populations is not particularly surprising, given that studies addressing these questions can differ in many ways that may significantly influence outcome, including the specific tests used, age of testing, culture, nutrition, performance motivation, etc. Information on how the effects of lead are influenced by other environmental conditions may be of particular use both to public health protection measures and to the development of efficacious intervention and remediation strategies.

In experimental assessments of learning as a potential contributor to IQ reductions, it is always critical to consider that lead-associated changes may not be learning deficits per se, but they could arise in response to such factors as altered motivation levels, defective sensory processing, or even alterations in motor functions [73]. A study in rats was explicitly designed to address that possibility using a multiple schedule of repeated learning (RL) and performance (P) contingencies, a paradigm originally established for use with human subjects [74] but since adapted for use in both rats [75] and mice [76–78]. In this multiple schedule format, RL and P contingencies alternated during a behavioral test session on the basis of time and/or number of reward deliveries, and the component in effect at any given time was indicated by an external signal, for example, color of an illuminated light. In the repeated learning or RL components of the schedule, rewards were provided after the correct completion of a sequence of three lever-press responses, with the correct three-response sequence changing with each successive experimental session. In the performance or P components of the test schedule, rewards were also provided after the correct completion of a three-lever response sequence, but the specific sequence was the same for all sessions. Thus, learning a new three-response sequence in each session was required during the RL component, whereas only completion of the same already learned three-member sequence was required during P components.

Under these conditions, equivalent sensory capabilities, motor function, and motivation are required in both the RL and P components, but learning is only required in the RL component. Under such conditions, learning-specific impairments would be characterized by selective reductions in accuracy in the RL component, with no changes in accuracy during the P component. Consistent with this interpretation, post-weaning lead-exposed rats demonstrated highly selective impairments that were restricted to the RL component [75].

Fundamentally, the RL component of the schedule is a discrimination learning task (based on spatial positioning). Given that the correct RL sequence changes with each successive test session, it constitutes a discrimination reversal in that it requires both extinction of the previously learned sequence from the prior session and concurrent learning of a new sequence. The intra-extra-dimensional (IED) Set Shift task from the automated touch-screen CANTAB battery (Cambridge Cognition) is a task with behavioral requirements similar to those operative in the multiple RL and P schedule in that it also involves discrimination reversal learning and was used to test for learning deficits in children [79]. The IED Set Shift test proceeds through nine stages, with the first seven being comprised of discrimination learning and reversal learning problems based on identically colored shapes, while the final two stages involve a shift of the reinforcement contingency to a discrimination task involving a previously experienced irrelevant dimension of the stimuli (extra-dimensional shift to white lines; stage 8) and its reversal (stage 9).

The most difficult components of the IED Set Shift task are the shifts that occur in the final two stages. Children from the Rochester cohort [79], whose lifetime average blood lead level was 7.2 µg/dL, exhibited lead-related impairments on performance of the more difficult stages in the Set Shift task in analyses that controlled for maternal ethnicity, HOME scores collected at 24 months of age and at 6 years of age, and prenatal smoking. Specifically, few children completed the extra-dimensional (ED) shift (stages 8 and 9), and there was a trend toward lower blood lead levels in those who shifted successfully: for each 1 µg/dL increase in blood lead, children were 1.2 times less likely to complete the ED shift. Additionally, higher blood lead levels were associated with a decrease in the number of stages completed. Considering only children who successfully completed the first seven stages of this task, higher blood lead levels were associated with an increased number of trials to reach criterion. These results may be consistent with the report by Stiles and Bellinger [80] of an association of childhood blood lead level with increased perseverative errors and perseverative responses on the Wisconsin Card Sort Test, a task that requires matching cards according to various rules (e.g., color, shape) with the rule changing over the course of the task in an unpredictable manner.

As with effects described above related to studies of lead exposure and IQ in children, the nature of the learning impairments seen in response to lead exposure appears to differ depending upon numerous other factors, such as exposure concentration, exposure duration, developmental period of exposure, and gender—information that can be highly relevant to mechanisms of neurotoxic effects. For example, in studies involving drinking water lead exposure of rats beginning at approximately two months of age, a six months exposure impaired spatial learning in a water maze in both sexes [81], whereas a shorter term exposure of 30–45 days was found to produce deficits only in males [82]. Lead effects on learning may also be highly context dependent: in female rats exposed to lead during gestation and lactation, reduced accuracy levels in the RL component were not seen for all sequences, but for very specific directions of spatial movement [83].

DO ATTENTION DEFICITS AND/OR HEARING LOSS SERVE AS BEHAVIORAL MECHANISMS OF LEAD-INDUCED LEARNING AND IQ DEFICITS?

Attention Deficit—Deficits in attention-related behaviors have been repeatedly linked to lead exposure in human studies [84–89]. Such deficits may be critical not only as a disruptive behavior per se, but also with respect to their potential to serve as a potential behavioral mechanism of lead-associated learning impairments and conduct disorders (see below) [90,91].

Attention represents a behavioral construct comprised of multiple components, and from a diagnostic perspective, attention deficit disorder includes three major categories: inattention, hyperactivity, and impulsivity/inability to manage delay of reinforcement, with hyperactivity/impulsivity often combined. While early studies assessing associations of lead with attention-related behaviors tended to rely on parental and/or teacher ratings, more recent findings from case–control studies suggest associations with clinically diagnosed attention deficit hyperactivity disorder (ADHD) [92,93], both hyperactivity and inattention categories, at blood lead values $< 5 \mu g/dL$. A recent meta-analysis examining 33 studies published between 1972 and 2010 of lead exposure and attention found associations between inattention and lead exposure, and combined hyperactivity/impulsivity with lead exposure (when analyses of lead levels by hair analyses were excluded) with deficits similar in magnitude to those cited for IQ [94].

Animal models have provided additional insight into the impact of lead exposure on attention, with hyperactivity (locomotor activity) studied the most intensively [86]. Over the years, there have been reports of reductions, increases, or no changes in locomotor activity in various animal models of lead exposure, with no one outcome dominating. The varied findings with respect to hyperactivity likely reflect the significantly different lead exposure protocols and other experimental conditions used across such studies and the nonlinearity of the brain dopamine systems mediating such behavior [95–97]. In that context, a systematic review of how lead exposure protocols relate to changes in locomotor activity has never been undertaken but could prove fruitful.

Animal studies have also provided information with respect to the potential involvement of alterations in sustained attention versus impulsivity. In animal studies, deficits in sustained attention have proven difficult to demonstrate [98] or have been of relatively modest magnitude [99]. However, changes in measures indicative of impulsivity have shown more pronounced effects, as seen in fixed interval (FI) schedule-controlled behavior and in a fixed ratio (FR) waiting-for-reward paradigm.

FI behavior was described by Darcheville et al. as a surrogate for impulsivity [100,101]. In his studies, children exhibiting high FI response rates reliably made impulsive choices in self-control paradigms (choosing smaller rewards to be received quickly rather than larger rewards to be received later), whereas children with low FI response rates made self-controlled choices (choosing larger rewards to be received later rather than small rewards to be received quickly). Changes in FI behavior have also been reported in boys with a clinical diagnosis of attention deficit disorder [102]. FI schedule-controlled behavior has been demonstrated to be highly sensitive to lead exposures [103]. Indeed, changes in FI performance have been observed in animals after lead exposure initiated at several different developmental periods and across species [103]. Lower blood lead levels have been associated with increases in response rates, whereas higher blood lead levels are associated with decreases in FI response rates, again, likely reflecting the underlying U-shaped dopamine curves that mediate FI behavior [97].

Further support for impulsivity as a potentially important behavioral mechanism for lead-associated cognitive deficits was found using an FR waiting-for-reward paradigm [104,105]. Under this multiple behavioral schedule, rats were first required to complete an FR of 50 lever presses which, when completed, initiated a component in which free reward deliveries occurred, but with the time intervals between each free reward delivery doubling, for example, 2, 4, 8s, etc. Thus, in the waiting-for-reward component, rats could simply inhibit responding and accrue free reward deliveries, but the waiting time continued to increase with each free reward delivered. Any lever press response that occurred during the free reward component reset the schedule to the FR 50, such that another 50 responses would be required to reinitiate the free reward deliveries initiating the shorter wait times. Several outcomes were measured, including the mean longest waiting time during the free reward delivery component, the total number of FR resets, as well as the mean number of responses required to produce a reward delivery, that is, the total number of FR responses divided by the total number of reinforcers earned in both the FR and free reward components.

Chronic post-weaning lead exposures of 0, 50, or 150 ppm in drinking water (corresponding to mean blood lead levels of <5, 10.8, and 28.5 µg/dL, respectively) resulted in a lead concentration-dependent decrease in the mean long-wait time, with reductions of approximately 33% at 50 ppm and 50–60% at 150 ppm, findings consistent with a type of impulsivity and inability to inhibit responding. Additionally, lead exposure significantly increased the number of FR resets, resulting in a corresponding lead-associated increase in the number of responses made per each reinforcer delivery.

Parallels to these collective findings from animal studies are now actually being reported in children [88,89]. In a study of 150 children between 8 and 17 years of age with maximal blood lead levels of 3.4 µg/dL, blood lead was associated with diagnostic symptoms (assessed by licensed clinicians blinded to blood lead levels) of hyperactivity–impulsivity, but not inattention–disorganization after control for covariates including income and gender. The lead association with hyperactivity–impulsivity was not accounted for by IQ; in contrast, hyperactivity was found to mediate lead association with IQ. A "stop" task of impulsivity was also administered in which subjects were instructed to interrupt a rapid response on 25% of trials. Poorer performance on this task was found to mediate the hyperactivity–impulsivity symptoms and did so independently of IQ, suggesting it as a mediating factor or behavioral mechanism for reduced IQ. Thus, as in the animal studies, impulsivity proved to be the most prominent outcome with respect to different components of impaired attention.

As of yet, no direct studies of impulsivity as measured using the well-validated "self-control" paradigm have been undertaken in relation to elevated blood lead in children. However, effects from rat studies demonstrating lead-induced changes in FI schedule-controlled behavior [106] and FR waiting-for-reward behavior [104], coupled with the most in-depth assessments of lead effects on behavior in children [88,89], as cited above, suggest it may be a key behavioral mechanism of lead-associated cognitive deficits. In further support of this assertion is the fact that lead exposure-related impairments in performance on a differential reinforcement of low rates (DRL) schedule have been found both in children [107] and in nonhuman primates [108]. The DRL schedule requires that responses be separated by a designated period of time, with premature responses punished by a reset of the time. DRL performance is often used as a measure of impulsivity in the context of inability to inhibit responding [109]. The work of Nicolescu et al. [87] provides support for this possibility in a study of 8–12-year-old Romanian children in which four different computer-based tests were administered in conjunction with parent and teacher ratings of attention-related behaviors. Notably, false alarm responses (i.e., responding in the absence of a signal) rather than response latencies were positively associated with lead, suggesting a greater impact on impulsivity than sustained attention. The fact that the observed effect sizes were shown to be considerably larger than those associated with IQ reduction led the authors to suggest that attention deficit could be the more basic deficit, that is, a behavioral mechanism underlying other cognitive impairments. As of yet, the extent to which the supralinearity of the concentration–effect curves relating lead exposure to IQ reduction is paralleled in attention-related behaviors is not clear.

Auditory Deficits—Lead exposure-associated alterations in auditory function have long been recognized [110,111] and could conceivably also contribute to impaired cognition [112,113], an outcome first suggested by Otto and Fox [114]. Early studies reported increases in hearing thresholds with increasing blood lead concentrations [110] and even at this early stage to the fact that such effects appeared to extend to blood lead levels below the then CDC level of concern of $10\,\mu g/dL$, as well as an apparent absence of any threshold blood lead level for the effect [115]; similar effects were reported in a later cohort of children 1–6 years of age from China based on measures of brainstem auditory-evoked potentials [116] and in 4- to 14-year-old children in Poland [117].

An interesting series of studies by Rothenberg and colleagues furthered such assessments with respect to defining underlying biological substrates. In one study, they reported that auditory-evoked brainstem response interpeak intervals I–V and III–V in children 5–7 years of age declined as blood lead levels rose from 1 to $8\,\mu g/dL$, but then increased as blood lead levels rose further from 8 to $30.5\,\mu g/dL$. Similar nonlinearities in interpeak intervals had previously been reported with a bisection point at $25\,\mu g/dL$ [118]. Rothenberg et al. [66] attributed the reductions below $8\,\mu g/dL$ to reductions in auditory path length, which could reflect as much as a $3.4\,mm$ difference between the cochlea nucleus and the ipsilateral medial superior olivary nucleus, while increasing effects above $8\,\mu g/dL$ likely reflected lead damage to myelin and/or synaptic transmission. The nonlinearity of these findings is potentially relevant not only to the repeated reports of such effects with lead exposure but also to the greater magnitude of IQ reductions at lower vs. higher blood lead levels.

A potential inconsistency of the literature assessing auditory function, however, involves reports from a series of studies of Andean children with quite high blood lead levels (mean $37.7\,\mu g/dL$) who do not show changes in distortion product otoacoustic emissions (reflecting outer hair cells of the inner ear) [119] or in auditory brainstem-mediated acoustic stapedius reflex measurements [120], despite deficits in auditory memory as assessed via a digit span test [121]. Given that effects in other studies have reported no apparent blood lead threshold and biphasic effects, it may be that the limited sample size of the lower range of blood lead levels of this particular population does not provide adequate sensitivity. Alternatively, individual differences in vulnerability may obscure population effects, or other specific features of this environment/cohort may mitigate these effects of lead.

Studies in animal models have been generally consistent with adverse auditory effects of lead exposure. These were initially reported in nonhuman primate studies that also confirmed the potential for individual differences in vulnerability to such effects of lead, as measured using detection of pure tone thresholds [122] and distortion product otoacoustic emissions [123]. Some of these studies indicated that such changes represented permanent effects of lead exposure [124]. Post-weaning exposures of mice yielding blood lead values of approximately 20–25 μg/dL produced increased auditory brainstem response thresholds in conjunction with morphological changes of the outer hair cells, including reduction in the tight junctions between endothelial and border cells [125]. A study of gestationally and lactationally exposed mice indicated primary effects on neuronal structural proteins, most prominently manifest as increased phosphorylation of both the medium- and high-weight neurofilaments within the auditory brainstem nuclei, as opposed to changes in myelin [126].

Of potential interest to the possibility that auditory deficits underlie the cognitive deficits associated with lead would be assessments of the correlations of auditory effects with IQ or other cognitive tests. Such evaluations either were insignificant or have generally, surprisingly, not been pursued. Dietrich et al. [127] reported deficits in both auditory processing abilities and cognitive deficits measured with the Kaufman Assessment Battery for children, but no correlations of these measures were described, and effects were no longer significant after controlling for home environment and maternal intelligence.

PERSISTENCE OF DEVELOPMENTAL LEAD NEUROTOXICITY

The extended period encompassed by several of the longitudinal cohort studies in children has permitted an assessment of the potential for persistence of the adverse consequences of lead well beyond early development, as well as of the potential emergence of impairments at a later age. Among such consequences is a recent report indicating the persistence of early deficits in IQ, with childhood blood lead concentrations measured at 6 months of age and 4 and 10 years of age still related to Full-Scale IQ in adults between 28 and 30 years of age. While intriguing, further studies are needed to examine such outcomes given the inability to control for maternal IQ in this study and its small sample size [128]. Lead also impairs academic achievement, first reported in 10-year-old children from the Boston cohort study [129] and more recently shown in elementary and junior high school children in a retrospective study from Detroit, demonstrating that high blood lead values (geometric mean value of 7.12 μg/dL) prior to age 6 were associated with poorer academic achievement in grades 3, 5, and 8 [130].

Far more attention has been afforded to studies of delinquent behaviors, antisocial behaviors, and conduct disorders, with the collective results suggesting a significant role for early lead exposure, studies largely prompted by a 1996 report by Needleman and colleagues demonstrating relationships between bone lead levels, a measure of cumulative exposure, and delinquent and antisocial behavior [131] in a retrospective cohort study of boys from 7 to 11 years of age. At 11 years of age, significant increases in somatic complaints and in delinquent, aggressive, internalizing, and externalizing behaviors as reported by parents were found with increasing lead exposure levels. Teachers likewise reported higher levels of these behaviors in boys with higher bone lead levels. Considered over the four-year period, higher lead subjects were also more likely to have worse scores on all items of the Child Behavior Check List. A subsequent case–control study [132] confirmed higher mean bone lead concentrations of adjudicated delinquents than controls in youths 12–18 years of age. In a longitudinal prospective study [133], both prenatal and postnatal blood lead exposures were associated with increased frequencies of antisocial and delinquent behaviors in both self- and parental reports. In the longitudinal cohort study from New Zealand [134], dentine lead levels at 6–9 years of age were associated with recorded violence/property convictions at ages 14–21 and self-reports of violent/property offending at ages 14–21, findings consistent with those of Needleman et al. [131]. In a time course analysis of changes in violent crime, marked similarities were found in the long-term changes in exposure to leaded gasoline and subsequent changes in violent crime assessed using a 20-year lag period [135]. This association of lead exposure in children with conduct problems was further supported by a meta-analysis that showed the magnitude of this association was very similar to that relating lead exposure to IQ reductions [136].

Thus, it appears that both early lead exposure (prenatal and postnatal blood lead, shed deciduous teeth) as well as cumulative lead exposure, as derived from later bone lead measures, show associations with delinquent behaviors. While these studies do not suggest a critical period of exposure for the association of lead with delinquent behavior, it was notable that educational underachievement was found to mediate the links between dentin lead and violent behavior in the New Zealand cohort [134], findings consistent with a body of prior research linking educational underachievement to antisocial behavior and delinquency [137,138]. Of direct relevance to the associations of lead exposure with attention deficit and impulsivity are studies that report specific links between impulsivity and later delinquency [139], with such effects shown to be greater in lower socioeconomic status conditions after controlling for grade level, gender, race, language, maternal education, and socioeconomic status [140]. Considered collectively, such findings raise the assertion of a pathway in which lead-associated early and persistent reductions in educational success may themselves contribute to impulsive-like behaviors that increase risk for antisocial and delinquent behaviors, particularly in low socioeconomic status communities that already preferentially experience the highest levels of lead exposure.

That early lead exposure can also produce morphological changes in the brain as children reach adolescence and adulthood has now also been shown. For example, annual mean blood lead levels from 3 to 6 years of age (mean of 13.1 µg/dL and maximum values of 23.1 µg/dL) were inversely associated with gray matter volume reductions, particularly in the prefrontal cortex in 21-year-old adults [141] as well as in associated brain metabolic markers [142], findings consistent with an earlier study with a smaller sample size and greater blood lead levels [143]. Additional adulthood effects have been reported in the Port Pirie Cohort study, particularly in females in relation to childhood average (to age seven) blood leads in the form of near-significant ($p=0.06$) increases in anxiety and phobias [144,145].

Several studies have now raised a potential association of lead exposure with schizophrenia, a mental illness that onsets in late adolescence/early adulthood that is of unknown etiology [146,147]. A neurodevelopmental hypothesis posits that schizophrenia is a developmental disorder that arises from interactions of genetic and environmental factors. A retrospective case–control study reported that increased levels of the lead biomarker delta-aminolevulinic acid were found in second-trimester serum samples in cases (corresponding blood lead values

>15 µg/dL) with an odds ratio of 2.43 ($p = 0.051$) after adjustment for covariates [148], findings that were subsequently replicated in a second study using data from a different cohort [149]. A subsequent review article provides additional mechanistic support for this potential association, citing work in mice mutant in the human form of the disrupted in schizophrenia 1 (DISC-1) gene [147]. The findings from both human and experimental animal studies and their potential public health implications should provide a basis for further evaluation of this potential relationship.

EN ROUTE TO MORE RELEVANT EVALUATIONS OF LEAD EXPOSURE AND PUBLIC HEALTH PROTECTION

In the human environment, lead exposure occurs in conjunction with many other risk factors for problems such as educational underachievement, attention deficit disorder, and juvenile delinquency. Importantly, some of these other risk factors may have the potential to modify the effects of lead exposure, either by enhancing or unmasking its toxicity (deleterious risk modifiers) or by mitigating its effects (protective factors). Bellinger, in fact, raised such a possibility as an explanation for the differences in the specific profile of lead effects on children's cognition [150]. By modifying the toxicity of lead exposure, that is, interacting with lead exposure, these other factors modify its public health consequences. For example, factors that protect against the effects of lead, for example, calcium or iron supplementation or higher socioeconomic status, could effectively lower the threat of harm from lead and/or increase the exposure level of lead associated with adverse effects. In contrast, factors that enhance or unmask effects of lead, for example, iron insufficiency or low socioeconomic status, carry the potential to increase the threat of harm from lead and effectively lower the exposure level associated with adverse effects. Further, such interactions may impact select subgroups within a population, for example, sex differences.

Because the ongoing multiplicity of risk factors constitutes real-world scenarios, it is imperative to understand the effects of toxicants in such relevant contexts rather than as isolated variables [151]. In the case of human epidemiological studies, such information is obtained from studying interactions of lead with other variables, an approach infrequently pursued. In the case of experimental animal models, this requires assessment of the effects of an environmental chemical in conjunction with other potential relevant risk factors. Important criteria for such potential interactive risk factors include both co-occurrence and shared biological substrates, thereby providing a biological basis for an interaction.

Most studied among such potential interactions with lead exposure has been the influence of stress, studies that began in animal models and have since expanded to human population studies. A study in rats combined maternal lead exposure (beginning 2 months prior to breeding and ending at weaning to insure significant body burden of lead at the time of pregnancy, consistent with human exposure) with or without prenatal stress (immobilization restraint on gestational days 16–17) was based on their shared biological targeting of both the mesocorticolimbic system of the brain (prefrontal cortex, nucleus accumbens, hippocampus) as well as the hypothalamic-pituitary-adrenal (HPA) axis [152]. It revealed multiple instances of enhanced and/or unmasked effects of lead exposure by prenatal stress, including alterations in dams (corticosterone levels, altered frontal cortex dopamine turnover, nucleus DOPAC,

HVA and serotonin, and striatal DOPAC), and sex-dependent differences in the profile of interactive effects in offspring that included altered corticosterone, frontal cortex dopamine, and DOPAC in females, and striatal serotonin in both males and females. This study has been followed by additional efforts that included interactions of lifelong lead exposure with stress, as well as direct stress challenges to offspring and its interactions with lead exposure [83,153–160].

Translational studies addressing the interactions of lead and stress followed. Consistent with findings in a rat model of demonstrating interactive alterations in HPA axis negative feedback in response to lead and prenatal stress [69], Gump et al. [161] demonstrated that low-level (<10 μg/dL) lead exposure altered the adrenocortical response to acute stress in children, resulting in delayed recovery of cortisol levels post-stress challenge. Subsequent studies confirmed that blood lead was a significant mediator of the association between socioeconomic status and the peripheral vascular response to acute stress challenge [162,163].

Several studies have expanded the assessment of combined lead and stress and its impact on children's neurodevelopment with somewhat mixed results to date. For example, positive interactions were seen by Kordas et al. [164], who reported that blood lead levels ≥5 μg/dL in either mothers or children between 13 and 55 months of age were associated with lower maternal perceptions of being skilled at discipline. Similar interactions of lower self-esteem with lead exposure on the Mental Development Index score of 397 mother–infant pairs from Mexico City (mean blood lead at 24 months of 6.4 μg/dL and range of 0.08–25.8 μg/dL) were seen by Surkan et al. [164] but were not statistically significant. However, Hubbs-Tait et al. [165] reported detrimental effects of blood lead values >2.5 μg/dL on the McCarthy Scales of Children's Ability that were further negatively exacerbated by permissive (e.g., lack of involvement) parenting.

Fewer studies have addressed the converse question, that is, can an enriched environment mitigate the neurodevelopmental toxicity of lead exposure. An early study by Bellinger [39] demonstrated the importance of socioeconomic status as a risk modifier for lead neurotoxicity, with developmental neurotoxicity of lead evident at lower levels in low socioeconomic status children within the second year of life than seen in children of higher socioeconomic status. Such findings could indicate that the positive behavioral experiences/environment of a higher socioeconomic status leads to a more resilient, less vulnerable phenotype, and/or that negative experiences associated with low SES can enhance the neurotoxicity of lead.

Early behavioral experience is well known to have a profound and enduring influence on subsequent behavior and brain function. Early child abuse, that is, negative behavioral experience or adversity, for example, has been shown to produce enduring changes in prefrontal cortical structure and function, corresponding brain dopamine function, and associated cognitive behaviors [166,167]. Cory-Slechta et al. [168] compared the effects of maternal lead exposure, prenatal stress, or the combination on brain monoamine levels after early negative vs. positive behavioral experience in mice. Positive behavioral experience was composed of daily behavioral test sessions in which food reward was available on an FI schedule of reinforcement. Negative behavioral experience was comprised of four exposures to a forced swim test in which mice were immersed for 5 min in a pool of water from which escape was not possible [168].

These differences in early behavioral experience were found to significantly alter the neurochemical consequences of lead, prenatal stress, and the combination in brain mesocortico-limbic regions of mice [168]. For example, lead exposure significantly reduced frontal cortex dopamine turnover in female offspring with negative behavioral experience, whereas no lead exposure, prenatal stress, or combined lead exposure and prenatal stress effects were observed in littermates with positive behavioral experience. Further, hippocampal glutamate levels were elevated by lead exposure after negative behavioral experience in females but reduced under conditions of positive behavioral experience. In males, a trend toward lead exposure-induced increases in midbrain dopamine and significant increases in its metabolite, homovanillic acid, in the midbrain was only seen in offspring that had negative behavioral experience. Understanding the role of early behavioral experience as a modifier for the developmental neurotoxicity of lead is important to the ability to establish effective therapeutic or behavioral interventions to ameliorate its effects, and it clearly warrants further attention.

As can be noted from the above text, differences in neurodevelopmental toxicity of lead as examined in animal models are almost always sex dependent, indicative of differential susceptibility of males and females. Given that brain differentiation begins early in gestation, sex differences in outcome would be expected rather than considered unusual. Importantly, sex differences are also likely to be context dependent, that is, depend upon outcome measures, etc. rather than an all-encompassing greater susceptibility of one sex vs. the other.

Unfortunately, even with the large human epidemiological literature on the developmental neurotoxicity of lead, the extent to which outcomes differ by sex is almost never mentioned, but it is more often used as a covariate and not examined for interaction effects. Ris et al. [169] reported greater effects of developmental lead exposure on both attention and visuoconstruction in adolescent males. Greater lead-associated reductions (mean cord blood lead values of 1.38 µg/dL) in Mental Development Index scores were found for boys at 36 months of age in a prospective cohort in Poland. A report from a longitudinal cohort studied for almost 30 years suggested near-significant increases in 25- to 29-year-old females in odds ratios after covariate adjustment (both p values = 0.06) for anxiety (1.2) and phobias (3.39) in relation to childhood average (to age 7) blood lead level [144,145]. Understanding sex differences in the developmental neurotoxicity of lead can be particularly useful to understanding mechanisms of neurotoxicity, to the appropriate development of therapeutic intervention strategies, and to public health protection.

AN INTEGRATED SUMMARY AND HYPOTHESIZED RELATIONSHIPS OF THE CONSEQUENCES OF DEVELOPMENTAL LEAD EXPOSURE

Figure 7.1 depicts a schematic summarizing the trajectory of the reported consequences of developmental lead exposure and hypothesized interrelationships among these consequences. Lead exposures occurring prenatally as well as postnatally and even considered cumulatively across early developmental exposure periods have all been linked with adverse outcomes. With respect to underlying CNS substrates, robust changes in dopamine and glutamate function of the mesocorticolimbic regions of the brain and changes in HPA axis are observed. In addition, auditory deficits are observed in human and experimental animal studies, likely as a result of direct effects on auditory system structures.

Lead has also been linked to attention-related behaviors and diagnosed deficit hyperactivity disorder, which could be a consequence of changes in any or all of these neurobiological

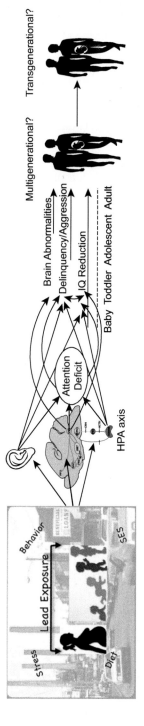

FIGURE 7.1 Schematic overview summarizing consequences of developmental lead exposure and their potential interrelationships. Exposures to lead ranging from in utero to young childhood have all been associated with adverse effects. Three neurobiological substrates of lead include auditory processing, brain mesocorticolimbic system dopamine/glutamate system function, and HPA axis function. Alterations in each of these three targets have been shown to influence attention, IQ/cognition, and delinquency/aggression—three behavioral outcomes also reliably associated with lead exposure. Attention deficit has been suggested to underlie both IQ/cognitive deficits and delinquency/aggression outcomes, and the latter two have also been shown to be associated. Depending upon other risk factors in the environment (e.g., stress, diet, socioeconomic status, behaviors, and other chemical exposures), the influence of lead on the various neurobiological substrates could be modulated with consequences for the corresponding behavioral outcomes, a scenario likely to explain differences across studies/populations in the profile of lead-associated behavioral deficits. As of now, the extent to which lead effects may be carried to the next generation (multigenerational) or to the offspring of the next generation (transgenerational) through epigenetic mechanisms has yet to be determined.

substrates. Changes in mesocorticolimbic function, for example, are well known to impact attention-related behaviors [170–172]. Not surprisingly, auditory deficits, in turn, can be tied to attention deficits [173–177]. HPA axis alterations also occur in attention deficit disorder [178–184]. Thus, all three of the potential neurobiological targets of lead influence attention-related behaviors. It may be that individually, none of the three are significantly enough impaired to lead to attention deficit, but their downstream consequences could merge to alter attention-related behavioral functions.

Further, all three of these neurobiological targets of lead exposures can be associated with delinquency and aggression. Links between brain mesocorticolimbic function, particularly dopamine, with aggression have been widely reported in human and animal models [171,185–188]. Delinquency/aggression can also be a co-morbidity of hearing impairment, that is, hearing impaired children are more prone to development of aggression and conduct disorder [189,190]. HPA axis alterations are reported in antisocial behavior [191], oppositional defiant disorder and conduct disorder [192,193], disruptive and antisocial behavior [194], and externalizing behaviors [195].

All three of lead's neurobiological substrates also influence IQ/cognition. Through the release of hormones and neurotransmitters, the HPA axis affects higher level cognition [196] as well as memory processes [197,198]. Auditory deficits can lead to learning impairments, especially in the absence of early intervention [112,113]. The critical role of mesocorticolimbic dopamine and the glutamate system in mediating executive function and cognition is extensively documented both in humans and in animal models [199–202]. Thus, all three neurobiological substrates of lead, that is, auditory processing, mesocorticolimbic system function, and the HPA axis, can act to deleteriously impact the three major behavioral features of lead exposure, reduction in IQ/cognition, attention deficit, and delinquency/aggression/conduct disorder, providing a basis from which additive or enhanced consequences of lead exposure may occur. Further, the extent to which lead exposure differentially influences these three neurobiological mediators, which may depend upon other extant risk or protective factors, may ultimately underlie the variability observed in the behavioral profile of lead exposure across studies.

Another arc in the circle that could influence the ultimate behavioral consequences of lead exposure is the known associations of attention deficit with delinquency and conduct disorder types of behaviors [203–206], as well as with IQ/cognitive function. As pointed out previously in the chapter, in the studies of Nigg et al., attention-related behavioral deficits were found to mediate the association of lead level with IQ [88,89]. Attention deficits could also serve as a behavioral mechanism of both the cognitive and delinquency consequences of developmental lead exposure. In fact, Nicolescu et al. [87] observed that effect sizes for attention deficit-related behaviors were considerably larger than those associated with IQ reduction. And, as a final linking of all lead-related outcomes, delinquent and conduct disorder behaviors are thought to involve impaired cognitive control over emotional behavior [207–209], linking IQ reductions to delinquency/aggression, with delinquency/aggression likely leading to even further reductions in academic achievement and cognition, a reciprocal interaction.

Figure 7.1 also raises questions as to where intervention strategies might be most efficacious, considering that primary prevention has been precluded as an option and chelation therapy should only be considered at blood lead levels associated with poisoning. The significant consequences engendered by prenatal exposure would certainly suggest screening pregnant women or women of childbearing age planning pregnancies as a type of prevention strategy. However, the toxicokinetics of lead, specifically its release from bone back into the blood during periods of calcium mobilization and lactation, complicates such an approach. For children, early tests for potential auditory loss and for attention-related behavioral deficits might be the most feasible.

References

[1] Needleman H. Lead poisoning. Annu Rev Med 2004;55:209–22.

[2] Nriagu JO. Lead and lead poisoning in antiquity. New York: John Wiley & Sons; 1983.

[3] Rosner D, Markowitz G. A 'gift of God'? the public health controversy over leaded gasoline during the 1920s. Am J Public Health 1985;75(4):344–52.

[4] Markowitz G, Rosner D. "Cater to the children": the role of the lead industry in a public health tragedy, 1900–1955. Am J Public Health 2000;90(1):36–46.

[5] Alexander FW, Delves HT. Deaths from acute lead poisoning. Arch Dis Child 1972;47(253):446–8.

[6] Barnako D. Childhood lead poisoning. JAMA 1972;220(13):1737–8.

[7] Craven EM, Metzger L, Walsh ES, Handler J. Lead poisoning in Wilmington, Delaware, 1974–1984. Del Med J 1989;61(3):135–6.

[8] Rodriguez JG, Sattin RW. Epidemiology of childhood poisonings leading to hospitalization in the United States, 1979–1983. Am J Prev Med 1987;3(3):164–70.

[9] Byers R, Lord E. Late effects of lead poisoning on mental development. Am J Dis Child 1943;66:471–94.

[10] Lippmann M. 1989 Alice Hamilton lecture. Lead and human health: background and recent findings. Environ Res 1990;51(1):1–24.

[11] Rabin R. Warnings unheeded: a history of child lead poisoning. Am J Public Health 1989;79(12):1668–74.

[12] Fewtrell LJ, Pruss-Ustun A, Landrigan P, Ayuso-Mateos JL. Estimating the global burden of disease of mild mental retardation and cardiovascular diseases from environmental lead exposure. Environ Res 2004;94(2):120–33.

[13] Lau WK, Liang P, Man YB, Chung SS, Wong MH. Human health risk assessment based on trace metals in suspended air particulates, surface dust, and floor dust from e-waste recycling workshops in Hong Kong, China. Environ Sci Pollut Res Int 2014;21(5):3813–25.

[14] Yang H, Huo X, Yekeen TA, Zheng Q, Zheng M, Xu X. Effects of lead and cadmium exposure from electronic waste on child physical growth. Environ Sci Pollut Res Int 2013;20(7):4441–7.

[15] Wang X, Miller G, Ding G, Lou X, Cai D, Chen Z, et al. Health risk assessment of lead for children in tinfoil manufacturing and e-waste recycling areas of Zhejiang Province, China. Sci Total Environ 2012;426:106–12.

[16] Liu J, Xu X, Wu K, Piao Z, Huang J, Guo Y, et al. Association between lead exposure from electronic waste recycling and child temperament alterations. Neurotoxicology 2011;32(4):458–64.

[17] Li Y, Xu X, Wu K, Chen G, Liu J, Chen S, et al. Monitoring of lead load and its effect on neonatal behavioral neurological assessment scores in Guiyu, an electronic waste recycling town in China. J Environ Monit 2008;10(10):1233–8.

[18] Greig J, Thurtle N, Cooney L, Ariti C, Ahmed AO, Ashagre T, et al. Association of blood lead level with neurological features in 972 children affected by an acute severe lead poisoning outbreak in Zamfara state, northern Nigeria. PLoS One 2014;9(4):e93716.

[19] Bashir M, Umar-Tsafe N, Getso K, Kaita IM, Nasidi A, Sani-Gwarzo N, et al. Assessment of blood lead levels among children aged </=5 years – Zamfara state, Nigeria, June–July 2012. Morb Mortal Wkly Rep 2014;63(15):325–7.

[20] Bartrem C, Tirima S, von Lindern I, von Braun M, Worrell MC, Mohammad Anka S, et al. Unknown risk: co-exposure to lead and other heavy metals among children living in small-scale mining communities in Zamfara state, Nigeria. Int J Environ Health Res 2013;24(4):304–19.

[21] Lo YC, Dooyema CA, Neri A, Durant J, Jefferies T, Medina-Marino A, et al. Childhood lead poisoning associated with gold ore processing: a village-level investigation-Zamfara state, Nigeria, October–November 2010. Environ Health Perspect 2012;120(10):1450–5.

[22] Safety IPoC. Inorganic lead. Environmental health criteria 165. Geneva: World Health Organization; 1995. p. 99–118.

[23] Needleman HL, Gunnoe C, Leviton A, Reed R, Peresie H, Maher C, et al. Deficits in psychologic and classroom performance of children with elevated dentine lead levels. N Engl J Med 1979;300:689–95.

[24] Baghurst PA, McMichael AJ, Wigg NR, Vimpani GV, Robertson EF, Roberts RJ, et al. Environmental exposure to lead and children's intelligence at the age of seven years. The Port Pirie Cohort Study. N Engl J Med 1992;327(18):1279–84.

[25] Bellinger DC, Needleman HL, Leviton A, Waternaux C, Rabinowitz MB, Nichols ML. Early sensory-motor development and prenatal exposure to lead. Neurobehav Toxicol Teratol 1984;6(5):387–402.

[26] Cooney GH, Bell A, McBride W, Carter C. Low-level exposures to lead: the Sydney lead study. Dev Med Child Neurol 1989;31(5):640–9.

[27] Dietrich KN, Krafft KM, Bornschein RL, Hammond PB, Berger O, Succop PA, et al. Low-level fetal lead exposure effect on neurobehavioral development in early infancy. Pediatrics 1987;80(5):721–30.

[28] Ernhart CB, Morrow-Tlucak M, Wolf AW, Super D, Drotar D. Low level lead exposure in the prenatal and early preschool periods: intelligence prior to school entry. Neurotoxicol Teratol 1989;11(2):161–70.

[29] Needleman HL, Schell A, Bellinger D, Leviton A, Allred EN. The long-term effects of exposure to low doses of lead in childhood. An 11-year follow-up report. N Engl J Med 1990;322(2):83–8.

[30] Rothenberg SJ, Poblano A, Garza-Morales S. Prenatal and perinatal low level lead exposure alters brainstem auditory evoked responses in infants. Neurotoxicology 1994;15(3):695–9.

[31] Wasserman G, Graziano JH, Factor-Litvak P, Popovac D, Morina N, Musabegovic A, et al. Independent effects of lead exposure and iron deficiency anemia on developmental outcome at age 2 years. J Pediatr 1992; 121(5 Pt 1):695–703.

[32] Canfield RL, Henderson Jr CR, Cory-Slechta DA, Cox C, Jusko TA, Lanphear BP. Intellectual impairment in children with blood lead concentrations below 10 microg per deciliter. N Engl J Med 2003;348(16):1517–26.

[33] Surkan PJ, Zhang A, Trachtenberg F, Daniel DB, McKinlay S, Bellinger DC. Neuropsychological function in children with blood lead levels <10 microg/dL. Neurotoxicology 2007;28(6):1170–7.

[34] Lanphear BP, Hornung R, Khoury J, Yolton K, Baghurst P, Bellinger DC, et al. Low-level environmental lead exposure and children's intellectual function: an international pooled analysis. Environ Health Perspect 2005;113(7):894–9.

[35] Huang PC, Su PH, Chen HY, Huang HB, Tsai JL, Huang HI, et al. Childhood blood lead levels and intellectual development after ban of leaded gasoline in Taiwan: a 9-year prospective study. Environ Int 2012;40:88–96.

[36] Lucchini RG, Zoni S, Guazzetti S, Bontempi E, Micheletti S, Broberg K, et al. Inverse association of intellectual function with very low blood lead but not with manganese exposure in Italian adolescents. Environ Res 2012;118:65–71.

[37] Budtz-Jorgensen E, Bellinger D, Lanphear B, Grandjean P. An international pooled analysis for obtaining a benchmark dose for environmental lead exposure in children. Risk Anal 2013;33(3):450–61.

[38] Schwartz J. Low-level lead exposure and children's IQ: a meta-analysis and search for a threshold. Environ Res 1994;65:42–55.

[39] Bellinger DC, Leviton A, Waternaux C, Needleman H, Rabinowitz M. Low level lead exposure, social class, and infant development. Neurotoxicol Teratol 1988;10:497–503.

[40] Wasserman GA, Liu X, Lolacono NJ, Factor-Litvak P, Kline JK, Popovac D, et al. Lead exposure and intelligence in 7-year-old children: the Yugoslavia Prospective Study. Environ Health Perspect 1997;105(9):956–62.

[41] Hansen ON, Trillingsgaard A, Beese I, Lyngbye T, Grandjean P. A neuropsychological study of children with elevated dentine lead level: assessment of the effect of lead in different socio-economic groups. Neurotoxicol Teratol 1989;11:205–13.

[42] Li Y, Dai Q, Jackson JC, Zhang J. Overweight is associated with decreased cognitive functioning among school-age children and adolescents. Obesity (Silver Spring) 2008;16(8):1809–15.

[43] Ohmann S, Popow C, Rami B, Konig M, Blaas S, Fliri C, et al. Cognitive functions and glycemic control in children and adolescents with type 1 diabetes. Psychol Med 2010;40(1):95–103.

[44] Liu X, Dietrich KN, Radcliffe J, Ragan NB, Rhoads GG, Rogan WJ. Do children with falling blood lead levels have improved cognition? Pediatrics 2002;110(4):787–91.

[45] Cory-Slechta DA, Pokora MJ. Behavioral manifestations of prolonged lead exposure initiated at different stages of the life cycle: I. Schedule-controlled responding. Neurotoxicology 1991;12(4):745–60.

[46] Cory-Slechta DA, Pokora MJ, Widzowski DV. Behavioral manifestations of prolonged lead exposure initiated at different stages of the life cycle: II. Delayed spatial alternation. Neurotoxicology 1991;12:761–76.

[47] Cory-Slechta DA, Flaugher CL, Evans SB, Pokora MJ, Greenamyre JT. Susceptibility of adult rats to lead-induced changes in NMDA receptor complex function. Neurotoxicol Teratol 1997;19:517–30.

[48] Lasley SM, Green MC, Gilbert ME. Influence of exposure period on in vivo hippocampal glutamate and GABA release in rats chronically exposed to lead. Neurotoxicology 1999;20(4):619–30.

[49] Bellinger DC, Leviton A, Waternaux C, Needleman HL, Rabinowitz M. Longitudinal analyses of prenatal and postnatal lead exposure and early cognitive development. N Engl J Med 1987;316:1037–43.

[50] Hu H, Tellez-Rojo MM, Bellinger D, Smith D, Ettinger AS, Lamadrid-Figueroa H, et al. Fetal lead exposure at each stage of pregnancy as a predictor of infant mental development. Environ Health Perspect 2006;114(11):1730–5.

[51] Bellinger DC, Sloman J, Leviton A, Rabinowitz M, Needleman HL, Waternaux C. Low-level lead exposure and children's cognitive function in the preschool years. Pediatrics 1991;87:219–27.

[52] Bellinger D, Leviton A, Allred E, Rabinowitz M. Pre- and postnatal lead exposure and behavior problems in school-aged children. Environ Res 1994;66(1):12–30.

[53] Leviton A, Bellinger D, Allred EN, Rabinowitz M, Needleman H, Schoenbaum S. Pre- and postnatal low-level lead exposure and children's dysfunction in school. Environ Res 1993;60(1):30–43.

[54] Chen A, Dietrich KN, Ware JH, Radcliffe J, Rogan WJ. IQ and blood lead from 2 to 7 years of age: are the effects in older children the residual of high blood lead concentrations in 2-year-olds? Environ Health Perspect 2005;113(5):597–601.

[55] Chen A, Cai B, Dietrich KN, Radcliffe J, Rogan WJ. Lead exposure, IQ, and behavior in urban 5- to 7-year-olds: does lead affect behavior only by lowering IQ? Pediatrics 2007;119(3):e650–8.

[56] Davis JM, Svendsgaard DJ. U-shaped dose-response curve-shaped dose-response curves: their occurrence and implications for risk assessment. J Toxicol Environ Health 1990;30:71–83.

[57] Alfano DP, Petit TL. Postnatal lead exposure and the cholinergic system. Physiol Behav 1985;34:449–55.

[58] Bjorklund H, Olson L, Seiger A, Hoffer B. Chronic lead and brain development: intraocular brain grafts as a method to reveal regional and temporal effects in the central nervous system. Environ Res 1980;22(1):224–36.

[59] Cory-Slechta DA, Weiss B, Cox C. Delayed behavioral toxicity of lead with increasing exposure concentration. Toxicol Appl Pharmacol 1983;71:342–52.

[60] Gross-Selbeck E, Gross-Selbeck M. A biphasic effect of lead on operant behaviour of rats induced by different exposure. Toxicol Lett Spec Issue 1980;1:128.

[61] Geist GR, Balko SW, Morgan ME, Angiak R. Behavioral effects following rehabilitation from postnatal exposure to lead acetate. Percept Mot Skills 1985;60:527–36.

[62] Grant LD, Kimmel CA, West GL, Martinez-Vargas CM, Howard JL. Chronic low-level lead toxicity in the rat. II. Effects of postnatal physical and behavioral development. Toxicol Appl Pharmacol 1980;56:42–58.

[63] Lilienthal H, Winneke G, Brockhaus A, Molik B, Schlipkoter H-W. Learning-set formation in rhesus monkeys pre- and postnatally exposed to lead. In: International conference: heavy metals in the environment. Edinburgh, United Kingdom: CEP Consultants; 1983. p. 901–3.

[64] Pocock SJ, Shaper AG, Ashby D, Delves T, Whitehead TP. Blood lead concentration, blood pressure, and renal function. Br Med J Clin Res Ed 1984;289(6449):872–4.

[65] Schwartz J, Angle C, Pitcher H. Relationship between childhood blood lead levels and stature. Pediatrics 1986;77(3):281–8.

[66] Rothenberg SJ, Poblano A, Schnaas L. Brainstem auditory evoked response at five years and prenatal and postnatal blood lead. Neurotoxicol Teratol 2000;22(4):503–10.

[67] Cheng Y, Willett WC, Schwartz J, Sparrow D, Weiss S, Hu H. Relation of nutrition to bone lead and blood lead levels in middle-aged to elderly men. The Normative Aging Study. Am J Epidemiol 1998;147(12):1162–74.

[68] Victery W. Evidence for effects of chronic lead exposure on blood pressure in experimental animals: an overview. Environ Health Perspect 1988;78:71–6.

[69] Rossi-George A, Virgolini MB, Weston D, Cory-Slechta DA. Alterations in glucocorticoid negative feedback following maternal Pb, prenatal stress and the combination: a potential biological unifying mechanism for their corresponding disease profiles. Toxicol Appl Pharmacol 2009;234(1):117–27.

[70] Kern M, Audesirk G. Stimulatory and inhibitory effects of inorganic lead on calcineurin. Toxicology 2000;150(1–3):171–8.

[71] Kern M, Wisniewski M, Cabell L, Audesirk G. Inorganic lead and calcium interact positively in activation of calmodulin. Neurotoxicology 2000;21(3):353–63.

[72] Bellinger DC. Very low lead exposures and children's neurodevelopment. Curr Opin Pediatr 2008;20(2):172–7.

[73] Cory-Slechta DA. Behavioral measures of neurotoxicity. Neurotoxicology 1989;10:271–95.

[74] Kelly TH, Fischman MW, Foltin RW, Brady JV. Response patterns and cardiovascular effects during response sequence acquisition by humans. J Exp Anal Behav 1991;56(3):557–74.

[75] Cohn J, Cox C, Cory-Slechta DA. The effects of lead exposure on learning in a multiple repeated acquisition and performance schedule. Neurotoxicology 1993;14:329–46.

[76] Brooks AI, Cory-Slechta DA, Murg SL, Federoff HJ. Repeated acquisition and performance chamber for mice: a paradigm for assessment of spatial learning and memory. Neurobiol Learn Mem 2000;74(3):241–58.

[77] Brooks AI, Cory-Slechta DA, Bowers WJ, Murg SL, Federoff HJ. Enhanced learning in mice parallels vector-mediated nerve growth factor expression in hippocampus. Hum Gene Ther 2000;11(17):2341–52.

[78] Brooks AI, Stein CS, Hughes SM, Heth J, McCray Jr PM, Sauter SL, et al. Functional correction of established central nervous system deficits in an animal model of lysosomal storage disease with feline immunodeficiency virus-based vectors. Proc Natl Acad Sci USA 2002;99(9):6216–21.

[79] Canfield RL, Gendle MH, Cory-Slechta DA. Impaired neuropsychological functioning in lead-exposed children. Dev Neuropsychol 2004;26(1):513–40.

[80] Stiles KM, Bellinger DC. Neuropsychological correlates of low-level lead exposure in school-age children: a prospective study. Neurotoxicol Teratol 1993;15:27–35.

[81] Mansouri MT, Naghizadeh B, Lopez-Larrubia P, Cauli O. Behavioral deficits induced by lead exposure are accompanied by serotonergic and cholinergic alterations in the prefrontal cortex. Neurochem Int 2013;62(3):232–9.

[82] Mansouri MT, Naghizadeh B, Lopez-Larrubia P, Cauli O. Gender-dependent behavioural impairment and brain metabolites in young adult rats after short term exposure to lead acetate. Toxicol Lett 2012;210(1):15–23.

[83] Cory-Slechta DA, Stern S, Weston D, Allen JL, Liu S. Enhanced learning deficits in female rats following lifetime Pb exposure combined with prenatal stress. Toxicol Sci 2010;117(2):427–38.

[84] Bellinger DC, Hu H, Titlebaum L, Needleman HL. Attentional correlates of dentin and bone lead levels in adolescents. Arch Environ Health 1994;49(2):98–105.

[85] Boucher O, Jacobson SW, Plusquellec P, Dewailly E, Ayotte P, Forget-Dubois N, et al. Prenatal methylmercury, postnatal lead exposure, and evidence of attention deficit/hyperactivity disorder among Inuit children in Arctic Quebec. Environ Health Perspect 2012;120(10):1456–61.

[86] Eubig PA, Aguiar A, Schantz SL. Lead and PCBs as risk factors for attention deficit/hyperactivity disorder. Environ Health Perspect 2010;118(12):1654–67.

[87] Nicolescu R, Petcu C, Cordeanu A, Fabritius K, Schlumpf M, Krebs R, et al. Environmental exposure to lead, but not other neurotoxic metals, relates to core elements of ADHD in Romanian children: performance and questionnaire data. Environ Res 2010;110(5):476–83.

[88] Nigg JT, Knottnerus GM, Martel MM, Nikolas M, Cavanagh K, Karmaus W, et al. Low blood lead levels associated with clinically diagnosed attention-deficit/hyperactivity disorder and mediated by weak cognitive control. Biol Psychiatry 2008;63(3):325–31.

[89] Nigg JT, Nikolas M, Mark Knottnerus G, Cavanagh K, Friderici K. Confirmation and extension of association of blood lead with attention-deficit/hyperactivity disorder (ADHD) and ADHD symptom domains at population-typical exposure levels. J Child Psychol Psychiatry 2010;51(1):58–65.

[90] Biederman J, Newcorn J, Sprich S. Comorbidity of attention deficit hyperactivity disorder with conduct, depressive, anxiety, and other disorders. Am J Psychiatry 1991;148(5):564–77.

[91] Cantwell DP, Baker L. Association between attention deficit-hyperactivity disorder and learning disorders. J Learn Disabil 1991;24(2):88–95.

[92] Kim S, Arora M, Fernandez C, Landero J, Caruso J, Chen A. Lead, mercury, and cadmium exposure and attention deficit hyperactivity disorder in children. Environ Res 2013;126:105–10.

[93] Kim Y, Cho SC, Kim BN, Hong YC, Shin MS, Yoo HJ, et al. Association between blood lead levels (<5 mug/dL) and inattention-hyperactivity and neurocognitive profiles in school-aged Korean children. Sci Total Environ 2010;408(23):5737–43.

[94] Goodlad JK, Marcus DK, Fulton JJ. Lead and attention-deficit/hyperactivity disorder (ADHD) symptoms: a meta-analysis. Clin Psychol Rev 2013;33(3):417–25.

[95] Goto Y, Otani S, Grace AA. The Yin and Yang of dopamine release: a new perspective. Neuropharmacology 2007;53(5):583–7.

[96] Monte-Silva K, Kuo MF, Thirugnanasambandam N, Liebetanz D, Paulus W, Nitsche MA. Dose-dependent inverted U-shaped effect of dopamine (D2-like) receptor activation on focal and nonfocal plasticity in humans. J Neurosci 2009;29(19):6124–31.

[97] Cory-Slechta DA, O'Mara DJ, Brockel BJ. Nucleus accumbens dopaminergic mediation of fixed interval schedule-controlled behavior and its modulation by low-level lead exposure. J Pharmacol Exp Ther 1998;286(3):794–805.

[98] Brockel BJ, Cory-Slechta DA. The effects of postweaning low-level lead exposure on sustained attention: a study of target densities, stimulus presentation rate and stimulus predictability. Neurotoxicology 1999;20:921–34.

[99] Morgan RE, Garavan H, Smith EG, Driscoll LL, Levitsky DA, Strupp BJ. Early lead exposure produces lasting changes in sustained attention, response initiation, and reactivity to errors. Neurotoxicol Teratol 2001;23(6):519–31.

[100] Darcheville JC, Riviere V, Wearden JH. Fixed-interval performance and self-control in children. J Exp Anal Behav 1992;57:187–99.

[101] Darcheville JC, Riviere V, Wearden JH. Fixed-interval performance and self-control in infants. J Exp Anal Behav 1993;60:239–54.

[102] Sagvolden T, Aase H, Zeiner P, Berger D. Altered reinforcement mechanisms in attention-deficit/hyperactivity disorder. Behav Brain Res 1998;94:61–71.

[103] Cory-Slechta DA. Comparative neurobehavioral toxicology of heavy metals. In: Suzuki T, editor. Toxicology of metals. New York: CRC Press; 1996. pp. 537–60.

[104] Brockel BJ, Cory-Slechta DA. Lead, attention, and impulsive behavior: changes in a fixed-ratio waiting-for-reward paradigm. Pharmacol Biochem Behav 1998;60:545–52.

[105] Brockel BJ, Cory-Slechta DA. Lead-induced decrements in waiting behavior: involvement of D2-like dopamine receptors. Pharmacol Biochem Behav 1999;63:423–34.

[106] Cory-Slechta DA, Weiss B. Alterations in schedule-controlled behavior of rodents correlated with prolonged lead exposure. In: Balster RL, editor. Behavioral pharmacology: the current status. New York: Alan R. Liss; 1985. pp. 487–501.

[107] Stewart PW, Sargent DM, Reihman J, Gump BB, Lonky E, Darvill T, et al. Response inhibition during Differential Reinforcement of Low Rates (DRL) schedules may be sensitive to low-level polychlorinated biphenyl, methylmercury, and lead exposure in children. Environ Health Perspect 2006;114(12):1923–9.

[108] Rice DC. Chronic low-lead exposure from birth produces deficits in discrimination reversal in monkeys. Toxicol Appl Pharmacol 1985;75:201–10.

[109] McClure FD, Gordon M. Performance of disturbed hyperactive and nonhyperactive children on an objective measure of hyperactivity. J Abnorm Child Psychol 1984;12:561–71.

[110] Schwartz J, Otto DA. Blood lead, hearing thresholds, and neurobehavioral development in children and youth. Arch Environ Health 1987;42:153–60.

[111] Schwartz J, Otto DA. Lead and minor hearing impairment. Arch Environ Health 1991;46:300–5.

[112] Hayes D. Improved health and development of children who are deaf and hard of hearing following early intervention. Ann Acad Med Singapore 2008;37(Suppl. 12):10–3.

[113] Verhaert N, Willems M, Van Kerschaver E, Desloovere C. Impact of early hearing screening and treatment on language development and education level: evaluation of 6 years of universal newborn hearing screening (ALGO) in Flanders, Belgium. Int J Pediatr Otorhinolaryngol 2008;72(5):599–608.

[114] Otto DA, Fox DA. Auditory and visual dysfunction following lead exposure. Neurotoxicology 1993;14:191–210.

[115] Otto DA, Robinson G, Bauman S, Schroeder S, Mushak P, Kleinbaum D, et al. 5 year follow-up study of children with low-to-moderate lead absorption: electrophysiological evaluation. Environ Res 1985;38:168–86.

[116] Zou C, Zhao Z, Tang L, Chen Z, Du L. The effect of lead on brainstem auditory evoked potentials in children. Chin Med J Engl 2003;116(4):565–8.

[117] Osman K, Pawlas K, Schutz A, Gazdzik M, Sokal JA, Vahter M. Lead exposure and hearing effects in children in Katowice, Poland. Environ Res 1999;80(1):1–8.

[118] Robinson G, Baumann S, Kleinbaum D, Barton C, Schroeder S, Mushak P, et al. Effects of low-to-moderate lead exposure on brainstem auditory evoked potentials in children. Copenhagen: World Health Organization; 1985.

[119] Buchanan LH, Counter SA, Ortega F. Environmental lead exposure and otoacoustic emissions in Andean children. J Toxicol Environ Health A 2011;74(19):1280–93.

[120] Counter SA, Buchanan LH, Ortega F, van der Velde J, Borg E. Assessment of auditory brainstem function in lead-exposed children using stapedius muscle reflexes. J Neurol Sci 2011;306(1–2):29–37.

[121] Counter SA, Buchanan LH, Ortega F. Neurocognitive screening of lead-exposed andean adolescents and young adults. J Toxicol Environ Health A 2009;72(10):625–32.

[122] Rice DC. Effects of lifetime lead exposure in monkeys on detection of pure tones. Fundam Appl Toxicol 1997;36(2):112–8.

[123] Lasky RE, Maier MM, Snodgrass EB, Hecox KE, Laughlin NK. The effects of lead on otoacoustic emissions and auditory evoked potentials in monkeys. Neurotoxicol Teratol 1995;17(6):633–44.

[124] Lilienthal H, Winneke G. Lead effects on the brain stem auditory evoked potential in monkeys during and after the treatment phase. Neurotoxicol Teratol 1996;18(1):17–32.

[125] Liu X, Zheng G, Wu Y, Shen X, Jing J, Yu T, et al. Lead exposure results in hearing loss and disruption of the cochlear blood-labyrinth barrier and the protective role of iron supplement. Neurotoxicology 2013;39:173–81.

[126] Jones LG, Prins J, Park S, Walton JP, Luebke AE, Lurie DI. Lead exposure during development results in increased neurofilament phosphorylation, neuritic beading, and temporal processing deficits within the murine auditory brainstem. J Comp Neurol 2008;506(6):1003–17.

[127] Dietrich KN, Succop PA, Berger OG, Keith RW. Lead exposure and the central auditory processing abilities and cognitive development of urban children: the Cincinnati lead study cohort at 5 years of age. Neurotoxicol Teratol 1992;14:51–6.

[128] Mazumdar M, Bellinger DC, Gregas M, Abanilla K, Bacic J, Needleman HL. Low-level environmental lead exposure in childhood and adult intellectual function: a follow-up study. Environ Health 2011;10:24.

[129] Bellinger DC, Stiles KM, Needleman HL. Low-level lead exposure, intelligence and academic achievement: a long-term follow-up study. Pediatrics 1992;90:855–61.

[130] Zhang N, Baker HW, Tufts M, Raymond RE, Salihu H, Elliott MR. Early childhood lead exposure and academic achievement: evidence from Detroit public schools, 2008–2010. Am J Public Health 2013;103(3):e72–7.

[131] Needleman HL, Riess JA, Tobin MJ, Biesecker GE, Greenhouse JB. Bone lead levels and delinquent behavior. J Am Med Assoc 1996;275(5):363–9.

[132] Needleman HL, McFarland C, Ness RB, Fienberg SE, Tobin MJ. Bone lead levels in adjudicated delinquents. A case control study. Neurotoxicol Teratol 2002;24(6):711–7.

[133] Dietrich KN, Ris MD, Succop PA, Berger OG, Bornschein RL. Early exposure to lead and juvenile delinquency. Neurotoxicol Teratol 2001;23(6):511–8.

[134] Fergusson DM, Boden JM, Horwood LJ. Dentine lead levels in childhood and criminal behaviour in late adolescence and early adulthood. J Epidemiol Community Health 2008;62(12):1045–50.

[135] Nevin R. How lead exposure relates to temporal changes in IQ, violent crime, and unwed pregnancy. Environ Res 2000;83(1):1–22.

[136] Marcus DK, Fulton JJ, Clarke EJ. Lead and conduct problems: a meta-analysis. J Clin Child Adolesc Psychol 2010;39(2):234–41.

[137] Hinshaw SP. Externalizing behavior problems and academic underachievement in childhood and adolescence: causal relationships and underlying mechanisms. Psychol Bull 1992;111(1):127–55.

[138] Fergusson DM, Horwood LJ. Early disruptive behavior, IQ, and later school achievement and delinquent behavior. J Abnorm Child Psychol 1995;23(2):183–99.

[139] White JL, Moffitt TE, Caspi A, Bartusch DJ, Needles DJ, Stouthamer-Loeber M. Measuring impulsivity and examining its relationship to delinquency. J Abnorm Psychol 1994;103(2):192–205.

[140] Lynam DR, Caspi A, Moffitt TE, Wikstrom PO, Loeber R, Novak S. The interaction between impulsivity and neighborhood context on offending: the effects of impulsivity are stronger in poorer neighborhoods. J Abnorm Psychol 2000;109(4):563–74.

[141] Brubaker CJ, Dietrich KN, Lanphear BP, Cecil KM. The influence of age of lead exposure on adult gray matter volume. Neurotoxicology 2010;31(3):259–66.

[142] Cecil KM, Dietrich KN, Altaye M, Egelhoff JC, Lindquist DM, Brubaker CJ, et al. Proton magnetic resonance spectroscopy in adults with childhood lead exposure. Environ Health Perspect 2011;119(3):403–8.

[143] Trope I, Lopez-Villegas D, Cecil KM, Lenkinski RE. Exposure to lead appears to selectively alter metabolism of cortical gray matter. Pediatrics 2001;107(6):1437–42.

[144] McFarlane AC, Searle AK, Van Hooff M, Baghurst PA, Sawyer MG, Galletly C, et al. Prospective associations between childhood low-level lead exposure and adult mental health problems: the Port Pirie Cohort Study. Neurotoxicology 2013;39:11–7.

[145] Searle AK, Baghurst PA, van Hooff M, Sawyer MG, Sim MR, Galletly C, et al. Tracing the long-term legacy of childhood lead exposure: a review of three decades of the Port Pirie Cohort Study. Neurotoxicology 2014;43:46–56.

[146] Cory-Slechta DA, McCoy L, Richfield EK. Time course and regional basis of Pb-induced changes in MK-801 binding: reversal by chronic treatment with the dopamine agonist apomorphine but not the D1 agonist SKF-82958. J Neurochem 1997;68:2012–23.

[147] Guilarte TR, Opler M, Pletnikov M. Is lead exposure in early life an environmental risk factor for Schizophrenia? Neurobiological connections and testable hypotheses. Neurotoxicology 2012;33(3):560–74.

[148] Opler MG, Brown AS, Graziano JH, Desai M, Zheng W, Schaefer C, et al. Prenatal lead exposure, delta-aminolevulinic acid, and schizophrenia. Environ Health Perspect 2004;112(5):548–52.

[149] Opler MG, Buka SL, Groeger J, McKeague I, Wei C, Factor-Litvak P, et al. Prenatal exposure to lead, delta-aminolevulinic acid, and schizophrenia: further evidence. Environ Health Perspect 2008;116(11):1586–90.

[150] Bellinger DC. Interpreting the literature on lead and child development: the neglected role of the "experimental system". Neurotoxicol Teratol 1995;17(3):201–12.

[151] Cory-Slechta DA. Studying toxicants as single chemicals: does this strategy adequately identify neurotoxic risk? Neurotoxicology 2005;26(4):491–510.

[152] Cory-Slechta DA, Virgolini MB, Thiruchelvam M, Weston DD, Bauter MR. Maternal stress modulates the effects of developmental lead exposure. Environ Health Perspect 2004;112(6):717–30.

[153] Cory-Slechta DA, Virgolini MB, Rossi-George A, Thiruchelvam M, Lisek R, Weston D. Lifetime consequences of combined maternal lead and stress. Basic Clin Pharmacol Toxicol 2008;102(2):218–27.

[154] Cory-Slechta DA, Virgolini MB, Liu S, Weston D. Enhanced stimulus sequence-dependent repeated learning in male offspring after prenatal stress alone or in conjunction with lead exposure. Neurotoxicology 2012;33(5):1188–202.

[155] Cory-Slechta DA, Merchant-Borna K, Allen J, Liu S, Weston D, Conrad K. Variations in the nature of behavioral experience can differentially alter the consequences of developmental exposures to lead, prenatal stress and the combination. Toxicol Sci 2012;131(1):194–205.

[156] Cory-Slechta DA, Weston D, Liu S, Allen JL. Brain hemispheric differences in the neurochemical effects of lead, prenatal stress, and the combination and their amelioration by behavioral experience. Toxicol Sci 2013;132(2):419–30.

[157] Rossi-George A, Virgolini MB, Weston D, Thiruchelvam M, Cory-Slechta DA. Interactions of lifetime lead exposure and stress: behavioral, neurochemical and HPA axis effects. Neurotoxicology 2011;32(1):83–99.

[158] Virgolini MB, Chen K, Weston DD, Bauter MR, Cory-Slechta DA. Interactions of chronic lead exposure and intermittent stress: consequences for brain catecholamine systems and associated behaviors and HPA axis function. Toxicol Sci 2005;87(2):469–82.

[159] Virgolini MB, Bauter MR, Weston DD, Cory-Slechta DA. Permanent alterations in stress responsivity in female offspring subjected to combined maternal lead exposure and/or stress. Neurotoxicology 2006;27(1):11–21.

[160] Virgolini MB, Rossi-George A, Lisek R, Weston DD, Thiruchelvam M, Cory-Slechta DA. CNS effects of developmental Pb exposure are enhanced by combined maternal and offspring stress. Neurotoxicology 2008;29(5):812–27.

[161] Gump BB, Stewart P, Reihman J, Lonky E, Darvill T, Parsons PJ, et al. Low-level prenatal and postnatal blood lead exposure and adrenocortical responses to acute stress in children. Environ Health Perspect 2008;116(2):249–55.

[162] Gump BB, Reihman J, Stewart P, Lonky E, Darvill T, Matthews KA. Blood lead (Pb) levels: a potential environmental mechanism explaining the relation between socioeconomic status and cardiovascular reactivity in children. Health Psychol 2007;26(3):296–304.

[163] Gump BB, Reihman J, Stewart P, Lonky E, Granger DA, Matthews KA. Blood lead (Pb) levels: further evidence for an environmental mechanism explaining the association between socioeconomic status and psychophysiological dysregulation in children. Health Psychol 2009;28(5):614–20.

[164] Kordas K, Ardoino G, Ciccariello D, Manay N, Ettinger AS, Cook CA, et al. Association of maternal and child blood lead and hemoglobin levels with maternal perceptions of parenting their young children. Neurotoxicology 2011;32(6):693–701.

[165] Hubbs-Tait L, Mulugeta A, Bogale A, Kennedy TS, Baker ER, Stoecker BJ. Main and interaction effects of iron, zinc, lead, and parenting on children's cognitive outcomes. Dev Neuropsychol 2009;34(2):175–95.

[166] Hart H, Rubia K. Neuroimaging of child abuse: a critical review. Front Hum Neurosci 2012;6:52.

[167] Burrus C. Developmental trajectories of abuse–an hypothesis for the effects of early childhood maltreatment on dorsolateral prefrontal cortical development. Med Hypotheses 2013;81(5):826–9.

[168] Cory-Slechta DA, Merchant-Borna K, Allen JL, Liu S, Weston D, Conrad K. Variations in the nature of behavioral experience can differentially alter the consequences of developmental exposures to lead, prenatal stress, and the combination. Toxicol Sci 2013;131(1):194–205.

[169] Ris MD, Dietrich KN, Succop PA, Berger OG, Bornschein RL. Early exposure to lead and neuropsychological outcome in adolescence. J Int Neuropsychol Soc 2004;10(2):261–70.

[170] Liston C, Malter Cohen M, Teslovich T, Levenson D, Casey BJ. Atypical prefrontal connectivity in attention-deficit/hyperactivity disorder: pathway to disease or pathological end point? Biol Psychiatry 2011;69(12):1168–77.

[171] Viggiano D, Vallone D, Ruocco LA, Sadile AG. Behavioural, pharmacological, morpho-functional molecular studies reveal a hyperfunctioning mesocortical dopamine system in an animal model of attention deficit and hyperactivity disorder. Neurosci Biobehav Rev 2003;27(7):683–9.

[172] Clark KL, Noudoost B. The role of prefrontal catecholamines in attention and working memory. Front Neural Circuits 2014;8:33.

[173] Ashitani M, Ueno C, Doi T, Kinoshita T, Tomoda K. Clinical features of functional hearing loss with inattention problem in Japanese children. Int J Pediatr Otorhinolaryngol 2011;75(11):1431–5.

[174] Dye MW, Bavelier D. Attentional enhancements and deficits in deaf populations: an integrative review. Restor Neurol Neurosci 2010;28(2):181–92.

[175] Parasnis I, Samar VJ, Berent GP. Deaf adults without attention deficit hyperactivity disorder display reduced perceptual sensitivity and elevated impulsivity on the Test of Variables of Attention (T.O.V.A.). J Speech Lang Hear Res 2003;46(5):1166–83.

[176] Chermak GD, Somers EK, Seikel JA. Behavioral signs of central auditory processing disorder and attention deficit hyperactivity disorder. J Am Acad Audiol 1998;9(1):78–84. quiz 5.

[177] Kelly D, Forney J, Parker-Fisher S, Jones M. The challenge of attention deficit disorder in children who are deaf or hard of hearing. Am Ann Deaf 1993;138(4):343–8.

[178] Isaksson J, Nilsson KW, Lindblad F. Early psychosocial adversity and cortisol levels in children with attention-deficit/hyperactivity disorder. Eur Child Adolesc Psychiatry 2013;22(7):425–32.

[179] Isaksson J, Nilsson KW, Nyberg F, Hogmark A, Lindblad F. Cortisol levels in children with attention-deficit/hyperactivity disorder. J Psychiatr Res 2012;46(11):1398–405.

[180] Scassellati C, Bonvicini C, Faraone SV, Gennarelli M. Biomarkers and attention-deficit/hyperactivity disorder: a systematic review and meta-analyses. J Am Acad Child Adolesc Psychiatry 2012;51(10):1003–19 e20.

[181] Palma SM, Fernandes DR, Muszkat M, Calil HM. The response to stress in Brazilian children and adolescents with attention deficit hyperactivity disorder. Psychiatry Res 2012;198(3):477–81.

[182] Ma L, Chen YH, Chen H, Liu YY, Wang YX. The function of hypothalamus-pituitary-adrenal axis in children with ADHD. Brain Res 2011;1368:159–62.

[183] Maldonado EF, Trianes MV, Cortes A, Moreno E, Escobar M. Salivary cortisol response to a psychosocial stressor on children diagnosed with attention-deficit/hyperactivity disorder: differences between diagnostic subtypes. Span J Psychol 2009;12(2):707–14.

[184] Kariyawasam SH, Zaw F, Handley SL. Reduced salivary cortisol in children with comorbid attention deficit hyperactivity disorder and oppositional defiant disorder. Neuro Endocrinol Lett 2002;23(1):45–8.

[185] Schluter T, Winz O, Henkel K, Prinz S, Rademacher L, Schmaljohann J, et al. The impact of dopamine on aggression: an [18F]-FDOPA PET Study in healthy males. J Neurosci 2013;33(43):16889–96.

[186] Yu Q, Teixeira CM, Mahadevia D, Huang Y, Balsam D, Mann JJ, et al. Dopamine and serotonin signaling during two sensitive developmental periods differentially impact adult aggressive and affective behaviors in mice. Mol Psychiatry 2014;19(6):688–98.

[187] Beiderbeck DI, Reber SO, Havasi A, Bredewold R, Veenema AH, Neumann ID. High and abnormal forms of aggression in rats with extremes in trait anxiety–involvement of the dopamine system in the nucleus accumbens. Psychoneuroendocrinology 2012;37(12):1969–80.

[188] Yanowitch R, Coccaro EF. The neurochemistry of human aggression. Adv Genet 2011;75:151–69.

[189] Theunissen SC, Rieffe C, Netten AP, Briaire JJ, Soede W, Schoones JW, et al. Psychopathology and its risk and protective factors in hearing-impaired children and adolescents: a systematic review. JAMA Pediatr 2014;168(2):170–7.

[190] Fellinger J, Holzinger D, Sattel H, Laucht M. Mental health and quality of life in deaf pupils. Eur Child Adolesc Psychiatry 2008;17(7):414–23.

[191] Schaefer JM, Fetissov SO, Legrand R, Claeyssens S, Hoekstra PJ, Verhulst FC, et al. Corticotropin (ACTH)-reactive immunoglobulins in adolescents in relation to antisocial behavior and stress-induced cortisol response. The TRAILS Study. Psychoneuroendocrinology 2013;38(12):3039–47.

[192] Hastings PD, Fortier I, Utendale WT, Simard LR, Robaey P. Adrenocortical functioning in boys with attention-deficit/hyperactivity disorder: examining subtypes of ADHD and associated comorbid conditions. J Abnorm Child Psychol 2009;37(4):565–78.

[193] Freitag CM, Hanig S, Palmason H, Meyer J, Wust S, Seitz C. Cortisol awakening response in healthy children and children with ADHD: impact of comorbid disorders and psychosocial risk factors. Psychoneuroendocrinology 2009;34(7):1019–28.

[194] van Goozen SH, Fairchild G. Neuroendocrine and neurotransmitter correlates in children with antisocial behavior. Horm Behav 2006;50(4):647–54.

[195] Snoek H, Van Goozen SH, Matthys W, Buitelaar JK, van Engeland H. Stress responsivity in children with externalizing behavior disorders. Dev Psychopathol 2004;16(2):389–406.

[196] Shansky RM, Lipps J. Stress-induced cognitive dysfunction: hormone-neurotransmitter interactions in the prefrontal cortex. Front Hum Neurosci 2013;7:123.

[197] Cornelisse S, Joels M, Smeets T. A randomized trial on mineralocorticoid receptor blockade in men: effects on stress responses, selective attention, and memory. Neuropsychopharmacology 2011;36(13):2720–8.

[198] Zoladz PR, Diamond DM. Linear and non-linear dose-response functions reveal a hormetic relationship between stress and learning. Dose-Response 2009;7:132–48.

[199] Puig MV, Miller EK. Neural substrates of dopamine D2 receptor modulated executive functions in the monkey prefrontal cortex. Cereb Cortex 2014; [epub ahead of print].

[200] Fried PJ, Rushmore 3rd RJ, Moss MB, Valero-Cabre A, Pascual-Leone A. Causal evidence supporting functional dissociation of verbal and spatial working memory in the human dorsolateral prefrontal cortex. Eur J Neurosci 2014;39(11):1973–81.

[201] Watanabe K, Funahashi S. Neural mechanisms of dual-task interference and cognitive capacity limitation in the prefrontal cortex. Nat Neurosci 2014;17(4):601–11.

[202] Yuan P, Raz N. Prefrontal cortex and executive functions in healthy adults: a meta-analysis of structural neuroimaging studies. Neurosci Biobehav Rev 2014;42C:180–92.

[203] Belcher JR. Attention deficit hyperactivity disorder in offenders and the need for early intervention. Int J Offender Ther Comp Criminol 2014;58(1):27–40.

[204] von Polier GG, Vloet TD, Herpertz-Dahlmann B. ADHD and delinquency–a developmental perspective. Behav Sci Law 2012;30(2):121–39.

[205] Sibley MH, Pelham WE, Molina BS, Gnagy EM, Waschbusch DA, Biswas A, et al. The delinquency outcomes of boys with ADHD with and without comorbidity. J Abnorm Child Psychol 2011;39(1):21–32.

[206] Bussing R, Mason DM, Bell L, Porter P, Garvan C. Adolescent outcomes of childhood attention-deficit/hyperactivity disorder in a diverse community sample. J Am Acad Child Adolesc Psychiatry 2010;49(6):595–605.

[207] Jantzer V, Haffner J, Parzer P, Roos J, Steen R, Resch F. The relationship between ADHD, problem behaviour and academic achievement at the end of primary school. Prax Kinderpsychol Kinderpsychiatr 2012;61(9):662–76.

[208] McLeod JD, Uemura R, Rohrman S. Adolescent mental health, behavior problems, and academic achievement. J Health Soc Behav 2012;53(4):482–97.

[209] Matthys W, Vanderschuren LJ, Schutter DJ. The neurobiology of oppositional defiant disorder and conduct disorder: altered functioning in three mental domains. Dev Psychopathol 2013;25(1):193–207.

Thyroid-Disrupting Chemicals as Developmental Neurotoxicants

David S. Sharlin

INTRODUCTION

Thyroid hormone (TH) signaling is critical for the development and maturation of the nervous system. It is well documented that untreated congenital thyroid disorders such as hypothyroidism result in a number of neurological deficits including a low intelligence

Environmental Factors in Neurodevelopmental and Neurodegenerative Disorders
http://dx.doi.org/10.1016/B978-0-12-800228-5.00008-X

quotient, auditory deficits, and motor impairments [1,2]. Similarly, developmental exposure to several specific neurotoxicants, such as polychlorinated biphenyls (PCBs), results in a similar set of neurocognitive deficits [3–5]. Considering these observations, it has been hypothesized that environmental chemicals may impact brain development by interfering with TH signaling [6,7].

This hypothesis has been tested in the human population and in experimental models by investigating measurements of thyroid function with toxicant load and neurodevelopmental outcomes. In these studies, it is routinely observed that end points of TH signaling are associated with disrupted neurodevelopment and neuropsychological outcomes following toxicant exposure. Moreover, it has become apparent that even mild perturbation in TH signaling can produce measurable deficits in very specific functions of TH action, and that the specific consequences of disrupted TH signaling depend on the precise developmental timing of the deficiency [8]. Thus, the quantity, length, and timing of toxicant exposure are correlated with the array and severity of developmental defects that are responsible for measurable neurological deficits.

Control of TH action in development is complex (see Figure 8.1), and thus several mechanisms have been developed to help explain deficits associated with perturbed TH signaling. However, two major mechanisms have been largely explored experimentally. The first mechanism is that thyroid toxicants produce adverse effects on the developing nervous system by creating a state of TH insufficiency. Consistent with this idea, a number of studies in humans and experimental animals have documented a relationship between toxicant exposure and reduced serum TH levels [9]. The second major presumption is that specific toxicants can disrupt TH-mediated brain development by directly interacting with thyroid hormone receptors (TRs). The goal of this chapter is to articulate the complexities of TH action in development and assimilate our current understanding of how thyroid toxicants may disrupt brain development. It begins by discussing the regulation of TH action and then examines the potential points of thyroid disruption with a discussion of representative thyroid-disrupting chemicals (TDCs). Finally, it concludes by describing the consequences of disruption of TH action by environmental chemicals in the developing brain, by focusing on well-understood TH-mediated processes such as neocortical development, gliogenesis and myelination, cerebellar maturation, and cochlear development.

REGULATION OF TH ACTION

TH Production and Levels in the Blood

Production and secretion of THs from the thyroid gland into the circulation is controlled largely by the pituitary gland, which in turn is controlled by the hypothalamus (Figure 8.1). Together, the hypothalamus–pituitary–thyroid (HPT) axis is a classic neuroendocrine system, which maintains, through negative feedback, circulating levels of TH within a narrow limit [10]. Hypothalamic control of the HPT axis is mediated by thyrotropin-releasing hormone (TRH) neurons located in the paraventricular nucleus of the hypothalamus (PVH). TRH released from the terminals of these neurons at the median

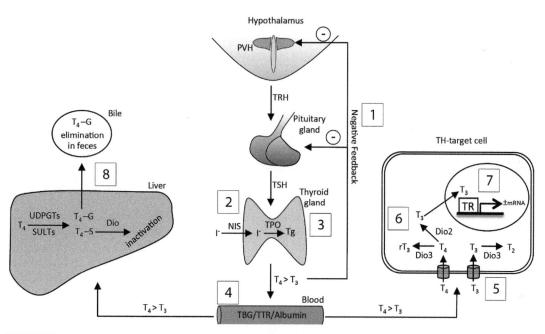

FIGURE 8.1 Regulation of thyroid hormone (TH) action and possible mechanisms of action of environmental chemicals that target the thyroid system. Under normal conditions, TH levels in the serum are maintained through the action of the hypothalamus–pituitary–thyroid (HPT) axis, TH binding to serum-binding proteins, local metabolism via deiodinase enzymes, and elimination and inactivation in the liver. Environmental toxicants can disrupt normal TH action through a number of mechanisms: (1) Disrupting negative feedback. (2) Blocking iodine uptake into the thyroid by inhibiting the Na/I symporter (NIS). (3) Blocking thyroglobulin (Tg) iodination by inhibiting thyroperoxidase (TPO) activity. (4) Competing for or inhibiting TH binding to serum-transfer proteins. (5) Inhibiting uptake or efflux of THs by interfering with membrane transporters. (6) Disrupting cellular TH availability by altering deiodinase enzymatic activities. (7) Directly perturbing TR function. (8) Altering TH clearance from the body through liver glucuronidation and sulfation. See text for details and a discussion on specific chemicals. TRH, thyrotropin-releasing hormone; TSH, thyroid-stimulating hormone; TBG, thyroxine-binding globulin; TTR, transthyretin; UDPGTs, UDP glucuronosyltransferases; SULTs, sulfotransferases; T_4-G, glucuronidated thyroxine; T_4-S, sulfonated thyroxine; TR, thyroid hormone receptor; Dio, deiodinase enzyme.

eminence stimulates the production and secretion of thyroid-stimulating hormone (TSH) from the anterior pituitary [11]. TSH then enters the circulation and regulates the activity of the thyroid gland. The hormones secreted from the thyroid gland, thyroxine (T_4) and triiodothyronine (T_3), are produced by coupling two iodinated tyrosyl residues that make up the precursor hormone thyroglobulin (Tg) [12]. T_4 is the most abundant product of the thyroid gland in circulation and is peripherally deiodinated to T_3 (discussed below).

The bioavailability of T_4 and T_3 is controlled in part by serum-binding proteins (i.e., thyroxine-binding globulin, transthyretin (TTR), and albumin), which contribute to maintaining the free fractions of THs (unbound hormone in serum) [13]. In addition, these binding proteins shield the hydrophobic tyrosine core from the aqueous environment of serum, increasing the half-lives of THs and contributing to the total (bound and free) carrying capacity of blood [14].

The regulation of TH clearance from serum is largely controlled by metabolism in the liver, where glucuronidation and sulfation of the phenolic hydroxyl group occurs via phase II and phase III enzymes [15]. Glucuronidated THs are rapidly cleared by excretion into the bile and subsequent removal through the feces. Sulfonated THs are usually deiodinated within hepatocytes, which results in permanent degradation and inactivation [16].

TRs in the Developing Nervous System

The majority of the biological actions of T_3 are mediated by the TRs. TRs are ligand-regulated transcription factors that mediate changes in the gene expression of target genes, and ultimately drive transcriptional programs that direct developmental processes. TRs are encoded by two distinct genes, *THRA* and *THRB* [17,18]. Through differential promoter usage and alternative splicing, these two genes produce four major receptor isoforms in humans and rodents: TRα1, TRα2, TRβ1, and TRβ2 [19,20]. TRα1, TRβ1, and TRβ2 are bona fide receptors that bind T_3 with nearly identical affinities [21,22]. TRα2 binds DNA, but contains an alternatively spliced C-terminal that changes the ligand-binding domain amino acid sequence and is unable to bind T_3 [23]. Therefore, TRα2 is not considered a bona fide TR and functions as a weak dominant-negative receptor.

Analysis of TR expression in human fetal whole brain tissue demonstrated the presence of the mRNAs for TRα1, TRα2, and TRβ1 as early as eight weeks of gestation [24], two weeks prior to the onset of fetal thyroid function [25]. The TRα1 and TRα2 isoforms increased in expression eightfold between 8 and 14 weeks of gestation, while the TRβ1 isoform increased at 14 weeks of gestation after remaining relatively steady between weeks 8 and 10. This study could not reliably detect TRβ2 mRNA in whole brain homogenates. However, a study by Kilby et al. [26] reported mRNA for all TR isoforms (TRα1, TRα2, TRβ1, and TRβ2) in homogenates from fetal neocortex during the second and third trimesters. Only TRα1 and TRα2 were detected in the developing cortex during the first trimester [26]. Little is known about the specific neural cell types that express TRs during development although several hypothalamic regions were immunopositive for TR proteins in adult postmortem tissue [27]. The general finding that TH receptors are present in the developing brain, even before fetal TH function, suggests that environmental chemicals that target the thyroid can directly impact fetal brain development.

Much of our understanding of the role of individual TRs is based on studies conducted in rodent models. Analysis of TR expression in the developing rodent nervous system indicated developmental and cell-specific expression profiles [28–30]. In general, TRα1 and TRα2 are expressed earlier and at greater levels in the murine brain than TRβ1 and TRβ2. Both TRα isoforms were detected in the neural tube at embryonic day 11.5 (E11.5). Robust overlapping expression of TRα1 and TRα2 were detected throughout the developing brain from E13.5 until adulthood [28]. Recently, detailed mapping of the TRα1 protein was completed using a transgenic mouse line that expresses a TRα1 protein fused to green fluorescent protein (GFP) as a chimera from the endogenous THRA gene [29]. This study demonstrated that TRα1 is expressed in most postmitotic neurons throughout the brain, with specific developmental patterns in some neurons. For example, early postmitotic purkinje cells of the cerebellum are TRα1-GFP positive, but negative for TRα1-GFP when assayed in the adult brain [29]. Interestingly, TRα1-GFP protein was detected in only a few glial cell lineages. Tanycytes that line the third ventricle were TRα1-GFP positive, but astrocytes in other regions were negative. Oligodendrocytes populating white matter tracts, the caudate putamen, the cerebral cortex, and the hippocampus were negative for TRα1-GFP. However, a few TRα1-GFP positive oligodendrocytes were observed in the hypothalamus. Detailed mapping of TRβ at the cellular level has yet to be reported, but in situ hybridization studies have reported that TRβ1 mRNA expression is first seen at E15.5 in the upper tegmental area. As development continues, TRβ1 mRNA is noted in the neocortex, caudate putamen, ammons horn, and dentate gyrus of the hippocampus and specific nuclei of the hypothalamus such as the PVH, with lower levels in a few other brain regions [28]. TRβ2 expression in the developing brain was highly restricted, with expression only noted in the developing striatum and hippocampus. However, robust TRβ2 mRNA is observed in the pituitary gland [28] and plays an important role in negative feedback that maintains normal serum TH levels [31,32].

Outside the brain, TR expression has been documented in several sensory organs. In the cochlea, in situ hybridization demonstrated that all four major TR isoforms are expressed in the cochlea. Temporally, in contrast to TR expression in the central nervous system (CNS) where TRα is the predominantly expressed isoform during development [28], TRβ is the predominant isoform expressed in the developing cochlea with TRβ1 expression greater than TRβ2 [33]. TRβ1 and TRβ2 isoforms have overlapping expression in several cochlear tissues including the greater and lesser epithelial ridges, outer sulcus epithelium, spiral limbus, and sensory epithelium with lower levels noted in the spiral ganglion [33]. In contrast, TRα1 expression is comparatively lower in TRβ-positive cochlear tissues, but with higher expression in the spiral ganglia. The expression of TRα2 in the cochlea was very similar to the expression pattern of TRα1, with the exception that TRα2 was also noted within the vestibular complex [33]. Interestingly, in the eye, TRβ2 was shown to be specifically expressed in immature cone photoreceptors [34,35]. The expression of the other TR isoforms in the retina is not clear.

Deiodinase Enzymes in the Developing Nervous System

Deiodinase enzymes regulate the intracellular availability of TH. Deiodinases are membrane-bound selenocysteine-containing enzymes that can amplify or reduce T_3 by the removal of an iodine atom. Within the nervous system, two deiodinases function to control TH bioavailability. The type II deiodinase (Dio2) locally generates T_3 from T_4, and the type III deiodinase (Dio3) prevents T_3 bioavailability by converting T_4 to reverse-T_3 (rT_3), or T_3 to diiodothyronine (T_2) [36] (Figure 8.2). Cellular expression of these enzymes during development controls tissue sensitivity and developmental processes such as cochlear development [37]. During early brain and sensory system development in rodents and humans, Dio3 activity and mRNA expression is high, restricting the bioavailability of T_3 [38–41]. Conversely, the activity and expression of Dio2 in early brain development is low, but increases transiently around the perinatal period [41–43]. A widely cited role for Dio2 in adult and developing nervous tissues is to function as an adaptive mechanism to protect the nervous system from perturbations in TH levels. This idea is largely based on the observation that as much as 80% of the T_3 present in the developing brain is derived from local deiodination [44,45], and that Dio2 activity is upregulated in response to low tissue TH levels [46]. However, whether changes in Dio2 activity actually offer protection and compensation during brain development has been questioned [47].

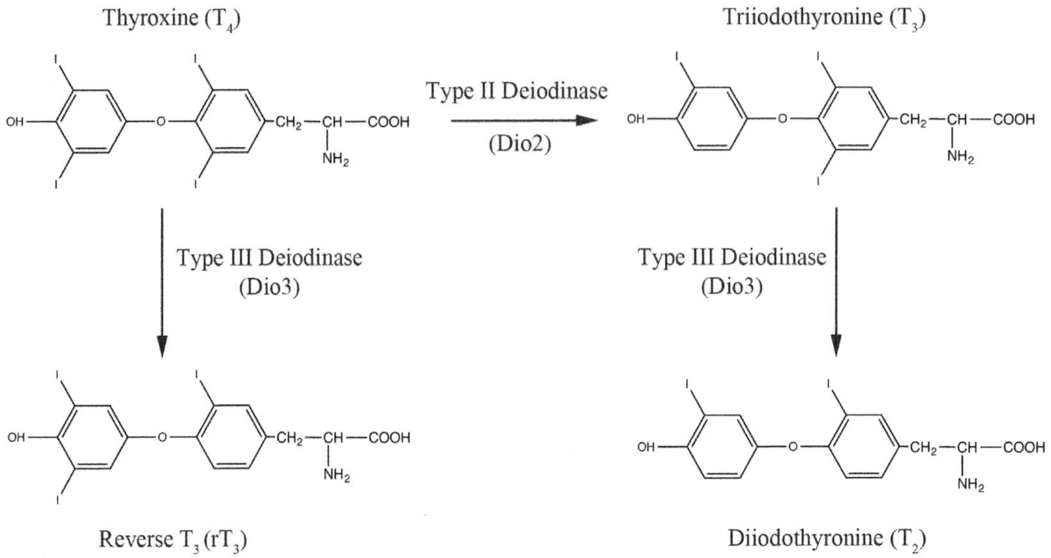

FIGURE 8.2   Overview of the deiodination of thyroxine (T_4) into triiodothyronine (T_3) by the type II deiodinase (Dio2) and the deiodination of T_4 to reverse-triiodothyronine (rT_3) and diiodothyronine (T_2) by the type II deiodinase (Dio33).

The cellular expressions of Dio2 and Dio3 suggest that TH signaling in the brain and sensory systems operates in a paracrine-like fashion. In the brain, Dio2 is expressed in gray matter astrocytes and tanycytes that line the third ventricle [43]. Dio3 is expressed in virtually all neurons [38]. This differential expression suggests that astrocytes take up T_4 from the extracellular space, convert it to T_3, and then release this newly generated T_3 for uptake by neurons. This paracrine-like mechanism is supported by an elaborate series of in vitro experiments [48]. A similar mechanism has been proposed in the developing cochlea, where T_3 is generated by Dio2-positive fibrocytes of the lateral wall and spiral limbus and then released for uptake by TRβ-positive tissues that include the greater epithelial ridge (GER) and hair cells [42,49].

TH Transporters in the Developing Nervous System

Both TR action at the receptor and metabolism by deiodinases are intracellular events that require the uptake of iodothyronines across the plasma by specific transmembrane transporters [50]. Although THs are lipid-soluble molecules, they do not readily cross the lipid bilayer by simple diffusion. THs are unable to cross the lipid bilayer by simple diffusion because of their intrinsic structure which is derived from the amino acid tyrosine. The alanine side chain of tyrosine contains a polar zwitter-ionic charge that prevents TH from passively diffusing through the hydrophobic core of the lipid bilayer [51]. Several families of transmembrane proteins have been demonstrated to mediate the uptake and efflux of THs, and is the focus of several reviews [52–56].

The physiological importance of TH transporters was clearly demonstrated with the identification that mutations in the MCT8 gene result in a form of severe X-linked mental retardation known as Allan–Herndon–Dudley syndrome [57–59]. The neurological deficits in these patients are consistent with reduced TH signaling in the developing brain. Interestingly, these patients have elevated serum T_3, low-normal T_4, and low-normal TSH [60,61]. Taken together, these findings support the idea that TH transporters, such as MCT8, are required for T_3 (and/ or T_4) entry into cells for normal TH signaling, an idea that is supported by studies using knockout animals [62–65].

REPRESENTATIVE TDCs AND MECHANISMS OF DISRUPTION

The dynamic relationship between maintaining circulating levels of THs and controlling TH action at the receptor, as discussed above, creates a complex setting where environmental chemicals that perturb TH action might result in a mosaic of effects that likely depend on the chemical(s) present, as well as the timing and brain region being investigated. The current front-line screens for thyroid disruptors in vivo—thyroid weight and histopathology, as well as hormone measurements (T_4, T_3, and TSH)—are not sufficient, given the complexity of TH action, indicating that additional end points need to be recruited to fully understand the potential of chemicals to disrupt development of the nervous system [66]. Described below and outlined in Figure 8.1 are potential points of thyroid disruption by neurotoxic chemicals. In addition, representative chemicals that target a specific point in the regulation of TH action are discussed.

Toxicant Effects on Thyroidal Synthesis of TH

As described above, the production and secretion of TH from the thyroid gland is controlled by the action of the HPT. TSH released by the pituitary gland stimulates the thyroid to increase the synthesis and release of TH. A major step in the synthesis of TH is the iodination of tyrosine residues on thyroglobulin (Tg) within the thyroid follicle [12]. This conjugation is mediated by the enzyme thyroperoxidase (TPO) which reacts first with thyroidal hydrogen peroxide to form an oxidized enzyme that then oxidizes iodide to form an enzyme-active iodide complex that is transferrable to Tg (a precursor hormone) [67]. Clearly, any chemical capable of inhibiting the action of TPO would directly result in a reduced efficiency of TH synthesis. This idea is demonstrated by the clinically relevant drug 6-propyl-2-thiouracil (a thioamide) that is used to treat patients with hyperthyroidism and is well documented to reduce circulating levels of TH in humans [68] and rodent models [69]. Isoflavones, classes of naturally occurring organic compounds found in soy protein (e.g., the "phytoestrogens" genistein, coumesterol, and daidzein), show significant TPO inhibition in experimental studies [70,71]. In humans, excessive consumption of soy leads to a transient increase in TSH levels, with significant correlation between daidzein and TSH [72]. Furthermore, goiter has been observed in infants fed a soy-based formula [73,74]. Whether diets rich in soy isoflavones result in neurotoxic effects by altering TH synthesis remains to be tested, but the idea has been speculated as a partial explanation for the sudden increase in autistic-like behaviors [75].

Critical to the conjugation of iodide to Tg is the uptake and sequestration of iodide into the follicular cells, the functional units of the thyroid gland. This uptake is mediated by the sodium/iodine symporter (NIS) [76]. Although NIS is required for TH synthesis [77], it is rather promiscuous in its ligand selectivity [78]. Perchlorate (ClO_4), a powerful oxidant used in a number of applications such as rocket fuels and fireworks [79], has become nearly ubiquitous in the human population [80] due to its environmental stability and widespread contamination. A number of studies have demonstrated that perchlorate significantly inhibits the uptake of iodide into the thyroid gland [78,81]. As a result, perchlorate is suspected to alter thyroid function by reducing iodide uptake and ultimately TH synthesis. Therefore, there is general concern that developmental processes, such as those discussed below, could be perturbed with perchlorate exposure, especially exposure in infants who are particularly sensitive to TH insufficiency. Notably, Blount et al. [80] found that urinary perchlorate levels were associated with serum TSH in women of the general population (not in men, however).

Toxicant Effects on Cellular Metabolism, Serum Transport, and Excretion of THs

As discussed above, the deiodinase enzymes regulate the cellular availability of TH in the developing nervous system. However, few studies have focused on the ability of environmental toxicants to disrupt TH metabolism by directly inhibiting the function of deiodinase enzymes. Considering that as much as 80% of the T_3 present in the developing brain is derived from local deiodination of T_4 by Dio2 [44], this is an area of research that warrants investigation. Moreover, it has been proposed that the developing brain possesses an effective compensatory mechanism to protect it from small to moderate reductions in circulating levels of TH. This concept is based on the observation that Dio2 is upregulated in the hypothyroid brain [46] and responds to changes in circulating levels of THs in a complex manner that involves the circulating levels of both T_3 and T_4 [82]. A number of studies have documented changes in deiodinase activities following exposure to environmental chemicals, such as PCBs, that reduce circulating levels of THs [83–85]. This increase in Dio2 activity is indicative of a local reduction in tissue TH, but whether such changes in Dio2 are actually protective to the developing nervous system remains unclear [47].

The ability of toxicants to reduce circulating levels of THs in developing animals has received a great deal attention due to the important role TH plays in brain maturation (discussed in detail below). Two main mechanisms have been identified that may contribute to toxicant-induced reduction in circulating THs. First, a number of chemicals—including dioxin-like chemicals [86], PCBs and their hydroxylated metabolites [87–90], polybrominated diphenylethers (PBDEs) and their hydroxylated metabolites [87,91], and bisphenol A (BPA) and its halogenated derivatives [91–94]—can displace TH from its serum-binding proteins, which would essentially make THs free for elimination. Of note is that some hydroxylated chlorinated biphenyls have an affinity for TTR that is greater than their affinity for T_4 [89,90]. Although this mechanism is plausible and deserves acknowledgment, TTR is not required for normal TH homeostasis [95–97]; therefore, it is not clear whether such displacement would result in adverse effects.

The second identified mechanism that may explain how certain environmental chemicals can reduce circulating levels of TH, ultimately altering TH-mediated brain development, is through the induction of the primary pathways that clear TH from circulation by glucuronic acid and sulfate conjugation [98]. Xenobiotics that reduce circulating levels of THs, such as PCBs and PBDEs, are known to induce the expression of phase I–III enzymes that are principal components in the detoxification and elimination of xenobiotics, hormones, and bile acids, perhaps by activating orphan nuclear receptors—constitutive androstane receptor and/or pregnane X receptor [87,99]. The induced enzymes—cytochrome p450s, UDP-glucuronyl transferase (UDPGTs), and sulfotranferases (SULTs)—produce glucuronidated and sulfated TH metabolites that are concentrated in the bile and subsequently excreted [16,100]. That toxicant exposure reduces circulating levels of THs suggests that thyroid disruptors might exert a neurotoxic effect on the developing brain by inducing a state of relative hypothyroidism.

Disruption of Receptor Function

The majority of the biological actions and developmental processes mediated by TH are the result of TR action. Considering this, there is general concern that specific environmental chemicals could alter TH-mediated brain development by directly altering TR function and subsequently TR target genes. Several known neurotoxic chemicals including PCBs, PBDEs, halogenated BPA, and triclosan (a ubiquitous bactericidal) are structurally similar to TH (Figure 8.3), which has led to the hypothesis that specific environmental chemicals might act as imperfect TH analogues [101].

A number of parent PCB congeners are reported to not bind TRs in vitro [102]. However, several hydroxylated PCB metabolites can bind to the TR with an IC_{50} as low as $5\,\mu M$ [103]. Consistent with these findings, Gauger et al. demonstrated that PCB105 and PCB118 can act as TR agonists in cell culture only after induction of cytochrome P4501A1 activation [104], presumably forming the same hydroxylated metabolite

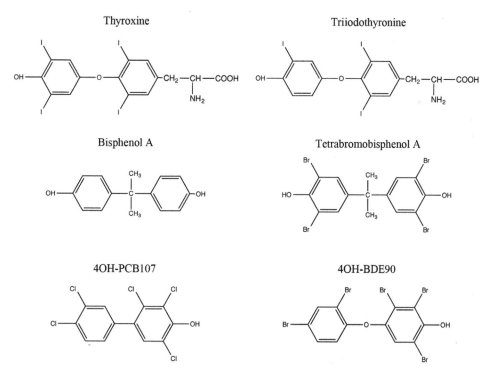

FIGURE 8.3 Structural comparison of thyroxine (T_4) and triiodothyronine (T_3) with chemicals known to disrupt thyroid hormone receptor function.

(4-hydroxy-2,3,3′,4′,5-pentachlorobiphenyl; 4-OH-PCB107). However, whether this common metabolite was formed in the assay remains to be demonstrated. Nonetheless, consistent with the idea that specific PCBs or hydroxylated metabolites can act as TR agonists in vitro, PCB exposure in vivo can increase the expression of known TH-responsive genes in the fetal and neonatal brain [102,105–107]. It should also be noted that not all reports indicate agonistic activity of PCBs. Several hydroxylated PCBs have been shown to inhibit T_3 action in in vitro assays [108,109].

Although PCBs are the best characterized thyroid toxicants at the TR level, BPA and its halogenated derivatives, as well as PBDEs, also have been shown to interact with TRs. Parent BPA as well as tetrabromobisphenol A and tetrachlorobisphenol A were documented to bind to TR and antagonize TR-mediated transactivation in cell culture reporter assays [110,111]. PBDEs were shown to bind TRs, with perhaps a selective preference for the TRβ over TRα isoform [112].

DISRUPTION OF TH ACTION IN THE DEVELOPING BRAIN

TH is essential for normal brain development. This idea is best exemplified by the severe neurological deficiencies observed with endemic cretinism and untreated congenital hypothyroidism. The neurological deficits that result from these conditions depend in part on the timing and severity of TH insufficiency and include mental retardation, spastic dysplasia, growth retardation, psychosocial behavioral abnormalities, and auditory deficits [1]. The variety of neurological phenotypes observed likely result from perturbing TH action during discrete developmental windows. Studies in experimental animals indicate that TH controls a number of processes during brain development, including neuronal migration and proliferation, and glial genesis and myelination (reviewed Ref. [113]). Below, I discuss the role of TH in these developmental processes and provide evidence suggesting that toxicants can disrupt the ability of TH to mediate these events.

TH and Neuronal Migration in the Developing Cerebral Cortex

The construction of the mammalian neocortex is a complex biological process that results in an orderly laminated structure. The spatial and temporal coordination of neuronal generation, migration, and differentiation are tightly regulated and required for normal brain function. Studies completed in pups from dams that had been rendered severely hypothyroid with the goitrogen methimazole demonstrated that TH is required for neocortical cell migration and proper cortical lamination [114,115]. Moreover, transient maternal hypothyroxinemia was sufficient to induce "blurring" of cortical layers [116], suggesting that neocortical cell migration is highly sensitive to perturbations in TH signaling. Goodman and Gilbert [117] reported a cluster of heterotopic cells within the corpus callosum in pups following mild-to-moderate TH insufficiency. The authors hypothesized that the heterotopic cells were the result of a migration defect. Recently it was demonstrated that neuronal migration in the neocortex requires reelin (a matrix-associated glycoprotein) and Notch signaling (Notch is a transmembrane receptor essential for stem cell maintenance) [118], and both of these pathways are targets of TH signaling in development [105,119]. Therefore, it is attractive to propose that defects in neuronal migration observed with developmental TH insufficiency are the result of altered reelin–Notch signaling.

Although it is clear that TH is essential for neocortical lamination, few studies have tested whether toxicants that target the thyroid can disrupt TH-mediated neocortical lamination. Bansal et al. [105] demonstrated that Hes1 and Hes5 mRNA expression, transcriptional targets of the activated Notch receptor, were upregulated in the neocortex of gestational day-16 fetuses following maternal exposure to the PCB mixture Aroclor 1254. The authors did not investigate cortical lamination in their study, but considering the recent report indicating the importance of Notch in cortical lamination, such investigation appears warranted. Recently, Naveau et al. [120] tested this idea more directly by exposing dams to Aroclor 1254 and then determining whether neuronal progenitor proliferation, cell cycle exit, and radial migration were disrupted in gestational day 17–20 pups. These authors reported that Aroclor 1254 significantly reduced total and free dam T_4, consistent with previous reports [107]. Furthermore, the authors reported that in utero PCB exposure resulted in increased cell cycle exit and delayed radial neuronal migration. However, no effects of PCBs were observed on cortical lamination as measured by Satb2 and Sox5 immunofluorescence. Considering that PCBs potentially disrupt TH-mediated brain developments through several mechanisms (discussed above), the observed "mosaic" of effects was not surprising and was consistent with previous reports [106]. Nonetheless, these reports indicate that early PCB exposure can disrupt neurogenesis, in part by altering TH signaling. Such effects would likely have lifelong consequences for neurocognitive outcomes and psychosocial behaviors by altering cortical function. Lastly, disrupted cortical neogenesis has also been reported in animals prenatally exposed to BPA [121,122]. Interestingly, it was further reported that the mRNA expression of TRα and TRβ in fetal forebrain exposed to BPA at E12 was significantly reduced, but TRα mRNA was increased at E14. Notch target genes Hes1 and Hes5 followed patterns of change in expression similar to those of TRα and TRβ, respectively.

TH Action in Gliogenesis and Myelination

It is well established that TH is important for the development of white matter tracts in experimental animals [123–126] and humans [127,128]. Specifically, TH controls oligodendrocyte precursor cell proliferation and survival [129], oligodendrocyte number [123], and oligodendrocyte differentiation [126,130]. Moreover, TH controls the balance of production of oligodendrocytes and astrocytes in major white matter tracts of the developing brain, potentially acting on a common precursor of these two cell types [126]. Considering that low TH during brain development results in permanent neurological deficits [1], it is reasonable to speculate that specific neurological deficits result in part from perturbed myelination.

Developmental PCB exposure in humans is associated with neurological deficits similar to those observed in children with congenital thyroid disease. Stewart et al. [131] reported that the size of the corpus callosum in children is a strong predictor of the strength of the association between PCB body burden and response inhibition. Response inhibition is considered a key neurological deficit in attention deficit hyperactivity disorder (ADHD). Interestingly, ADHD is associated with both congenital hypothyroidism and PCB exposure [132,133]. Furthermore, magnetic resonance imaging demonstrated that the anterior regions of the corpus callosum were significantly smaller in adolescent congenitally hypothyroid children identified by neonatal screening and treated [134]. This finding suggests that the anterior portions of the corpus callosum relay on TH during the late gestation/early postnatal period for normal development. Thus, toxicants that perturb TH signaling during brain development may result in altered corpus callosum development.

Consistent with this idea, developmental PCB exposure in experimental animals has been shown to reduce oligodendrocyte numbers in the corpus callosum, an effect consistent with hypothyroidism [135] (Figure 8.4). However, in contrast to the effect of hypothyroidism [126], PCB exposure did not increase the number of astrocytes in the corpus callosum [135]. Consistent with these findings, a single injection of 2,3,7,8-tetrachlorodibenzo-*p*-dioxin (which is known to reduce circulating THs) to pregnant rat dams on gestational day 18 is sufficient to induce a permanent reduction in the mRNA for myelin basic protein (MBP) in the cerebellum [136]. Whether this reduction is a result of fewer oligodendrocytes is unknown. Interestingly, Miller et al. [137] found that developmental exposure to a PCB mixture (Fox River Mix) induces hypothyroxinemia and sex-specific changes in glial proteins of the cerebellum. The authors reported that 18 mg/kg/day resulted in an increase in total MBP in males on postnatal day 42, but a decrease in females. Furthermore, a decrease in glial fibrillary acidic protein was observed in males on postnatal day 21 and 42, but an increase in females. It is not immediately clear why these results differ from those previously reported; however, it should be noted that myelination occurs in a caudal to rostral fashion [138], and thus timing

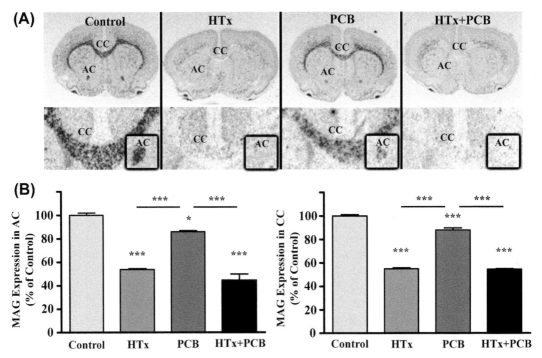

FIGURE 8.4   Effects of developmental hypothyroidism and PCB exposure on the oligodendrocyte marker MAG mRNA in the P15 rat brain. (A) Representative film autoradiographs obtained after in situ hybridization. Lower panels are higher magnifications of the corpus callosum (CC) and anterior commissure (AC; inset) of upper panels. (B) Quantification of density measurements depicted as bar graphs, showing the relative abundance of MAG mRNA in the AC (left) and CC (right). These data demonstrate the ability of PCB exposure to disrupt myelination in the juvenile rat brain. *Reproduced with permission from Ref. [135].*

of both hypothyroxinemia and myelination might play an important factor. Decabromodiphenyl ether (a fully brominated PBDE) was reported to induce mild hypothyroidism and reduce CNPase-positive oligodendrocytes in the cingulate deep cortex of the cerebrum in rat offspring after maternal exposure from mid-gestation through lactation [139]. Taken together, these studies support the idea that toxicant-induced reductions in serum THs can result in altered glial development and myelination.

In cell-culture systems utilizing purified primary oligodendrocyte precursor cells, TH promotes oligodendrocyte differentiation, and this timing mechanism is dependent on the presence of TRs [140]. Fritsche et al. [141] demonstrated that a specific PCB congener, PCB-118, can induce a dose-dependent increase in the number of oligodendrocytes in vitro, an effect that can be blocked with TR antagonist NH3 (Figure 8.5). These observations suggest that PCB exposure might alter white matter track development by directly interacting with TRs. Seiwa et al. [142] reported that BPA could inhibit a TH-mediated oligodendrocyte differentiation in vitro by blocking TRβ. Thus, PCB and PBDE exposure may interfere with TH action in developing white matter by disrupting receptor function in addition to their ability to reduce serum THs.

TH Action in Cerebellar Maturation

A major brain region dependent on TH for normal development is the cerebellum [143]. Hypothyroidism during the perinatal period in rodents results in two major structural defects. First, TH insufficiency delays granule cell proliferation and migration from the external granule layer (EGL) to the internal granule layer, resulting in a transient persistence of the EGL [144]. Second, developmental hypothyroidism results in stunted Purkinje cell dendritic

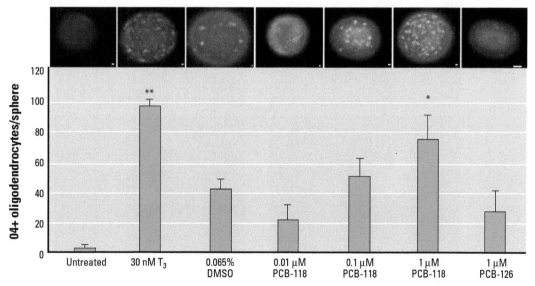

FIGURE 8.5 Induction of O4-positive (+) oligodendrocytes by T_3 or PCB-118 using in vitro differentiation of human neural progenitor cells. Photographs show O4-positive oligodendrocytes present in neural spheres. Histogram represents the count of cells per sphere. These data indicate that PCB-118 has a TH-like effect on the differentiation of oligodendrocytes in vitro. *Reproduced with permission from Ref. [141].*

growth and arborization [145,146]. Considering that these developmental events occur largely postnatally and role of TRs in cerebellar development is fairly well understood [147], the cerebellum offers an ideal system to investigate mechanistic aspects of TDCs, and several investigators have utilized this system.

Bansal and Zoeller [106] utilized cerebellar development in an attempt to discriminate between PCB effects on TH signaling in the developing brain by their ability to reduce circulating TH or by interact directly with the TR. The authors reasoned that if PCBs exert a direct action on the TR, then this effect should be observed in animals with low TH (i.e., ameliorate the effects of low TH). Using Purkinje cell protein 2 (PCP-2) mRNA expression (TH target gene in the cerebellum [148]) as a metric of TH action, the authors reported that PCB exposure could rescue the effect of hypothyroidism on PCP-2 expression. In addition, PCB treatment partially ameliorated the effect of hypothyroidism on persistence of the EGL during cerebellar development. These observations support the idea that PCBs can act as direct TR agonists in the developing cerebellum, although not all TH-dependent end points measured support this conclusion, indicating that the effects of PCBs on TH-directed brain development are highly complex [106,149].

Two different hydroxylated PCBs with high structural similarity to TH, 4-OH-2′,3,3′,4′,5′-pentachlorobiphenyl, and 4-OH-2′,3,3′,4′,5,5′-hexachlorobiphenyl were found to inhibit T_4-dependent Purkinje cell dendritic arborization at relatively low concentrations in primary cell cultures [109] (Figure 8.6). However, in vivo, no effect on Purkinje cell dendritic arborization was observed

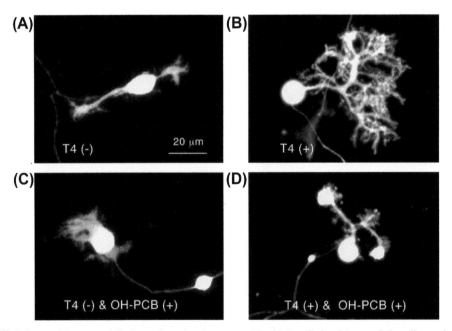

FIGURE 8.6   Visualization of T_4-dependent development of Purkinje cell dendrites and the effects of adding a hydroxylated PCB to cell culture media. Purkinje cells were stained with an anticalbindin antibody. (A) No T_4 addition. (B) Addition of T_4 only. (C) Addition of 4-OH-2′,3,3′,4′,5′-pentachlorobiphenyl (4-OH-PCB107; $5×10^{-11}$ M) only. (D) Addition of T_4 and 4-OH-PCB107. These results indicate that 4-OH-PCB107 inhibits T_4-dependent Purkinje cell dendritic arborization. *Reproduced with permission from Ref. [109].*

following gestational and lactational exposure to Aroclor 1254 [150]. Nonetheless, a number of studies have demonstrated motor dysfunction following developmental PCB exposure that can be partially attributed to altered cerebellar function induced by perturbing TR action [151].

Ibhazehiebo et al. [152] investigated whether specific parent PBDE congeners and hydroxylated metabolites were able to disrupt TH-mediated transcription and Purkinje cell arborization in vitro. The authors reported that several parent PBDE congeners, but not hydroxylated metabolites, were able to inhibit TR transactivation. Specifically, BDE209 ($2,2',3,3',4,4',5,5',6,6'$-DecaBDE) was capable of inhibiting T_3 transactivation of a luciferase reporter construct at a dose of 10^{-11} M. Interestingly, the inhibitory action of BDE209 was observed with the F2 (chicken lysozyme) and DR4 (consensus repeat separated by four base pairs) TH response elements (TREs), but the PAL TRE (consensus repeat separated by zero base pairs). Moreover, the BDE209 was more effective at inhibiting TR transactivation of TRβ than of TRα. These findings suggest that disruption of TR action in vivo by PBDEs could potentially be gene specific and dependent on target gene TRE and TR presence in the cell. Lastly, these authors reported that BDE209, but not BDE47, could significantly inhibit T_4-mediatd Purkinje cell arborization. This finding is consistent with the observation that BDE47 did not inhibit T_3-induced transactivation.

Although BPA and its halogenated derivatives have been shown to bind to TRs in vitro (discussed above), few studies have investigated whether BPA can disrupt TH-mediated brain development. BPA has been documented to bind to and antagonize T_3 activation of the TR, potentially uncoupling negative feedback in vivo [111,153]. Gestational and lactational exposure to BPA was reported to result in increased thickness of cerebellar EGL in postnatal day 11 pups [154]. Although this effect was not attributed to altered TH action, this finding is consistent with inhibition of TH action [144]. In contrast, BPA increased Purkinje dendritic growth, a TH-stimulated event [155]. The molecular basis for these differential effects of BPA on TH-mediated development of the cerebellum is not immediately clear, but it potentially could be related to differential affinities of BPA for TRs, and potential changes in circulating levels of THs following developmental BPA exposure [153]. Nonetheless, these reports support the idea that BPA and its halogenated derivatives can disrupt TH-mediated cerebellar development and that disruption will not simply follow a hypo- or hyperthyroid-like phenotype.

TH AND DEVELOPMENT OF THE AUDITORY SYSTEM

Auditory deficits associated with specific developmental thyroid disorders in humans are well documented, such as congenital hypothyroidism [156–158] and resistance to TH caused by mutations in the THRB gene [159–161]. Our mechanistic understanding of the role of TH in development of the auditory system has greatly advanced, in part by correlating the functional deficits observed in humans with congenital thyroid disorders, with those observed in studies conducted on model species [2]. Moreover, these studies have indicated that the developing auditory system is highly sensitive to perturbations in TH signaling.

Characterization of the relationship between the degree of hypothyroxinemia (low serum T_4 with normal serum T_3 and TSH) and hearing function demonstrated that a

50% reduction in serum T_4 was sufficient to permanently reduce auditory function [162] (Figure 8.7). Considering this sensitivity and our recently increased understanding of the mechanistic role of TH in directing development of the auditory system, the auditory system offers a unique set of functional and mechanistic end points for investigating whether environmental contaminants can disrupt TH-mediated auditory system development.

TH and Development of the Auditory System

Experimental rodent studies have indicated that TH plays a key role in the maturation of many tissues within the auditory system during a critical period that extends from the late gestation period to approximately two weeks after birth [2]. In humans, this developmental window corresponds to the mid-first to late-second trimesters [163]. Studies have indicated that a T_3 signal is required for both maturation of the outer ear auricle [164,165] and middle ear cavitation and ossicle development [166,167], but the best-understood role for TH in the maturation auditory system is within the cochlea. At birth in rodents, the mechanosensory hair cells and support cells within the cochlea that facilitate auditory function are largely specified [168], but anatomically, the cochlear structure is in an immature state. During the first two postnatal weeks, T_3 promotes the terminal differentiation of the cochlea by directing maturation and remodeling of the sensory epithelium prior to the start of cochlear function [2].

FIGURE 8.7 The correlation between log T_4 concentration during the early postnatal period and low-frequency hearing loss in adult rats. Data were compiled from several experiments to determine whether there was relationship between the degree of postnatal hypothyroxinemia and hearing loss. The solid line represents the correlation, while the thinner lines represent the 95% confidence intervals. Data in the insert exclude the extreme upper right value. These important data indicate a strong correlation between early postnatal serum T_4 concentrations and auditory thresholds, and suggest that early serum T_4 could be predictive for adverse neurological outcomes following developmental exposure to thyroid-disrupting chemicals. *Reproduced with permission from Ref. [162].*

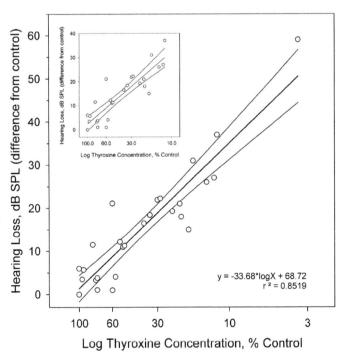

A major role for TH in maturation of the cochlea is to mediate regression of the GER (the GER also known as Kölliker's organ) [169], a transient epithelial cell structure located medially to the sensory hair cells (Figure 8.8). Although the functions of the GER are not fully understood, it is accepted that GER cells secrete several important sulfonated glycoproteins that contribute to the formation of the tectorial membrane [170–172]. More recently, GER cells were identified as the sources of adenosine triphosphate (ATP) that induce the spontaneous electrical activity in inner hair cells that is required for neuronal survival and refinement of the tonotopic map in the brain [173,174]. In euthyroid rodents, the GER regresses during the first two weeks of postnatal development before the onset of hearing, creating the inner sulcus that allows proper suspension of the tectorial membrane [175]. Low TH during the perinatal period results in a delayed GER regression and subsequent overproduction of tectorial proteins [176]. Delayed TH-mediated remodeling is associated with permanent auditory deficits [177].

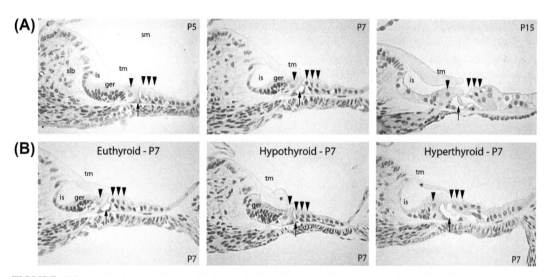

FIGURE 8.8 Postnatal remodeling in the rodent cochlea, and the effects of hypothyroidism and hyperthyroidism on remodeling. (A) Histological sections of the cochlea at developmental ages of P5, P7, and P15. Each image shows a midbasal turn of the cochlea. Two major remodeling events are histologically evident during this two-week period: (1) Regression of the greater epithelial ridge (GER) forming the inner sulcus (is) and (2) opening of the tunnel of corti (indicated by the arrow). Arrowheads indicate the single inner row and three outer rows of hair cells. Cells of the GER, which secrete extracellular matrix proteins to form the tectorial membrane (tm), slowly disappear as development proceeds. The tunnel of corti, a triangular fluid-filled space, separates the inner and outer hair cells and is dependent on pillar cell maturation (asterisk) (slb, spiral limbus; sm, scala media). (B) The role of T_3 and on postnatal remodeling in the rodent cochlea. Each image shows a midbasal turn of the cochlea at postnatal day 7 (P7) from a hypothyroid, euthyroid, or hypothyroid animal. Note that regression of GER cells is delayed in the hypothyroid cochlea and advanced in the hyperthyroid cochlea, compared with euthyroid control animals. This delay or advancement in GER cell regression results in an enlarged (swollen) or thin tectorial membrane (tm), respectively (asterisk). In addition, opening of the tunnel of corti (arrow) is delayed in the hypothyroid cochlea and advanced in the hyperthyroid cochlea, compared with the euthyroid cochlea. Note that under either hypothyroid or hyperthyroid conditions, cochlea hair cells (arrowheads) are present in correct numbers at this developmental age (is, inner sulcus).

Thyroid Disruptors as Inducers of Auditory Dysfunction

Considering the importance of TH in development of the auditory system, several investigators have used functional auditory testing to determine whether environmental chemicals that target the thyroid system can disrupt auditory function. Crofton and colleagues have completed the most comprehensive set of experiments that clearly indicate that PCB-induced reductions in TH result in permanent auditory deficits restricted to the low frequency range [178–181]. Using a cross-fostering paradigm, they demonstrated that the PCB-induced auditory deficits were of postnatal origin and resulted from lactational exposure [178]. Furthermore, it was proposed that PCB-induced hearing deficits are due in part to hair cell loss in the cochlea of exposed animals [182]. Although these findings are not fully consistent with the known effects of developmental hypothyroidism, the authors could partially rescue the auditory deficits by providing exogenous T_4 [180]. Taken together, these results suggest that PCBs present in dam milk reduce pup THs perinatally, altering TH-mediated cochlear development during the critical window of development [2]. Similarly, Schantz and colleagues, using a slightly different battery of functional auditory assessments in rats, have confirmed auditory deficits following developmental PCB exposure, but extended their studies to show similar auditory deficits in animals exposed to PBDEs alone or in conjunction with PCBs [183–186].

Although these groups have clearly demonstrated that exposure to thyroid toxicants results in auditory deficits consistent with insufficient TH, several unknowns remain. First, neither of these groups have investigated whether TH-mediated cochlear development (discussed above) is perturbed. This is an important unknown to resolve, because although PCBs and PBDEs can reduce circulating THs, these chemicals are also known to induce TH-like effects [107] in the face of low TH. Consider that excess TH advances GER remodeling and reduces auditory function [39], the TH-like activities of environmental chemicals could potentially advance cochlear development and induce deafness. Thus, measurement of auditory function alone, following developmental exposure to TDCs, is not sufficient to reach well-informed conclusions.

Considering these experimental studies linking toxicant exposure to perturbed TH action and auditory dysfunction, it is surprising that only a few epidemiological studies have been conducted to test whether exposure to specific environmental contaminants is associated with auditory dysfunction. Trnovec et al. [187] measured serum PCB concentrations and otoacoustic emissions (OAEs: can be used to test auditory functions) in 574 12-year-old Slovakian children. Cross-sectional analysis was then used to associate the resulting thresholds with serum PCBs. Several classes of PCBs were significantly associated with lower OAE thresholds at frequencies of 1000 and 1500 Hz. Recently, Min et al. [188] utilized data from the *National Health and Nutrition Examination Survey 1999–2004* to investigate the association between 11 specific PCB congeners in serum and hearing impairment, defined as a >25 dB hearing threshold at 0.5, 1, 2, and 4 kHz in adults older than 20 years of age. The calculated odds ratios for hearing impairment were significantly elevated for several PCB congeners (PCB170, PCB180, PCB187, PCB194, PCB196). Thus, PCB exposure appears to be associated with hearing dysfunction in the general population and highly exposed individuals. Although these results are not directly linked to perturbed TH action, they are consistent with controlled laboratory experimental studies and suggest a cochlear site of action, as would be expected based on the known roles of TH in cochlear development.

CONCLUDING REMARKS

TH is essential for development of the nervous system. A growing body of literature clearly indicates that synthetic chemicals released into the environment can alter TH signaling. Therefore, there is general concern that TH signaling during critical windows of development could result in permanent neurological deficits. Animal studies have documented that known TH-sensitive developmental events in the brain are altered following exposure to specific toxicants such as PCBs, PBDEs, and BPA. These reports have found that TDCs interact with the developing nervous in complex ways, interfering with TH signaling through several mechanisms. As a result, end points following exposure are mosaic, with some developmental events consistent with hypothyroidism, while others appear to be TH like. Moreover, recent studies have documented that the developing brain is exquisitely sensitive to perturbations in TH, and may not have the capacity to adapt to these perturbations as was once thought. Considering this, future evaluation of TDCs will require careful analysis and likely require more than just determining whether a specific chemical can reduce circulating levels of TH.

References

[1] Zoeller RT, Rovet J. Timing of thyroid hormone action in the developing brain: clinical observations and experimental findings. J Neuroendocrinol 2004;16(10):809–18.

[2] Ng L, Kelley MW, Forrest D. Making sense with thyroid hormone–the role of T(3) in auditory development. Nat Rev Endocrinol 2013;9(5):296–307.

[3] Huisman M, Koopman-Esseboom C, Lanting CI, Van der Paauw CG, Tuinstra LG, Fidler V, et al. Neurological condition in 18-month-old children perinatally exposed to polychlorinated biphenyls and dioxins. Early Hum Dev 1995;43:165–76.

[4] Longnecker MP, Rogan WJ, Lucier G. The human health effects of DDT (dichlorodiphenyl-trichloroethane) and PCBs (polychlorinated biphenyls) and an overview of organochlorines in public health. Annu Rev Public Health 1997;18:211–44.

[5] Lundqvist C, Zuurbier M, Leijs M, Johansson C, Ceccatelli S, Saunders M, et al. The effects of PCBs and dioxins on child health. Acta Paediatr 2006;95(453):55–64.

[6] Colborn T. Neurodevelopment and endocrine disruption. Environ Health Perspect 2004;112(9):944–9.

[7] Howdeshell KL. A model of the development of the brain as a construct of the thyroid system. Environ Health Perspect 2002;110(Suppl. 3):337–48.

[8] Zoeller RT, Rovet J. Timing of thyroid hormone action in the developing brain – clinical observations and experimental findings. J Neuroendocrinol 2004;16:809–18.

[9] Zoeller RT, Crofton KM. Thyroid hormone action in fetal brain development and potential for disruption by environmental chemicals. Neurotoxicology 2000;21(6):.

[10] Andersen S, Pedersen KM, Bruun NH, Laurberg P. Narrow individual variations in serum T(4) and T(3) in normal subjects: a clue to the understanding of subclinical thyroid disease. J Clin Endocrinol Metab 2002;87(3):1068–72.

[11] Cohen RN, Wondisford F. Chemistry and biosynthesis of thyrotropin. In: Braverman LE, Cooper D, editors. The thryoid: a fundamental and clinical text. 10 ed. Philadelphia: Lippincott Williams and Wilkins; 2012. pp. 149–62.

[12] Kopp PA. Thyroid hormone synthesis. In: Braverman LE, Cooper D, editors. The thyroid: a fundamental and clinical text. 10th ed. Philadelphia: Lippincott Williams and Wilkins; 2013. pp. 48–73.

[13] Kohrle J. Thyroid hormone metabolism and action in the brain and pituitary. Acta Med Austriaca 2000;27(1):1–7.

[14] Schussler GC. The thyroxine-binding proteins. Thyroid Off J Am Thyroid Assoc 2000;10(2):141–9.

[15] Wu S-y, Visser TJ, Veterans Administration Medical Center (Long Beach Calif.), Southwest Regional Medical Education Center (U.S.). Thyroid hormone metabolism: molecular biology and alternate pathways. Boca Raton: CRC Press; 1994.

[16] Visser TJ. Pathways of thyroid hormone metabolism. Acta Med Austriaca 1996;23(1–2):10–6.

[17] Sap J, Munoz A, Damm K, Goldberg Y, Ghysdael J, Lentz A, et al. The c-erb-A protein is a high-affinity receptor for thyroid hormone. Nature 1986;324(6098):635–40.

[18] Weinberger C, Thompson CC, Ong ES, Lebo R, Gruol DJ, Evans RM. The c-erb-A gene encodes a thyroid hormone receptor. Nature 1986;324(6098):641–6.

[19] Zhang J, Lazar MA. The mechanism of action of thyroid hormones. Annu Rev Physiol 2000;62:439–66.

[20] Jones I, Srinivas M, Ng L, Forrest D. The thyroid hormone receptor beta gene: structure and functions in the brain and sensory systems. Thyroid 2003;13(11):1057–68.

[21] Schwartz HL, Strait KA, Ling NC, Oppenheimer JH. Quantitation of rat tissue thyroid hormone binding receptor isoforms by immunoprecipitation of nuclear triiodothyronine binding capacity. J Biol Chem 1992;267(17):11794–9.

[22] Oppenheimer JH, Schwartz HL, Strait KA. Thyroid hormone action 1994: the plot thickens. Eur J Endocrinol/Eur Fed Endocr Soc 1994;130(1):15–24.

[23] Lazar MA, Hodin RA, Darling DS, Chin WW. Identification of a rat c-erbA alpha-related protein which binds deoxyribonucleic acid but does not bind thyroid hormone. Mol Endocrinol 1988;2(10):893–901.

[24] Iskaros J, Pickard M, Evans I, Sinha A, Hardiman P, Ekins R. Thyroid hormone receptor gene expression in first trimester human fetal brain. J Clin Endocrinol Metab 2000;85(7):2620–3.

[25] Shepard TH. Onset of function in the human fetal thyroid: biochemical and radioautographic studies from organ culture. J Clin Endocrinol Metab 1967;27(7):945–58.

[26] Kilby MD, Gittoes N, McCabe C, Verhaeg J, Franklyn JA. Expression of thyroid receptor isoforms in the human fetal central nervous system and the effects of intrauterine growth restriction. Clin Endocrinol 2000;53(4):469–77.

[27] Alkemade A, Vuijst CL, Unmehopa UA, Bakker O, Vennstrom B, Wiersinga WM, et al. Thyroid hormone receptor expression in the human hypothalamus and anterior pituitary. J Clin Endocrinol Metab 2005;90(2):904–12.

[28] Bradley DJ, Towle HC, Young 3rd WS. Spatial and temporal expression of alpha- and beta-thyroid hormone receptor mRNAs, including the beta 2-subtype, in the developing mammalian nervous system. J Neurosci 1992;12(6):2288–302.

[29] Wallis K, Dudazy S, van Hogerlinden M, Nordstrom K, Mittag J, Vennstrom B. The thyroid hormone receptor alpha1 protein is expressed in embryonic postmitotic neurons and persists in most adult neurons. Mol Endocrinol 2010;24(10):1904–16.

[30] Forrest D, Hallbook F, Persson H, Vennstrom B. Distinct functions for thyroid hormone receptors alpha and beta in brain development indicated by differential expression of receptor genes. EMBO J 1991;10(2):269–75.

[31] Ng L, Hurley JB, Dierks B, Srinivas M, Salto C, VennstromB, et al. A thyroid hormone receptor that is required for the development of green cone photoreceptors. Nat Genet 2001;27(1):94–8.

[32] Abel ED, Boers ME, Pazos-Moura C, Moura E, Kaulbach H, Zakaria M, et al. Divergent roles for thyroid hormone receptor beta isoforms in the endocrine axis and auditory system. J Clin Invest 1999;104(3):291–300.

[33] Bradley DJ, Towle HC, Young 3rd WS. Alpha and beta thyroid hormone receptor (TR) gene expression during auditory neurogenesis: evidence for TR isoform-specific transcriptional regulation in vivo. Proc Natl Acad Sci USA 1994;91(2):439–43.

[34] Ng L, Ma M, Curran T, Forrest D. Developmental expression of thyroid hormone receptor beta2 protein in cone photoreceptors in the mouse. Neuroreport 2009;20(6):627–31.

[35] Sjoberg M, Vennstrom B, Forrest D. Thyroid hormone receptors in chick retinal development: differential expression of mRNAs for alpha and N-terminal variant beta receptors. Development 1992;114(1):39–47.

[36] St Germain DL, Galton VA, Hernandez A. Minireview: defining the roles of the iodothyronine deiodinases: current concepts and challenges. Endocrinology 2009;150(3):1097–107.

[37] Ng L, Goodyear RJ, Woods CA, Schneider MJ, Diamond E, Richardson GP, et al. Hearing loss and retarded cochlear development in mice lacking type 2 iodothyronine deiodinase. Proc Natl Acad Sci USA 2004;101(10):3474–9.

[38] Tu HM, Legradi G, Bartha T, Salvatore D, Lechan RM, Larsen PR. Regional expression of the type 3 iodothyronine deiodinase messenger ribonucleic acid in the rat central nervous system and its regulation by thyroid hormone. Endocrinology 1999;140(2):784–90.

[39] Ng L, Hernandez A, He W, Ren T, Srinivas M, Ma M, et al. A protective role for type 3 deiodinase, a thyroid hormone-inactivating enzyme, in cochlear development and auditory function. Endocrinology 2009;150(4):1952–60.

[40] Ng L, Lyubarsky A, Nikonov SS, Ma M, Srinivas M, Kefas B, et al. Type 3 deiodinase, a thyroid-hormone-inactivating enzyme, controls survival and maturation of cone photoreceptors. J Neurosci 2010;30(9):3347–57.

[41] Kester MH, Martinez de Mena R, Obregon MJ, Marinkovic D, Howatson A, Visser TJ, et al. Iodothyronine levels in the human developing brain: major regulatory roles of iodothyronine deiodinases in different areas. J Clin Endocrinol Metab 2004;89(7):3117–28.

[42] Campos-Barros A, Amma LL, Faris JS, Shailam R, Kelley MW, Forrest D. Type 2 iodothyronine deiodinase expression in the cochlea before the onset of hearing. Proc Natl Acad Sci USA 2000;97(3):1287–92.

[43] Guadano-Ferraz A, Obregon MJ, St Germain DL, Bernal J. The type 2 iodothyronine deiodinase is expressed primarily in glial cells in the neonatal rat brain. Proc Natl Acad Sci USA 1997;94(19):10391–6.

[44] Larsen PR, Silva JE, Kaplan MM. Relationships between circulating and intracellular thyroid hormones: physiological and clinical implications. Endocr Rev 1981;2(1):87–102.

[45] Arrojo EDR, Fonseca TL, Werneck-de-Castro JP, Bianco AC. Role of the type 2 iodothyronine deiodinase (D2) in the control of thyroid hormone signaling. Biochim Biophys Acta 2013;1830(7):3956–64.

[46] Silva JE, Larsen PR. Comparison of iodothyronine 5'-deiodinase and other thyroid-hormone-dependent enzyme activities in the cerebral cortex of hypothyroid neonatal rat. Evidence for adaptation to hypothyroidism. J Clin Invest 1982;70(5):1110–23.

[47] Sharlin DS, Gilbert ME, Taylor MA, Ferguson DC, Zoeller RT. The nature of the compensatory response to low thyroid hormone in the developing brain. J Neuroendocrinol 2010;22(3):153–65.

[48] Freitas BC, Gereben B, Castillo M, Kallo I, Zeold A, Egri P, et al. Paracrine signaling by glial cell-derived triiodothyronine activates neuronal gene expression in the rodent brain and human cells. J Clin Invest 2010;120(6):2206–17.

[49] Sharlin DS, Visser TJ, Forrest D. Developmental and cell-specific expression of thyroid hormone transporters in the mouse cochlea. Endocrinology 2011;152(12):5053–64.

[50] Friesema EC, Jansen J, Milici C, Visser TJ. Thyroid hormone transporters. Vitam Horm 2005;70:137–67.

[51] Friesema EC, Docter R, Moerings EP, Verrey F, Krenning EP, Hennemann G, et al. Thyroid hormone transport by the heterodimeric human system L amino acid transporter. Endocrinology 2001;142(10):4339–48.

[52] Friesema EC, Jansen J, Visser TJ. Thyroid hormone transporters. Biochem Soc Trans 2005;33(Pt 1):228–32.

[53] van der Deure WM, Peeters RP, Visser TJ. Molecular aspects of thyroid hormone transporters, including MCT8, MCT10, and OATPs, and the effects of genetic variation in these transporters. J Mol Endocrinol 2010;44(1):1–11.

[54] Visser TJ. Thyroid hormone transporters. Horm Res 2007;68(Suppl. 5):28–30.

[55] Visser WE, Friesema EC, Jansen J, Visser TJ. Thyroid hormone transport by monocarboxylate transporters. Best Pract Res Clin Endocrinol Metab 2007;21(2):223–36.

[56] Visser WE, Friesema EC, Jansen J, Visser TJ. Thyroid hormone transport in and out of cells. Trends Endocrinol Metab 2008;19(2):50–6.

[57] Dumitrescu AM, Liao XH, Best TB, Brockmann K, Refetoff S. A novel syndrome combining thyroid and neurological abnormalities is associated with mutations in a monocarboxylate transporter gene. Am J Hum Genet 2004;74(1):168–75.

[58] Friesema EC, Grueters A, Biebermann H, Krude H, von Moers A, Reeser M, et al. Association between mutations in a thyroid hormone transporter and severe X-linked psychomotor retardation. Lancet 2004;364(9443):1435–7.

[59] Schwartz CE, Stevenson RE. The MCT8 thyroid hormone transporter and Allan-Herndon-Dudley syndrome. Best Pract Res Clin Endocrinol Metab 2007;21(2):307–21.

[60] Heuer H, Visser TJ. Minireview: pathophysiological importance of thyroid hormone transporters. Endocrinology 2009;150(3):1078–83.

[61] Heuer H, Visser TJ. The pathophysiological consequences of thyroid hormone transporter deficiencies: insights from mouse models. Biochim Biophys Acta 2012;1830:3974–78.

[62] Dumitrescu AM, Liao XH, Weiss RE, Millen K, Refetoff S. Tissue-specific thyroid hormone deprivation and excess in monocarboxylate transporter (mct) 8-deficient mice. Endocrinology 2006;147(9):4036–43.

[63] Trajkovic M, Visser TJ, Mittag J, Horn S, Lukas J, Darras VM, et al. Abnormal thyroid hormone metabolism in mice lacking the monocarboxylate transporter 8. J Clin Invest 2007;117(3):627–35.

[64] Trajkovic-Arsic M, Visser TJ, Darras VM, Friesema EC, Schlott B, Mittag J, et al. Consequences of monocarboxylate transporter 8 deficiency for renal transport and metabolism of thyroid hormones in mice. Endocrinology 2010;151(2):802–9.

[65] Mayerl S, Müller J, Bauer R, Richert S, Kassmann CM, Darras VM, et al. Transporters MCT8 and OATP1C1 maintain murine brain thyroid hormone homeostasis. J Clin Invest 2014;124(5):1987–99.

[66] Zoeller RT, Tyl RW, Tan SW. Current and potential rodent screens and tests for thyroid toxicants. Crit Rev Toxicol 2007;37(1–2):55–95.

[67] Nunez J, Pommier J. Formation of thyroid hormones. Vitam Horm 1982;39:175–229.

[68] Sato A, Koizumi Y, Kanno Y, Yamada T. Inhibitory effect of large doses of propylthiouracil and methimazole on an increase of thyroid radioiodine release in response to thyrotropin. Proc Soc Exp Biol Med 1976;152(1):90–4.

[69] Crofton KM, Zoeller RT. Mode of action: neurotoxicity induced by thyroid hormone disruption during development–hearing loss resulting from exposure to PHAHs. Crit Rev Toxicol 2005;35(8–9):757–69.

[70] Doerge DR, Sheehan DM. Goitrogenic and estrogenic activity of soy isoflavones. Environ Health Perspect 2002;110(Suppl. 3):349–53.

[71] Divi RL, Doerge DR. Inhibition of thyroid peroxidase by dietary flavonoids. Chem Res Toxicol 1996;9(1):16–23.

[72] Hampl R, Ostatnikova D, Celec P, Putz Z, Lapcik O, Matucha P. Short-term effect of soy consumption on thyroid hormone levels and correlation with phytoestrogen level in healthy subjects. Endocr Regul 2008;42(2–3):53–61.

[73] Ripp JA. Soybean-induced goiter. Am J Dis Child 1961;102:106–9.

[74] Labib M, Gama R, Wright J, Marks V, Robins D. Dietary maladvice as a cause of hypothyroidism and short stature. BMJ 1989;298(6668):232–3.

[75] Roman GC. Autism: transient in utero hypothyroxinemia related to maternal flavonoid ingestion during pregnancy and to other environmental antithyroid agents. J Neurol Sci 2007;262(1–2):15–26.

[76] Dohan O, De la Vieja A, Paroder V, Riedel C, Artani M, Reed M, et al. The sodium/iodide symporter (NIS): characterization, regulation, and medical significance. Endocr Rev 2003;24(1):48–77.

[77] De la Vieja A, Ginter CS, Carrasco N. Molecular analysis of a congenital iodide transport defect: G543E impairs maturation and trafficking of the Na+/I− symporter. Mol Endocrinol 2005;19(11):2847–58.

[78] Wolff J. Perchlorate and the thyroid gland. Phamacol Rev 1998;50:89–105.

[79] National Academies Press (U.S.), National Research Council (U.S.). Division on earth and life studies. Health implications of perchlorate ingestion. Washington, DC: National Academies Press; 2005.

[80] Blount BC, Valentin-Blasini L, Osterloh JD, Mauldin JP, Pirkle JL. Perchlorate exposure of the US Population, 2001–2002. J Expo Sci Environ Epidemiol 2007;17(4):400–7.

[81] Urbansky ET. Perchlorate as an environmental contaminant. Environ Sci Pollut Res Int 2002;9(3):187–92.

[82] Burmeister LA, Pachucki J, Germain DLS. Thyroid hormones inhibit type 2 iodothyronine deiodinase in the rat cerebral cortex by both pre- and posttranslational mechanisms. Endocrinology 1997;138:5231–7.

[83] Morse DC, Groen D, Veerman M, van Amerongen CJ, Koëter HB, Smits van Prooije AE, et al. Interference of polychlorinated biphenyls in hepatic and brain thyroid hormone metabolism in fetal and neonatal rats. Toxicol Appl Pharmacol 1993;122(1):27–33.

[84] Morse DC, Wehler EK, Wesseling W, Koeman JH, Brouwer A. Alterations in rat brain thyroid hormone status following pre- and postnatal exposure to polychlorinated biphenyls (Aroclor 1254). Toxicol Appl Pharmacol 1996;136(2):269–79.

[85] Viluksela M, Raasmaja A, Lebofsky M, Stahl BU, Rozman KK. Tissue-specific effects of 2,3,7,8-tetrachlorodibenzo-p-dioxin (TCDD) on the activity of 5′-deiodinases I and II in rats. Toxicol Lett 2004;147(2):133–42.

[86] McKinney JD, Chae K, Oatley SJ, Blake CC. Molecular interactions of toxic chlorinated dibenzo-p-dioxins and dibenzofurans with thyroxine binding prealbumin. J Med Chem 1985;28(3):375–81.

[87] Hallgren S, Darnerud PO. Polybrominated diphenyl ethers (PBDEs), polychlorinated biphenyls (PCBs) and chlorinated paraffins (CPs) in rats-testing interactions and mechanisms for thyroid hormone effects. Toxicology 2002;177(2–3):227–43.

[88] Darnerud PO, Morse D, Klasson-Wehler E, Brouwer A. Binding of a 3,3′, 4,4′-tetrachlorobiphenyl (CB-77) metabolite to fetal transthyretin and effects on fetal thyroid hormone levels in mice. Toxicology 1996;106(1–3): 105–14.

[89] Chauhan KR, Kodavanti PR, McKinney JD. Assessing the role of ortho-substitution on polychlorinated biphenyl binding to transthyretin, a thyroxine transport protein. Toxicol Appl Pharmacol 2000;162(1):10–21.

[90] Cheek AO, Kow K, Chen J, McLachlan JA. Potential mechanisms of thyroid disruption in humans: interaction of organochlorine compounds with thyroid receptor, transthyretin, and thyroid-binding globulin. Environ Health Perspect 1999;107(4):273–8.

[91] Meerts IA, van Zanden JJ, Luijks EA, van Leeuwen-Bol I, Marsh G, Jakobsson E, et al. Potent competitive interactions of some brominated flame retardants and related compounds with human transthyretin in vitro. Toxicol Sci 2000;56(1):95–104.

[92] Cao J, Guo LH, Wan B, Wei Y. In vitro fluorescence displacement investigation of thyroxine transport disruption by bisphenol A. J Environ Sci (China) 2011;23(2):315–21.

[93] Marchesini GR, Meimaridou A, Haasnoot W, Meulenberg E, Albertus F, Mizuguchi M, et al. Biosensor discovery of thyroxine transport disrupting chemicals. Toxicol Appl Pharmacol 2008;232(1):150–60.

[94] Yamauchi K, Ishihara A, Fukazawa H, Terao Y. Competitive interactions of chlorinated phenol compounds with 3,3',5-triiodothyronine binding to transthyretin: detection of possible thyroid-disrupting chemicals in environmental waste water. Toxicol Appl Pharmacol 2003;187(2):110–7.

[95] Palha JA. Transthyretin as a thyroid hormone carrier: function revisited. Clin Chem Lab Med 2002;40(12):1292–300.

[96] Palha JA, Fernandes R, de Escobar GM, Episkopou V, Gottesman M, Saraiva MJ. Transthyretin regulates thyroid hormone levels in the choroid plexus, but not in the brain parenchyma: study in a transthyretin-null mouse model. Endocrinology 2000;141(9):3267–72.

[97] Palha JA, Hays MT, Morreale de Escobar G, Episkopou V, Gottesman ME, Saraiva MJ. Transthyretin is not essential for thyroxine to reach the brain and other tissues in transthyretin-null mice. Am J Physiol 1997;272 (3 Pt 1):E485–93.

[98] Kelly GS. Peripheral metabolism of thyroid hormones: a review. Altern Med Rev 2000;5(4):306–33.

[99] Kretschmer XC, Baldwin WS. CAR and PXR: xenosensors of endocrine disrupters? Chem Biol Interact 2005;155(3):111–28.

[100] Visser TJ, Rutgers M, de Herder WW, Rooda SJ, Hazenberg MP. Hepatic metabolism, biliary clearance and enterohepatic circulation of thyroid hormone. Acta Med Austriaca 1988;15(Suppl. 1):37–9.

[101] McKinney JD, Waller CL. Polychlorinated biphenyls as hormonally active structural analogues. Environ Health Perspect 1994;102:290–7.

[102] Gauger KJ, Kato Y, Haraguchi K, Lehmler HJ, Robertson LW, Bansal R, et al. Polychlorinated biphenyls (PCBs) exert thyroid hormone-like effects in the fetal rat brain but do not bind to thyroid hormone receptors. Environ Health Perspect 2004;112(5):516–23.

[103] Kitamura S, Jinno N, Suzuki T, Sugihara K, Ohta S, Kuroki H, et al. Thyroid hormone-like and estrogenic activity of hydroxylated PCBs in cell culture. Toxicology 2005;208(3):377–87.

[104] Gauger KJ, Giera S, Sharlin DS, Bansal R, Iannacone E, Zoeller RT. Polychlorinated biphenyls 105 and 118 form thyroid hormone receptor agonists after cytochrome P4501A1 activation in rat pituitary GH3 cells. Environ Health Perspect 2007;115(11):1623–30.

[105] Bansal R, You SH, Herzig CT, Zoeller RT. Maternal thyroid hormone increases HES expression in the fetal rat brain: an effect mimicked by exposure to a mixture of polychlorinated biphenyls (PCBs). Brain Res Dev Brain Res 2005;156(1):13–22.

[106] Bansal R, Zoeller RT. Polychlorinated biphenyls (Aroclor 1254) do not uniformly produce agonist actions on thyroid hormone responses in the developing rat brain. Endocrinology 2008;149(8):4001–8.

[107] Zoeller RT, Dowling AL, Vas AA. Developmental exposure to polychlorinated biphenyls exerts thyroid hormone-like effects on the expression of RC3/neurogranin and myelin basic protein messenger ribonucleic acids in the developing rat brain. Endocrinology 2000;141(1):181–9.

[108] Iwasaki T, Miyazaki W, Takeshita A, Kuroda Y, Koibuchi N. Polychlorinated biphenyls suppress thyroid hormone-induced transactivation. Biochem Biophys Res Commun 2002;299(3):384–8.

[109] Kimura-Kuroda J, Nagata I, Kuroda Y. Hydroxylated metabolites of polychlorinated biphenyls inhibit thyroid-hormone-dependent extension of cerebellar Purkinje cell dendrites. Brain Res Dev Brain Res 2005;154(2):259–63.

[110] Kitamura S, Jinno N, Ohta S, Kuroki H, Fujimoto N. Thyroid hormonal activity of the flame retardants tetrabromobisphenol A and tetrachlorobisphenol A. Biochem Biophys Res Commun 2002;293(1):554–9.

[111] Moriyama K, Tagami T, Akamizu T, Usui T, Saijo M, Kanamoto N, et al. Thyroid hormone action is disrupted by bisphenol A as an antagonist. J Clin Endocrinol Metab 2002;87(11):5185–90.

[112] Talsness CE. Overview of toxicological aspects of polybrominated diphenyl ethers: a flame-retardant additive in several consumer products. Environ Res 2008;108(2):158–67.

[113] Bernal J. Thyroid hormone receptors in brain development and function. Nat Clin Pract Endocrinol Metab 2007;3(3):249–59.

[114] Berbel P, Auso E, Garcia-Velasco JV, Molina ML, Camacho M. Role of thyroid hormones in the maturation and organisation of rat barrel cortex. Neuroscience 2001;107(3):383–94.

[115] Lucio RA, Garcia JV, Ramon Cerezo J, Pacheco P, Innocenti GM, Berbel P. The development of auditory callosal connections in normal and hypothyroid rats. Cereb Cortex 1997;7(4):303–16.

[116] Lavado-Autric R, Auso E, Garcia-Velasco JV, Arufe Mdel C, Escobar del Rey F, Berbel P, et al. Early maternal hypothyroxinemia alters histogenesis and cerebral cortex cytoarchitecture of the progeny. J Clin Invest 2003;111(7):1073–82.

[117] Goodman JH, Gilbert ME. Modest thyroid hormone insufficiency during development induces a cellular malformation in the corpus callosum: a model of cortical dysplasia. Endocrinology 2007;148(6):2593–7.

I. NEURODEVELOPMENTAL DISORDERS

[118] Hashimoto-Torii K, Torii M, Sarkisian MR, Bartley CM, Shen J, Radtke F, et al. Interaction between Reelin and Notch signaling regulates neuronal migration in the cerebral cortex. Neuron 2008;60(2):273–84.

[119] Alvarez-Dolado M, Ruiz M, Del Rio JA, Alcantara S, Burgaya F, Sheldon M, et al. Thyroid hormone regulates reelin and dab1 expression during brain development. J Neurosci 1999;19(16):6979–93.

[120] Naveau E, Pinson A, Gerard A, Nguyen L, Charlier C, Thome JP, et al. Alteration of rat fetal cerebral cortex development after prenatal exposure to polychlorinated biphenyls. PLoS ONE, 2014;9:e91903.

[121] Nakamura K, Itoh K, Yaoi T, Fujiwara Y, Sugimoto T, Fushiki S. Murine neocortical histogenesis is perturbed by prenatal exposure to low doses of bisphenol A. J Neurosci Res 2006;84(6):1197–205.

[122] Nakamura K, Itoh K, Sugimoto T, Fushiki S. Prenatal exposure to bisphenol A affects adult murine neocortical structure. Neurosci Lett 2007;420(2):100–5.

[123] Schoonover CM, Seibel MM, Jolson DM, Stack MJ, Rahman RJ, Jones SA, et al. Thyroid hormone regulates oligodendrocyte accumulation in developing rat brain white matter tracts. Endocrinology 2004;145(11):5013–20.

[124] Berbel P, Guadano-Ferraz A, Angulo A, Ramon Cerezo J. Role of thyroid hormones in the maturation of interhemispheric connections in rats. Behav Brain Res 1994;64(1–2):9–14.

[125] Gravel C, Hawkes R. Maturation of the corpus callosum of the rat: I. Influence of thyroid hormones on the topography of callosal projections. J Comp Neurol 1990;291(1):128–46.

[126] Sharlin DS, Tighe D, Gilbert ME, Zoeller RT. The balance between oligodendrocyte and astrocyte production in major white matter tracts is linearly related to serum total thyroxine. Endocrinology 2008;149(5):2527–36.

[127] Annunziata P, Federico A, D'Amore I, Corona RM, Guazzi GC. Impairment of human brain development: glycoconjugate and lipid changes in congenital athyroidism. Early Hum Dev 1983;8(3–4):269–78.

[128] Gika AD, Siddiqui A, Hulse AJ, Edward S, Fallon P, McEntagart ME, et al. White matter abnormalities and dystonic motor disorder associated with mutations in the SLC16A2 gene. Dev Med Child Neurol 2010;52(5):475–82.

[129] Jones SA, Jolson DM, Cuta KK, Mariash CN, Anderson GW. Triiodothyronine is a survival factor for developing oligodendrocytes. Mol Cell Endocrinol 2003;199(1–2):49–60.

[130] Baxi EG, Schott JT, Fairchild AN, Kirby LA, Karani R, Uapinyoying P, et al. A selective thyroid hormone β receptor agonist enhances human and rodent oligodendrocyte differentiation. Glia 2014;62(9):1513–29.

[131] Stewart P, Fitzgerald S, Reihman J, Gump B, Lonky E, Darvill T, et al. Prenatal PCB exposure, the corpus callosum, and response inhibition. Environ Health Perspect 2003;111(13):1670–7.

[132] Rovet J, Alvarez M. Thyroid hormone and attention in school-age children with congenital hypothyroidism. J Child Psychol Psychiatry 1996;37(5):579–85.

[133] Sagiv SK, Thurston SW, Bellinger DC, Tolbert PE, Altshul LM, Korrick SA. Prenatal organochlorine exposure and behaviors associated with attention deficit hyperactivity disorder in school-aged children. Am J Epidemiol 2010;171(5):593–601.

[134] Rovet JF, Skocic J, Pinzon N. Corpus callosum morphology in adolescents with congenital hypothyroidism. In: 79th annual meeting of the American Thyroid Association. Chicago, IL: Mary Ann Liebert, Inc.; 2008. p. S16–7.

[135] Sharlin DS, Bansal R, Zoeller RT. Polychlorinated biphenyls exert selective effects on cellular composition of white matter in a manner inconsistent with thyroid hormone insufficiency. Endocrinology 2006;147(2):846–58.

[136] Fernandez M, Paradisi M, D'Intino G, Del Vecchio G, Sivilia S, Giardino L, et al. A single prenatal exposure to the endocrine disruptor 2,3,7,8-tetrachlorodibenzo-p-dioxin alters developmental myelination and remyelination potential in the rat brain. J Neurochem 2010;115(4):897–909.

[137] Miller VM, Kahnke T, Neu N, Sanchez-Morrissey SR, Brosch K, Kelsey K, et al. Developmental PCB exposure induces hypothyroxinemia and sex-specific effects on cerebellum glial protein levels in rats. Int J Dev Neurosci Off J Int Soc Dev Neurosci 2010;28(7):553–60.

[138] Verity AN, Campagnoni AT. Regional expression of myelin protein genes in the developing mouse brain: in situ hybridization studies. J Neurosci Res 1988;21(2–4):238–48.

[139] Fujimoto H, Woo GH, Inoue K, Takahashi M, Hirose M, Nishikawa A, et al. Impaired oligodendroglial development by decabromodiphenyl ether in rat offspring after maternal exposure from mid-gestation through lactation. Reprod Toxicol 2011;31(1):86–94.

[140] Billon N, Tokumoto Y, Forrest D, Raff M. Role of thyroid hormone receptors in timing oligodendrocyte differentiation. Dev Biol 2001;235(1):110–20.

[141] Fritsche E, Cline JE, Nguyen NH, Scanlan TS, Abel J. Polychlorinated biphenyls disturb differentiation of normal human neural progenitor cells: clue for involvement of thyroid hormone receptors. Environ Health Perspect 2005;113(7):871–6.

[142] Seiwa C, Nakahara J, Komiyama T, Katsu Y, Iguchi T, Asou H. Bisphenol A exerts thyroid-hormone-like effects on mouse oligodendrocyte precursor cells. Neuroendocrinology 2004;80(1):21–30.

[143] Koibuchi N, Iwasaki T. Regulation of brain development by thyroid hormone and its modulation by environmental chemicals. Endocr J 2006;53(3):295–303.

[144] Morte B, Manzano J, Scanlan T, Vennstrom B, Bernal J. Deletion of the thyroid hormone receptor alpha 1 prevents the structural alterations of the cerebellum induced by hypothyroidism. Proc Natl Acad Sci USA 2002;99(6):3985–9.

[145] Hajos F, Patel AJ, Balazs R. Effect of thyroid deficiency on the synaptic organization of the rat cerebellar cortex. Brain Res 1973;50(2):387–401.

[146] Heuer H, Mason CA. Thyroid hormone induces cerebellar Purkinje cell dendritic development via the thyroid hormone receptor alpha1. J Neurosci Off J Soc Neurosci 2003;23(33):10604–12.

[147] Koibuchi N. The role of thyroid hormone on functional organization in the cerebellum. Cerebellum 2013;12(3):304–6.

[148] Zou L, Hagen SG, Strait KA. Identification of thyroid hormone response elements in rodent Pcp-2, a developmentally regulated gene of cerebellar Purkinje cells. J Biol Chem 1994;269:13346–52.

[149] Darras VM. Endocrine disrupting polyhalogenated organic pollutants interfere with thyroid hormone signalling in the developing brain. Cerebellum 2008;7(1):26–37.

[150] Roegge CS, Morris JR, Villareal S, Wang VC, Powers BE, Klintsova AY, et al. Purkinje cell and cerebellar effects following developmental exposure to PCBs and/or MeHg. Neurotoxicol Teratol 2006;28(1):74–85.

[151] Roegge CS, Schantz SL. Motor function following developmental exposure to PCBS and/or MEHG. Neurotoxicol Teratol 2006;28(2):260–77.

[152] Ibhazehiebo K, Iwasaki T, Kimura-Kuroda J, Miyazaki W, Shimokawa N, Koibuchi N. Disruption of thyroid hormone receptor-mediated transcription and thyroid hormone-induced Purkinje cell dendrite arborization by polybrominated diphenyl ethers. Environ Health Perspect 2011;119(2):168–75.

[153] Zoeller RT, Bansal R, Parris C. Bisphenol-A, an environmental contaminant that acts as a thyroid hormone receptor antagonist in vitro, increases serum thyroxine, and alters RC3/neurogranin expression in the developing rat brain. Endocrinology 2005;146(2):607–12.

[154] Mathisen GH, Yazdani M, Rakkestad KE, Aden PK, Bodin J, Samuelsen M, et al. Prenatal exposure to bisphenol A interferes with the development of cerebellar granule neurons in mice and chicken. Int J Dev Neurosci Off J Int Soc Dev Neurosci 2013;31(8):762–9.

[155] Shikimi H, Sakamoto H, Mezaki Y, Ukena K, Tsutsui K. Dendritic growth in response to environmental estrogens in the developing Purkinje cell in rats. Neurosci Lett 2004;364(2):114–8.

[156] DeLong GR, Stanbury JB, Fierro-Benitez R. Neurological signs in congenital iodine-deficiency disorder (endemic cretinism). Dev Med Child Neurol 1985;27(3):317–24.

[157] Rovet J, Walker W, Bliss B, Buchanan L, Ehrlich R. Long-term sequelae of hearing impairment in congenital hypothyroidism. J Pediatr 1996;128(6):776–83.

[158] Leger J, Ecosse E, Roussey M, Lanoe JL, Larroque B. Subtle health impairment and socioeducational attainment in young adult patients with congenital hypothyroidism diagnosed by neonatal screening: a longitudinal population-based cohort study. J Clin Endocrinol Metab 2011;96(6):1771–82.

[159] Brucker-Davis F, Skarulis MC, Pikus A, Ishizawar D, Mastroianni MA, Koby M, et al. Prevalence and mechanisms of hearing loss in patients with resistance to thyroid hormone. J Clin Endocrinol Metab 1996;81(8):2768–72.

[160] Ferrara AM, Onigata K, Ercan O, Woodhead H, Weiss RE, Refetoff S. Homozygous thyroid hormone receptor beta-gene mutations in resistance to thyroid hormone: three new cases and review of the literature. J Clin Endocrinol Metab 2012;97(4):1328–36.

[161] Refetoff S, DeWind LT, DeGroot LJ. Familial syndrome combining deaf-mutism, stuppled epiphyses, goiter and abnormally high PBI: possible target organ refractoriness to thyroid hormone. J Clin Endocrinol Metab 1967;27(2):279–94.

[162] Crofton KM. Developmental disruption of thyroid hormone: correlations with hearing dysfunction in rats. Risk Anal 2004;24(6):1665–71.

[163] Moore JK, Linthicum Jr FH. The human auditory system: a timeline of development. Int J Audiol 2007;46(9):460–78.

[164] Sprenkle PM, McGee J, Bertoni JM, Walsh EJ. Development of auditory brainstem responses (ABRs) in Tshr mutant mice derived from euthyroid and hypothyroid dams. J Assoc Res Otolaryngol 2001;2(4):330–47.

[165] Sprenkle PM, McGee J, Bertoni JM, Walsh EJ. Consequences of hypothyroidism on auditory system function in Tshr mutant (hyt) mice. J Assoc Res Otolaryngol 2001;2(4):312–29.

[166] Cordas EA, Ng L, Hernandez A, Kaneshige M, Cheng SY, Forrest D. Thyroid hormone receptors control developmental maturation of the middle ear and the size of the ossicular bones. Endocrinology 2012;153(3):1548–60.

[167] Fraser JS. The pathology of deaf-mutism. Proc R Soc Med 1932;25(6):861–78.

[168] Kelley MW. Regulation of cell fate in the sensory epithelia of the inner ear. Nat Rev Neurosci 2006;7(11):837–49.

[169] Uziel A, Gabrion J, Ohresser M, Legrand C. Effects of hypothyroidism on the structural development of the organ of corti in the rat. Acta Otolaryngol 1981;92:469–80.

[170] Legan PK, Rau A, Keen JN, Richardson GP. The mouse tectorins. Modular matrix proteins of the inner ear homologous to components of the sperm-egg adhesion system. J Biol Chem 1997;272(13):8791–801.

[171] Rau A, Legan PK, Richardson GP. Tectorin mRNA expression is spatially and temporally restricted during mouse inner ear development. J Comp Neurol 1999;405(2):271–80.

[172] Killick R, Richardson GP. Antibodies to the sulphated, high molecular mass mouse tectorin stain hair bundles and the olfactory mucus layer. Hear Res 1997;103(1–2):131–41.

[173] Tritsch NX, Yi E, Gale JE, Glowatzki E, Bergles DE. The origin of spontaneous activity in the developing auditory system. Nature 2007;450(7166):50–5.

[174] Leao RN, Sun H, Svahn K, Berntson A, Youssoufian M, Paolini AG, et al. Topographic organization in the auditory brainstem of juvenile mice is disrupted in congenital deafness. J Physiol 2006;571(Pt 3):563–78.

[175] Hinojosa R. A note on development of Corti's organ. Acta Otolaryngol 1977;84(3–4):238–51.

[176] Rusch A, Ng L, Goodyear R, Oliver D, Lisoukov I, Vennstrom B, et al. Retardation of cochlear maturation and impaired hair cell function caused by deletion of all known thyroid hormone receptors. J Neurosci 2001;21(24):9792–800.

[177] Uziel A, Rabie A, Marot M. The effect of hypothyroidism on the onset of cochlear potentials in developing rats. Brain Res 1980;182(1):172–5.

[178] Crofton KM, Kodavanti PR, Derr-Yellin EC, Casey AC, Kehn LS. PCBs, thyroid hormones, and ototoxicity in rats: cross-fostering experiments demonstrate the impact of postnatal lactation exposure. Toxicol Sci 2000;57(1):131–40.

[179] Crofton KM, Rice DC. Low-frequency hearing loss following perinatal exposure to 3,3′,4,4′,5-pentachlorobiphenyl (PCB 126) in rats. Neurotoxicol Teratol 1999;21(3):299–301.

[180] Goldey ES, Crofton KM. Thyroxine replacement attenuates hypothyroxinemia, hearing loss, and motor deficits following developmental exposure to Aroclor 1254 in rats. Toxicol Sci 1998;45(1):94–105.

[181] Goldey ES, Kehn LS, Lau C, Rehnberg GL, Crofton KM. Developmental exposure to polychlorinated biphenyls (Aroclor 1254) reduces circulating thyroid hormone concentrations and causes hearing deficits in rats. Toxicol Appl Pharmacol 1995;135(1):77–88.

[182] Crofton KM, Ding D, Padich R, Taylor M, Henderson D. Hearing loss following exposure during development to polychlorinated biphenyls: a cochlear site of action. Hear Res 2000;144(1–2):196–204.

[183] Lasky RE, Widholm JJ, Crofton KM, Schantz SL. Perinatal exposure to Aroclor 1254 impairs distortion product otoacoustic emissions (DPOAEs) in rats. Toxicol Sci 2002;68(2):458–64.

[184] Poon E, Powers BE, McAlonan RM, Ferguson DC, Schantz SL. Effects of developmental exposure to polychlorinated biphenyls and/or polybrominated diphenyl ethers on cochlear function. Toxicol Sci 2011;124(1):161–8.

[185] Powers BE, Poon E, Sable HJ, Schantz SL. Developmental exposure to PCBs, MeHg, or both: long-term effects on auditory function. Environ Health Perspect 2009;117(7):1101–7.

[186] Powers BE, Widholm JJ, Lasky RE, Schantz SL. Auditory deficits in rats exposed to an environmental PCB mixture during development. Toxicol Sci 2006;89(2):415–22.

[187] Trnovec T, Sovcikova E, Pavlovcinova G, Jakubikova J, Jusko TA, Hustak M, et al. Serum PCB concentrations and cochlear function in 12-year-old children. Environ Sci Technol 2010;44(8):2884–9.

[188] Min JY, Kim R, Min KB. Serum polychlorinated biphenyls concentrations and hearing impairment in adults. Chemosphere 2014;102:6–11.

Environmental Factors in Neurodevelopmental Disorders: Summary and Perspectives

Lucio G. Costa, Michael Aschner

It has been estimated that neurodevelopmental disorders may affect greater than 15% of all births. Some specific diagnoses, such as attention deficit hyperactivity disorder and autism spectrum disorders (ASD) seem to have undergone exponential increases in recent years (see Chapter 1). In many instances, subtle decreases of brain functions or other behavioral abnormalities may be clinically unnoticeable, yet not less relevant from a societal perspective. There has been an ongoing debate on the etiological factors that contribute to such neurodevelopmental disorders. In several cases, genetic background is known to play a central and most relevant role. Perhaps 30–40% of all neurodevelopmental disorders have a genetic causation; Rett syndrome and trisomy 21 syndrome are two of the best known, but there are hundreds if not thousands of such diseases. Yet, nongenetic environmental factors should be responsible for the remaining 60–70% of neurodevelopmental disorders. It is well known that in utero or early postnatal exposure to certain chemicals is strongly associated with specific neurodevelopmental disorders. Some, such as ethanol (Chapter 3), methylmercury (Chapter 6), and lead (Chapter 7) are highlighted in this volume. However, several more have been associated with alterations in brain development, in experimental in vitro, animal, and epidemiological studies. It has been indicated that over 200 substances are highly toxic to the human nervous system, including several inorganic and organic metal compounds, organic solvents, pesticides, and a wide variety of organic chemicals [1]. While some may affect the brain directly, either in utero or during the early postnatal period, others may have indirect effects; for example, by interfering with hormonal systems. Indeed, several compounds known as endocrine disruptors may be developmental neurotoxicants. For example, Chapter 8 discusses the importance of thyroid hormones in normal brain development, and highlights how disruptions of this hormonal system by various chemicals lead to significant central nervous system (CNS) impairment. Another important consideration relates to the fact that environmental factors are not necessarily synonymous with one or more chemical entities; indeed other "environmental factors" may have much relevance

193

as well. Chapter 4 discusses how prenatal infection plays a most relevant role in the etiology of ASD, schizophrenia, and perhaps other CNS disorders, possibly by activating pro-inflammatory processes. Oxidative and neuroinflammatory processes also appear to be responsible for the adverse effects of air pollution (particularly traffic-related air pollution) on the developing nervous system (Chapter 5), as evidenced in recent studies. It has been suggested by many, and there is evidence in this regard, that a key role in neurodevelopmental disorders may be played by gene–environment interactions, in which one or more "susceptibility genes" increase the likelihood for certain disorders to appear, but only after a certain environmental exposure. Ken Olden, former Director of the National Institute of Environmental Health Sciences (NIEHS), described the relationship between genes and the environment in an interesting way: "genes load the gun, the environment pulls the trigger"; indeed a loaded gun by itself poses no harm, but when the trigger is pulled, the potential for harm is released [2].

There is no doubt that environmental factors directly or indirectly, through interactions with specific genetic predispositions or with epigenetic mechanisms, may affect nervous system development and contribute to neurodevelopmental disorders. Despite the magnitude of the existing situation, the number of developmental neurotoxicants and other factors that negatively affect brain development may be even larger; hence, there is possibly a current underestimation of the real situation worldwide [1]. While some exposures can be clearly and easily avoided (e.g., alcohol during pregnancy), for others (e.g., some food contaminants and air pollutants) the task is harder. What can be done, however, is to further characterize and delineate the dose–response effects of certain exposures on the developing nervous system, so that even if unavoidable, such exposures are kept below specific limits. A better understanding of the cellular and molecular mechanisms that underlie adverse effects on the developing nervous system would also be useful, in order to develop nutritional and/or pharmacological interventions to counteract such actions. Finally, an important strategy would be that of preventing the further use and release into the environment of substances that may affect the developing nervous system. This may be accomplished by implementing more rigorous screening approaches for novel entities, utilizing standard toxicological approaches, and using new rapid screening tools that are becoming available.

References

[1] Grandjean P, Landrigan PJ. Neurobehavioural effects of developmental neurotoxicity. Lancet Neurol 2014;13:330–8.
[2] Costa LG, Eaton DL. Gene-environment interactions. Fundamentals of ecogenetics. Hoboken, NJ: John Wiley & Sons; 2006. p. 556.

NEURODEGENERATIVE DISORDERS

Overview of Neurodegenerative Disorders and Susceptibility Factors in Neurodegenerative Processes

Ruth E. Musgrove, Sarah A. Jewell, Donato A. Di Monte

CLASSIFICATION

A wide variety of clinical and pathological conditions can generally be classified as "neurodegenerative diseases." Typical examples are diseases such as Alzheimer's disease (AD), Parkinson's disease (PD), Huntington's diseases (HD), amyotrophic lateral sclerosis (ALS), and spinocerebellar ataxia; in each case, a progressive degenerative process cripples neuronal function, disables synaptic transmission/brain circuitry, and ultimately leads to neuronal death and brain atrophy. The prevalence of these disorders varies quite significantly. AD is the most prevalent, affecting greater than five million people in the United States alone [1]. PD is the second most common, with an estimated 600,000 to one million US patients [2].

Other neurodegenerative diseases are relatively rare; for example, HD affects only approximately 40,000 individuals in the United States [3].

Historically, neurodegenerative diseases have been recognized and defined based on relatively late pathological and clinical manifestations, examples being the widespread presence of amyloid plaques in the postmortem AD brain and the occurrence of frank dementia in AD patients. Major efforts over the past two decades have instead focused on the identification and characterization of early events in the disease process associated with neuronal dysfunction rather than neuronal death (e.g., loss of synaptic plasticity), and with subtle clinical signs (e.g., mild cognitive impairment). Several considerations justify this new interest. A clearer understanding of early degenerative processes would be fundamental to unraveling temporal and causal relationships between molecular, cellular, and pathological changes at disease onset and during disease development. For example, it could shed light upon mechanisms of selective vulnerability by which distinct neuronal populations are targeted during the development of each neurodegenerative disease. Investigations into early pathogenetic events will likely contribute to the identification of disease biomarkers that could be used for diagnostic purposes and for assessing pathological and clinical progression. Finally, a primary yet unmet goal in the field of neurodegenerative diseases is the implementation of preventive public health measures and development of therapeutic strategies capable of halting or at least slowing the disease process. Disease-modifying intervention (i.e., therapeutic approaches that not only affect symptoms but are capable of counteracting the disease process itself) would be most effective if initiated as early as possible, and should therefore target incipient neurodegenerative changes.

The shift of focus from neuronal death toward neuronal dysfunction and network abnormalities bears important implications, not only from pathogenetic and therapeutic standpoints but also from a nosological one. It is foreseeable that in the near future the boundary between "classic" neurodegenerative diseases and other neurological disorders will become less stringent, and be replaced by a more continuous spectrum of diseases that albeit distinct from each other, share similar pathways and mechanisms of neuronal perturbation. An intriguing example of mechanism-based disease associations is given by recent findings linking AD to epilepsy. AD patients are at higher risk for seizures, and the occurrence of seizures in AD patients is associated with more severe symptoms, more rapid disease progression, and greater neurodegenerative changes at autopsy [4,5]. Experimental studies are also consistent with aberrant neuronal firing being a feature shared by AD and epilepsy; in the former, increased network excitability may represent an early pathogenetic event causally linked to synaptic deficit, and ultimately cognitive dysfunction [6].

MECHANISTIC CLUES

The importance of studying early manifestations of neurodegenerative diseases is also evident when it comes to disease risk factors, since environmental conditions and genetic variations could modulate clinical and pathological signs of these diseases, particularly during their initial development. A recent example is the finding of a relationship between occupational exposure to solvents and cognitive deficits, which may represent early manifestations of or predisposal to the development of AD-like neurodegenerative changes [7]. A corollary to these emerging concepts is that a great deal of interest over the past decade has been directed toward the identification of molecular mechanisms likely to be involved in early pathogenetic processes. It is not surprising, in this respect, that both experimental and pathological investigations have focused on an intriguing feature common to most neurodegenerative diseases, that is, the intra- or extracellular accumulation of disease-specific proteins into insoluble aggregates. Proteins include (1) β-amyloid and tau that form senile plaques and neurofibrillary tangles, respectively, in AD; (2) α-synuclein, which is a primary component of Lewy bodies and Lewy neurites in PD; (3) polyglutamine-rich huntingtin, which is accumulated in HD; (4) infectious prion protein (PrP^{Sc}) that triggers the formation of plaques in Creutzfeldt–Jakob disease (CJD); and (5) TDP-43 (trans-activator regulatory DNA-binding protein 43), which is often aggregated in ALS and frontotemporal dementia.

Protein accumulation and inclusion formation could result from a variety of mechanisms acting separately, additively, or synergistically. Intrinsic biochemical properties of disease-specific proteins (e.g., their structural conformation), which may be affected by genetic and toxic factors, have been shown to modulate protein tendencies to aggregate. For example, changes in α-synuclein conformation and protein aggregation can be promoted by (1) mutations in the α-synuclein gene (SNCA) linked to familial forms of parkinsonism, such as an alanine-to-threonine substitution in position 53 of the protein's amino acid sequence [8,9], and (2) α-synuclein interactions with metals or pesticides [10,11].

An imbalance between protein expression and neuronal mechanisms of protein degradation could also favor aggregation reactions by increasing intracellular concentrations of disease-associated proteins; this is because the propensity to aggregate is enhanced at higher protein concentrations [9,12]. It is therefore noteworthy that several lines of evidence from experimental investigations and postmortem pathological observations support the hypothesis that an impairment of degradation mechanisms (e.g., the ubiquitin–proteasome and autophagic pathways) plays a role in protein accumulation and aggregation in AD, PD, HD, and other neurodegenerative disorders [13–15]. Furthermore, aging (an unequivocal risk factor for human neurodegenerative diseases) is characterized by abnormal clearance of proteins and protein aggregates that could ultimately contribute to neuronal overload and predispose to further pathology [16,17].

A relatively novel concept in the pathogenesis of neurodegenerative diseases has been derived from the prion field and bears important implications from the standpoint of environmental exposures. Misfolding of prion proteins, replication of the misfolded conformation, templated protein assembly, and interneuronal transmission of pathological aggregates are key features that underlie the infectivity of prions and their ability to cause progressive neurodegenerative pathology. A similar sequence of events has been proposed to explain the toxicity of other proteins (e.g., β-amyloid, tau, and α-synuclein) associated with neurodegenerative diseases, and could possibly be triggered by a yet unidentified environmental pathogen [18,19]. The tendency of proteins such as β-amyloid or α-synuclein to aggregate, as discussed above, is not itself evidence of "prion-like" behavior. However, these disease-associated proteins possess other important features that were originally attributed to prion particles—they can spread from one neuron to another and diffuse throughout the brain via anatomically interconnected pathways [20–23] (Figure 10.1). Misfolded aggregated proteins

FIGURE 10.1 (A) As initially proposed by Braak and colleagues [28], α-synuclein pathology in PD progressively spreads from the medulla oblongata toward higher brain regions, reaching the midbrain and ultimately the cerebral cortex. (B) A similar pattern of caudorostral protein spreading has been reproduced in animal models (see schematic view of a rat brain) [23], suggesting α-synuclein propagation via anatomically interconnected pathways.

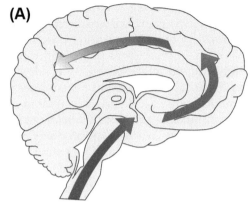

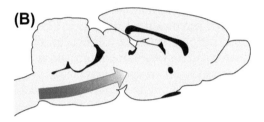

can also act as corruptive templates (or seeds) capable of recruiting unaltered soluble proteins, thus initiating a self-perpetuating process of pathological amplification [24–26].

Prion-like properties of disease-associated proteins have mostly been documented experimentally. Their involvement in pathogenetic processes is supported, however, by key post-mortem observations in diseased human brain. For instance, interneuronal protein spreading via axonally connected regions could well underlie the stereotypical pattern of pathological disease progression that is characteristically seen in the brain of AD or PD patients [27,28] (Figure 10.1). A note of caution should be added, by emphasizing that in our opinion, prion-like properties of proteins involved in neurodegenerative diseases do not necessarily imply that AD, PD, or HD should now be considered prion diseases. The latter are defined as transmissible spongiform encephalopathies caused by infectious prion particles. Proteins like β-amyloid and α-synuclein may be capable of propagating from cell to cell; no evidence to date, however, indicates that they can be transmitted from human to human as disease-causing infectious agents [29].

COMPLEX DISORDERS

Neurodegenerative diseases are multifaceted heterogeneous disorders. From a clinical standpoint, they are characterized by a wide spectrum of symptomatic manifestations that not only distinguish one disease from another but also can vary among patients affected by the same neurodegenerative disorder. PD patients, for example, can share similar motor abnormalities, while at the same time being affected by a diverse range of nonmotor symptoms such as cognitive impairment, depression, and autonomic dysfunction. Remarkable clinical variability characterizes diseases that arise from similar pathogenetic mechanisms. Within the group of prion diseases, rapidly progressive dementia is a primary symptom of CJD, whereas cerebellar ataxia and inability to sleep are typical of kuru and fatal familial insomnia, respectively [30]. The onset and clinical progression of neurodegenerative diseases can also vary significantly. In the instance of HD, the length of cytosine–adenine–guanine trinucleotide repeats within the *Huntingtin* gene affects prognosis, as a longer expansion results in an earlier age of onset and possibly a more rapid symptomatic progression [31]. From a pathological standpoint, inter- and intradisease heterogeneity is reflected by an assortment of neurodegenerative changes that include the intriguing feature of mixed pathologies. Tau and α-synuclein aggregates are pathological hallmarks of AD and PD, respectively. They can also coexist, however, in the brains of patients with typical or atypical (e.g., PD with dementia) forms of these diseases. Copathology is not a rare occurrence, can involve proteins other than tau and α-synuclein, and is often associated with severe clinical manifestations and an unfavorable prognosis [32].

The terms "complex disorders" are often used in reference to neurodegenerative diseases. Strictly, this designation is meant to indicate the multifactorial etiology of AD, PD, and other neurodegenerative disorders involving intricate interactions between inherited and environmental factors. Neurodegenerative diseases can be triggered by a single genetic or environmental cause, examples being (1) mutations in the *Huntingtin* gene causing HD [31], (2) mutations in the genes encoding amyloid precursor protein (APP) or presenilin responsible for early-onset, autosomal-dominant AD [33], (3) mutations in *SNCA* or other genes (e.g., *Park2/parkin* or *Park7/DJ-1* genes) underlying familial forms of parkinsonism [34], and (4) exposure to the neurotoxicant 1-methyl-4-phenyl-1,2,3,6-tetrahydropyridine causing a parkinsonian syndrome [35]. However, monogenic or monoexposure forms of AD, PD, and other neurodegenerative disorders (e.g., ALS) only explain a small percentage of cases. Most likely, sporadic neurodegenerative diseases are multifactorial, justifying their definition as complex disorders.

Evidence accumulated over the past decade is consistent with the concept of genetic and environmental factors modulating the risk of neurodegenerative diseases, such as PD and AD, against the background of age-related increases in susceptibility. Possible mechanisms involved in this age-dependent vulnerability have been mentioned above (see section Mechanistic Clues). It is noteworthy to add that at the time of clinical diagnosis, most neurodegenerative disorders have already been developing for several years and possibly decades. Thus, although aging should continue to be considered an important risk factor, neurodegenerative diseases may not necessarily be classified as illnesses of old age; rather, they are likely to begin in middle-aged (and sometimes younger) individuals and are characterized by a progressive worsening of clinical manifestations in aging-affected patients.

Research efforts over the past decade have led to the identification of genetic variants capable of modifying the risk for neurodegenerative disorders, including *APOE* (apolipoprotein E) polymorphisms (e.g., the *APOE* ε4 allele) in AD, and variations in the Rep1 promoter region of *SNCA* in PD (see below) [33,36]. Advances have also been made in our understanding of the role of environmental factors in pathogenetic processes; particularly interesting are findings concerning the influence of dietary habits and other lifestyle patterns (e.g., exercising) on disease development and progression. Relatively less is known about how genetic, environmental, and age-related risk factors interact, making this topic a primary focus for future investigations at the clinical, epidemiological, and experimental levels.

A complete list of novel findings on risk factors in neurodegenerative disorders would be beyond the scope of this chapter. We will therefore focus next on a few examples that (1) underscore the relevance of lifestyle modifiers, (2) provide evidence of gene–environment interactions, and (3) reveal the emergence of exciting research avenues for further investigating relationships between the environment and human neurodegenerative diseases.

LIFESTYLE MODIFIERS OF NEURODEGENERATIVE DISEASE RISK

Every human is unique not only in genetic constitution, but in the environmental exposures encountered over the course of a lifetime. In broad terms, these environmental interactions encompass lifestyle factors, generally understood as ways of living that can impact health in positive or negative ways. Lifestyle factors are important considerations because they offer insight into the biological mechanisms of disease risk (or resilience) and—unlike nonmodifiable factors such as age, gender, and genotype—provide opportunities to apply effective disease prevention methods. Such intervention could be aimed at protecting healthy people from developing disease in the first place, as well as limiting the progression and impact of disease in its early stages (i.e., primary and secondary prevention).

One intriguing example of lifestyle interaction with neurodegenerative disease has emerged from strong epidemiological evidence that smokers have a significantly reduced (approximately halved) risk of PD compared with nonsmokers [37–39]. The mechanisms underlying this consistently reported finding have not been defined. Since the effect is dependent upon dose (both average and total lifetime exposure) and time since cessation [40], a causal mechanism is likely, in which smoking results in exposure to neuroprotective agent(s) and/or triggers compensatory mechanisms. It has been hypothesized that nicotine is one of the key tobacco components that could mediate a disease-preventing effect [41,42]. Interestingly, nicotine may also provide relief from the dyskinesias that often arise after long-term PD therapy with L-dopa [43]. Clinical trials are now underway to assess the efficacy of nicotine as both a preventive agent and one to be used in the clinical management of advanced disease.

In the case of dementia caused by AD (and mixed pathology from cerebrovascular disease), there is ample evidence to support lifestyle-based intervention efforts; in fact, it has been estimated that up to half of all AD cases may be attributable to modifiable risk factors [44]. Many factors, including hypertension, diabetes, obesity, dyslipidemia, smoking, low cognitive reserve, and physical inactivity contribute to disease risk; these factors often coexist and may exert combined effects [45]. Despite some ominous societal trends in the prevalence of these conditions, there is epidemiological evidence of decreasing dementia incidence [46,47]. A comparison of dementia incidence within two enrollment cohorts (starting in 1990 vs 2000) within the population-based Rotterdam study demonstrated that age-adjusted dementia incidence rates were consistently lower in the later cohort [47]. The later cohort also had less brain atrophy and small-blood-vessel disease in magnetic-resonance brain scans.

These changes were attributable at least in part to increased use of antithrombotic and lipid-lowering medications.

Another interesting question is whether lifestyle-based preventive measures could counteract strong genetic risk of disease. Evidence in support of this possibility concerns the *APOE* ε4 allele in AD and is discussed in section Gene–Environment Interactions.

GENE–ENVIRONMENT INTERACTIONS

It has long been hypothesized that in neurodegenerative disorders, environmental factors could contribute to disease pathogenesis against a background of genetic susceptibility, and vice versa. Growing evidence from both epidemiological and experimental investigations supports this hypothesis. In AD, several studies have investigated interactions between environmental factors and the most prevalent genetic modulator of AD risk, that is, variations in the *APOE* gene. In PD, these investigations have focused mostly on the synergistic effects that genetic variations and environmental insults may have on α-synuclein expression and pathology.

Apolipoprotein E

Apolipoprotein E (APOE) is a member of the lipoprotein family of cholesterol transporters. Within the central nervous system (CNS), it mediates functions that are critical for neuronal development, plasticity, and repair [48,49]. The three most prevalent APOE isoforms are differentially encoded by the *APOE* ε2, ε3, and ε4 allele variants [48]. The *APOE* ε4 variant is the most prevalent modulator of AD risk. Inheritance of this allele is associated with enhanced susceptibility, ranging from an approximate threefold to ~15-fold increase in heterozygote and homozygote individuals, respectively [50,51]. In contrast, ε2 allele inheritance is negatively correlated with disease risk [52]. Several mechanisms could account for the ability of APOE to influence AD risk. For example, APOE is capable of binding soluble β-amyloid peptide. This binding property, which is most pronounced for APOE2 and least efficient with APOE4, could facilitate β-amyloid degradation and removal [48,53].

As mentioned above, an important aspect of APOE-associated disease risk concerns the possibility that enhanced vulnerability conferred by the ε4 allele may be modulated by lifestyle factors. For example, several observational studies have suggested that regular exercise/fitness correlates with better cognitive performance in *APOE* ε4 carriers, to an even greater extent than in noncarriers [54,55]. The neuroprotective effect of physical activity is also suggested by studies employing functional and structural magnetic resonance imaging of the brain [56,57]. One recent longitudinal study evaluated leisure-time physical activity and change in hippocampus volume (a biomarker for AD progression), over a relatively brief time frame of 18 months in cognitively intact adults who were either ε4 carriers or noncarriers. The hippocampal volume decreased by 3% in the ε4 carriers with low physical activity, but was maintained in those with moderately vigorous frequency and intensity of exercise [56]. Thus, although physical activity is known to preserve cognitive performance in general, it may be particularly important in forestalling the consequences of genetic risk for dementia. These studies highlight the importance of including a biologically based stratification in the design of human studies, to identify priority cohorts for public health intervention.

α-Synuclein

SNCA mutations that are associated with familiar parkinsonism include single-point mutations resulting in "qualitative" changes in protein sequence, as well as multiplication mutations that lead to increased expression of the normal protein [58]. Genome-wide association studies have also identified *SNCA* as a major PD risk locus [59]. In particular, genetic variation in the *SNCA* promoter region, such as variability in the length of the dinucleotide repeat sequence REP1, have been reported to modulate PD risk, most likely by affecting α-synuclein levels in the brain, blood, and other tissues [36,60]. Taken together, these observations not only underscore the prominent role of α-synuclein in human parkinsonism (both familial and sporadic), but also indicate a relationship between enhanced neuronal levels of α-synuclein and pathogenetic processes. Enhanced protein expression is causally related to familial parkinsonism in patients with *SNCA* multiplication mutations; conversely, it represents a disease risk factor in individuals who carry polymorphic *SNCA* variants.

The pathogenetic relevance of increased α-synuclein levels, as indicated by genetic findings, bears a number of broader implications. First, experimental work has revealed that exposure to toxic agents is capable of triggering neuronal upregulation of α-synuclein in a variety of animal models [61,62]. It is therefore possible that one of the mechanisms by which toxic environmental challenges enhance neuronal vulnerability to α-synuclein-related pathology is through an elevation of protein expression. Secondly, investigations in both humans and experimental animals have shown a progressive increase in α-synuclein expression as a function of age within neuronal populations highly vulnerable to neurodegenerative processes of PD (e.g., dopaminergic neurons in the substantia nigra pars compacta (SNpc)) [63,64]. As discussed above, higher α-synuclein concentrations could trigger pathological changes such as protein aggregation and interneuronal protein spreading, thus suggesting the following scenario: genetic, environmental, or age-related PD risk factors could act additively or synergistically by raising the intraneuronal levels of α-synuclein, thus promoting a protein gain of toxic/pathological function.

Epidemiological evidence is consistent with this scenario, supporting gene–environment interactions centered on α-synuclein. Exposure to pesticides and traumatic brain injury are two conditions often (though not always) found to be associated with increased PD risk by population-based studies [65–67]. When recent investigations have correlated these environmental challenges with specific genotypes based on the allele length of the *SNCA* REP1 region, data have revealed remarkable interactions. Confirming earlier investigations, Gatto and colleagues [68] reported that the short repeat length REP1 259 genotype conferred protection against PD, while the long repeat REP1 263 genotype was associated with increased disease risk. Interestingly, study results also indicated gene–environment interactions between *SNCA* variations and pesticide exposure. In particular, pesticide exposure counteracted the protective effect of short REP1, whereas long REP1 increased pesticide susceptibility, particularly in younger-onset patients.

Goldman et al. [69] did not find an overall association between head injury and PD risk. However, when individuals in this case–control study were interviewed regarding head injuries as well as genotyped for Rep1, a significant positive association between brain trauma and long REP1 became evident (odds ratio 3.5). We believe that genetic, pathological, experimental, and epidemiological data on α-synuclein represent an exciting example of convergent research through interdisciplinary approaches. Stemming from genetic and clinical evidence (e.g., REP1 variability), experimental work has validated mechanisms of likely pathogenetic relevance (e.g., the role of enhanced α-synuclein expression), and epidemiological investigations have focused on specific gene–environment interactions underlying PD susceptibility.

EPIGENETICS

As defined by the National Institutes of Health's *Roadmap Epigenomics Project*, "epigenetics is an emerging frontier of science that involves the study of changes in the regulation of gene activity and expression that are not dependent on gene sequence" (http://www.roadmapepigenomics.org/overview). Epigenetic modifications regulate protein synthesis and cellular phenotypes in a heritable yet reversible fashion. There is an obvious relationship between epigenetics and environment, and epigenetic alterations can indeed result from a variety of environmental conditions such as micronutrient availability, exposure to xenobiotics, stress, and social interactions [70–72]. It has been proposed that epigenetic modifications, including DNA methylation, histone changes, and micro-RNA–associated gene silencing [73], play important roles in the pathogenesis of human diseases. Pertinent to the topic of this chapter, initial evidence also supports their involvement in neurodegenerative processes [70,72,74,75]. The relationship between epigenetics, environment, and neurodegenerative disorders represents a promising area of clinical and experimental research that could have a direct impact on personalized, preventive, and curative medicine. The following examples of DNA methylation and histone modifications were chosen to underscore this important concept.

DNA Methylation

DNA methylation represses gene transcription by physically interfering with transcription factor-binding sites [75]. This process is mediated by DNA methyltransferases that catalyze the addition of methyl groups to the cysteine residue of CpG dinucleotides [76]. An interesting model of interactions among environmental exposures, DNA methylation, and neurodegenerative processes has been described as a consequence of lead (Pb) administration to rodent and nonhuman primates. Pb exposure during development (in rats) or early life (in primates) caused an upregulation of AD-related proteins (e.g., APP and β-site APP cleaving enzyme 1 (BACE1)), intracellular accumulation of β-amyloid, and evidence of AD-like pathology (amyloid plaques) in aged animals [77,78]. These effects were attributed to Pb-mediated downregulation of DNA methyltransferase leading to hypomethylation of the APP promoter (this promoter region is rich in CpG dinucleotides and therefore highly prone to methylation [79,80]). Methylation of the APP promoter and transcription of APP mRNA are inversely correlated [81,82], making it possible that Pb-mediated APP hypomethylation ultimately leads to increased APP expression and higher levels of β-amyloid. Overall, this model supports the hypothesis that environmental exposures and consequent epigenetic imprinting in early life may have long-lasting effects and influence susceptibility to neurodegenerative processes, such as those involving β-amyloid in AD.

Histone Modifications

Another mechanism (besides DNA methylation) for regulation of gene transcription involves posttranslational modifications of chromatin proteins, in particular histones. The spatial organization of histones, which is affected by their acetylation, regulates the physical accessibility of genes along the chromosome. The degree of histone acetylation is controlled by the opposing activities of histone acetylases and histone deacetylases (HDACs) [83]. Hyperacetylated histones are associated with chromosomal regions available for transcription, whereas hypoacetylation leads to chromatin condensation and impaired transcription [73,83]. Several lines of evidence support an important role of histone acetylation in neurodegenerative disease [75,84]. Particularly intriguing is the work of Fischer and colleagues [85], who utilized a transgenic mouse model characterized by severe synaptic and neuronal loss in the forebrain. In this model, environmental enrichment was able to restore learning by promoting growth of new dendrites and synapses. This effect was associated with enhanced histone acetylation in the hippocampus and cortical regions, and could be reproduced by pharmacological treatment with HDAC inhibitors.

As a noteworthy follow-up to this study, these investigators also showed that histone acetylation at a specific site (lysine 12 of histone 4, H4K12) modulates memory processes in the aging brain [86]. In young mice, increased hippocampal histone acetylation paralleled learning, whereas in older animals, memory disturbances were associated with lack of H4K12 acetylation and impaired hippocampal gene expression. Thus, epigenetic changes that could be influenced by environmental conditions (e.g., enriched environment) and targeted by drugs appear capable of protecting and reestablishing neuronal networks critical for learning and memory. By doing so, histone modifications may contribute to the onset and progression of cognitive dysfunction characteristic of human neurodegenerative diseases (see above in section Classification).

CONCLUSIONS

The public health and financial burdens associated with human neurodegenerative diseases are already substantial and expected to increase in the next few decades [87]. Prolonged life expectancy of the general population will result in enhanced incidence of age-related diseases such as AD and PD unless preventive intervention becomes available. This trend has been one of the drivers for the growing interest in neurodegenerative diseases from governmental institutions, research funding agencies, and other for-profit (e.g., pharmaceutical companies) and nonprofit (e.g., patient advocacy groups) organizations. Within this context of greater awareness and more pronounced research efforts, the issue of environmental factors modulating neurodegenerative processes warrants special attention, if for no other reason than its direct impact on the design of public health measures for disease prevention (see above in section Lifestyle Modifiers of Neurodegenerative Disease Risk).

Conventionally, our view of how environmental factors may contribute to pathogenetic processes has mostly focused on the role of toxic agents (e.g., industrial chemicals, aerial pollutants, and food poisons) that could enhance disease susceptibility by damaging cells and perturbing tissue function. The importance of such toxic insults in neurodegenerative processes should not be underestimated. On the other hand, the role of the environment in the pathogenesis of neurodegenerative diseases should also be seen from a wider perspective and using a more comprehensive definition of environmental determinants. As reviewed above, important factors and mechanisms likely to enhance or lower disease risk include lifestyle habits (e.g., food consumption and exercising), use of medications (e.g., cholesterol-lowering drugs), and epigenetic modifications. Other intriguing determinants, such as microbiomal changes in the nasal cavity or gastrointestinal tract, and activation of the inflammasome protein complex, have recently been suggested by clinical evidence and/or experimental studies [88–90]. These new findings underscore the wide range of means by which disease pathogenesis could be affected by environmental modifiers. They are also a reflection of research perspectives that are rapidly evolving and will likely have significant impacts on our understanding of neurodegenerative diseases and our approaches to disease prevention and treatment.

References

[1] Thies W, Bleiler L. Alzheimer's disease facts and figures. Alzheimers Dement 2013;9:208–45.
[2] Mayeux R. Epidemiology of neurodegeneration. Annu Rev Neurosci 2003;26:81–104.
[3] Pringsheim T, Wiltshire K, Day L, Dykeman J, Steeves T, Jette N. The incidence and prevalence of Huntington's disease: a systematic review and meta-analysis. Mov Disord 2012;27:1083–91.
[4] Palop JJ, Mucke L. Epilepsy and cognitive impairments in Alzheimer disease. Arch Neurol 2009;66:435–40.
[5] Vossel KA, Beagle AJ, Rabinovici GD, Shu H, Lee SE, Naasan G, et al. Seizures and epileptiform activity in the early stages of Alzheimer disease. JAMA Neurol 2013;70:1158–66.
[6] Sanchez PE, Zhu L, Verret L, Vossel KA, Orr AG, Cirrito JR, et al. Levetiracetam suppresses neuronal network dysfunction and reverses synaptic and cognitive deficits in an Alzheimer's disease model. Proc Natl Acad Sci USA 2012;109:e2895–90.
[7] Sabbath EL, Gutierrez LA, Okechukwu CA, Singh-Manoux A, Amieva H, Goldberg M, et al. Time may not fully attenuate solvent-associated cognitive deficits in highly exposed workers. Neurology 2014;82:1716–23.
[8] Conway KA, Harper JD, Lansbury PT. Accelerated in vitro fibril formation by a mutant α-synuclein linked to early-onset Parkinson disease. Nat Med 1998;4:1318–20.
[9] Li J, Uversky VN, Fink AL. Effect of familial Parkinson's disease point mutations A30P and A53T on the structural properties, aggregation, and fibrillation of human α-synuclein. Biochemistry 2001;40:11604–13.
[10] Uversky VN, Li J, Fink AL. Metal-triggered structural transformations, aggregation, and fibrillation of human α-synuclein. A possible molecular NK between Parkinson's disease and heavy metal exposure. J Biol Chem 2001;276:44284–96.
[11] Manning-Bog AB, McCormack AL, Li J, Uversky VN, Fink AL, Di Monte DA. The herbicide paraquat causes up-regulation and aggregation of α-synuclein in mice. J Biol Chem 2002;277:1641–4.
[12] Uversky VN. Neuropathology, biochemistry, and biophysics of α-synuclein aggregation. J Neurochem 2007;103:17–37.
[13] Cuervo AM, Wong ES, Martinez-Vicente M. Protein degradation, aggregation, and misfolding. Mov Disord 2010;25:S49–54.
[14] Harris H, Rubinsztein DC. Control of autophagy as a therapy for neurodegenerative disease. Nat Rev Neurol 2011;8:108–17.
[15] Tanaka K, Matsuda N. Proteostasis and neurodegeneration: the roles of proteasomal degradation and autophagy. Biochim Biophys Acta 2014;1843:197–204.
[16] Koga H, Kaushik S, Cuervo AM. Protein homeostasis and aging: the importance of exquisite quality control. Ageing Res Rev 2011;10:205–15.

[17] Tan CC, Yu JT, Tan MS, Jiang T, Zhu XC, Tan L. Autophagy in aging and neurodegenerative diseases: implications for pathogenesis and therapy. Neurobiol Aging 2014;35:941–57.

[18] Brundin P, Melki R, Kopito R. Prion-like transmission of protein aggregates in neurodegenerative diseases. Nat Rev Mol Cell Biol 2010;11:301–7.

[19] Jucker M, Walker LC. Self-propagation of pathogenic protein aggregates in neurodegenerative diseases. Nature 2013;501:45–51.

[20] Clavaguera F, Bolmont T, Crowther RA, Abramowski D, Frank S, Probst A, et al. Transmission and spreading of tauopathy in transgenic mouse brain. Nat Cell Biol 2009;11:909–13.

[21] Eisele YS, Obermüller U, Heilbronner G, Baumann F, Kaeser SA, Wolburg H, et al. Peripherally applied Aβ-containing inoculates induce cerebral β-amyloidosis. Science 2010;330:980–2.

[22] Luk KC, Kehm VM, Zhang B, O'Brien P, Trojanowski JQ, Lee VM. Intracerebral inoculation of pathological α-synuclein initiates a rapidly progressive neurodegenerative α-synucleinopathy in mice. J Exp Med 2012;209:975–86.

[23] Ulusoy A, Rusconi R, Pérez-Revuelta BI, Musgrove RE, Helwig M, Winzen-Reichert B, et al. Caudo-rostral brain spreading of α-synuclein through vagal connections. EMBO Mol Med 2013;5:1051–9.

[24] Guo JL, Covell DJ, Daniels JP, Iba M, Stieber A, Zhang B, et al. Distinct α-synuclein strains differentially promote tau inclusions in neurons. Cell 2013;154:103–17.

[25] Langer F, Eisele YS, Fritschi SK, Staufenbiel M, Walker LC, Jucker M. Soluble Aβ seeds are potent inducers of cerebral β-amyloid deposition. J Neurosci 2011;31:14488–95.

[26] Masuda-Suzukake M, Nonaka T, Hosokawa M, Oikawa T, Arai T, Akiyama H, et al. Prion-like spreading of pathological α-synuclein in brain. Brain 2013;136:1128–38.

[27] Braak H, Braak E. Neuropathological stageing of Alzheimer-related changes. Acta Neuropathol 1991;82:239–59.

[28] Braak H, Del Tredici K, Rüb U, de Vos RA, Jansen Steur EN, Braak E. Staging of brain pathology related to sporadic Parkinson's disease. Neurobiol Aging 2003;24:197–211.

[29] Irwin DJ, Abrams JY, Schonberger LB, Leschek EW, Mills JL, Lee VM, et al. Evaluation of potential infectivity of Alzheimer and Parkinson disease proteins in recipients of cadaver-derived human growth hormone. JAMA Neurol 2013;70:462–8.

[30] Jackson WS. Selective vulnerability to neurodegenerative disease: the curious case of prion protein. Dis Model Mech 2014;7:21–9.

[31] Ross CA, Tabrizi SJ. Huntington's disease: from molecular pathogenesis to clinical treatment. Lancet Neurol 2011;10:83–98.

[32] Irwin DJ, Lee VM, Trojanowski JQ. Parkinson's disease dementia: convergence of α-synuclein, tau and amyloid-β pathologies. Nat Rev Neurosci 2013;14:626–36.

[33] Guerreiro RJ, Gustafson DR, Hardy J. The genetic architecture of Alzheimer's disease: beyond APP, PSENs and APOE. Neurobiol Aging 2012;33:437–56.

[34] Trinh J, Farrer M. Advances in the genetics of Parkinson disease. Nat Rev Neurol 2013;9:445–54.

[35] Di Monte DA, Lavasani M, Manning-Bog AB. Environmental factors in Parkinson's disease. Neurotoxicology 2002;23:487–502.

[36] Maraganore DM, de Andrade M, Elbaz A, Farrer MJ, Ioannidis JP, Krüger R, et al. Collaborative analysis of α-synuclein gene promoter variability and Parkinson disease. JAMA 2006;296:661–7.

[37] Bronstein J, Carvey P, Chen H, Cory-Slechta D, DiMonte D, Duda J, et al. Meeting report: consensus statement-Parkinson's disease and the environment: collaborative on health and the environment and Parkinson's Action Network (CHE PAN) conference 26–28 June 2007. Environ Health Perspect 2009;117:117–21.

[38] Ritz B, Ascherio A, Checkoway H, Marder KS, Nelson LM, Rocca WA, et al. Pooled analysis of tobacco use and risk of Parkinson disease. Arch Neurol 2007;64:990–7.

[39] Tanner CM, Goldman SM, Aston DA, Ottman R, Ellenberg J, Mayeux R, et al. Smoking and Parkinson's disease in twins. Neurology 2002;58:581–8.

[40] van der Mark M, Nijssen PC, Vlaanderen J, Huss A, Mulleners WM, Sas AM, et al. A case-control study of the protective effect of alcohol, coffee, and cigarette consumption on Parkinson disease risk: time-since-cessation modifies the effect of tobacco smoking. PLoS One 2014;9:e95297.

[41] Kawamata J, Suzuki S, Shimohama S. α7 nicotinic acetylcholine receptor mediated neuroprotection in Parkinson's disease. Curr Drug Targets 2012;13:623–30.

[42] Quik M, O'Neill M, Perez XA. Nicotine neuroprotection against nigrostriatal damage: importance of the animal model. Trends Pharmacol Sci 2007;28:229–35.

[43] Xie CL, Pan JL, Zhang SF, Gan J, Liu ZG. Effect of nicotine on L-dopa-induced dyskinesia in animal models of Parkinson's disease: a systematic review and meta-analysis. Neurol Sci 2014;35:653–62.

[44] Barnes DE, Yaffe K. The projected effect of risk factor reduction on Alzheimer's disease prevalence. Lancet Neurol 2011;10:819–28.

[45] Mangialasche F, Kivipelto M, Solomon A, Fratiglioni L. Dementia prevention: current epidemiological evidence and future perspective. Alzheimers Res Ther 2012;4:6.

[46] Rocca WA, Petersen C, Knopman DS, Hebert LE, Evans DA, Hall KS, et al. Trends in the incidence and prevalence of Alzheimer's disease, dementia, and cognitive impairment in the United States. Alzheimers Dement 2011;7:80–93.

[47] Schrijvers EM, Verhaaren BF, Koudstaal PJ, Hofman A, Ikram MA, Breteler MM. Is dementia incidence declining?: Trends in dementia incidence since 1990 in the Rotterdam Study. Neurology 2012;78:1456–63.

[48] Kim J, Basak JM, Holtzman DM. The role of apolipoprotein E in Alzheimer's disease. Neuron 2009;63:287–303.

[49] Mahley RW, Huang Y. Apolipoprotein E sets the stage: response to injury triggers neuropathology. Neuron 2012;76:871–85.

[50] Bettens K, Sleegers K, Van Broeckhoven C. Genetic insights in Alzheimer's disease. Lancet Neurol 2013;12:92–104.

[51] Corder EH, Saunders AM, Strittmatter WJ, Schmechel DE, Gaskell PC, Small GW, et al. Gene dose of apolipoprotein E type 4 allele and the risk of Alzheimer's disease in late onset families. Science 1993;261:921–3.

[52] Corder EH, Saunders AM, Risch NJ, Strittmatter WJ, Schmechel DE, Gaskell PCJ, et al. Protective effect of apolipoprotein E type 2 allele for late onset Alzheimer disease. Nat Genet 1994;7:180–4.

[53] Strittmatter WJ, Weisgraber KH, Huang DY, Dong LM, Salvesen GS, Pericak-Vance M, et al. Binding of human apolipoprotein E to synthetic amyloid β peptide: isoform-specific effects and implications for late-onset Alzheimer disease. Proc Natl Acad Sci USA 1993;90:8098–102.

[54] Etnier JL, Caselli RJ, Reiman EM, Alexander GE, Sibley BA, Tessier D, et al. Cognitive performance in older women relative to ApoE-ε4 genotype and aerobic fitness. Med Sci Sports Exerc 2007;39:199–207.

[55] Purnell C, Gao S, Callahan CM, Hendrie HC. Cardiovascular risk factors and incident Alzheimer disease: a systematic review of the literature. Alzheimer Dis Assoc Disord 2009;23:1–10.

[56] Smith JC, Nielson KA, Woodard JL, Seidenberg M, Durgerian S, Hazlett KE, et al. Physical activity reduces hippocampal atrophy in elders at genetic risk for Alzheimer's disease. Front Aging Neurosci 2014;6:61.

[57] Smith JC, Nielson KA, Woodard JL, Seidenberg M, Verber MD, Durgerian S, et al. Does physical activity influence semantic memory activation in amnestic mild cognitive impairment? Psychiatry Res 2011;193:60–2.

[58] Ross OA, Braithwaite AT, Skipper LM, Kachergus J, Hulihan MM, Middleton FA, et al. Genomic investigation of α-synuclein multiplication and parkinsonism. Ann Neurol 2008;63:743–50.

[59] Simón-Sánchez J, Schulte C, Bras JM, Sharma M, Gibbs JR, Berg D, et al. Genome-wide association study reveals genetic risk underlying Parkinson's disease. Nat Genet 2009;41:1308–12.

[60] Fuchs J, Tichopad A, Golub Y, Munz M, Schweitzer KJ, Wolf B, et al. Genetic variability in the SNCA gene influences α-synuclein levels in the blood and brain. FASEB J 2008;22:1327–34.

[61] Mak SK, McCormack AL, Langston JW, Kordower JH, Di Monte DA. Decreased α-synuclein expression in the aging mouse substantia nigra. Exp Neurol 2009;220:359–65.

[62] Purisai MG, McCormack AL, Langston WJ, Johnston LC, Di Monte DA. α-Synuclein expression in the substantia nigra of MPTP-lesioned non-human primates. Neurobiol Dis 2005;20:898–906.

[63] Chu Y, Kordower JH. Age-associated increases of α-synuclein in monkeys and humans are associated with nigrostriatal dopamine depletion: Is this the target for Parkinson's disease? Neurobiol Dis 2007;25:134–49.

[64] McCormack AL, Mak SK, Di Monte DA. Increased α-synuclein phosphorylation and nitration in the aging primate substantia nigra. Cell Death Dis 2012;3:e315.

[65] Di Monte DA. The environment and Parkinson's disease: is the nigrostriatal system preferentially targeted by neurotoxins? Lancet Neurol 2003;2:531–8.

[66] Lee PC, Bordelon J, Bronstein J, Ritz B. Traumatic brain injury, paraquat exposure, and their relationship to Parkinson disease. Neurology 2012;79:2061–6.

[67] Tanner CM. Advances in environmental epidemiology. Mov Disord 2010;25:S58–62.

[68] Gatto NM, Rhodes SL, Manthripragada AD, Bronstein J, Cockburn M, Farrer M, et al. α-Synuclein gene may interact with environmental factors in increasing risk of Parkinson's disease. Neuroepidemiology 2010;35:191–5.

[69] Goldman SM, Kamel F, Ross GW, Jewell SA, Bhudhikanok GS, Umbach D, et al. Head injury, α-synuclein Rep1, and Parkinson's disease. Ann Neurology 2012;71:40–8.

[70] Babenko O, Kovalchuk I, Metz GA. Epigenetic programming of neurodegenerative diseases by an adverse environment. Brain Res 2012;444:96–111.

[71] Feil R, Fraga MF. Epigenetics and the environment: emerging patterns and implications. Nat Rev Genet 2012;13:97–109.

[72] Kwok JB. Role of epigenetics in Alzheimer's and Parkinson's disease. Epigenomics 2010;2:671–82.

[73] Egger G, Liang G, Aparicio A, Jones PA. Epigenetics in human disease and prospects for epigenetic therapy. Nature 2004;429:457–63.

[74] Gapp K, Woldemichael BT, Bohacek J, Mansuy IM. Epigenetic regulation in neurodevelopment and neurodegenerative diseases. Neuroscience 2014;264:99–111.

[75] Marques S, Outeiro TF. Epigenetics in Parkinson's and Alzheimer's diseases. Subcell Biochem 2013;61:507–25.

[76] Portela A, Esteller M. Epigenetic modifications and human disease. Nat Biotechnol 2010;28:1057–68.

[77] Basha MR, Wei W, Bakheet SA, Benitez N, Siddiqi HK, Ge Y, et al. The fetal basis of amyloidogenesis: exposure to lead and latent overexpression of amyloid precursor protein and β-amyloid in the aging brain. J Neurosci 2005;25:823–9.

[78] Wu J, Basha MR, Brock B, Cox DP, Cardozo-Pelaez F, McPherson CA, et al. Alzheimer's disease (AD)-like pathology in aged monkeys after infantile exposure to environmental metal lead (Pb): evidence for a developmental origin and environmental link for AD. J Neurosci 2008;28:3–9.

[79] Lahiri DK, Maloney B, Basha MR, Ge YW, Zawia NH. How and when environmental agents and dietary factors affect the course of Alzheimer's disease: the "LEARn" model (latent early-life associated regulation) may explain the triggering of AD. Curr Alzheimer Res 2007;4:219–28.

[80] Querfurth HW, Jiang J, Xia W, Selkoe DJ. Enhancer function and novel DNA binding protein activity in the near upstream βAPP gene promoter. Gene 1999;232:125–41.

[81] Bakulski KM, Rozek LS, Dolinoy DC, Paulson HL, Hu H. Alzheimer's disease and environmental exposure to lead: the epidemiologic evidence and potential role of epigenetics. Curr Alzheimer Res 2012;9:563–73.

[82] Rogaev EI, Lukiw WJ, Lavrushina O, Rogaeva EA, St George-Hyslop PH. The upstream promoter of the β-amyloid precursor protein gene (APP) shows differential patterns of methylation in human brain. Genomics 1994;22:340–7.

[83] Fischer A, Sananbenesi F, Mungenast A, Tsai LH. Targeting the correct HDAC(s) to treat cognitive disorders. Trends Pharmacol Sci 2010;31:605–17.

[84] Hahnen E, Hauke J, Tränkle C, Eyüpoglu IY, Wirth B, Blümcke I. Histone deacetylase inhibitors: possible implications for neurodegenerative disorders. Expert Opin Investig Drugs 2008;17:169–84.

[85] Fischer A, Sananbenesi F, Wang X, Dobbin M, Tsai LH. Recovery of learning and memory is associated with chromatin remodelling. Nature 2007;447:178–82.

[86] Peleg S, Sananbenesi F, Zovoilis A, Burkhardt S, Bahari-Javan S, Agis-Balboa RC, et al. Altered histone acetylation is associated with age-dependent memory impairment in mice. Science 2010;328:753–6.

[87] Dorsey ER, George BP, Leff B, Willis AW. The coming crisis: obtaining care for the growing burden of neurodegenerative conditions. Neurology 2013;80:1989–96.

[88] Bhattacharjee S, Lukiw WJ. Alzheimer's disease and the microbiome. Front Cell Neurosci 2013;7:153.

[89] Gabrielli M, Bonazzi P, Scarpellini E, Bendia E, Lauritano EC, Fasano A, et al. Prevalence of small intestinal bacterial overgrowth in Parkinson's disease. Mov Disord 2011;26:889–92.

[90] Heneka MT, Kummer MP, Stutz A, Delekate A, Schwartz S, Vieira-Saecker A, et al. NLRP3 is activated in Alzheimer's disease and contributes to pathology in APP/PS1 mice. Nature 2013;493:674–8.

Environmental Neurotoxins Linked to a Prototypical Neurodegenerative Disease

Peter S. Spencer, C. Edwin Garner, Valerie S. Palmer, Glen E. Kisby

Environmental Factors in Neurodevelopmental and Neurodegenerative Disorders
http://dx.doi.org/10.1016/B978-0-12-800228-5.00011-X

INTRODUCTION

ALS-PDC: A Prototypical Neurodegenerative Disease

The Western Pacific amyotrophic lateral sclerosis and parkinsonism–dementia complex (ALS-PDC) is a fatal neurodegenerative disease that has affected three genetically distinct populations on the islands of Guam (Chamorro), Honshu (Japanese), and New Guinea (Papuan). The progressive temporal reduction in ALS-PDC prevalence in all three affected populations, coupled with the absence of known genetic markers of related neurodegenerative diseases, indicates a disappearing exposure to one or more environmental factors. The culpable agent(s) must be of natural origin because the disease was present in New Guinea prior to the introduction of human-made materials. Epidemiological studies of migrants to and from Guam demonstrate that a latent period of years or decades intervenes between environmental exposure and clinical appearance of this acquired disease. The neuropathology of ALS-PDC is dominated by neurofibrillary tangles reminiscent of Alzheimer's disease (AD). Indeed, whereas progressive AD-like dementia is today a characteristic clinical expression of ALS-PDC, in former years the disease was expressed clinically as atypical parkinsonism with dementia (P–D) and, before that, amyotrophic lateral sclerosis (ALS). The declining incidence and changing clinical presentation of ALS-PDC has been associated with an increasing age of onset. Thus, over the past 50–100 years, ALS-PDC has changed its clinical expression from a disease dominated by motor neuron degeneration in young adults (i.e., ALS) to a condition akin to AD in the elderly, the reason for which is unknown but may reflect declining exposure to the culpable environmental agents. Western Pacific ALS-PDC is therefore a prototypical environmental neurodegenerative disease that combines in a single condition the elements of aging and central nervous system (CNS) disorders that are generally considered distinct. Understanding the etiology and molecular pathogenesis underlying ALS-PDC may therefore yield valuable information about environmental triggers of related brain disorders, including sporadic ALS and noninherited forms of AD.

Western Pacific ALS-PDC is a progressive disease associated with amyotrophy, parkinsonism, and dementia. The amyotrophy of ALS arises from degeneration of spinal and brain stem motor neurons, resulting in weakness, wasting, and paralysis of the arms and legs, swallowing, and breathing difficulty. The central motor pathway arising from frontal cortical motor neurons that project along the spinal cord also undergoes degeneration. Those with atypical parkinsonism have a stooped posture, slowed movement, unsteady gait, tremulous hands, and impaired memory. Those with AD-like dementia (D), the third form of

ALS-PDC, become forgetful, confused, irritable, suspicious, and may hallucinate. Some have overlapping clinical forms of the disease. The neuropathology of ALS-PDC is dominated by neurofibrillary tangles that are formed by hyperphosphorylation of the microtubule protein tau, which causes it to aggregate in insoluble forms known as paired helical filaments (PHF). Brain disorders with this type of neuropathology, known as tauopathies, include a number of other neurodegenerative diseases, notably AD. Importantly, whereas several tauopathies arise from mutations in the *tau* gene, these have *not* been found in ALS-PDC.

ALS-PDC has disproportionately affected families both in Guam and Kii-Japan, leading investigators to seek evidence for an inherited genetic etiology. However, evidence does not support an inherited disease. Epidemiological studies conducted over the past 50+ years have demonstrated that ALS-PDC has undergone a declining prevalence in the manner of a disappearing environmental brain disease. The declining prevalence of ALS-PDC in three genetically disparate groups (Chamorro, Japanese, Papuan), coupled with evidence that migrants to Guam (Filipinos, Caucasians) have succumbed to the disease after adopting the Chamorro lifestyle, indicate the dominant operation of nongenetic factors. Taken together, available evidence points clearly and unequivocally to a primary or exclusive environmental etiology, which is common to all three geographic isolates of this disease. Exposure to the culpable environmental agent(s) must have declined as the three affected populations have slowly abandoned their traditional lifestyles and certain cultural practices.

Cycads: A Plausible Environmental Trigger of ALS-PDC

First discovered among Chamorros in Guam, later among Japanese in Honshu Island, and then among Auyu and Jaqai linguistic groups of southeastern West New Guinea (now the extreme eastern part of Indonesia), ALS-PDC has affected communities that have used raw or incompletely detoxified cycad seed (*Cycas* spp.) for medicine and/or food, respectively. The cause of a poorly defined neuromuscular disease in grazing animals, cycads associated with ALS-PDC harbor a number of neurotoxic agents including two with developmental neurotoxic properties. The principal agent methylazoxymethanol (MAM) (Figure 11.1), the genotoxic aglycone of cycad azoxyglycosides (cycasin, macrozamin), is epidemiologically associated with ALS-PDC on Guam. MAM produces specific guanine DNA lesions in the postnatal rodent brain that disrupt neurodevelopment [1–4] and, in adult animals, modulate cell signal pathways linked to neurodegeneration and neurodevelopment [5]. The second agent is a free amino acid, β-*N*-methylamino-L-alanine (L-BMAA), a weak excitotoxin and cyanobacterial product that has ALS-PDC-relevant neurotoxic properties in developing rodents and young adult primates [6–9] (Figure 11.2). Although cycad chemicals are not the proven cause of ALS-PDC, they are by far the most plausible etiological candidates.

While the neurotoxic properties of L-BMAA and MAM are the focus of this chapter, the reader should note that cycads harbor other agents of possible significance to the etiology of ALS-PDC [10]. One such unidentified neurotoxic agent with a molecular weight >1000 was found in the retentate of aqueous extracts of the cycad *Encephalartos altensteinii* [11]. Yagi and colleagues [12,13] conducted several studies demonstrating that cycad plant beta-glucosidases

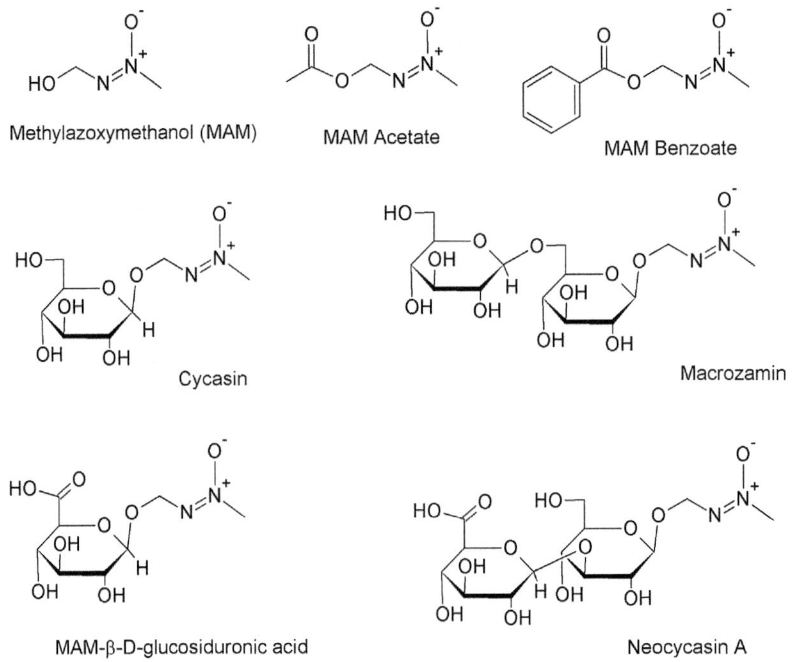

FIGURE 11.1   β-*N*-Methylamino-L-alanine (BMAA) and its structural relationship to L-alanine.

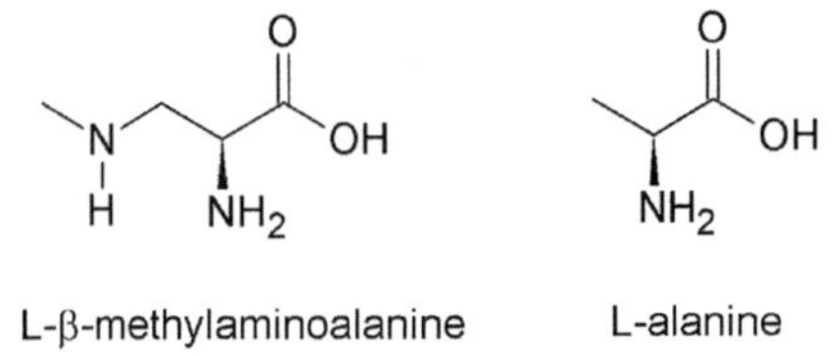

L-β-methylaminoalanine   L-alanine

FIGURE 11.2   Cycad azoxyglycosides (cycasin, macrozamin, neocyasin A), methylazoxymethanol (MAM) and some synthetic derivatives. *Adapted from Woo Y-T, Lai DY. Nitrosamine congener alkyloxymethanol-derived alkylating agents: cycasin and related compounds. McLean, Virginia: Sci Appl Internat Corp; 1986. EPA Contract #68–02–3948.*

can cause the synthesis of higher molecular weight azoxyglycosides through transglucosylation (e.g., neocycasins H, I, J) of either cycasin or macrozamin, the chief azoxyglycosides in most cycad species (Figure 11.1). It is unknown whether the higher molecular weight substances found in *E. altensteinii* [11] were the trisaccharide neocycasins G, H, J or, possibly, other higher molecular weight azoxyglycosides. A thorough digestion of the polysaccharides in various cycad species is required to confirm or refute their presence. The issue is important in determining total exposure to MAM, the common highly toxic and neurotoxic azoxyglycoside metabolite.

Nonprotein amino acids in *Cycas revoluta* other than L-BMAA include: α-aminoadipic acid, β-alanine, α-, β-, and γ-aminobutyric acid, citrulline, hydroxyargine, cycasindene (3-[3′-amino-indenyl-2′]-alanine), and cycasthioamide (*N*-[glycinyl-alaninyl-11-thio]-5-one-pipecolic acid) [14].

β-*N*-Oxalylamino-L-alanine (L-BOAA) [15], a potent excitotoxin that is responsible for the induction of a self-limiting, nonprogressive upper motor neuron disease (lathyrism) [16], has also been reported but not confirmed. Furthermore, it should be noted that endogenous MAM reacts with physiological amino acids such as aspartate and glutamate to form neuroexcitotoxic species [17,18].

Although disappearing traditional practices that expose populations to cycad toxins are the most plausible causes of ALS-PDC, this conclusion is not without controversy because a complete animal model of the human disease has yet to be demonstrated. This is not surprising given the long period of time between human exposure and disease expression (years or decades), a phenomenon that suggests the involvement of a "slow toxin" that utilizes novel molecular mechanisms to induce neuronal degeneration [19,20]. Of the many possibilities, two have been extensively studied to date: (1) MAM-induced persistent DNA damage that activates pathways leading to neurodevelopment and neurodegeneration, (2) triggered initially by the incorporation of L-BMAA into neuroproteins that misfold and malfunction [21].

ALS-PDC: A DECLINING OR DISAPPEARING ENVIRONMENTAL DISEASE

Mariana Islands: Guam and Rota

Highly prevalent neurodegenerative disease among Chamorro residents of Guam was discovered by medical science in the aftermath of World War II, although the disease (*lytico-bodig*) had long been present on the island. Incidence studies revealed rates for ALS ~100 times higher than in the continental United States [22,23]. Parkinsonism usually accompanied by dementia (P–D) was also high prevalent among Chamorros [24], as were subjects with both ALS and P–D. Subsequent clinical and neuropathological studies demonstrated ALS and P–D to be variants of a single, apparently unique brain disease [25]. Familial and genetic surveys of the high incidence of ALS and P–D among the Chamorro people of Guam were initiated in 1958 with the establishment of a case-control registry. Whereas a dominantly inherited disease was initially proposed [26], it is now clear the etiology is primarily if not exclusively environmental in origin. An early clue was the declining incidence of both ALS and P–D between 1958 and 1983 [27]. Another study found that rates of ALS and P–D had dropped by 50% between 1965 and 1972 [28]. Others showed declining ALS incidence from the early 1950s through the next 35 years and declining P–D incidence from the early 1960s to the 1980s [29].

Although siblings and offspring of ALS cases tended to have a higher risk than control relatives, the difference was not significant, incompatible with Mendelian inheritance and, as early as 1975, indicative of the operation of unknown environmental factors [30]. Nevertheless, interest is continuing in the possibility of a genetic contribution to the etiology of neurodegenerative disease on Guam. A study of 22 subjects with parkinsonism–dementia and 19 controls failed to find a single gene locus [31]. The expanded *C9orf72* repeat alleles (an ALS genotype found in western populations) and *LRRK2* mutations (an autosomal dominant Parkinson's disease mutation) were not detected among Chamorros with ALS-PDC [32]. Somatic mtDNA deletions in the entire mtDNA genome were also undetectable in 23 ALS or 42 P–D Guam subjects [33] A search for mutations in *tau* proved negative [34,35] although a single polymorphism was later identified as a possible weak susceptibility factor for Guam ALS-PDC [36]. More recently, three single-nucleotide polymorphism sites (snps) have been reported that are presumed to influence the risk of neurodegenerative disease in 11 affected Chamorro families [37,38].

The presence of these snps in some Chamorros with the neurodegenerative disease should be viewed in light of evidence that diverse genotypes (Chamorro, Japanese, and Papuan) are susceptible to western Pacific ALS-PDC. A dominant environmental etiology is evident given the declining prevalence of ALS-PDC in all three affected populations. This is supported by the observation that non-Chamorro migrants to Guam (Filipinos and US mainland Caucasians) who adopted the Chamorro lifestyle developed ALS-PDC one year to three decades after arrival on the island [30,39,40]. This latency period between exposure and disease onset is comparable to the experience of Chamorros who spent their childhood and adolescence in Guam and who developed ALS 1–34 years after migrating from Guam to the US mainland and elsewhere [41]. Based on data from Chamorro migrants, Román [42] estimated that the critical age of exposure was during their period of adolescence on Guam. Early studies showed that preference for traditional Chamorro food was the only 1 of 23 variables that was significantly associated with an increased risk of P–D [43]. Later, Borenstein and colleagues [44] concluded that eating traditional food (cycad *fadang*) in young adulthood was the highest attributable risk for PDC, "Guam dementia" (GD), and mild cognitive impairment (MCI). Based on our data showing that the concentration of cycasin was 10-fold higher than that of L-BMAA in Chamorro flour samples collected from Guam villagers [45], Zhang and colleagues [46] demonstrated that cycasin, but not L-BMAA content, was significantly correlated with historical village-based rates for ALS and P–D among males and females. These important conclusions were reached ~40 years after the late Marjorie Whiting (http://www.libraries.psu.edu/psul/icik/aboutMGW.html) first proposed the cycad hypothesis, when she noted that Chamorro people attributed *lytico*, many cases of which were identified as ALS, to their practice of using cycad seed for food [47]. The cycad hypothesis lay dormant until 1987 when Spencer [6,7,19] proposed that Guam ALS-PDC is a long-latency neurotoxic disorder caused by "slow" toxin(s) in food. This remarkable but disappearing disease has been extant on Guam for at least 150 years [48].

Kii Peninsula, Honshu Island, Japan

A second focus of Guam-like neurodegenerative disease exists in the southern coastal mountainous region of the Kii peninsula, including the Hohara district in Mie Prefecture and the Kozagawa district in Wakayama Prefecture. Descriptions of "Muro disease" (endemic ALS) in folk literature date from the 17th century and medical literature from 1911 [49]. The first epidemiological survey in the 1960s identified these two high-incidence foci of ALS [50]. ALS patients occasionally showed parkinsonian features or dementia, and neuropathological studies revealed numerous neurofibrillary tangles (NFTs) of the type seen in Guam ALS-PDC and in AD [51–53]. Tau protein, a main component of neurofibrillary tangles, consisted of 3R and 4R tau isoforms phosphorylated at 18 sites [54]. Age-adjusted incidence rates during 1950 and 2000 showed that ALS had gradually declined for 50 years and that PDC had increased in the 1990s essentially replacing ALS, a changing pattern of disease reminiscent of Guam in the 1970s [49,55]. ALS prevalence continued to be elevated relative to worldwide figures, and the incidence even increased among women between 2000 and 2008 in the southern coastal region of Kozagawa and Kushimoto, including the small island of Oshima [57,58]. As in Guam, individuals who migrate from the high-risk areas may develop the disease one to four decades later [49].

A family history of ALS-PDC was recorded in more than 70% of Japanese patients with ALS-PDC, but gene analyses failed to demonstrate mutations of *SOD1*, *parkin*, *α-synuclein*, *tau*, *progranulin*, *TDP-43*, *OPTN*, and other genes related to dementia, parkinsonism, and motor neuron disease [59,60]. Three of 15 ALS patients from the Kozagawa area were found to have expanded alleles, but these were absent in 14 ALS and 16 PDC patients from the Hohara disease focus in Mie Prefecture [61]. Nevertheless, Japanese investigators continue to propose a genetic underpinning for ALS-PDC, despite the absence of evidence and recognition that familial disease can have environmental origins. Interest continues in Japan in a decades-old hypothesis (beyond the scope of this chapter) that proposes an etiological link between low levels of calcium and magnesium and an excess of aluminum and manganese in drinking water, abnormal mineral metabolism, and ALS [52,49,56–58,62]. The applicability to ALS-PDC patients of our discovery [63] in 1987 of an association between ALS in Hohara and use of cycad seed as a tonic has been questioned [64].

Papua, Indonesia

The third pocket of ALS-PDC involves the Auyu and Jaqai people of Papua, Indonesia, in the southwest lowlands of the western part of the island of New Guinea. Although the neuropathological identity is unknown, the clinical picture and epidemiological trends mirror those of the Guam and Kii-Japan isolates. The focus in Papua was discovered in the 1960s by Gajdusek [65,66] who considered a noninfectious environmental etiology probable and the culpable agent(s) of natural origin because the disease was extant prior to the introduction of human-made materials. Mineral deficiency, hyperparathyroidism, and metal neurotoxicity arising from reliance on drinking water from springs and shallow wells were advanced as the likely cause. However, our research in the epicenter of the Papua focus in 1987 and 1990 revealed a decline in ALS and a reversal in the relative prevalence of ALS and parkinsonism in the absence of any changes in the population's water supply [67]. We also discovered that cycad seed (termed *kurru*) gametophyte was used as a poultice to treat large skin lesions, a practice that exposes subcutaneous tissues to cycad toxins continuously for several days until the lesion is healed [68]. (Cycad poultices were also used as a traditional Chamorro method to heal skin wounds on Guam). Unfortunately, etiology was not addressed in a recent extensive epidemiological survey (2001–2002) that confirmed the temporal decline in ALS prevalence throughout the Papua focus, with the continued presence of overlapping ALS, parkinsonism, and cognitive impairment as in Guam and Kii-Japan [69].

ALS-PDC: LINKS TO CYCAD EXPOSURE

Whiting [70] was the first to associate cycads with neurodegenerative disease among Chamorros on Guam [47]. She noted that Chamorro folklore long ago linked the practice of handling and consuming cycad seed to *lytico*, cases of which correspond to ALS. The neurotoxic properties of the cycad plant and two of its chemical components (MAM, L-BMAA) are established, but a causal relationship between either compound and ALS-PDC remains *sub judice*. The long latent period between exposure and clinical disease expression, measured in decades, poses particular problems for the experimentalist who seeks to model ALS-PDC in laboratory animals. A further complication is the carcinogenic property of cycasin and its aglycone (MAM), a discovery that initially sidelined scientific interest in the neurotoxic properties of these compounds. However, we now appreciate the possibility that the molecular mechanisms of MAM-induced carcinogenicity and developmental neurotoxicity may be linked through DNA damage, consequent perturbations of cellular systems biology, and epigenetic mechanisms. This subject and the relationship between AD and certain cancers are discussed elsewhere [71].

The acute toxic effects of ingesting cycad flour or sago are attributable to the hepatotoxic action of MAM. Dogs and other mammalian species also display acute cycad toxicosis [72]. Guam children who ate food prepared from cycad seed flour sometimes fell acutely ill [69], presumably because Chamorro flour contains residual cycasin/MAM [45]. Similarly, 12–40 h after ingestion of incompletely detoxified cycad flour, nausea and vomiting develop suddenly, convulsions and hepatomegaly then appear, and the subject loses consciousness prior to death [73]. [Efforts to follow-up Okinawan and Floridian victims of cycad toxicosis to determine their long-term clinical outcome were unsuccessful (Spencer and Palmer, unpublished)]. Aboriginal Australians go to extraordinary lengths to detoxify cycad seed and consume the resulting material without any acute or, as far as is known, long-term adverse effect. Fermented seed of *C. revoluta* is also used to prepare *Nari miso* in Okinawa Prefecture and Amami islands in Japan, references to which can be found in recipes available on the World Wide Web.

The acute and long-term effects of cycad plant ingestion on neuromuscular function have been recorded for more than a century from animal farms in Australia, Japan, USA (Puerto Rico), and elsewhere [74]. As in humans, acute cycadism causes hepatotoxicity, enterotoxicity, and death. Rapid twitching of the eyelids, nostrils, lips, jaw muscles, with periodic tremors of the body are reported in sheep, and muscle fasciculation has been seen in heifers [75–79]. Such phenomena are consistent with excitotoxicity, but not necessarily from L-BMAA. Delayed effects of cycad consumption are characterized by a progressive and irreversible paralysis of the hind limbs. Initially, in cattle, there is a staggering weaving gait, with crossing of the legs, incoordination, and ataxia. More severe forms are characterized by posterior motor weakness, dragging of the hind limbs, and, occasionally, a stringhalt-like action of the hocks reminiscent of equine lathyrism. Weakness and neurogenic muscle atrophy in the hind limbs may be prominent. Function of the bladder, anus, and tail is unimpaired [80]. Feeding fresh *Cycas* leaves produced this locomotor disorder in 85 days, and goats displayed "spheroids in the spinal white matter" several months after a single intraruminal administration of cycad leaves (species unknown, probably *C. revoluta*). Importantly, goats experimentally dosed with purified cycasin developed hepatic and neurologic lesions similar to those seen in cattle after grazing on the leaves and seed of cycads [81]. Neuropathological examination of bovine cycadism revealed distal degeneration of descending spinal tracts, readily observed from T7 to L6, with similar involvement of the fasciculus gracilis and dorsal spinocerebellar tracts, most prominent in the cervical region [82]. The picture may be complicated by an additional spongiform myelopathy that has been attributed to hepatic encephalopathy [79]. Taken in concert, cycads induce a cycasin-associated progressive neuromuscular disease (cycadism) in ruminants, but there is insufficient neuropathological data to determine the relevance of the veterinary disease to western Pacific ALS-PDC.

Primate studies designed to assess the effects of cycad consumption on the nervous system were first carried out by the neuropathologist Darub Dastur (*dec.*) in the 1960s [83–85], by which time the identity of cycasin and its hepatotoxic properties were known [86,87]. Macaques were fed on different cycad seed preparations, including unwashed seed, boiled seed (which inactivates the plant's β-glycosidase that cleaves MAM from cycasin), and seed that had been water-soaked (one day) and water-washed (seven to eight days) Chamorro-style, which resulted in seed material containing <1 mg of detectable cycasin [74,83,84]. Six of nine animals fed cycad materials for 26 days to 2 years displayed hepatic insufficiency attributable to cycasin. CNS pathology was evident in all macaques, especially in animals fed washed and soaked cycad seed. After 9 months on the latter diet, one of these animals (aged only seven months at onset) exhibited weight loss, progressive emaciation, and moderate weakness and wasting, especially of the right arm, with mild generalized loss of hair and thinning of the skin [85]. The neurological signs in this primate closely resemble ALS. At necropsy two months later, Dastur reported a fairly severe loss and chromatolysis of the remainder of the anterior horn cells of both the cervical and lumbar spinal cord. The cerebral cortex was thinned, especially in the frontal lobes. The cerebellar cortex and brain stem were well preserved, and the hippocampus and substantia nigra appeared unaffected. Animals fed boiled cycad seed showed a moderate depletion of cortical neurons, chromatolytic anterior horn cells, and motor neurons and axons swollen with 160 kDa neurofilaments [74,84], a neuropathological picture reminiscent of the giant axonal swellings seen in ALS [88]. An older animal fed washed cycad seed for two years (amounting to a total of 1454 g) showed severe hepatic insufficiency at death (aged ~11 years), better preservation of spinal motor neurons and striking changes in the cerebral cortex and white matter. The larger pyramidal neurons, including Betz cells, were not as severely depleted in the younger animal, and argyrophilic, amyloid-free neuritic plaques formed in the deep cerebral cortex and subcortical white matter, especially the frontal lobes, along with depleted and degenerated axons [74]. The findings are consistent with the neuritic plaques in ALS-PDC.

Subequent primate studies focused on the carcinogenic properties of cycad [89]. Animals were administered cycad flour, cycasin, and/or MAM starting at one day of age. The cycad flour was given for the first 7–10 months, with the highest dose of 1 g/day on alternate days or alternate weeks, such that the highest dosage given to animals living past eight months was 53 g, a mere 3.6% of the dose given to Dastur's older animal. A group of 10 monkeys received MAM acetate by weekly ip injections (3–10 mg/kg). Six of these animals developed tumors after receiving an average of 6.14 g (range, 3.58–9.66 g) of MAM acetate for an average of 75 months (range, 50–89 months). Four of the monkeys developed hepatocellular carcinomas, and two had multiple primary tumors including hepatocellular carcinomas, renal carcinomas, squamous cell carcinomas of the esophagus, and adenocarcinomas of the small intestine [89]. Unfortunately, none of these animals was assessed with neurological or neuropathological techniques.

ALS-PDC: LINKS TO CYCASIN

Occurrence in Cycads

Use of cycad seed has been epidemiologically associated with western Pacific ALS/PDC among Chamorros of Guam, Japanese of the Kii peninsula of Honshu Island, and the Auyu and Jaqai linguistic groups of Papua (west New Guinea), Indonesia [20,64,68,90]. Although cycad seed contains many possible neurotoxic chemicals, the three most studied groups are the azoxyglycosides (e.g., cycasin, macrozamin), excitotoxic amino acids (i.e., L-BMAA), and the β-sitosterols (vide infra) [91]. Whereas raw cycad seed contains small concentrations of L-BMAA (0.1%), cycasin—the glucoside of the potent DNA alkylating agent methylazoxy-methanol (MAM)—accounts for up to 4% w/w of *Cycas* seed [92]. MAM and its glycosides are thus the principal toxic moieties of cycads [93]. In home-processed cycad flour samples obtained from 17 different Chamorro families in southern Guam, the concentration of cycasin was found on average 10-fold higher than that of L-BMAA [45,94]. The largest concentrations of cycasin were found in cycad flour samples from villages with a high reported prevalence of ALS/PDC. Based on these results, ingestion of cycad-derived food would result in estimated human exposure to milligram amounts of cycasin per day. Strong correlation exists between the cycasin content of cycad flour and the historical age-adjusted incidence of both ALS ($p < 0.00002$) and P–D ($p < 0.000005$) among both male and female villagers of Guam [42,46,95]. In contrast, the correlation between the L-BMAA concentration in the cycad flour samples and Guam ALS-PDC was not significant. Thus, cycasin is delivered in much larger doses than L-BMAA when cycad flour is used as food (Guam), and when the raw seed is used as an oral tonic (Kii-Japan), or for wound repair (Guam, West Papua).

Several other reasons exist to suspect an etiological relationship between the cycad geno-toxins cycasin/MAM and western Pacific ALS-PDC. First, treatment of postnatal mice with MAM arrests cerebellar development [96] leading to the production of ectopic, multinu-cleated Purkinje-like cells comparable to those reported in Guam and Kii ALS brains [97]. In contrast, similar pathological changes have *not* been reported in postnatal rats after treatment with either low (50–200 mg/kg) or high doses (460–600 mg/kg) of L-BMAA [9,98]. Second, Guam ALS-PDC neurons with tau inclusions also contain mitotic markers [99,100]; such changes result from disruption of neuronal development, as occurs in neonatal rodents treated with cycasin or MAM [1,101,102]. Moreover, pathological forms of tau have been detected in the hippocampus of MAM-treated neonatal mice [91,103], but *not* in the hippo-campus of L-BMAA-treated neonatal mice [9,98]. These findings are consistent with emerging evidence indicating that neurons in AD and other tauopathies exhibit binucleation, cell cycle disturbances, and aneuploidy [104–110]. It has also been suggested that cycad genotoxins may be responsible for the very high prevalence of glucose intolerance and diabetes mellitus among Chamorros with this neurodegenerative disease [19,111]. Cycasin and MAM impair both rodent and human β-islet function and may cause death of these pancreatic cells in vitro. Cytotoxic mechanisms appear to include DNA alkylation, nitric oxide generation [111,112], and, in neurons, perturbed DNA repair ([113]; Delaney and colleagues, unpublished data).

Individuals with ALS-PDC and cycad-exposed animals also develop nonneurological changes (e.g., skin and bone), which have been summarized elsewhere [19]. Animals grazing on cycad leaves lose their horns and hooves in the manner of a molt, in which dermal regen-eration is active [74]. In parallel fashion, cycad seed materials have been shown to speed skin repair in rodents, which is consistent both with their therapeutic use to treat dermal injuries in Guam and Papua, and with the resistance to bed sores of bedridden ALS patients on Guam and elsewhere. In sporadic ALS, but not Guam ALS, this phenomenon has been linked to the presence of small-diameter collagen fibers and copious ground substance [114]. Inter-estingly, a large number of skin lesions (including tumors) was observed in rats 28 months after treatment with the cycad genotoxin MAM [115]. Chamorros historically have had a high prevalence of diaphyseal aclasis, which is visible as nonmalignant bony protuberances near the growing ends of long bones in the upper extremities [19]. These observations raise the possibility that a component within cycad (i.e., cycasin or MAM) activates molecular processes associated with epidermal growth. Unpublished observations showing unusual skull bone thickness among Chamorros with ALS or PD are consistent with this idea.

Cycasin/MAM: Chemistry and Molecular Mechanisms

Cycasin is a member of a family of naturally occurring azoxyglycosides in cycad plants, which comprise nine genera and approximately 100 species. Genera include: *Bowenia*, *Ceratozamia*, *Cycas*, *Dioon*, *Encephalartos*, *Macrozamia*, *Microcycas*, *Stangeria*, and *Zamia*. *Macrozamia* spp. and *Cycas* spp. contain the primeveroside macrozamin and/or the glucoside cycasin, the predominant azoxyglycosides in cycad species, and macrozamin can undergo conversion to cycasin [116] (Figure 11.1). *Cycas revoluta* also contains neocycasin A (β-laminaribioside of MAM) and B (β-gentiobioside of MAM) plus a xylose derivative of macrozamin [117].

Cycasin is synthesized from MAM by the cycad plant enzyme UDP-glucosyltransferase [118]. The concentration of cycasin in cycad seed ranges from 2 to 4% (w/w) [13] and it readily undergoes hydrolysis under acidic conditions (e.g., stomach) to yield methanol, formaldehyde, nitrogen, and glucose [119]. Cycasin is metabolized by plant, bacterial, or mammalian β-glucosidases (hydrolysis of β-glycosyl linkage) to liberate the aglycone MAM [92,120]. β-Glucosidases play important roles in diverse aspects of plant physiology including the synthesis and activation of chemical defense compounds (e.g., azoxyglycosides) against herbivores and pathogens [121,122]. During periods of drought, the β-glucosidase activity in plants increases to mobilize saccharides and possibly chemical defense compounds to sustain respiration and protect the plant from predators, respectively [123]. The β-glucosidase in cycads is also capable of forming di- and trisaccharides through the transglucosylation of cycasin or macrozamin to a number of neocycasins [124]. The synthesis of saccharides by cycad plant β-glucosidase may be a mechanism for storing more stable forms of the toxic azoxyglycosides during times of environmental stress. Washing, boiling, fermenting, and storage of cycad flour remove substantial amounts of the azoxyglycosides and other toxins (e.g., MAM, L-BMAA), as well as the β-glucosidase. However, we and others found significant amounts of cycasin (up to 75 µg/g) in traditionally washed cycad flour used for foodstuffs [45,125].

Methylazoxymethanol (MAM) is the active toxic metabolite of cycasin and all other cycad plant azoxyglycosides. MAM spontaneously breaks down into reactive molecules such as methyldiazonium ions and carbon-centered free radicals (potent alkylating agents), which methylate DNA in the O^6-, N7-, and C8-positions of guanine [91,126–128]. The accumulation of these DNA lesions is responsible for the teratogenic, mutagenic, hepatotoxic, carcinogenic, and neurotoxic properties of MAM [89,91,129]. N7-mG and O^6-mG are the predominant MAM-induced DNA lesions; these are repaired by the base-excision and direct reversal (i.e., O^6-methylguanine methyltransferase, MGMT) repair pathways, respectively [130]. MAM alkylates brain-tissue DNA to produce N7-methylguanine (N7-mG) and O^6-methylguanine (O^6-mG) DNA lesions [92] when administered to rodents either in utero [131,132] or after birth [1,3]. These DNA lesions also accumulate and may persist in the brain of fetal rats or neonatal mice after a single injection of MAM or related alkylating agents [1,92,133] because their repair in the rodent brain varies across different regions and is significantly less efficient than in peripheral organs (e.g., liver) [134]. O^6-mG DNA lesions have greater persistence in rodent brains [1,92,133], and they are highly mutagenic at the level of transcription in human cells, leading to an altered protein load, especially in cells with significantly reduced MGMT [135].

Cycasin/MAM: Developmental Neurotoxicity

Exposure to the cycad plant during childhood or adolescence, critical periods of normal human brain development, has been shown to increase the risk of developing MCI, Guam PD, and GD [44]. Thus, early-life exposure to cycad azoxyglycosides may be an important contributor to both the neurodevelopmental and neurodegenerative changes observed in the brain of individuals with ALS-PDC [97,136]. Because the neurotoxic properties of cycasin and its active metabolite MAM are poorly understood, initial studies by our group examined the distribution, uptake, and metabolism of cycasin and its active metabolite MAM in rodents. Unlike L-BMAA, cycasin and MAM readily cross the blood–brain regulatory interface with ~5% of either genotoxin detected in brain tissue [92]. Depending on the route of exposure (topical or oral), cycasin gains entry into the brain by a high-affinity brain endothelial [137] or intestinal [138] Na$^+$-dependent glucose transporter (SGLT). However, cycasin showed a 100-fold greater affinity for the brain endothelial SGLT, suggesting that the genotoxin rapidly enters brain tissue. Cycasin is metabolized to the toxic aglycone MAM by both β-glucosidase in the gut and brain, but brain activity of this enzyme is 500-fold higher than that found in the gut or any other rodent organ [92]. Thus, cycasin is efficiently transported by the brain Na$^+$/glucose cotransporter and readily metabolized by the brain β-glucosidase, which is evidence that these CNS systems are responsible for MAM's neurotoxic properties in animals and possibly humans [84,92,137].

We next explored the relationship between early-life exposure to cycasin and western Pacific ALS-PDC by examining the response of the neonatal and young adult murine brain to MAM, the genotoxic metabolite of cycasin. Parallel studies were conducted in neuronal and astrocyte cell cultures from neonatal mice to compare the response of different CNS cell types to MAM. Unlike most organs, the fetal and young adult human brain has a low or absent capacity to repair alkylation-induced DNA damage [139]. Neonatal and young adult MGMT DNA-repair knock-out mice (i.e., $Mgmt^{-/-}$) are therefore appropriate animal models to examine the response of the underdeveloped human brain to MAM. Another reason for using neonatal mice is that previous studies demonstrated that postnatal exposure (days 1–5) to MAM leads to misalignment of Purkinje cells, ectopic and multinucleated granule cells, pathological changes comparable to those observed in the cerebellum of individuals with ALS-PDC [97,136].

Global gene expression profiling [1] and proteomic studies [140] of the brain of neonatal C57BL/6 mice given a single injection of MAM (21.5 mg/kg) showed that the cycad genotoxin produces an early and persistent increase in DNA damage, and disrupts the expression of genes and proteins that regulate the neuronal cytoskeleton, protein degradation, and mitochondrial metabolism. Global gene expression profiling of young postmitotic cerebellar neurons and astrocytes treated in vitro with MAM also showed that immature neurons are more vulnerable than astrocytes, and this vulnerability is associated with the accumulation of DNA lesions and distinct alterations in gene expression [2]. The preferential modulation of genes involved in diverse functions such as cell signaling, transcriptional regulation, differentiation, and the stress and immune response suggests that MAM targets distinct neuronal networks in the developing but not in the mature brain. This is supported by the insensitivity of adult neurons to MAM [101,141]. Because these cellular pathways are also perturbed in neurodegenerative disease, and the under-developed human brain is particularly inefficient in repairing alkyl DNA lesions (i.e., O^6-mG) [139,142], MAM may have targeted one or more of these CNS pathways by a DNA damage-mediated mechanism. Although O^6-mG DNA lesions are cytotoxic and mutagenic (in cycling cells), their persistence in immature neurons [133] could lead to a long-term disruption of CNS function that culminates in an enhanced predisposition to late-life neurodegenerative disorders. Therefore, the DNA damage produced by MAM in the undeveloped human brain may constitute an important cause of the prominent tau and synuclein pathology observed in western Pacific ALS-PDC [44,143]. The ability of MAM to induce latent tau phosphorylation (serine 199) in the hippocampus of young adult mice that overexpress normal human tau (i.e., htau) after neona-tal administration of the cycad genotoxin [91,103], and latent heavy deposits of α-synuclein (i.e., inclusions) in the cerebellum of MAM-treated neonatal mice [140], are consistent with this notion.

How neuronal DNA damage produced by cycad genotoxins contributes to the neuro-developmental and neurodegenerative changes observed in western Pacific ALS-PDC has not been investigated. One possibility is that MAM disrupts the expression of genes that regulate neurodevelopment or induce neurodegeneration either through DNA damage [3,5] and/or epigenetic-mediated mechanisms [71,91,144]. To address this question, we compared the response of the neonatal brain of DNA repair-proficient (C57BL6, background strain), -deficient ($Mgmt^{-/-}$) and -overexpressing ($Mgmt^{Tg+}$) mice to MAM or related genotoxins, namely dimethyl sulfate (DMS), and the monofunctional or bifunctional chloroethylating agents chloroethylamine (CEA) and nitrogen mustard (HN2) [1,3]. MAM methylates DNA to produce N7-mG and O^6-mG DNA lesions, whereas HN2 rapidly and irreversibly alkylates guanine (G) and adenine (A) (i.e., N7-alkylG and 3-alkylA lesions, respectively) of DNA to produce monoadducts, intrastrand cross-links, and interstrand cross-links. DMS is a meth-ylating agent that produces primarily N7-mG and 3-methyladenine DNA lesions, whereas CEA, the monofunctional analog of nitrogen mustard, produces predominantly N7-alkylG DNA lesions, but not cross-links. By comparing the response of the neonatal brain to MAM and related alkylating agents among these DNA-repair genotypes, one can determine if a specific DNA lesion is responsible for the neuropathological and associated neurobehavioral changes induced by this cycad genotoxin. Neurodevelopment and motor function were more severely affected by MAM in $Mgmt^{-/-}$ mice than after treatment with the other alkylating agents, and these effects were less pronounced in wild-type mice [3]. The reduced number of $Mgmt^{Tg+}$ neurons undergoing apoptosis after MAM treatment and the preservation of cer-ebellar morphology and motor function in similarly treated $Mgmt^{Tg+}$ mice, provide strong evidence that MGMT plays an important role in protecting the immature brain from cycad genotoxin-induced injury. These findings demonstrate that the neurotoxic effects of MAM occur primarily through the production of specific DNA lesions (i.e., O^6-mG), a result com-parable to the influence of 8-oxoguanine DNA lesions on methamphetamine-induced neu-rodevelopmental deficits [145]. Such a mechanism may explain the developmental cellular abnormalities (e.g., ectopic cells) observed in the cerebellum of Chamorro and Japanese indi-viduals with ALS-PDC [97,136].

An enriched environment produces long-lasting effects on rodents by epigenetic mechanisms [146]. The ability of environmental enrichment to reverse the sensorimotor deficits and improve learning in young adult rodents after injection of neonatal mice with MAM [147–149] suggests that the cycad genotoxin induces neurobehavioral changes in part by an epigenetic mechanism, such as histone methylation. To address this question, Mackowiak and colleagues [144] studied the methylation of histone proteins in the brain of adult rats after a single in utero (E17) injection of MAM. An examination of the medial prefrontal cortex of young adult rats revealed a pronounced effect on the methylation of histone proteins (i.e., H3K4 and H3K9). H3K9me2 levels were decreased before puberty (post-natal day [PND] 15 and PND 45), and H3K4me3 levels were significantly decreased when rats reached adulthood (PND 60 and PND 70). Thus, fetal exposure to MAM reduced H3K9me2 and H3K4me3 levels (markers of histone H3 methylation) in the prefrontal cortex before and after puberty, respectively. These histone H3 methylation changes were also associated with altered levels of enzymes that methylate (ASH2L) and demethylate (LSD1, JARID1c) histone proteins. Because epigenetic changes (i.e., histone lysine methylation and altered levels of methyltransferases and demethylases) occur in the adolescent rodent brain after in utero MAM administration, this cycad toxin might disrupt neurodevelopment by inducing both DNA damage and epigenetic changes.

MAM also appears to interfere with the normal development of differentiating postmitotic neurons, causing retraction and persistently shortened axons [150]. MAM increases glutamate-stimulated neuronal *tau* mRNA expression and cell loss in neuronal cultures, raising the possibility that MAM produces long-term perturbation in neuronal responses to physiological levels of glutamate [113]. Additionally, protein kinase C activation, which stimulates the release of secreted amyloid precursor protein (APPs) in several cell lines, is permanently hyperactivated in the cerebral cortex and hippocampus of MAM-treated rats. Synaptosomes derived from these brain areas show decreased membrane-bound APP concentration and a concomitant increase in soluble fractions [151].

Cyasin/MAM Neurotoxicity in Young Adults

A prospective 15-year epidemiological study of ALS-PDC among Chamorros on Guam and Rota showed that of 23 selected variables, only the preference for traditional Chamorro food was significantly associated with an increased risk of PD [43]. Twenty years later, another epidemiological study showed that exposure of *young* adults to the picking, processing, and eating of cycad food products (*fadang*) was associated with the highest attributable risk for MCI, GD, and P–D [44]. However, no association was observed between the picking, processing, or eating of *fadang* among adults and the risk for MCI, GD, or P–D. These observations suggest the developing human brain is especially vulnerable to cycad genotoxins. We explored this hypothesis further by examining the response of the young adult murine brain to the cycad genotoxin MAM. As in the neonatal studies (vide supra), both DNA repair-proficient (C57BL/6) and *Mgmt$^{-/-}$* mice were used to determine whether DNA damage also plays an important role in the response of the young adult brain to cycad genotoxins.

Global gene expression profiling was used to compare the extent of MAM-induced DNA damage (i.e., O^6-mG) with the response of the young adult brain of both DNA repair-proficient mice (C57BL6/J) and DNA repair-deficient (*Mgmt$^{-/-}$*) animals [5]. The brains of young adult *Mgmt$^{-/-}$* mice treated with a single systemic dose of MAM showed significantly higher levels of O^6-mG than the brains of comparably treated C57BL/6 wild-type mice. The DNA lesions in the brain of MAM-treated young adult $Mgmt^{-/-}$ mice remained elevated (up to seven days posttreatment), an indication that MAM-induced DNA damage is persistent, and MGMT plays an important role in protecting the young adult brain from alkylation-induced DNA damage [133,152,153]. The O^6-mG levels were linked to changes in the expression of genes in several cell-signaling pathways (i.e., tumor protein p53 [TP53], nuclear factor κB [NFκB], mitogen-activated protein kinase [MAPK]) associated with cancer, human neurological disease, and neurodevelopmental and skin disorders. These data are consistent with the established developmental, neurotoxic, and carcinogenic properties of MAM in rodents [151–154]. The prominent modulation of "cancer genes" in the "tumor-insensitive" brains of MAM-treated young adult animals suggests that perturbations of these genes in the undeveloped brain have consequences other than cancer [71]. They also support the hypothesis that early-life exposure to the MAM-glucoside (cycasin) has an etiological association with western Pacific ALS-PDC.

Additional experiments by our group were conducted to determine if MAM induces latent transcriptional changes in the brains of $Mgmt^{-/-}$ vs. wild-type mice [155]. Brain transcriptional changes six months after single systemic treatment with MAM were comparable to those seen at seven days post-MAM treatment, and there was elevation of MAPK and increased caspase-3 activity, both of which are involved in tau aggregation and neurofibrillary tangle formation typical of ALS-PDC and AD. Additionally, the transcriptional profile of the $Mgmt^{-/-}$ mouse brains was dominated by the presence of 28 genes involved in olfactory transduction, which suggested the presence of an MAM-induced change in olfaction status. Although caution is merited when comparing rodent and human data, olfactory dysfunction is among the first signs of neurodegenerative disease, and it is especially marked in Guam ALS-PDC [156,157]. The inability to distinguish the nature of olfactory dysfunction among Guam P–D, AD [158], and ALS patients [157] suggests a common neurologic substrate and underlines the close relationship between ALS-PDC and the more familiar neurodegenerative disorders seen in other parts of the world.

ALS-PDC: LINKS TO L-BMAA

L-BMAA: Occurrence

The amino acid α-amino-β-methylaminopropionic acid, discovered by the late Arthur Bell [159] and renamed β-N-methylamino-L-alanine (L-BMAA) by Spencer in 1987 [160], is a non-protein amino acid with stereospecific acute neurotoxic properties found within seeds and other tissues of certain species of cycads. L-BMAA is also produced by cyanobacteria in freshwater, marine, and terrestrial ecosystems. The amino acid was originally isolated from the seeds of *C. circinalis* (later renamed *Cycas micronesica*) [159] and characterized by high-voltage electrophoresis and nuclear magnetic resonance spectroscopy [161]. Although initial animal studies revealed the excitotoxic properties of the amino acid [162], interest waned until we demonstrated in macaques that repeated oral doses of L-BMAA induced a motorsystem disorder [7] distinct from that induced by L-BOAA (β-N-oxalylamino-L-alanine, the neurotoxin in *Lathyrus sativus* responsible for lathyrism) and with clinical and neuropathological features recalling those of ALS-PDC [7,16,160]. Large doses of L-BMAA and L-BOAA were used to induce neurological changes within a reasonable experimental period (weeks/months). Animals treated with L-BMAA developed muscle weakness, bradykinesia, and tremor, improved markedly with L-DOPA, and displayed histological evidence of neuronal chromatolysis and early degenerative changes in anterior horn cells and cortical motor neurons, respectively. Although, the primate disorder did not progress once dosing was stopped, these experimental findings (which should be confirmed and extended) re-created interest in the cycad hypothesis for ALS-PDC and stimulated a substantial body of subsequent research that continues expansively. Initial criticism that doses of L-BMAA used to treat the macaques bore no relationship to human exposure on Guam [42,94,163] was later challenged and muted by the work of Paul Cox, who proposed that L-BMAA was biomagnified in flying foxes, a previously unidentified Chamorro food item [164].

Analysis of the leaves and seeds of a variety of cycads showed the content of L-BMAA to vary among species, with the highest concentration in *Cycas* spp. (up to 1.8 mg/g) [163]. The L-BMAA content in cycad flour obtained from 17 Chamorro residents in Guam and the adjacent island of Rota ranged from 0 to 18.39 µg/g [45]. A detailed review of the L-BMAA content in a variety of cycad species and other plants is available [165]. The origin of L-BMAA within cycad tissues is not well understood. Brenner and colleagues [166] proposed a hypothetical biosynthetic pathway based on their analysis of the *Cycas* genome (Figure 11.3).

Murch and colleagues [167] found that axenic cultures of cyanobacteria isolated from collaroid cycad roots also contained ~240 times more "bound" than "free" L-BMAA (72 mg/kg vs. 0.3 mg/kg). Collaroid roots of *C. micronesica* contained 2 mg L-BMAA per kg of the bound form. In non-collaroid roots without any cyanobacteria, L-BMAA was undetectable suggesting that cyanobacteria colonies within the collaroid roots of cycads produce L-BMAA, which in turn is absorbed by the plant and concentrated within the seed and other parts of the plant. Other parts of the cycad plant also contain neurotoxic amino acid: leaf, 738 mg L-BMAA per kg; outer seed layers: 48 mg/kg; the sarcotesta 89 mg/kg; and the female gametophyte 81 mg/kg. Subsequently, cycads grown in the absence of cyanobacteria have been shown to produce L-BMAA as they grow, findings that challenge the hypothesis that cyanobacteria are the sole source of L-BMAA in *C. micronesica* [10,168].

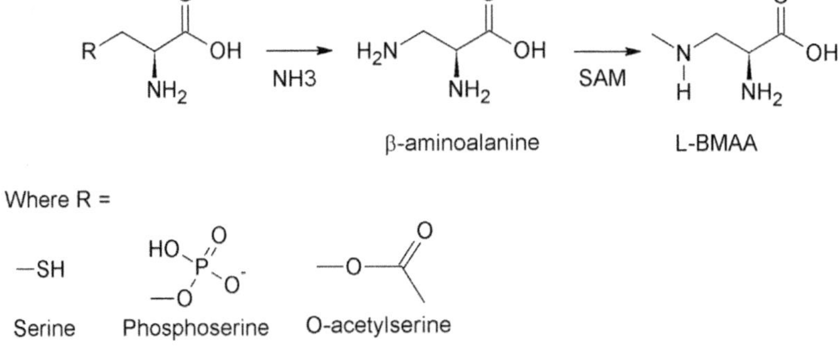

FIGURE 11.3   Proposed L-BMAA biosynthesis pathway in *Cycas* spp. SAM=S-adenosylmethionine. *Adapted from [166].*

Concern for possible biomagnification of L-BMAA in the food chain is based in part on observations that L-BMAA in cycad tissue is protein-bound [94,167] and thus may be incorporated into protein [21]. Observations of increased levels of free and "bound" L-BMAA found in the Guamanian ecosystem from plants to flying foxes (now extirpated from the island, the fruit bat was prized as a food source among native Guamanians) suggested that dietary exposure via bioconcentrated L-BMAA could deliver potentially neurotoxic doses to humans [167]. A 10,000-fold increase in L-BMAA concentrations was estimated in flying foxes that consumed cycad seed. Based on this observation, BMAA was speculated to induce protein misfolding and truncation as a function of misincorporation during mammalian protein synthesis [167,169]. It is known that nonproteogenic amino acids can insert into peptides [159,170] and examples exist of rare tRNA mischarging [171,172]. The detection of L-BMAA in cyanobacterial blooms has raised the possibility of a broader impact on human health through exposure to L-BMAA in water and the ingestion of fish and shellfish [173]. Free and protein-associated L-BMAA has also been detected in blue-green algae (primarily from cyanobacteria) from a variety of fresh, brackish, and oceanic water sources in the United States and other countries [174–177]. The analytical methodology used to measure trace levels of L-BMAA (free and protein-bound) in both complex plant and animal tissue samples is, however, hotly debated [178–181]. The potential for detecting artifacts by some of the current analytical methods makes comparisons of measurements among laboratories extremely difficult [181]. Until the analytical methodology for detecting L-BMAA is standardized, caution is warranted regarding speculation on the broader impact of this neurotoxin on human health (vide infra).

L-BMAA: Chemistry and Molecular Mechanisms

L-BMAA, an analog of alanine, shares similar chemical properties with more common amino acids (see Figure 11.2). Its distribution and potential uptake into tissues, and thus its neurotoxic potential, are dependent upon the pK_a of its ionizable groups [182]. Neutral amino acids, such as alanine, are zwitterions at physiological pH 7.4, and they have a net zero charge. L-BMAA, like lysine and ornithine, is a basic amino acid that contains an additional secondary amino group. Basic amino acids have a protonated side chain at neutral pH and, therefore, a net positive charge.

Vega and colleagues [161] determined that the pK_1 (acid), pK_2 (β), and pK_3 (α) for L-BMAA were 2.1, 6.5, and 9.8, respectively. Similar values of 6.63 and 9.76 for pK_2 and pK_3, respectively, were estimated by Nunn and associates [183]. Swedish National Food Laboratories reported an isoelectric point of 8.09 [165]. At physiological pH (7.4), the main fraction of L-BMAA will be a zwitterion with a neutral net charge. The pK values discussed above are referred to as "macroscopic" or "apparent" pK values. Using chemical modeling software, the charge state of each group shows that there are three forms at pH 7.4: doubly protonated (14%), α-protonated (6%), and β-protonated (79%) (Figure 11.4), which agree with previously published data [183].

In physiological solutions, the uncharged amine(s) of an amino acid is at equilibrium with bicarbonate, forming a carbamate [184]. A significant proportion of primary and secondary amines in biological samples exists as carbamates, and this may impact their activity and structure [185–187]. Using ^{1}H-NMR, Nunn and O'Brian [188] were the first to detect the formation of an L-BMAA α-carbamate. Subsequently, ^{13}C-NMR studies showed the formation of both an α- and a β-carbamate of L-BMAA [189] (see Figure 11.5). Under simulated physiological

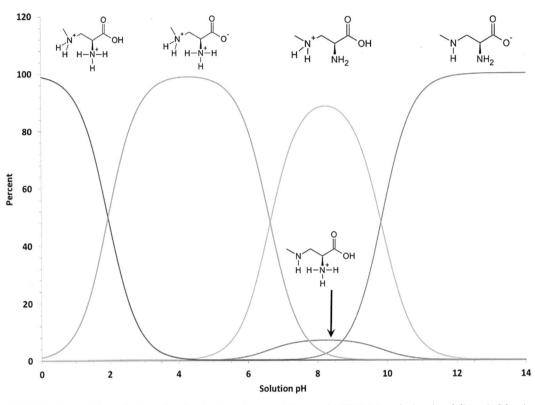

FIGURE 11.4   Effect of pH on the distribution of charged forms of L-BMAA in solution (modeling via Marvin 14.7.7, Chemaxon.com).

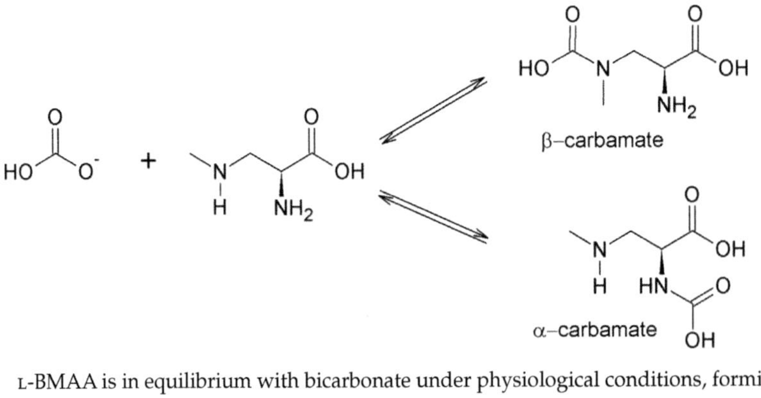

FIGURE 11.5   L-BMAA is in equilibrium with bicarbonate under physiological conditions, forming carbamates.

conditions (pH 7.4, 37 °C, and 23 mM bicarbonate), the rate of formation of the α-carbamate was approximately two times greater than that of the β-carbamate. At equilibrium, the mole fractions of L-BMAA existing as α-carbamate and β-carbamate were 0.22 and 0.09, respectively [189]. A 2:1 proportion for the α-carbamate over the β-carbamate forms of L-BMAA may reflect the relative values of the pK_as of the amino groups. Chen and colleagues [190] demonstrated that for amino acids with pK_as near 9.5, the rate of carbamate formation was very low relative to those with lower pK_a values. Although L-BMAA is known to interact with NMDA receptors (*N*-methyl-D-aspartate) [191–193], formation of the carbamate is required for measurable receptor activation. Binding of β-carbamino-L-BMAA to the NMDA receptor was shown in vitro to resemble that of the activated NMDA-receptor complex [194,195].

A proposed hypothetical mechanism of chronic neurotoxicity in humans is based on the accumulation of L-BMAA in neuroproteins following environmental exposure and the subsequent slow release over time of L-BMAA from brain protein following its catabolism [167]. Neuronal degeneration would arise following hyperstimulation of glutamate receptors and/or the formation of abnormal neuroproteins. However, primates that were treated chronically with L-BMAA showed no evidence of glutamatergic neurotoxicity [7], which makes the hyperstimulation hypothesis unlikely.

L-BMAA: Distribution, Uptake, and Metabolism

An understanding of the biochemical processes related to the distribution, uptake, protein binding, and potential persistence of L-BMAA and its metabolites in brain and peripheral tissues, preferably from animal models, may provide important clues to the potential impact of L-BMAA exposure on human health. All toxicological studies conducted to date with L-BMAA have used the "free" amino acid or one of its salts. No studies exist as of this writing that have followed the absorption or fate of L-BMAA associated strongly or bound covalently within the macromolecules of plant or animal tissues and then subsequently consumed.

Free L-BMAA is absorbed intact and is systemically bioavailable. In pharmacokinetic studies, L-BMAA has been shown to have high oral bioavailability in rodents and nonhuman primates [94,163]. Following a single oral dose of L-BMAA (100 mg/kg) to rats, the amino acid was rapidly absorbed (T_{max} ~5 min), detected in plasma and other tissues [163] and the bioavailability calculated as 80–93%. Using radiolabeled [1,2-^{14}C]-L-BMAA, the absolute oral absorption was determined to exceed 90% in both rats and mice across a broad dose range [196,197]. Plasma and brain concentrations of L-BMAA and its metabolic equivalents were linear and directly proportional to the dose at oral doses up to 400 mg/kg in rats [163,196,197]. The oral bioavailability of L-BMAA in nonhuman primates at doses of 2 mg/kg was 72–83%. The presence or absence of food had no impact on the oral bioavailability of L-BMAA [198]. Taken together, these findings suggest that, across species, nearly all of an oral dose of "free" L-BMAA is absorbed intact and is systemically available.

We showed and others confirmed that L-BMAA is an excitotoxic amino acid with acute neurotoxic properties that are attenuated by glutamate receptor antagonists. L-BMAA has effects on both ionotropic and metabotropic glutamate receptors [191–195,199]. The activation of glutamate receptors by L-BMAA (a membrane-mediated event), however, does not adequately explain how this cycad neurotoxin induces a chronic motorsystem disease in primates fed L-BMAA. In the early 1990s, we began investigating whether the chronic neurotoxic effects of L-BMAA are due to uptake by nervous tissue and then either incorporation into protein or biotransformation into a toxic species. We showed that L-BMAA is taken up by mouse cortical explants and synaptosomes in a time-dependent manner [200]. These studies were confirmed several years later by demonstrating that [^{14}C]-L-BMAA was taken up by both cortical explants and in-mouse cortical synaptosomes treated with [^{3}H]-L-BMAA [91], whereas the uptake of [^{14}C]-L-BMAA into cortical astrocytes was 5-fold lower. Thus, L-BMAA appears to be preferentially taken up by neurons. The uptake of L-BMAA into mouse cortical explants and synaptosomes seems to occur through a high-affinity transport system for amino acid neurotransmitters (e.g., GABA, proline, glutamate) [201]. Radiolabeled L-BMAA was also found to be released from cortical synaptosomes. The rapid uptake, accumulation, and release of L-BMAA by brain tissue, but not of the structural analog L-BOAA, suggested that the subchronic neurotoxic effects of L-BMAA in primates were related to its intracellular actions. Such a mechanism would explain the slow evolution of motor neuron changes and the lack of excitotoxic damage in the frontal cortex of heavily dosed primates that exhibited signs of motorsystem disease in the presence of low serum and cerebrospinal levels of L-BMAA [20,202].

In rodents, L-BMAA is subject to N-demethylation via the cytochrome P450s, releasing formaldehyde and 2,3-diaminopropionic acid [201,203]. The desmethyl metabolite 2,3-diaminopropionic acid is a known glutamate receptor-mediated neurotoxic agent [204]. The extent of conversion of L-BMAA into 2,3-diaminopropionic acid in vivo has not been accurately determined in mass balance studies. Within 8h of administration of intraperitoneal [U-^{14}C]-2,3-diaminopropionic acid to rats, nearly 60% of the dose is rapidly converted to $^{14}CO_2$ via pyruvic acid pathways with concomitant incorporation of radioactivity into tissue proteins [205]. Nunn and Ponnusamy [203] demonstrated that 2,3-diaminopropionic acid and methylamine are produced by rat liver and kidney homogenates after incubation with L-BMAA (10 nM), but not by brain homogenates. Methylamine is metabolized further by vascular semicarbazide-sensitive amine oxidase to formaldehyde, which is then converted to CO_2 [206,207]. The product of methylamine loss, the aldehyde 2-amino-3-oxopropionic acid, yields aminomalonic acid if further oxidized. Aminomalonic acid is unstable in physiological solutions and spontaneously decarboxylates, yielding glycine and CO_2 [208]. Aminomalonic acid can be incorporated into proteins via mischarge of t-RNA [171,172]. Incorporation of aminomalonic acid into proteins increases the capacity of the protein to chelate ions, such as calcium, causing protein precipitation [209]. Incubation of L-[1,2-^{14}C]-BMAA with rat brain homogenates released $^{14}CO_2$ at a rate approximately two times that of liver [210]. This suggests that liver and brain both possess the capacity to metabolize the "backbone" of L-BMAA in addition to N-demethylation. In rats and mice, $^{14}CO_2$ was the main route of L-[1,2-^{14}C]-BMAA excretion (>60% eliminated within ~12h as $^{14}CO_2$) regardless of dose [196,197].

Hashmi and Anders [211] showed that L-BMAA is also a substrate for L-amino oxidase. This enzyme, which is highly expressed in mammalian liver and kidney mitochondria [212,213], deaminates L-BMAA releasing equimolar amounts of H_2O_2 and methylaminopyruvate. Methylaminopyruvate is subject to spontaneous decarboxylation to yield methylglycine (i.e., sarcosine). Mitochondrial sarcosine dehydrogenase readily converts sarcosine to glycine [214,215]. The production of glycine or serine from the "backbone" of L-BMAA would yield significant amounts of CO_2 [216,217] resulting in measurable BMAA equivalents in rodent liver and kidney proteins if L-[1,2-^{14}C]-BMAA were administered to a rodent (see Figure 11.6).

L-BMAA: Evidence for Incorporation into Proteins

No studies are known to have followed the absorption or fate of L-BMAA bound covalently within macromolecules in plant or animal tissues and then subsequently consumed. Based on the proposed detection of L-BMAA in brain tissue from ALS-PDC patients [217], a contested observation [180,218], a prevailing hypothesis is that L-BMAA is absorbed by the

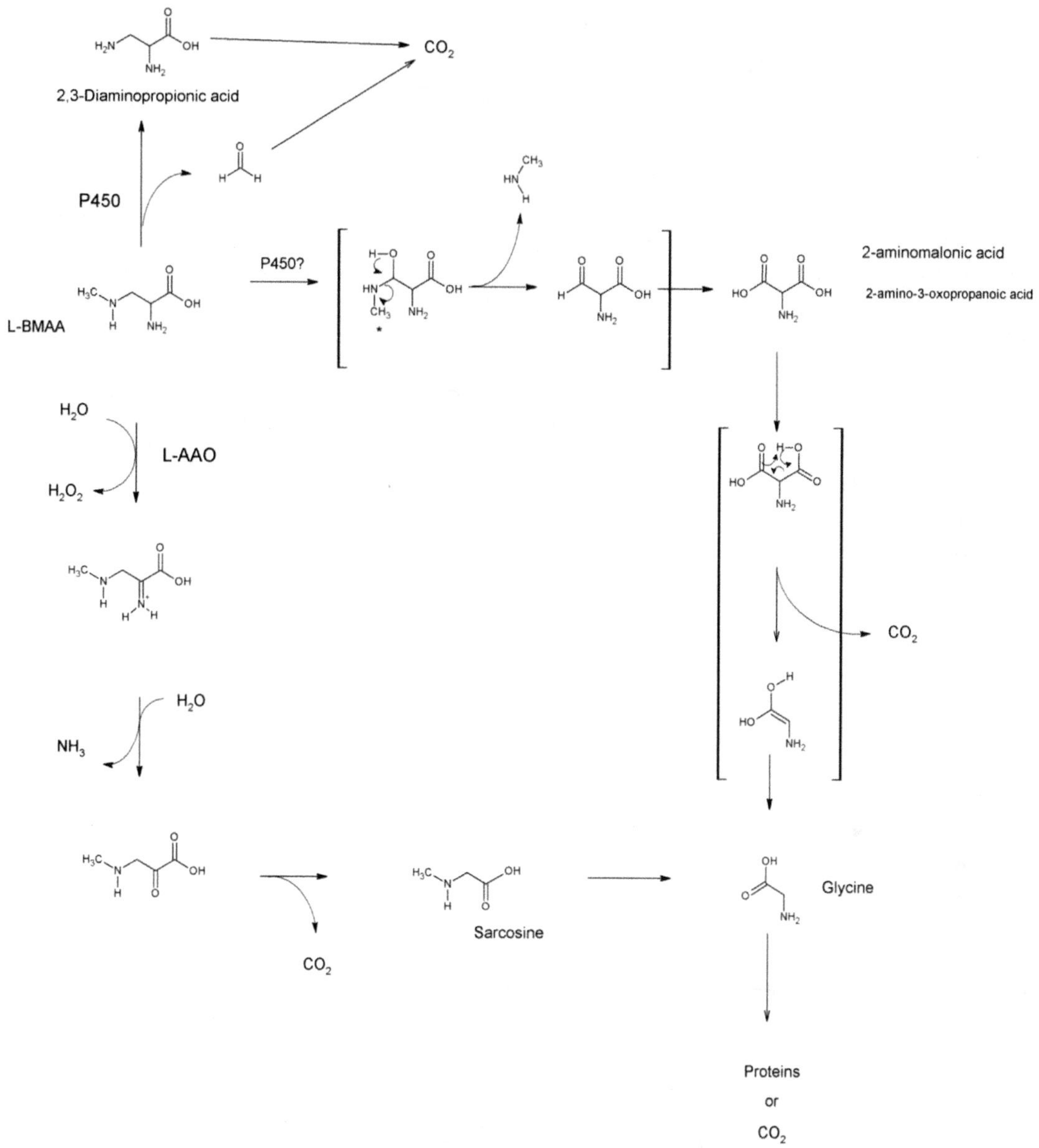

FIGURE 11.6 Proposed metabolic fate of L-BMAA.

gut, taken up by the CNS, and then incorporated directly into brain-tissue proteins [217]. Additionally, exposures to L-BMAA equivalents are proposed to accumulate in brain proteins and be released as "free" neurotoxic levels of L-BMAA, following protein catabolism [176]. Though their analytical methodology is strongly questioned due to nonspecificity [219], these studies have raised awareness about other environmental sources of L-BMAA that may be available to populations not exposed to cycads.

Current data suggest that, following oral absorption, "free" L-BMAA crosses the blood–brain regulatory interface or "barrier" (BBB). Kisby and colleagues [17] reported that "free" L-BMAA was detected in the cerebrospinal fluid of orally dosed monkeys, and in brain-tissue homogenates from rats administered a single intraperitoneal dose of L-BMAA. Subsequently, Duncan and associates [163,220] demonstrated that "free" L-BMAA levels in the brain of rats were proportional to those in plasma at oral doses up to 400 mg/kg. However, both groups of investigators reported unexpectedly low levels of "free" L-BMAA in the brain (<0.1% of dose). More recently, Snyder and colleagues [221,222] used 2DGC-TOF MS to detect L-BMAA (limit of detection 0.7 ppb) in brain homogenates following oral L-BMAA administration to mice. Evidence also exists that brain uptake of L-BMAA is a carrier-mediated event. L-BMAA uptake into adult Sprague–Dawley rat brain was measured using an in situ brain perfusion technique. These studies showed that L-BMAA is transported across the BBB by the large neutral amino acid carrier (L-system), and its transport is critically sensitive to competition from plasma neutral amino acids [163,220]. Pretreatment of mouse cortical explants (2–3 weeks in vitro) with the nonselective γ-aminobutyric acid-uptake inhibitor, nipecotic acid (100 μM and 1.0 mM), blocked L-BMAA uptake and L-BMAA-induced excitotoxicity, respectively [201]. These studies indicate that L-BMAA is taken up by brain tissue, a process that explains the subchronic neurotoxic effects of this amino acid in orally dosed primates [7,91,160].

Our recent studies with radiolabeled L-BMAA have supported findings from the earlier transport studies. After oral administration of L-[1,2-^{14}C]-BMAA at various doses, radioactivity in the brain was less than 0.2% of the administered dose for both rats and mice [196,197]. After intravenous administration of L-[1,2-^{14}C]-BMAA (3.8 mg/kg) to mice, Xie and colleagues [223] reported recoveries of <1% of the total dose in the brain. Concomitant with a rapid decline of ^{14}C in the plasma, the total radioactivity in the murine brain reached peak levels ~1.5 h after intravenous dosing and then declined with a long terminal phase. The long terminal phase was associated with an increasing percentage of radioactivity recoverable in TCA-precipitated brain protein. Thus, the elimination of radioactive equivalents in brain proteins was the principal driver for the slow decline of radioactivity in the total tissue. Garner and associates [197] gave single and daily oral administration of L-[1,2-^{14}C]-BMAA to rats and mice for 10 days and found that the majority of the ^{14}C radioactivity was recovered in peripheral tissues (liver, adipose muscle, and skin), with <0.01% dose from brain. At 1, 5, and 10 days of dosing, a majority of the ^{14}C radioactivity in the brain was recovered in the protein fraction after exhaustive extraction. The critical caveat with all these studies is that only total radioactivity is measured; *none* distinguishes between metabolites and the parent compound, L-BMAA. Thus, these studies fail to characterize the nature of the L-BMAA equivalents within brain proteins.

Dunlop and colleagues [224] incubated human MRC-5 fibroblasts with ^{3}H-L-BMAA in culture medium depleted in amino acids and observed a time-dependent increase in radioactivity in cell lysates, a portion of which was associated with cell proteins. Co-incubation of MRC-5 cells with ^{3}H-BMAA, along with the protein synthesis inhibitor cyclohexamide or serine, significantly reduced the amount of radioactivity in proteins precipitated from washed cells. The authors interpreted this effect as evidence of the mischarge of serine tRNA with L-BMAA, because incubation with up to 250 μM serine resulted in a concentration-dependent reduction in radiolabeled content of precipitated protein. Because the uptake of L-BMAA into cells is a transport-mediated mechanism [201,220], a potential source of artifact in this cytosolic incorporation experiment is competition at the level of transporter-mediated amino acid uptake. The uptake of L-serine occurs through several transport systems [225,226], including Na+-dependent transporters such as the neutral amino acid transporter system ASC and system A, and Na+-independent transporters such as system L, which have been shown to transport L-BMAA [220]. Additionally, cyclohexamide inhibits amino acid uptake in multiple systems [227,228]. Regardless of the mechanism, L-BMAA was detected when protein hydrosylates from cells were analyzed via 6-aminoquinolyl-N-hydroxysuccinimidyl carbamate (AQC) derivitization followed by UPLC-MS/MS with characterization via transitions specific for L-BMAA.

Karlsson and colleagues [229] administered two successive subcutaneous doses of L-[methyl-14C]-BMAA HCl (up to 460 mg/kg) to 10-day-old male rats and collected liver and brain tissue through 28 days post dose. Free and protein-associated L-BMAA was determined via AQC derivitization followed by UPLC-MS/MS characterization via selective reaction monitoring of daughter ions specific for L-BMAA. The UHPLC–MS/MS analysis demonstrated a dose-dependent increase of protein-associated L-BMAA in the brain and liver of neonatal rats. Higher levels of protein-associated L-BMAA in the liver samples were considered due to a higher rate of protein synthesis in the liver compared with the brain. Twenty-eight weeks after dosing, the protein-associated radioactivity and L-BMAA returned to background. Thus, the protein-associated effects of L-BMAA are transient, in contrast to the more persistent effects (DNA lesions) seen after cycasin or MAM treatment of rodents.

The recent radiotracer studies using 14C-labeled L-BMAA indicate that the uptake, and potential incorporation into proteins and metabolites of this unusual amino acid in CNS tissues does occur (Figure 11.7), but the level and degree require further study. In rat studies,

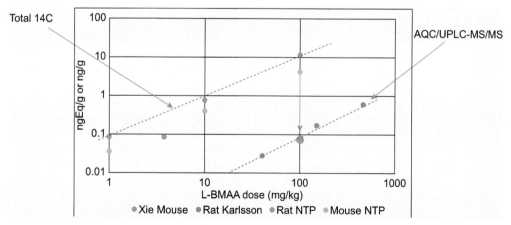

FIGURE 11.7 Concentration of free or hydrolysable "protein bound" L-BMAA in brain tissue[a] or total ng-equivalents in brain protein 24 h following the administration of L-BMAA to rats[b] or mice[c]. Estimated total recovery of L-BMAA-derived radioactivity in the rat and mouse brain was <1% of total residual 14C equivalents at all doses. Note logarithmic scale of ordinate axis. [a]Karlsson et al. [229]: L-BMAA in rat brain determined after oral administration. Analytical method: AQC derivitization + UPLC-MS/MS; LOQ 0.02 ng/mg tissue. [b]Garner et al. [197]: Total 14C in rat brain determined following oral administration. [c]Xie et al. [223]: Total 14C in mouse brain determined following intravenous administration (3.8 mg/kg). Total radioactivity in these studies is expressed as nanogram equivalents of L-[1,2-14C]-BMAA per mg tissue.

we showed that >80% of the brain radioactivity measured 24h after a single oral dose of L-[1,2-^{14}C]-BMAA was tightly associated with protein [91,196,197,200]. This observation is consistent with data reported by Xie and associates [223], who showed that murine brain radioactivity 24h after dosing with ^{14}C-labeled L-BMAA was mostly associated with precipitable protein. Taking the ^{14}C and UPLC-MS/MS data together, it is clear that the total equivalents of extractable and hydrolysable brain levels of L-BMAA were log-linear and proportional across a broad dose range.

These data show that the ratio of total L-BMAA residuals to hydrolysable L-BMAA associated with brain tissue is >100:1. Furthermore, the ratio of extractable L-BMAA to hydrolysable L-BMAA was 300- to 400-fold in rat brain [229]. That is, there may be as much as a 100-fold excess of L-BMAA and/or L-BMAA-derived metabolites for each mole of L-BMAA that is associated with tissue protein. However, great potential exists for residual-free L-BMAA within tissues to contaminate protein extracts [230], especially as the sensitivity of the detection method is in the *femtomole* range [178,180]. L-BMAA binds tightly to neuromelanin [231], which is abundant in the rodent CNS [232,233]. Further L-BMAA forms tight complexes with metals such as copper and zinc [183]. Thus, L-BMAA may be retained within neuroproteins during precipitation procedures currently used. None of the current LC-MS/MS methods uses more than a single or double wash of the brain protein precipitates. Classical protocols regarding the analysis of the covalent binding of metabolites to proteins typically suggest repeated washes (five times or more) or Soxhlet extraction to avoid artifacts caused by the tight noncovalent binding of metabolites to proteins [234–237].

Intense debate continues regarding the analytical methods used to detect trace levels of L-BMAA in protein complexes found in cyanobacterial, animal, and human matrices, both "free" and those associated with protein [178–181]. Recent studies, however, question the potential bias of nonspecific derivatization analytical methods toward significant overestimation of L-BMAA levels in archival human tissue samples and cyanobacterial samples due to a variety of artifacts [178,238]. Adding to the confusion is the lack of standardization of methods. Few or no laboratories have analyzed the same set of samples and compared results. Thus, the potential for analytical artifacts still exists, making comparisons of the data across laboratories difficult if not impossible [181].

L-BMAA research would benefit greatly from a cooperative effort to distribute matched samples of human, plant, and cyanobacterial origin among laboratories, with the goal of developing a standardized approach to the analysis of the neurotoxic amino acid. Tissue samples from animals treated with defined doses of L-BMAA should be used as reference samples to standardize the extraction methods, which would eliminate artifacts during sample work-up. Until the analytical and extraction methods are standardized across laboratories, caution should be exercised with regard to broader assessments of L-BMAA exposure, bioaccumulation, and risk to human health.

L-BMAA: Developmental Neurotoxicity

Initial studies of L-BMAA in adult animals demonstrated poor transport into the brain by the large neutral amino acid transporter [163,220]. Consequently, high doses of L-BMAA (>100mg/kg) are required to produce toxic levels (250μM) of L-BMAA in the rodent brain. Several years later, the uptake and distribution of L-BMAA into the adult murine brain was investigated by two independent groups using radiolabeled L-BMAA [231]. Karlsson and colleagues [231] gave adult mice a single intraperitoneal injection of [^{3}H]-L-BMAA (7.3 µg/kg) before examining the animals by whole-body autoradiography. The highest levels of radioactivity were found in the pigmented tissue of the eye, the corticomedullary region of kidney, liver, and hair follicles, but brain and spinal cord levels were low. Moreover, the regional distribution of radioactivity within the brain was homogenous with no specific uptake and binding in discrete regions, which the authors confirmed by liquid scintillation counting. These findings are consistent with previous reports of the poor uptake of L-BMAA into the brain and spinal cord of adult rats [163]. Four years later, Xie and colleagues [223] used radiolabeled L-BMAA (i.e., intravenous [^{14}C]-L-BMAA, 3.8 mg/kg) to show that uptake of this neurotoxic amino acid is also poor in the adult murine brain. Brain uptake accounted for less than 1% of the administered radiolabeled L-BMAA, a value comparable to that found by Duncan (0.08%), who administered "cold" L-BMAA to adult rats. Collectively, these autoradiographic studies demonstrate that L-BMAA is poorly transported into brain tissue, which would explain why large doses of L-BMAA are required to induce neurotoxicity in adult rats or primates [7,239,240]. By contrast, L-BMAA uptake into the fetal and neonatal murine CNS has been shown to be high and differentially localized within various brain regions (hippocampus, striatum) after administering a comparable dose of [^{3}H]-L-BMAA to either timed-pregnant (GD 14) or 9–10-day-old neonatal mice [241]. These studies show that the developing brain is much more permeable to L-BMAA than the mature brain, findings that are consistent with the increased risk of ALS-PDC for children or adolescents who were exposed to cycads [44]. Based on these findings, immature rodents are a better animal model than adult rodents to explore the chronic neurotoxic effects of the cyanobacterial and cycad toxin L-BMAA.

The fetal and neonatal periods of brain development are sensitive periods with respect to exposure to various xenobiotics [242], including L-BMAA. Purdie and colleagues [243] were one of the first groups to examine the effect of L-BMAA on neurodevelopment. Zebrafish embryos were exposed to L-BMAA at concentrations of ~300 µM for five days with and without added carbonate. This exposure induced a range of neuromuscular (e.g., clonus-like convulsions) and developmental abnormalities (e.g., spinal deformities) that were directly related to the disruption of glutamatergic signaling pathways. In that same year, Karlsson and associates [241,244] reported that neonatal rats (PND9–10) given two subcutaneous injections of either 200 or 600 mg/kg L-BMAA within a few days after dosing developed acute behavioral deficits, including impaired locomotor activity and hyperactivity, that disappeared by PND24–27. When the L-BMAA-treated neonatal rats reached young adulthood (10–12 weeks old), they displayed impaired learning but no memory deficits, evidence supporting the zebrafish data implicating L-BMAA as a developmental neurotoxin. Karlsson and colleagues continued using the neonatal rodent model to determine if the BMAA-induced neurobehavioral abnormalities were associated with neurodegenerative changes [8,9]. As in previous studies, L-BMAA (200 mg/kg, or 600 mg/kg) was injected subcutaneously at PND9 and PND10, but a lower dose (50 mg/kg) was added [8]. Low doses of BMAA (50 and 200 mg/kg) caused spatial learning and memory deficits in adult rats, but no morphological changes. Neuronal cell death in the hippocampus was only observed at the higher dose of L-BMAA. These results indicate that the long-term effects of L-BMAA on cognitive function are also associated with neuropathological changes, but these effects are dose-dependent. Subsequent experiments sought to determine if the learning and memory impairment observed in adult rats treated neonatally with L-BMAA [8,241,244] was associated with long-term protein changes within the hippocampus [9]. Brain-tissue sections of six-month-old rats from L-BMAA-treated neonatal animals (PND9–10) were examined by matrix-assisted laser desorption ionization imaging mass spectrometry to directly image proteins in brain-tissue sections and correlate with associated histological changes. Long-term changes included decreased expression of hippocampal proteins involved in energy metabolism and intracellular signaling at a dose (150 mg/kg) that produced no histopathological lesions. At higher doses of L-BMAA (460 mg/kg), increases occurred in the expression of S100β, histones, calcium- and calmodulin-binding proteins, and guanine nucleotide-binding proteins. At this dose of L-BMAA, Karlsson and colleagues also observed neuronal degeneration, cell loss, calcium deposits, and astrogliosis in the adult hippocampus. However, α-synuclein immunostaining of the CA1 pyramidal neurons was only marginally increased within the hippocampus. More importantly, immunostaining for neurofilament or phosphotau in the hippocampus of BMAA-treated rats did *not* differ from vehicle-treated rats. They also observed a similar effect of L-BMAA (460 mg/kg) on the pattern of α-synuclein and tau expression in the hippocampus of adult rats using a combination of immunohistochemical and proteomic methods [98]. Thus, high doses of L-BMAA did *not* induce the in vivo accumulation of pathological proteins (phosphotau, α-synuclein) that predominate in the ALS-PDC brain. In contrast, MAM has been shown to induce the latent accumulation of α-synuclein [140] and phosphotau (serine 199) [91] in the brains of young adult mice after neonatal administration of the cycad neurotoxin. In sum, the Karlsson group has demonstrated that L-BMAA induces subtle, sometimes dose-dependent, effects on many processes during development of the hippocampus, but their L-BMAA neonatal animal model did not produce the neuropathological hallmarks of ALS-PDC. At this time, therefore, the MAM neonatal animal model appears to be a better animal model of ALS-PDC.

Additional studies by the Karlsson group were conducted using a proteomic approach to evaluate neuropeptide changes in the striatum of L-BMAA-treated neonatal mice. Using a previous treatment protocol, they found that 25 different peptides were changed in the striatum of L-BMAA-treated neonatal rats [245]. L-BMAA induced a dose-dependent increase of virus growth factor (VGF)-derived peptides and, at the highest dose, changes were observed in the relative levels of cholecystokinin, chromogranin, secretogranin, melanin-concentrating hormone, somatostatin, and cortistatin-derived peptides. In female pups, they also showed an increase in the relative level of peptides derived from the proenkephalin-A and protachykinin-1 precursors, including substance P and neurokinin A. Because several of these peptides play critical roles in the development and survival of neurons, these neuropeptide changes are possible mediators of the L-BMAA-induced neurobehavioral changes. Moreover, some neuropeptide changes suggest potential sex-related differences in L-BMAA susceptibility.

ALS-PDC AND PHYTOSTEROLS

In addition to MAM and L-BMAA, there has been research interest in a possible role for cycad phytosterol glycosides in the etiology of ALS-PDC. Dietary plant phytosterols are structurally related to cholesterol, and they appear to play an important role in the regulation of serum cholesterol and to exhibit anticancer properties [246]. Interest in phytosterols arose from behavioral and neuropathological changes in mice fed washed cycad flour chips lacking "appreciable quantities of cycasin, MAM, or BMAA" [247]. Shaw and colleagues reported that cycad-derived β-sitosterol β-D-glucoside (BSSG) was acutely toxic to cells in the adult rat cerebral cortex [247]. Additionally, mice-fed pellets containing BSSG for 15 weeks showed deficient leg extension associated with motor neuron loss in the lumbar and thoracic spinal cord [248]. In a more recent publication, α-synuclein and ubiquitin aggregates plus progressive degeneration of dopaminergic cells in the substantia nigra (but not motor neurons in spinal cord) were observed in rats fed for 16 weeks with cycad seed that had been washed by a protocol designed to "remove all traces of BMAA, cycasin, and MAM" [249]. Respondents to this paper pointed out that the method used to wash cycad flour does not remove cycasin (or other azoxyglycosides), and it underestimates the concentration of L-BMAA, especially because it does not release amino acid associated with protein [250,251]. Moreover, previous studies showed that washed cycad flour not only contains significant amounts of cycasin (up to 75 μg/g), but possibly other much larger concentrations of azoxyglycosides (e.g., neocycasins) [45,125]. Murch and colleagues [167] also found that after removing free amino acids from cycad samples (including flour), L-BMAA concentrations increased 180-fold after hydrolyzing the remaining insoluble fraction. In reply, Shaw and colleagues noted that L-BMAA was "an unlikely toxin candidate for our rodent model of cycad flour-induced parkinsonism and that a contribution from cycasin and/or MAM could not be ruled out" [252,253]. Thus, further work is required to fully characterize the toxic compound(s) in both cycad seed and washed cycad flour.

These results notwithstanding, the proposal that the sterol glycoside BSSG is an etiologic agent for ALS-PDC is surprising, given the massive weight of evidence against this proposition. β-Sitosterol (BSS) and BSSG are the major phytosterols in higher plants—many of which are used by human for food—and especially in oils therefrom. Roasted peanuts, for example, contain 64–114 mg/100 g, 78–83% of which is in the form of BSS. Both BSS and BSSG are found in the tissue, plasma, and feces of healthy individuals. Importantly, individuals with phytosterolemia, a rare autosomal recessive genetic disorder leading to phytosterol accumulation, develop premature coronary artery disease and tendon xanthoma, not neurological disease [254].

BSS and BSSG also possess anti-inflammatory, antipyretic, antineoplastic, and blood sugar-controlling and immune-modulating effects [255,256]. Numerous experimental studies with BSSG have reported no adverse effects. BSS and BSSG (which would be converted to BSS by gut glycosidases) are used medicinally for the treatment of various chronic conditions, notably benign prostatic hypertrophy [257]. Klippel and colleagues [258] reported efficacy and no adverse effects in a multicenter, randomized, double-blind and placebo-controlled clinical trial of 177 patients treated daily for six months with 130 mg free BSS. Additionally, a July 2014 analysis based on reports from the US Food and Drug Administration and social media retrieved no side effects among those taking BSSG for medicinal purposes. In sum, it seems improbable that these sterol glycosides have a role in the etiology of ALS-PDC.

BIOMEDICAL RELEVANCE BEYOND THE WESTERN PACIFIC REGION

L-BMAA and Cyanobacterial Neurotoxins

Cox and colleagues [176] reported that L-BMAA is present in all known groups of cyanobacteria, including cyanobacterial symbionts and free-living cyanobacteria. Subsequent reports are numerous and beyond the scope of this chapter. In a nutshell, cyanobacteria in soil, freshwater, and marine bodies are proposed as a possible source of L-BMAA exposure [259,260] and thus a worldwide public health concern, especially aquatic species that can bioaccumulate the amino acid and are used for food [174]. To put this in perspective, it may be noted that cyanobacteria also produce genotoxins (e.g., microcystins) that are capable of inducing DNA damage in both human and rodent cells [261,262]. Like cycasin/MAM, cyanobacterial-derived microcystin is commonly characterized as a carcinogen and hepatotoxin, but substantial evidence exists that microcystins also induce neurotoxic effects. Microcystins have been shown to cross the BBB and accumulate in the brains of rodents [263–265], induce neurodevelopmental changes (i.e., reduced brain size) [266], and memory loss [267]. More recently, Li and colleagues [268] used a proteomic approach to determine the putative neurotoxic mechanism of microcystins. The hippocampus of microcystin-treated rats showed alterations in proteins involved in the cytoskeleton, neurodegenerative disease (i.e., septin 5, α-internexin, α-synuclein), oxidative stress, apoptosis, and energy metabolism. Western blotting studies showed that microcystins also reduced serine/threonine protein phosphatases and increased tau phosphorylation. Microcystin-treated mice also had impaired learning and memory and associated loss of hippocampal neurons (i.e., CA1). These recent findings suggest that microcystins are at least as important as L-BMAA as potential human health hazards. Equally important is the need for more research on human exposure to L-BMAA and microcystins in cyanobacteria, as well as L-BMAA-containing cycad products and their relationship to the risk of neurodegenerative disease.

Cycasin, Nitrosamines, and Alzheimer's Disease (AD)

Although the proposed link between human exposure to cycad toxins and western Pacific ALS-PDC is controversial, equally provocative suggestions have raised the possibility that related chemicals in the North American environment may be linked to sporadic Parkinson's disease and AD. Epidemiological studies published several years ago by de La Monte and colleagues [269] noted strong parallels between age-adjusted increases in the death rate from AD, Parkinson's disease, and diabetes mellitus, and progressive increases in human exposure to nitrates, nitrites, and nitrosamines through processed and preserved foods, as well as fertilizers. By contrast, other diseases, including HIV-AIDS, cerebrovascular disease, and leukemia did not exhibit those trends. De la Monte and colleagues proposed that the increase in exposure to these environmental chemicals plays a critical role in the cause, development, and the effects of these diseases through an insulin-resistant mechanism. Just as the cancer research community decades ago called for a reduction in exposure to nitrite preservatives in food [270], the question of a relationship between exposure to such compounds and neurodegenerative disease is now an issue. The basis for focusing on nitrite preservatives is their transformation in the gut to genotoxic nitrosamines, which share chemical relations with methylazoxymethanol (MAM).

N-Nitrosamines are potent carcinogens that occur widely in the environment, and they are formed endogenously in the stomach following the reaction of ingested nitrates or nitrites with secondary amines (e.g., proteins) [271]. Human exposure to nitrosamines or their precursors, nitrates and nitrites, can occur as well from exogenous sources, such as diet [272],

TABLE 11.1 Common Sources of N-Nitrosamines

Source	N-Nitrosamine[a]	Concentration
FOOD/BEVERAGES		
Bacon	DMN, DEN, NPYR	1–40 ppb
Salami	DMN	10–80 ppb
Luncheon meat	DMN, DEN	1–4 ppb
Hamburger	DMN	15–25 ppb
Chinese marine-salted fish	DMN DEN	0.05–21 ppm
Smoked salmon	DMN	0–26 ppb
Cheese	DMN	1–4 ppb
Wheat flour	DEN	0–10 ppb
Beer	DMA	
COMMERCIAL PRODUCTS		
Latex gloves	DMN	37–329 ppb
	DEN	<10 µg/kg
Cigarettes	DMN, DEN, NPYR	0–180 ng/cig.
Pesticide formulations	DMN	300 ppb–640 ppm
Rubber nipples	DMN, DEN	<100–200 µg/kg
Rubber toys	DMN	<25 µg/kg
Rubber balloons	DMN	<150 µg/kg
	DEN	>30 µg/kg
INDUSTRIAL EXPOSURE		
Leather tanneries	DMN	0.1–47 µg/m^3
Foundries, USA	DMN	0.062–2.4 µg/m^3
Rubber, curing, salt bath	DMN, DEN	
Rubber, curing, injection molding	DMN	

[a] *Abbreviations: DMN, N-nitrosodimethylamine (NDMA); DEN, N-nitrosodiethylamine (NDEA); NPYR, N-nitrosopyrrolidine; DMA, N-nitrosodimethylamine.*
Thomson MA, Green CR, Rasma BB, Holtshouser JL, Hinderer RK, Papciak RJ, et al. N-Nitrosamines. In: Clayton GD, Clayton FE, editors. Patty's industrial hygiene and toxicology. 4th ed. vol. 2 (Pt. A). New York: Wiley; 1993. p. 663–709.

drinking water [273], occupation (e.g., rubber industry) [274], or the environment (see Table 11.1). Nitrosamines, like MAM, are potent alkylating agents that methylate DNA to produce N7-mG and O^6-mG DNA lesions [275]. The antineoplastic agent streptozotocin, a monofunctional nitrosourea derivative isolated from *Streptomyces achromogenes*, is also a potent alkylating and highly genotoxic agent known to directly methylate DNA. Because antineoplastic agents impair cognition

[276] and the effects are persistent [277], these agents might also induce both structural and functional changes in the immature brain through a DNA damage-mediated mechanism.

SUMMARIZING THE EVIDENCE: ALS-PDC AND CYCAD TOXINS

Many unanswered questions remain regarding western Pacific ALS-PDC, notably its etiology and the relevance of the causative agent(s) to health worldwide. The following summarizes our interpretation of salient knowledge to date.

1. The disease has declined in all three geographic isolates consistent with diminishing exposure to a noninfectious environmental factor of natural origin.
2. Disease acquisition is most probable during childhood and/or adolescence.
3. Clinical disease expression may occur decades after environmental exposure.
4. ALS-PDC is a single pathological entity related to other tauopathies, and tau (as well as ubiquitin, Aβ, α−synuclein, and TDP-43) are involved in disease pathogenesis [278].
5. Disease expression could be a dose–response to the environmental risk factor(s), in which ALS>P–D>D> subclinical neurofibrillary pathology.
6. Exposure to neurotoxic cycad materials is a plausible but unproven etiology.
7. Cycads contain several azoxyglycosides and "uncommon" amino acids, some of which have not been assessed for neurotoxic potential.
8. In the case of azoxyglycosides, focus has been placed on cycasin (the probable cause of cycad-induced neuromuscular disease in grazing animals) and its aglycone methylazoxymethanol (MAM, common to all azoxyglycosides).
9. Cycasin and MAM are well-established developmental neurotoxins in rodents and hepatotoxins in humans.
10. MAM, like nitrosamines, alkylates a range of target molecules, including nucleic and amino acids.
11. The developmental neurotoxicity of MAM (cerebellar dysgenesis) is related to unrepaired toxin-induced brain DNA damage associated with the enzyme activity of the DNA-repair enzyme O^6-MGMT.
12. MAM-induced DNA damage (O^6-MGMT) in young adult mice activates intracellular pathways linked to human neurodegenerative diseases, neurodevelopment, and cancers.
13. In the case of cycad amino acids, focus has been placed on L-BMAA, a weak, stereospecific developmental and adult neurotoxin that, in the presence of carbamate, has excitotoxic and seizuregenic activity in rodents.
14. L-BMAA is taken up by the rodent brain much more in neonates than in adults.
15. L-BMAA has been reported (but not confirmed) to substitute for serine in neuroproteogenesis.
16. Chronic oral treatment of young adult primates to nonseizuregenic concentrations of L-BMAA induces cortical and spinal motor neuron pathology coupled with nonprogressive neurological signs reminiscent of ALS-PDC.
17. L-BMAA has been reported in cyanobacteria and diatoms [279], which can enter marine organisms used by humans for food.
18. Cycad phytosterols are improbable candidates for ALS-PDC.

Western Pacific ALS-PDC is a neurodegenerative disease of unquestionable importance in relation to understanding the causes of neuropathologically related brain disorders across the world, including AD [280,281]. Although the environmental cause of ALS-PDC has yet to be identified, the association with cycad exposure, and the neurotoxic properties of MAM and L-BMAA, justify a search for these compounds and chemically related species elsewhere. Although the search for L-BMAA has begun with a vengeance, little attention has been given to MAM and MAM-like compounds. This is unfortunate because compounds chemically related to MAM (nitrosamines, hydrazines) are present worldwide in food, the workplace, and the general environment [91,282].

If food and/or medicinal use of cycads are causally related to ALS-PDC, which cycad chemicals are of particular concern? In the early 1990s, we placed a spotlight on the neurotoxic properties of L-BMAA in molecular, cellular, rodent, and primate systems, and the possible relevance of exposure to this amino acid in the etiopathogenesis of ALS-PDC. These observations have stimulated a wide array of research studies, from basic research to public health. Although the importance of these studies should not be underestimated, it is concerning that a paucity of research interest has been focused on the MAM-mediated neurotoxic properties of cycad azoxyglucosides and their potential etiological role in ALS-PDC. MAM is an unequivocal developmental neurotoxin and its glycone (cycasin) induces cycad-like neuromuscular disease in ruminants. Moreover, cycasin, but not L-BMAA, is epidemiologically associated with ALS-PDC among Chamorros on Guam. Thus, MAM and its chemical relatives are perhaps even more important than L-BMAA as subjects for research attention in relation to the etiology of western Pacific ALS-PDC and related disorders worldwide.

Acknowledgment

The authors thank Todd Needs, OSM-2 student at Western University of Health Sciences, for assistance with the references.

References

[1] Kisby GE, Standley M, Lu X, O'Malley J, Lin B, Muniz J, et al. Molecular networks perturbed in a developmental animal model of brain injury. Neurobiol Dis 2005;19:108–18.

[2] Kisby GE, Olivas A, Standley M, Lu X, Pattee P, O'Malley J, et al. Genotoxicants target distinct molecular networks in neonatal neurons. Environ Health Perspect 2006;114:1703–12.

[3] Kisby GE, Olivas A, Park T, Churchwell M, Doerge D, Samson LD, et al. DNA repair modulates the vulnerability of the developing brain to alkylating agents. DNA Repair 2009;8:400–12.

[4] Kisby GE, Moore H, Spencer PS. Animal models of brain maldevelopment induced by cycad plant genotoxins. Birth Defects Res C Embryo Today 2013;99:247–55.

[5] Kisby GE, Fry RC, Lasarev MR, Bammler TK, Beyer RP, Churchwell M, et al. The cycad genotoxin MAM modulates brain cellular pathways involved in neurodegenerative disease and cancer in a DNA damage-linked manner. PLoS One 2011;6:e20911.

[6] Spencer PS, Nunn PB, Hugon J, Ludolph A, Roy DN. Motorneurone disease on Guam: possible role of a food neurotoxin. Lancet 1986;1:965.

[7] Spencer PS, Nunn PB, Hugon J, Ludolph AC, Ross SM, Roy DN, et al. Guam amyotrophic lateral sclerosis-parkinsonism-dementia linked to a plant excitant neurotoxin. Science 1987;237:517–22.

[8] Karlsson O, Roman E, Berg AL, Brittebo EB. Early hippocampal cell death, and late learning and memory deficits in rats exposed to the environmental toxin BMAA (β-N-methylamino-L-alanine) during the neonatal period. Behav Brain Res 2011;219:310–20.

[9] Karlsson O, Berg AL, Lindström AK, Hanrieder J, Arnerup G, Roman E, et al. Neonatal exposure to the cyano-bacterial toxin BMAA induces changes in protein expression and neurodegeneration in adult hippocampus. Toxicol Sci 2012;130:391–404.
[10] Snyder LR, Marler TE. Rethinking cycad metabolite research. Commun Integr Biol 2011;4:86–8.
[11] Louw WK, Oelofsen W. Carcinogenic and neurotoxic components in the cycad Encephalartos altensteineii Lehm. (Family Zamiaceae). Toxicon 1975;13:447–52.
[12] Yagi F, Tadera K. Substrate specificity and transglucosylation catalyzed by cycad beta-glucosidase. Biochim Biophys Acta 1996;1289:315–21.
[13] Yagi F. Azoxyglycoside content and beta-glycosidase activities in leaves of various cycads. Phytochemistry 2004;65:3243–7.
[14] Pan M, Mabry TJ, Beale JM, Mamiya BM. Nonprotein amino acids from Cycas revoluta. Phytochemistry 1997;45:517–9.
[15] Wilson JM, Khabazian I, Wong MC, Seyedalikhani A, Bains JS, Pasqualotto BA, et al. Behavioral and neurologi-cal correlates of ALS-parkinsonism dementia complex in adult mice fed washed cycad flour. Neuromolecular Med 2002;1:207–21.
[16] Spencer PS, Roy DN, Ludolph A, Hugon J, Dwivedi MP, Schaumburg HH. Lathyrism: evidence for role of the neuroexcitatory aminoacid BOAA. Lancet 1986;2:1066–7.
[17] Kisby GE, Roy DN, Spencer PS. Determination of beta-N-methylamino-L-alanine (BMAA) in plant (Cycas cir-cinalis L.) and animal tissue by precolumn derivatization with 9-fluorenylmethyl chloroformate (FMOC) and reversed-phase high-performance liquid chromatography. J Neurosci Methods 1988;26:45–54.
[18] Kisby GE, Spencer PS. Neurotoxic amino acids from the cycad carcinogen methylazoxymethanol. In: Rose FC, Norris FH, editors. ALS. New advances in toxicology and epidemiology. London: Smith-Gordon; 1990. p. 35–9.
[19] Spencer PS. Guam ALS/parkinsonism-dementia: a long-latency neurotoxic disorder caused by "slow toxin(s)" in food? Can J Neurol Sci 1987;14:347–57.
[20] Spencer PS, Kisby GE, Ludolph AC. Slow toxins, biologic markers, and long-latency neurodegenerative dis-ease in the Western Pacific region. Neurology 1991;41(Suppl. 2):62–6.
[21] Spencer PS, Fry RC, Palmer V, Kisby GE. Western Pacific ALS-PDC: a prototypical neurodegenerative disorder linked to DNA damage and aberrant proteogenesis? Front Neurol 2012;3:180. http://dx.doi.org/10.3389/fneur.2012.00180.
[22] Arnold A, Edgren DC, Paladino VS. Amyotrophic lateral sclerosis: fifty cases observed on Guam. J Nerv Ment Dis 1953;117:135–9.
[23] Kurland LT, Mulder DW. Epidemiologic investigations of amyotrophic lateral sclerosis. 1. Preliminary report on geographic distribution, with special reference to the Mariana Islands, including clinical and pathologic observations. Neurology 1954;4:355–78. 4, 438–48.
[24] Mulder DW, Kurland LT, Iriarte LLG. Neurologic diseases on the island of Guam. US Armed Forces Med J 1954;5:1724–39.
[25] Hirano A, Kurland LT, Krooth RS, Lessel S. Parkinsonism-dementia complex, an endemic disease on the Island of Guam. Brain 1961;84:642–61.
[26] Kurland LT, Mulder DW. Epidemiologic investigations of amyotrophic lateral sclerosis. 2. Familial aggrega-tions indicative of dominant inheritance. Neurology 1955;5:182–96. 5, 249–268.
[27] Plato CC, Garruto RM, Fox KM, Gajdusek DC. Amyotrophic lateral sclerosis and parkinsonism-dementia on Guam: a 25-year prospective case-control study. Am J Epidemiol 1986;124:643–56.
[28] Reed DM, Brody JA. Amyotrophic lateral sclerosis and parkinsonism-dementia on Guam, 1945-1972. I. Descrip-tive epidemiology. Am J Epidemiol 1975;101:287–301.
[29] Garruto RM, Yanagihara R, Gajdusek DC. Disappearance of high-incidence amyotrophic lateral sclerosis and parkinsonism-dementia on Guam. Neurology 1985;35:193–8.
[30] Reed DM, Torres JM, Brody JA. Amyotrophic lateral sclerosis and parkinsonism-dementia on Guam, 1945-1972. II. Familial and genetic studies. Am J Epidemiol 1975;101:302–10.
[31] Morris HR, Steele JC, Crook R, Wavrant-De Vrièze F, Onstead-Cardinale L, Gwinn-Hardy K, et al. Genome-wide analysis of the parkinsonism-dementia complex of Guam. Arch Neurol 2004;61:1889–97.
[32] Dombroski BA, Galasko DR, Mata IF, Zabetian CP, Craig UK, Garruto RM, et al. C9orf72 hexanucleotide repeat expansion and Guam amyotrophic lateral sclerosis-parkinsonism-dementia complex. JAMA Neurol 2013;70:742–5.

[33] Reiff DM, Spathis R, Chan CW, Vilar MG, Sankaranarayanan K, Lynch D, et al. Inherited and somatic mitochondrial DNA mutations in Guam amyotrophic lateral sclerosis and parkinsonism-dementia. Neurol Sci 2011; 32:883–92.

[34] Pérez-Tur J, Buée L, Morris HR, Waring SC, Onstead L, Wavrant-De Vrièze F, et al. Neurodegenerative diseases of Guam: analysis of TAU. Neurology 1999;53:411–3.

[35] Kowalska A, Konagaya M, Sakai M, Hashizume Y, Tabira T. Familial amyotrophic lateral sclerosis and parkinsonism-dementia complex—tauopathy without mutations in the tau gene? Folia Neuropathol 2003;41:59–64.

[36] Poorkaj P, Tsuang D, Wijsman E, Steinbart E, Garruto RM, Craig UK, et al. TAU as a susceptibility gene for amyotrophic lateral sclerosis-parkinsonism dementia complex of Guam. Arch Neurol 2001;58:1871–8.

[37] Sundar PD, Yu CE, Sieh W, Steinbart E, Garruto RM, Oyanagi K, et al. Two sites in the MAPT region confer genetic risk for Guam ALS/PDC and dementia. Hum Mol Genet 2007;16:295–306.

[38] Sieh W, Choi Y, Chapman NH, Craig UK, Steinbart EJ, Rothstein JH, et al. Identification of novel susceptibility loci for Guam neurodegenerative disease: challenges of genome scans in genetic isolates. Hum Mol Genet 2009;18:3725–38.

[39] Reed DM, Labarthe D, Stallones R. Health effects of westernization and migration among Chamorros. Am J Epidemiol 1970;92:94–112.

[40] Garruto RM, Gajdusek DC, Chen KM. Amyotrophic lateral sclerosis and parkinsonism-dementia among Filipino migrants to Guam. Ann Neurol 1981;10:341–50.

[41] Garruto RM, Gajdusek DC, Chen KM. Amyotrophic lateral sclerosis among Chamorro migrants from Guam. Ann Neurol 1980;8:612–9.

[42] Román GC. Neuroepidemiology of amyotrophic lateral sclerosis: clues to aetiology and pathogenesis. J Neurol Neurosurg Psychiat 1966;61:131–7.

[43] Reed D, Labarthe D, Chen KM, Stallones R. A cohort study of amyotrophic lateral sclerosis and parkinsonism-dementia on Guam and Rota. Am J Epidemiol 1987;125:92–100.

[44] Borenstein AR, Mortimer JA, Schofield E, Wu Y, Salmon DP, Gamst A, et al. Cycad exposure and risk of dementia, MCI, and PDC in the Chamorro population of Guam. Neurology 2007;68:1764–71.

[45] Kisby GE, Ellison M, Spencer PS. Content of the neurotoxins cycasin (methylazoxymethanol beta-D-glucoside) and BMAA (beta-N-methylamino-L-alanine) in cycad flour prepared by Guam Chamorros. Neurology 1992;42:1336–40.

[46] Zhang ZX, Anderson DW, Mantel N, Román GC. Motor neuron disease on Guam: geographic and familial occurrence 1956–1985. Acta Neurol Scand 1996;94:51–9.

[47] Whiting M. Identification of toxic elements of cycads. In: Toxicity of cycads; implications for neurodegenerative diseases and cancer, transcripts of the four Cycad Conferences. New York: Third World Medical Research Foundation; 1988.

[48] Steele J, Guzman T. Observations about amyotrophic lateral sclerosis and the parkinsonism-dementia complex of Guam with regard to epidemiology and etiology. Can J Neurol Sci 1987;14:358–62.

[49] Yoshida S, Uebayashi Y, Kihira T, Kohmoto J, Wakayama I, Taguchi S, et al. Epidemiology of motor neuron disease in the Kii Peninsula of Japan, 1989–1993: active or disappearing focus? J Neurol Sci 1998;155:146–55.

[50] Kuzuhara S. ALS-parkinsonism-dementia complex of the Kii Peninsula of Japan (Muro disease). Historical review, epidemiology and concept. Rinsho Shinkeigaku 2007;47:962–5.

[51] Yase Y. Neurologic disease in the Western Pacific islands, with a report on the focus of amyotrophic lateral sclerosis found in the Kii Peninsula, Japan. Am J Trop Med Hyg 1970;19:55–66.

[52] Yase Y, Yoshida S, Kihira T, Wakayama I, Komoto J. Kii ALS dementia. Neuropathology 2001;21:105–9.

[53] Kokubo Y, Kuzahara S. Neurological and neuropathological studies of amyotrophic lateral sclerosis/parkinsonism-dementia complex in the Kii Peninsula of Japan. Rinsho Shinkeigaku 2001;41:769–74.

[54] Kokubo Y. Clinical aspects, imaging and neuropathology of Kii ALS/PDC. Rinsho Shinkeigaku 2007;47:966–9.

[55] Kuzuhara S. Muro disease: amyotrophic lateral sclerosis/parkinsonism-dementia complex in Kii Peninsula of Japan. Brain Nerve 2011;63:119–29.

[56] Kihira T, Yoshida S, Murata K, Ishiguti H, Kondo T, Kohmoto J, et al. Changes in the incidence and clinical features of ALS in the Koza, Kozagawa, and Kushimoto area of the Kii Peninsula—from the 1960s to the 2000s (follow-up study). Brain Nerve 2010;62:72–80.

[57] Kihira T, Yoshida S, Kondo T, Iwai K, Wada S, Morinaga S, et al. An increase in ALS incidence on the Kii Peninsula, 1960-2009: a possible link to change. Amyotroph Lateral Scler 2010;13:347–50.

[58] Kihira T, Okamoto K, Yoshida S, Kondo T, Iwai K, Wada S, et al. Environmental characteristics and oxidative stress of inhabitants and patients with amyotrophic lateral sclerosis in a high-incidence area on the Kii Peninsula, Japan. Intern Med 2013;52:1479–86.

[59] Tomiyama H, Kokubo Y, Sasaki R, Li Y, Imamichi Y, Funayama M, et al. Mutation analyses in amyotrophic lateral sclerosis/parkinsonism-dementia complex of the Kii Peninsula, Japan. Mov Disord 2008;23:2344–8.

[60] Kaji R, Izumi Y, Adachi Y, Kuzuhara S. ALS-parkinsonism-dementia complex of Kii and other related diseases in Japan. Parkinsonism Relat Disord 2012;18(Suppl. 1):S190–1.

[61] Ishiura H, Takahashi Y, Mitsui J, Yoshida S, Kihira T, Kokubo Y, et al. C9ORF72 repeat expansion in amyotrophic lateral sclerosis in the Kii Peninsula of Japan. Arch Neurol 2012;69:1154–8.

[62] Yoshida S, Yase Y, Iwata S, Mizumoto Y, Gajdusek DC. Trace-elemental study on amyotrophic lateral sclerosis (ALS) and parkinsonism-dementia (PD) in the Kii Peninsula of Japan and Guam. Rinsho Shinkeigaku 1987;27:79–87.

[63] Spencer PS, Ohta M, Palmer VS. Cycad use and motor neurone disease in Kii Peninsula of Japan. Lancet 1987;19:1462–3.

[64] Iwami O, Niki Y, Watanabe T, Ikeda M. Motor neuron disease on the Kii Peninsula of Japan: cycad exposure. Neuroepidemiology 1993;12:307–12.

[65] Gajdusek DC. Motor-neuron diseases in natives of New Guinea. New Engl J Med 1963;268:474–6.

[66] Gajdusek DC. A focus of high incidence amyotrophic lateral sclerosis and parkinsonism and dementia syndromes in a small population of Auyu and Jakai people of southern West New Guinea. In: Tsubaki T, Toyokura Y, editors. Amyotrophic lateral sclerosis. Tokyo: Japan Medical Research Foundation Publ. No. 8, University of Tokyo Press; 1979. p. 287–303.

[67] Spencer PS, Palmer VS, Ludolph AC. On the decline and etiology of high-incidence motor system disease in West Papua (southwest New Guinea). Mov Disord 2005;20(Suppl. 12):S119–26.

[68] Spencer PS, Palmer VS, Herman A, Asmedi A. Cycad use and motor neurone disease in Irian Jaya. Lancet 1987;2:1273–4.

[69] Okumiya K, Wada T, Fujisawa M, Ishine M, Garcia Del Saz E, Hirata Y, et al. Amyotrophic lateral sclerosis and parkinsonism in Papua, Indonesia: 2001–2012 survey results. BMJ Open 2014;4:e004353.

[70] Whiting MG. Toxicity of cycads. Econ Bot 1963;17:271–302.

[71] Spencer P, Fry RC, Kisby GE. Unraveling 50-year-old clues linking neurodegeneration and cancer to cycad toxins: are microRNAs common mediators? Front Genet 2012;3:192. http://dx.doi.org/10.3389/fgene.2012.00192.

[72] Albretsen JC, Khan SA, Richardson JA. Cycad palm toxicosis in dogs: 60 cases (1987–1997). J Am Vet Med Assoc 1998;213:99–101.

[73] Hirono I, Kachi H, Kato T. A survey of acute toxicity of cycads and mortality rate from cancer in the Miyako Islands, Okinawa. Acta Pathol Jpn 1970;20:327–37.

[74] Spencer PS, Dastur D. Neurolathyrism and neurocycadism. In: Dastur DK, Shahani M, Bharucha EP, editors. Neurological sciences: an overview of current problems. Section VI. Tropical neurology and neurotoxicology. New Delhi: Interprint; 1989. p. 309–18.

[75] Hall WTK, McGavin MD. Clinical and neuropathological changes in cattle eating the leaves of *Macrozamia lucida* or *Bowenia serrulata* (family Zamiaceae). Pathol Vet 1968;5:26–34.

[76] Hooper PT. Cycad poisoning in Australia—etiology and pathology. In: Keeler RF, Van Kampen KR, James LF, editors. Effects of poisonous plants on livestock. New York, NY: Academic Press; 1978. p. 337–47.

[77] Hooper PT, Best SM, Campbell A. Axonal dystrophy in the spinal cords of cattle consuming the cycad palm. Aust Vet J 1974;50:146–9.

[78] Hooper PT. Cycad poisoning. In: Keeler RF, Tu AT, editors. Handbook of natural toxins. Plant and fungal toxins, vol. 1. NY: Marcell Dekker; 1983. p. 463.

[79] Reams RY, Janovitz EB, Robinson FR, Sullivan JM, Rivera Casanova C, Más E. Cycad (*Zamia puertoriquensis*) toxicosis in a group of dairy heifers in Puerto Rico. J Vet Diagn Invest 1993;5:488–94.

[80] Spencer PS. Linking cycad to the etiology of Western Pacific amyotrophic lateral sclerosis. In: Rose FC, Norris FH, editors. ALS. New advances in toxicology and epidemiology. London: Smith-Gordon; 1990. p. 29–34.

[81] Shimizu T, Yasuda N, Kono I, Yagi F, Tadera K, Kobayahsi A. Hepatic and spinal lesions in goats chronically intoxicated with cycasin. Jpn J Vet Sci 1986;48:1291–5.

[82] Jubb KVF, Kennedy PC, Palmer N. Pathology of domestic animals. 3rd ed. Orlando: Academic Press; 1985.

[83] Palekar RS, Dastur DK. Cycasin content of *Cycas circinalis*. Nature 1965;206:1363–5.

[84] Dastur DK, Palekar RS. Effect of boiling and storing on cycasin content of *Cycas circinalis* L. Nature 1966;210:841–3.

[85] Dastur DK. Cycad toxicity in monkeys: clinical, pathological, and biochemical aspects. Fed Proc 1964;2:1368–9.

[86] Nishida K, Kobayashi A, Nagahama T. Studies on cycasin, a new toxic glycoside of *Cycas revoluta*, thumb. Part II. Hydrolysis of cycasin with cycad-emulsion. Bull Agric Chem Soc Jpn 1955;19:172–7.

[87] Matsumoto H, Strong FM. The occurrence of methylazoxymethanol in *Cycas circinalis* L. Arch Biochem Biophys 1963;101:299–310.

[88] Xiao S, McLean J, Robertson J. Neuronal intermediate filaments and ALS: a new look at an old question. Biochim Biophys Acta 2006;1762:1001–12.

[89] Sieber SM, Correa P, Dalgard DW, McIntire KR, Adamson RH. Carcinogenicity and hepatotoxicity of cycasin and its aglycone methylazoxymethanol acetate in nonhuman primates. J Natl Cancer Inst 1980;65:177–89.

[90] Spencer PS. Are neurotoxins driving us crazy? Planetary observations on the causes of neurodegenerative diseases of old age. In: Russell RW, Flattau PE, Pope AM, editors. Behavioral measures of neurotoxicity. Washington, DC: National Academy Press; 1990. p. 11–36.

[91] Kisby GE, Spencer PS. Is neurodegenerative disease a long-latency response to early-life genotoxin exposure? Int J Environ Res Public Health 2011;8:3889–921.

[92] Kisby GE, Kabel H, Hugon J, Spencer P. Damage and repair of nerve cell DNA in toxic stress. Drug Metab Rev 1999;31:589–618.

[93] Beaton JM. Dangerous harvest: investigation in the late prehistoric occupation of southeast central Queensland [Ph.D. thesis]. Canberra: Australian National University; 1977.

[94] Duncan MW, Steele JC, Kopin IJ, Markey SP. 2-Amino-3-(methylamino)-propanoic acid (BMAA) in cycad flour: an unlikely cause of amyotrophic lateral sclerosis and parkinsonism-dementia of Guam. Neurology 1990;40:767–72.

[95] Román GC, Zhang Z-X, Ellenberg JH. The Neuroepidemiology of Parkinson's disease. In: Ellenberg JH, Koller WC, Langston JW, editors. Etiology of Parkinson's disease. New York: Marcel Dekker; 1995. p. 203–43.

[96] Jones MZ, Gardner E. Pathogenesis of methylazoxymethanol-induced lesions in the postnatal mouse cerebellum. J Neuropathol Exp Neurol 1976;35:413–44.

[97] Shiraki H, Yase Y. ALS in Japan. In: Vinken PJ, Bruyn GW, editors. Handbook of clinical neurology, vol. 22. System disorders and atrophy, Part 2. New York: American Elsevier; 1975. p. 353–419.

[98] Karlsson O, Berg AL, Hanrieder J, Arnerup G, Lindström AK, Brittebo EB. Intracellular fibril formation, calcification, and enrichment of chaperones, cytoskeletal, and intermediate filament proteins in the adult hippocampus CA1 following neonatal exposure to the nonprotein amino acid BMAA. Arch Toxicol 2015;89:423–36.

[99] Husseman JW, Nochlin D, Vincent I. Mitotic activation: a convergent mechanism for a cohort of neurodegenerative diseases. Neurobiol Aging 2000;21:815–28.

[100] Stone JG, Siedlak SL, Tabaton M, Hirano A, Castellani RJ, Santocanale C, et al. The cell cycle regulator phosphorylated retinoblastoma protein is associated with tau pathology in several tauopathies. J Neuropathol Exp Neurol 2011;70:578–87.

[101] Sullivan-Jones P, Ali SF, Gough B, Holson RR. Postnatal methylazoxymethanol: sensitive periods and regional selectivity of effects. Neurotoxicol Teratol 1994;16:631–7.

[102] Ciani E, Frenquelli M, Contestabile A. Developmental expression of the cell cycle and apoptosis controlling gene, Lot1, in the rat cerebellum and in cultures of cerebellar granule cells. Brain Res Dev Brain Res 2003;142:193–202.

[103] Kisby GE, Renslow P, Ryan A, Beam M, Woltjer R. The cycad genotoxin methylazoxymethanol (MAM) induces brain tissue DNA damage and accelerates tau pathology in htau mice. Soc Neurosci Abstr 2011. #351.25/V2.

[104] Vincent I, Pae CI, Hallows JL. The cell cycle and human neurodegenerative disease. Prog Cell Cycle Res 2003;5:31–41.

[105] McShea A, Lee HG, Petersen RB, Casadesus G, Vincent I, Linford NJ, et al. Neuronal cell cycle re-entry mediates Alzheimer disease-type changes. Biochim Biophys Acta 2007;1772:467–72.

[106] Zhu X, Siedlak SL, Wang Y, Perry G, Castellani RJ, Cohen ML, et al. Neuronal binucleation in Alzheimer disease hippocampus. Neuropathol Appl Neurobiol 2008;34:457–65.

[107] Chano T, Okabe H, Hulette CM. RB1CC1 insufficiency causes neuronal atrophy through mTOR signaling alteration and involved in the pathology of Alzheimer's diseases. Brain Res 2007;1168:97–105.

[108] Bajic V, Mandusic V, Stefanova E, Bozovic A, Davidovic R, Zivkovic L, et al. Skewed X-chromosome inactivation in women affected by Alzheimer's disease. J Alzheimers Dis 2015;43:1251–9.

[109] Arendt T. Cell cycle activation and aneuploid neurons in Alzheimer's disease. Mol Neurobiol 2012;46:125–35.

[110] Katsel P, Tan W, Fam P, Purohit DP, Haroutunian V. Cycle checkpoint abnormalities during dementia: a plausible association with the loss of protection against oxidative stress in Alzheimer's disease. PLoS One 2013;8:e68361.

[111] Eizirik DL, Spencer P, Kisby GE. Potential role of environmental genotoxic agents in diabetes mellitus and neurodegenerative diseases. Biochem Pharmacol 1996;51:1585–91.

[112] Eizirik DL, Kisby GE. Cycad toxin-induced damage of rodent and human pancreatic beta-cells. Biochem Pharmacol 1995;50:355–65.

[113] Esclaire F, Kisby G, Spencer P, Milne J, Lesort M, Hugon J. The Guam cycad toxin methylazoxymethanol damages neuronal DNA and modulates tau mRNA expression and excitotoxicity. Exp Neurol 1999;155:11–21.

[114] Ono S, Waring SC, Kurland LL, Katrina-Craig U, Petersen RC. Guamanian neurodegenerative disease: ultrastructural studies of skin. J Neurol Sci 1997;146:35–40.

[115] Notman J, Tan QH, Zedeck MS. Inhibition of methylazoxymethanol-induced intestinal tumors in the rat by pyrazole with paradoxical effects on skin and kidney. Cancer Res 1982;4:1774–80.

[116] Tate ME, Delaere IM, Jones GP, Tierkink ERT. Crystal and molecular structure of (Z)-β-O-D-glucopyranosyloxy-NNO-azoxymethane. Aust J Chem 1995;48:1059.

[117] Nagahama T, Numata T, Nishida K. Some new azoxyglycosides of *Cycas revoluta* Thumb. II. Neocycasin B and macrozamin. Bull Agr Chem Soc Jpn 1959;23:556–7.

[118] Tadera K, Yagi F, Arima M, Kobayashi A. Formation of cycasin from methylazoxymethanol by UDP-glucosyltransferase from leaves of Japanese cycad. Agric Biol Chem 1985;49:2827–8.

[119] Woo YT, Lai DY, Arcos JC, Argus MF. Natural, metal, fiber and macromolecular carcinogens. Section 5.3.2.2 Cycasin and related compounds. In: Arcos JC, Wolf G, editors. Chemical induction of cancer: structural basis and biological mechanisms. San Diego: Academic Press; 1988. p. 178.

[120] Laqueur GL. Carcinogenic effects of cycad meal and cycasin, methylazoxymethanol glycoside, in rats and effects of cycasin in germfree rats. Fed Proc 1964;23:1386–8.

[121] Morant AV, Jørgensen K, Jørgensen C, Paquette SM, Sánchez-Pérez R, Møller BL, et al. Beta-glucosidases as detonators of plant chemical defense. Phytochemistry 2008;69:1795–813.

[122] Ketudat-Cairns JR, Esen A. β-Glucosidases. Cell Mol Life Sci 2010;67:3389–405.

[123] Patro L, Mohapatra PK, Biswal UC, Biswal B. Dehydration induced loss of photosynthesis in Arabidopsis leaves during senescence is accompanied by the reversible enhancement in the activity of cell wall β-glucosidase. J Photochem Photobiol B 2014;137:49–54.

[124] Yagi F. Formation of 3 new azoxyglycosides by transglucosylation by cycad beta-glucosidase. Agric Biol Chem 1985;49:2985–90.

[125] Wells WW, Yang MG, Bolzer W, Mickelsen O. Gas-liquid chromatographic analysis of cycasin in cycad flour. Anal Biochem 1968;25:325–9.

[126] Matsumoto H, Higa HH. Studies on methylazoxymethanol, the aglycone of cycasin: methylation of nucleic acids in vitro. Biochem J 1966;98:20C.

[127] Shank RD, Magee PN. Similarities between the biochemical actions of cycasin and dimethylnitrosamine. Biochem J 1967;105:521.

[128] Nagasawa HT, Shirota FN, Matsumoto H. Decomposition of methylazoxymethanol, the aglycone of cycasin, in D_2O. Nature 1972;236:234–5.

[129] Kobayashi A, Matsumoto H. Studies on methylazoxymethanol, the aglycone of cycasin. Isolation, biological, and chemical properties. Arch Biochem Biophys 1965;110:373–80.

[130] Samson LD. The repair of DNA alkylation damage by methyltransferases and glycosylases. Essays Biochem 1992;27:69–78.

[131] Nagata Y, Matsumoto H. Studies on methylazoxymethanol: methylation of nucleic acids in fetal rat brain. Proc Soc Exp Biol Med 1969;132:383–5.

[132] Matsumoto H, Spatz M, Laqueur GL. Quantitative changes with age in the DNA content of methylazoxymethanol-induced microencephalic rat brain. J Neurochem 1972;19:297–306.

[133] Kleihues P, Bucheler J. Long-term persistence of O^6-methylguanine in rat brain DNA. Nature 1977;269:625–6.

[134] Kreklau EL, Liu N, Li Z, Cornetta K, Erickson LC. Comparison of single- versus double-bolus treatments of O(6)-benzylguanine for depletion of O(6)-methylguanine DNA methyltransferase (MGMT) activity in vivo: development of a novel fluorometric oligonucleotide assay for measurement of MGMT activity. J Pharmacol Exp Ther 2001;297:524–30.

[135] Burns JA, Dreij K, Cartularo L, Scicchitano DA. O^6-methylguanine induces altered proteins at the level of transcription in human cells. Nucleic Acids Res 2010;38:8178–87.

[136] Yase Y. The pathogenesis of amyotrophic lateral sclerosis. Lancet 1972;2:292–6.

[137] Matsuoka T, Nishizaki T, Kisby GE. Na+-dependent and phlorizin-inhibitable transport of glucose and cycasin in brain endothelial cells. J Neurochem 1998;70:772–7.

[138] Hirayama B, Hazama A, Loo DF, Wright EM, Kisby GE. Transport of cycasin by the intestinal Na+/glucose cotransporter. Biochim Biophys Acta 1994;1193:151–4.

[139] Silber JR, Blank A, Bobola MS, Mueller BA, Kolstoe DD, Ojemann GA, et al. Lack of the DNA repair protein O6-methylguanine-DNA methyltransferase in histologically normal brain adjacent to primary human brain tumors. Proc Natl Acad Sci USA 1996;93:6941–6.

[140] Kisby GE, Standley M, Park T, Olivas A, Fei S, Jacob T, et al. Proteomic analysis of the genotoxicant methyl-azoxymethanol (MAM)-induced changes in the developing cerebellum. J Proteome Res 2006;5:2656–65.

[141] Ferguson SA, Paule MG, Holson RR. Functional effects of methylazoxymethanol-induced cerebellar hypoplasia in rats. Neurotoxicol Teratol 1996;18:529–37.

[142] Bobola MS, Blank A, Berger MS, Silber JR. O6-methylguanine-DNA methyltransferase deficiency in developing brain: implications for brain tumorigenesis. DNA Repair (Amst) 2007;6:1127–33.

[143] Yang W, Woltjer RL, Sokal I, Pan C, Wang Y, Brodey M, et al. Quantitative proteomics identifies surfactant-resistant alpha-synuclein in cerebral cortex of parkinsonism-dementia complex of Guam but not Alzheimer's disease or progressive supranuclear palsy. Am J Pathol 2007;171:993–1002.

[144] Mackowiak M, Bator E, Latusz J, Mordalska P, Wedzony K. Prenatal MAM administration affects histone H3 methylation in postnatal life in the rat medial prefrontal cortex. Eur Neuropsychopharmacol 2014;24:271–89.

[145] Wong AW, McCallum GP, Jeng W, Wells PG. Oxoguanine glycosylase 1 protects against methamphetamine-enhanced fetal brain oxidative DNA damage and neurodevelopmental deficits. J Neurosci 2008;228:9047–54.

[146] Arai JA, Feig LA. Long-lasting and transgenerational effects of an environmental enrichment on memory formation. Brain Res Bull 2011;85:30–5.

[147] Guo N, Yoshizaki K, Kimura R, Suto F, Yanagawa Y, Osumi N. A sensitive period for GABAergic interneurons in the dentate gyrus in modulating sensorimotor gating. J Neurosci 2013;33:6691–704.

[148] Ueda S, Yoshimoto K, Kadowaki T, Hirata K, Sakakibara S. Improved learning in microencephalic rats. Congenit Anom (Kyoto) 2010;50:58–63.

[149] Wallace CS, Reitzenstein J, Withers GS. Diminished experience-dependent neuroanatomical plasticity: evidence for an improved biomarker of subtle neurotoxic damage to the developing rat brain. Environ Health Perspect 2003;111:1294–8.

[150] Hoffman JR, Boyne LJ, Levitt P, Fischer I. Short exposure to methylazoxymethanol causes a long-term inhibition of axonal outgrowth from cultured embryonic rat hippocampal neurons. J Neurosci Res 1996;46:349–59.

[151] Bigl V, Rossner S. Amyloid precursor protein processing in vivo–insights from a chemically-induced constitutive overactivation of protein kinase C in Guinea pig brain. Curr Med Chem 2003;10:871–82.

[152] Buecheler J, Kleihues P. Excision of O6-methylguanine from DNA of various mouse tissues following a single injection of N-methyl-nitrosourea. Chem Biol Interact 1977;16:325–33.

[153] Likhachev AJ, Alexandrov VA, Anisimov VN, Bespalov VG, Korsakov MV, Ovsyannikov AI, et al. Persistence of methylated purines in the DNA of various rat fetal and maternal tissues and carcinogenesis in the offspring following a single transplacental dose of N-methyl-N-nitrosourea. Int J Cancer 1983;31:779–84.

[154] Balduini W, Cimino M, Lombardelli G, Abbracchio MP, Peruzzi G, Cecchini T, et al. Microencephalic rats as a model for cognitive disorders. Clin Neuropharmacol 1986;9:S8–18.

[155] Kisby GE, Palmer VS, Lasarev MR, Fry R, Iordanov M, Magun E, et al. Does the cycad genotoxin MAM implicated in Guam ALS-PDC induce disease-relevant changes in mouse brain that includes olfaction? Commun Integr Biol 2011;4:731–4.

[156] Doty RL, Perl DP, Steele JC, Chen KM, Pierce Jr JD, Reyes P, et al. Odor identification deficient of the parkinsonism-dementia complex of Guam: equivalence to that of Alzheimer's and idiopathic Parkinson's disease. Neurology 1991;41:77–80.

[157] Ahlskog JE, Waring SC, Petersen RC, Esteban-Santillan C, Craig UK, O'Brien PC, et al. Olfactory dysfunction in Guamanian ALS, parkinsonism, and dementia. Neurology 1998;51:1672–7.

[158] Doty RL, Perl DP, Steele JC, Chen KM, Pierce Jr JD, Reyes P, et al. Olfactory dysfunction in three neurodegenerative diseases. Geriatrics 1991;46S:47–51.

[159] Bell EA. The discovery of BMAA, and examples of biomagnification and protein incorporation involving other non-protein amino acids. Amyotroph Lateral Scler 2009;10(Suppl. 2):21–5. http://dx.doi.org/10.3109/17482960903268700.

[160] Spencer PS, Hugon J, Ludolph A, Nunn PB, Ross SM, Roy DN, et al. Discovery and partial characterization of primate motor-system toxins. Ciba Found Symp 1987;126:221–38.

[161] Vega A, Bell EA, Nunn PB. The preparation of L- and D-α-amino-β-methylaminopropionic acid acids and the identification of the compound isolated from *Cycas circinalis* as the L-isomer. Phytochemistry 1968;7:1885–7.

[162] Polsky FI, Nunn PB, Bell EA. Distribution and toxicity of alpha-amino-beta-methylaminopropionic acid. Fed Proc 1972;31:1473–5.

[163] Duncan MW, Villacreses NE, Pearson PG, Wyatt L, Rapoport SI, Kopin IJ, et al. 2-Amino-3-(methylamino)-propanoic acid (BMAA) pharmacokinetics and blood–brain barrier permeability in the rat. J Pharmacol Exp Ther 1991;258:27–35.

[164] Cox PA, Banack SA, Murch SJ. Biomagnification of cyanobacterial neurotoxins and neurodegenerative disease among the Chamorro people of Guam. Proc Natl Acad Sci 2003;100:13380–3.

[165] Anon. Analysis, occurrence, and toxicity of beta-methylaminoalanine (BMAA). A risk for the consumer? TemaNord, vol. 561. Copenhagen: Nordic Council of Ministers; 2007.

[166] Brenner ED, Stevenson DW, McCombie RW, Katari MS, Rudd SA, Mayer KF, et al. Expressed sequence tag analysis in *Cycas*, the most primitive living seed plant. Genome Biol 2003;4:R78. Epub 2003 November 18.

[167] Murch SJ, Cox PA, Banack SA. A mechanism for slow release of biomagnified cyanobacterial neurotoxins and neurodegenerative disease in Guam. Proc Natl Acad Sci USA 2004;101:12228–31.

[168] Marler TE, Snyder LR, Shaw CA. *Cycas micronesica* (Cycadales) plants devoid of endophytic cyanobacteria increase in beta-methylamino-L-alanine. Toxicon 2010;56:563–8.

[169] Banack SA, Johnson HE, Cheng R, Cox PA. Production of the neurotoxin BMAA by a marine cyanobacterium. Mar Drugs 2007;5:180–96.

[170] Sting AR, Seebach D. In: Paquette LA, editor. Encyclopedia of reagents for organic synthesis, 1. Hoboken, NJ: Wiley; 1995. p. 308–9.

[171] Fowden L, Lea PJ, Bell EA. The nonprotein amino acids of plants. Adv Enzymol Relat Areas Mol Biol 1979;50:117–75.

[172] Copley SD, Frank E, Kirsch WM, Koch TH. Detection and possible origins of aminomalonic acid in protein hydrolysates. Anal Biochem 1992;201:152–7.

[173] Blunt JW, Copp BR, Munro MH, Northcote PT, Prinsep MR. Marine natural products. Nat Prod Rep 2005;22:15–61.

[174] Brand LE, Pablo J, Compton A, Hammerschlag N, Mash DC. Cyanobacterial blooms and the occurrence of the neurotoxin, β-*N*-methylamino-L-alanine (BMAA), in south Florida aquatic food webs. Harmful Algae 2010;9:620–35.

[175] Jonasson S, Eriksson J, Berntzon L, Spacil Z, Ilag LL, Ronnevi LO, et al. Transfer of a cyanobacterial neurotoxin within a temperate aquatic ecosystem suggests pathways for human exposure. Proc Natl Acad Sci USA 2010;107:9252–7.

[176] Cox PA, Banack SA, Murch SJ, Rasmussen U, Tien G, Bidigare RR, et al. Diverse taxa of cyanobacteria produce β-*N*-methylamino-L-alanine, a neurotoxic amino acid. Proc Natl Acad Sci USA 2005;102:5074–8.

[177] Metcalf JS, Banack SA, Lindsay J, Morrison LF, Cox PA, Codd GA. Co-occurrence of beta-*N*-methylamino-L-alanine, a neurotoxic amino acid with other cyanobacterial toxins in British waterbodies, 1990–2004. Environ Microbiol 2008;10:702–8.

[178] Krüger T, Oelmüller R. The origin of β-*N*-methylamino-L-alanine (BMAA): cycads and/or cyanobacteria? J Endocytobio Cell Res 2012;22:29–36.

[179] Cohen SA. Analytical techniques for the detection of a-amino- β-methylaminopropionic acid. Analyst 2012;137. http://dx.doi.org/10.1039/c2an16250d.

[180] Snyder LR, Cruz-Aguado R, Sadilek M, Galasko D, Shaw CA, Montine TJ. Lack of cerebral BMAA in human cerebral cortex. Neurology 2009;72:1360–1.

[181] Faassen EJ. Presence of the neurotoxin BMAA in aquatic ecosystems: what do we really know? Toxins (Basel) 2014;6:1109–38.

[182] Lehninger AL, Nelson DL, Cox MM. Principles of biochemistry. 2nd ed. New York: Worth; 1993.

[183] Nunn PB, O'Brien P, Pettit LD, Pyburn SI. Complexes of zinc, copper, and nickel with the nonprotein amino acid L-alpha-amino-beta-methylaminopropionic acid: a naturally occurring neurotoxin. J Inorg Biochem 1989;37:175–83.

[184] Stadie WC, O'Brien H. The carbamate equilibrium: I. The equilibrium of amino acids, carbon dioxide, and carbamates in aqueous solution; with a note on the Ferguson-Roughton carbamate method. J Biol Chem 1936;112:723–58.

[185] Morrow JS, Keim P, Gurd FRN. CO_2 adducts of certain amino acids, peptides and sperm whale myoglobin studied by carbon 13 and proton nuclear magnetic resonance. J Biol Chem 1974;249:7484–94.

[186] Schaefer WH. Reaction of primary and secondary amines to form carbamic acid glucuronides. Curr Drug Metab 2006;7:873–81.

[187] Terrier P, Douglas DJ. Carbamino group formation with peptides and proteins studied by mass spectrometry. J Am Soc Mass Spectrom 2010;21:1500–5.

[188] Nunn PB, O'Brien P. The interaction of beta-N-methylamino-L-alanine with bicarbonate: an 1H-NMR study. FEBS Lett 1989;251:31–5.

[189] Myers TG, Nelson SD. Neuroactive carbamate adducts of beta-N-methylamino-L-alanine and ethylenediamine. Detection and quantitation under physiological conditions by 13C NMR. J Biol Chem 1990;25:10193–5.

[190] Chen JG, Sandberg M, Weber SG. Chromatographic method for the determination of conditional equilibrium constants for the carbamate formation reaction from amino acids and peptides in aqueous solution. J Am Chem Soc 1993;115:7343–50.

[191] Ross SM, Seelig M, Spencer PS. Specific antagonism of excitotoxic action of 'uncommon' amino acids assayed in organotypic mouse cortical cultures. Brain Res 1987;425:120–7.

[192] Ross SM, Spencer PS. Specific antagonism of behavioral action of "uncommon" amino acids linked to motor-system diseases. Synapse 1987;1:248–53.

[193] Cucchiaroni ML, Viscomi MT, Bernardi G, Molinari M, Guatteo E, Mercuri NB. Metabotropic glutamate receptor 1 mediates the electrophysiological and toxic actions of the cycad derivative beta-N-methylamino-L-alanine on substantia nigra pars compacta DAergic neurons. J Neurosci 2010;30:5176–88.

[194] Weiss JH, Choi DW. Beta-N-methylamino-L-alanine neurotoxicity: requirement for bicarbonate as a cofactor. Science 1988;241:973–5.

[195] Mroz EA, Weiss JW, Choi DW. Neuroactive carbamate adducts of beta-N-methylamino-L-alanine and ethylenediamine. Detection and quantitation under physiological conditions by 13C NMR. J Biol Chem 1990;265:10193–5.

[196] Wegerski C, Sanders JM, Doyle-Eisele M, McDonald J. Disposition and metabolism of β-N-methylamino-L-alanine in Sprague Dawley rats and b6c3f1 mice. Toxicologist 2012;126:405.

[197] Garner C, Wegerski CJ, Doyle-Eisele M, Lucak J, Waidyanatha S, McDonald JD, et al. Accumulation of β-N-methylamino-L-alanine in tissues following repeat oral administration to Harlan Sprague-Dawley rats. Toxicologist 2013;132:107.

[198] Duncan MW, Markey SP, Weick BG, Pearson PG, Ziffer H, Hu Y, et al. 2-Amino-3-(methylamino)propanoic acid (BMAA) bioavailability in the primate. Neurobiol Aging 1992;13:333–7.

[199] Weiss JH, Koh JY, Choi DW. Neurotoxicity of beta-N-methylamino-L-alanine (BMAA) and beta-N-oxalylamino-L-alanine (BOAA) on cultured cortical neurons. Brain Res 1989;497:64–71.

[200] Kisby GE, Ross SM, Spencer PS, Gold BG, Nunn PB, Roy DN. Cycasin and BMAA: candidate neurotoxins for western Pacific amyotrophic lateral sclerosis/Parkinsonism-dementia complex. Neurodegeneration 1992;1:73–82.

[201] Kisby GE, Nottingham V, Kayton R, Roy DN, Spencer PS. Brain metabolism of β-N-methylamino-L-alanine (BMAA) and protection of excitotoxicity by GABA-uptake inhibitors. Soc Neurosci 1992;18:82.

[202] Spencer PS, Ross SM, Kisby G, Roy DN. Western Pacific amyotrophic lateral sclerosis: putative role of cycad toxins. In: Hudson AJ, editor. Amyotrophic lateral sclerosis: concepts in pathogenesis and etiology. Toronto: University of Toronto Press; 1990. p. 263–95.

[203] Nunn PB, Ponnusamy M. Beta-N-methylaminoalanine (BMAA): metabolism and metabolic effects in model systems and in neural and other tissues of the rat in vitro. Toxicon 2009;54:85–94.

[204] Weiss JH, Christine CW, Choi DW. Bicarbonate dependence of glutamate receptor activation by beta-N-methylamino-L-alanine: channel recording and study with related compounds. Neuron 1989;3:321–6.

[205] Vijayalakshmi KR, Hariharan K, Rao DR. Metabolism of 2,3-diaminopropionate in the rat. Proc Indian Acad Sci 1987;87:257–9.

[206] Boor PJ, Trent MB, Lyles GA, Tao M, Ansari GA. Methylamine metabolism to formaldehyde by vascular semicarbazide-sensitive amine oxidase. Toxicology 1992;73:251–8.

[207] Deng Y, Boomsma F, Yu PH. Deamination of methylamine and aminoacetone increases aldehydes and oxidative stress in rats. Life Sci 1998;63:2049–58.

[208] Thanassi JW. Aminomalonic acid. Spontaneous decarboxylation and reaction with 5-deoxypyridoxal. Biochemistry 1970;9:525–32.

[209] Van Buskirk JJ, Kirsch WM, Kleyer DL, Barkley RM, Koch TH. Aminomalonic acid: identification in *Escherichia coli* and atherosclerotic plaque. Proc Natl Acad Sci USA 1984;81:722–5.

[210] Kisby GE, Garner CE, Wegerski C, Doyle-Eisele M, Lucak JF, McDonald J, et al. Distribution, transport, and metabolism of BMAA in the rodent brain. Soc Neurosci 2012. 442.12/E52.

[211] Hashmi M, Anders MW. Enzymatic reaction of β-methylaminoalanine with L-amino acid oxidase. Biochim Biophys Acta 1991;1074:36–9.

[212] Wellner D, Meister A. Studies on the mechanism of action of L-amino acid oxidase. J Biol Chem 1961;236:2357–64.

[213] Nakano M, Danowski TS, Weitzel DR. Crystalline mammalian L-amino acid oxidase from rat kidney mitochondria. J Biol Chem 1966;241:2075–83.

[214] Sato M, Ohishi N, Yagi K. Identification of a covalently bound flavoprotein in rat liver mitochondria with sarcosine dehydrogenase. Biochem Biophys Res Commun 1979;87:706–11.

[215] Porter DH, Cook RJ, Wagner C. Enzymatic properties of dimethylglycine dehydrogenase and sarcosine dehydrogenase from rat liver. Arch Biochem Biophys 1985;243:396–407.

[216] Kikuchi G. The glycine cleavage system: composition, reaction mechanism, and physiological significance. Mol Cell Biochem 1973;1:169–87.

[217] Murch SJ, Cox PA, Banack SA, Steele JC, Sacks OW. Occurrence of beta-methylamino-L-alanine (BMAA) in ALS/PDC patients from Guam. Acta Neurol Scand 2004;110:267–9.

[218] Montine TJ, Li K, Perl GP, Glasko D. Lack of beta-methylamino-L-alanine in brain from controls, AD, or Chamorros with PDC. Neurology 2005;65:768–789.

[219] Krüger T, Mönch B, Oppenhäuser S, Luckas B. LC-MS/MS determination of the isomeric neurotoxins BMAA (beta-N-methylamino-L-alanine) and DAB (2,4-diaminobutyric acid) in cyanobacteria and seeds of *Cycas revoluta* and *Lathyrus latifolius*. Toxicon 2010;55:547–57.

[220] Smith QR, Nagura H, Takada Y, Duncan MW. Facilitated transport of the neurotoxin, beta-N-methylamino-L-alanine, across the blood–brain barrier. J Neurochem 1992;58:1330–7.

[221] Snyder LR, Cruz-Aguado R, Sadilek M, Galasko D, Shaw CA, Montine TJ. Parkinson-dementia complex and development of a new stable isotope dilution assay for BMAA detection in tissue. Toxicol Appl Pharmacol 2009;240:180–8.

[222] Snyder LR, Hoggard JC, Montine TJ, Synovec RE. Development and application of a comprehensive two-dimensional gas chromatography with time-of-flight mass spectrometry method for the analysis of L-beta-methylamino-alanine in human tissue. J Chromatogr A 2010;1217:4639–47.

[223] Xie X, Basile M, Mash DC. Cerebral uptake and protein incorporation of cyanobacterial toxin β-N-methylamino-L-alanine. Neuroreport 2013;24:779–84.

[224] Dunlop RA, Cox PA, Banack PA, Rodgers KJ. The non-protein amino acid BMAA is misincorporated into human proteins in place of L-serine causing protein misfolding and aggregation. PLoS One 2013;8:e75376.

[225] Yamamoto T, Nishizaki I, Furuya S, Hirabayashi Y, Takahashi K, Okuyama S, et al. Characterization of rapid and high-affinity uptake of L-serine in neurons and astrocytes in primary culture. FEBS Lett 2003;548:69–73.

[226] Takarada T, Balcar VJ, Baba K, Takamoto A, Acosta GB, Takano K, et al. Uptake of [^{3}H]-L-serine in rat brain synaptosomal fractions. Brain Res 2003;983:36–47.

[227] Blum JJ. Effects of cycloheximide and actinomycin D on the amino acid transport system of Tetrahymena. J Cell Physiol 1982;111:104–10.

[228] Baran DT, Lichtman MA, Peck WA. alpha-Aminoisobutyric acid transport in human leukemic lymphocytes: in vitro characteristics and inhibition by cortisol and cycloheximide. J Clin Invest 1972;51:2181–9.

[229] Karlsson O, Jiang L, Andersson M, Ilag LL, Brittebo EB. Protein association of the neurotoxin and non-protein amino acid BMAA (β-N-methylamino-L-alanine) in the liver and brain following neonatal administration in rats. Toxicol Lett 2014;226:1–5.

[230] Wells DE. Extraction, clean-up and recoveries of persistent trace organic contaminants from sediment and biota samples. In: Dijkstra A, Massart DL, Kaufman L, editors. Evaluation and optimization of laboratory methods and analytical procedures. New York: Elsevier; 1978.

[231] Karlsson O, Berg C, Brittebo EB, Lindquist NG. Retention of the cyanobacterial neurotoxin beta-N-methylamino-L-alanine in melanin and neuromelanin-containing cells–a possible link between Parkinson-dementia complex and pigmentary retinopathy. Pigment Cell Melanoma Res 2009;22:120–30.

[232] DeMattei M, Levi AC, Fariello RG. Neuromelanic pigment in substantia nigra neurons of rats and dogs. Neurosci Lett 1996;72:37–42.

[233] Breckenridge JM, Zahm DS, Westfall TC, MacArthur H. Dopaminochrome induces neuroinflammation in the substantia nigra of Sprague Dawley rats. FASEB J 2010;24 (Suppl) 1053.8.

[234] Zhang Z, Chen Q, Li Y, Doss GA, Dean BJ, Ngui JS, et al. In vitro bioactivation of dihydrobenzoxathin selective estrogen receptor modulators by cytochrome P450 3A4 in human liver microsomes: formation of reactive iminium and quinone type metabolites. Chem Res Toxicol 2005;18:675–85.

[235] Kramer RA, McMenamin MG, Boyd MR. Differential distribution and covalent binding of two labeled forms of methyl-CCNU in the Fischer 344 rat. Cancer Chemother Pharmacol 1985;14:150–5.

[236] Hanzlik RP. Biological reactive intermediates III: Chemistry of covalent binding: studies with bromobenzene and thiobenzamide. Adv Exp Med Biol 1986;197:31–40.

[237] Pantarotto C, Blonda C. Covalent binding to proteins as a mechanism of chemical toxicity. Arch Toxicol 1984;7:208–18.

[238] Faassen EJ, Gillissen F, Lürling M. A comparative study on three analytical methods for the determination of the neurotoxin BMAA in cyanobacteria. PLoS One 2012;7:e36667.

[239] Rakonczay Z, Matsuoka Y, Giacobini E. Effects of L-beta-N-methylamino-L-alanine (L-BMAA) on the cortical cholinergic and glutamatergic systems of the rat. J Neurosci Res 1991;29:121–6.

[240] Matsuoka Y, Rakonczay Z, Giacobini E, Naritoku D. L-beta-methylamino-alanine-induced behavioral changes in rats. Pharmacol Biochem Behav 1993;44:727–34.

[241] Karlsson O, Lindquist NG, Brittebo EB, Roman E. Selective brain uptake and behavioral effects of the cyanobacterial toxin BMAA (beta-N-methylamino-L-alanine) following neonatal administration to rodents. Toxicol Sci 2009;109:286–95.

[242] Rice D, Barone Jr S. Critical periods of vulnerability for the developing nervous system: evidence from humans and animal models. Environ Health Perspect 2000;108:511–33.

[243] Purdie EL, Samsudin S, Eddy FB, Codd GA. Effects of the cyanobacterial neurotoxin beta-N-methylamino-L-alanine on the early-life stage development of zebrafish (Danio rerio). Aquat Toxicol 2009;95:279–84.

[244] Karlsson O, Roman E, Brittebo EB. Long-term cognitive impairments in adult rats treated neonatally with beta-N-methylamino-L-alanine. Toxicol Sci 2009;112:185–95.

[245] Karlsson O, Kultima K, Wadensten H, Nilsson A, Roman E, Andrén PE, et al. Neurotoxin-induced neuropeptide perturbations in striatum of neonatal rats. J Proteome Res 2013;12:1678–90.

[246] Jones PJ, AbuMweis SS. Phytosterols as functional food ingredients: linkages to cardiovascular disease and cancer. Curr Opin Clin Nutr Metab Care 2009;12:147–51.

[247] Khabazian I, Bains JS, Williams DE, Cheung J, Wilson JM, Pasqualotto BA, et al. Isolation of various forms of sterol beta-D-glucoside from the seed of Cycas circinalis: neurotoxicity and implications for ALS-parkinsonism dementia complex. J Neurochem 2002;82:516–28.

[248] Tabata RC, Wilson JMB, Shaw CA. Chronic exposure to dietary sterol glucosides is neurotoxic to motor neurons and induces an ALS-PDC phenotype. Neuromolecular Med 2008;10:24–39.

[249] Shen WB, McDowell KA, Siebert AA, Clark SM, Dugger NV, Valentino KM, et al. Environmental neurotoxin-induced progressive model of parkinsonism in rats. Ann Neurol 2010;68:7–80.

[250] Spencer PS, Lasarev MR, Palmer VS, Kisby GE. Neurotoxic cycad components and Western Pacific ALS/PDC. Ann Neurol 2010;68:975–6. [Letter].

[251] Banack SA, Nunn PB, Cheng R, Bradley WG. Washed cycad flour contains beta-N-methylamino-L-alanine and may explain parkinsonism symptoms. Ann Neurol 2011;69:423. [Letter].

[252] McDowell KA, Shen WB, Siebert AA, Clark SM, Jinnah HA, Sztalryd C, et al. Reply to Banack et al., 2011. Ann Neurol 2011;68:423–4. [Letter].

[253] McDowell KA, Shen WB, Siebert AA, Clark SM, Jinnah HA, Sztalryd C, et al. Reply to Spencer et al., 2010. Ann Neurol 2010;68:975–6. [Letter].

[254] Patel MD, Thompson PD. Phytosterols and vascular disease. Atherosclerosis 2006;186:12–9.

[255] Bouic PJ, Etsebeth S, Liebenberg RW, Albrecht CF, Pegel K, Van Jaarsveld PP. beta-Sitosterol and beta-sitosterol glusoside stimulate human peripheral blood lymphocyte proliferation: implications for their use as an immunomodulatory vitamin combination. Int J Immunopharmacol 1996;18:693–700.

[256] Ju YH, Clausen LM, Allred KF, Almade AL, Hellerich WG. Beta-Sitosterol, beta-sitosterol glucoside, and a mixture of beta-sitosterol and beta-sitosterol glucoside modulate the growth of estrogen-responsive breast cancer cells in vitro and in ovariectomized athymic mice. J Nutr 2004;204:1145–51.

[257] Anon. Plant sterols and sterolins. Altern Med Rev 2001;6:203–6.

[258] Klippel KF, Hilti DM, Schipp B. A multicentric, placebo-controlled, double-blind clinical trial of beta-sitosterol (phytosterol) for the treatment of benign prostatic hyperplasia. German BPH-Phyto Study group. Br J Urol 1997;80:427–32.

[259] Aráoz R, Molgó J, Tandeau de Marsac N. Neurotoxic cyanobacterial toxins. Toxicon 2010;56:813–28.

[260] Merel S, Walker D, Chicana R, Snyder S, Baurès E, Thomas O. State of knowledge and concerns on cyanobacterial blooms and cyanotoxins. Environ Int 2013;59:303–27.

[261] Zegura B, Straser A, Filipič M. Genotoxicity and potential carcinogenicity of cyanobacterial toxins - a review. Mutat Res 2011;727:16–41.

[262] Sieroslawska A. Assessment of the mutagenic potential of cyanobacterial extracts and pure cyanotoxins. Toxicon 2013;74:76–82.

[263] Ding XS, Li XY, Duan HY, Chung IK, Lee JA. Toxic effects of microcystis cell extracts on the reproductive system of male mice. Toxicon 2006;48:973–9.

[264] Meriluoto JA, Nygård SE, Dahlem AM, Eriksson JE. Synthesis, organotropism and hepatocellular uptake of two tritium-labeled epimers of dihydromicrocystin-LR, a cyanobacterial peptide toxin analog. Toxicon 1990;28:1439–46.

[265] Wang Q, Xie P, Chen J, Liang G. Distribution of microcystins in various organs (heart, liver, intestine, gonad, brain, kidney and lung) of Wistar rat via intravenous injection. Toxicon 2008;52:721–7.

[266] Falconer IR, Smith JV, Jackson AR, Jones A, Runnegar MT. Oral toxicity of a bloom of the cyanobacterium *Microcystis aeruginosa* administered to mice over periods up to 1 year. J Toxicol Environ Health 1988;24:291–305.

[267] Maidana M, Carlis V, Galhardi FG, Yunes JS, Geracitano LA, Monserrat JM, et al. Effects of microcystins over short- and long-term memory and oxidative stress generation in hippocampus of rats. Chem Biol Interact 2006;159:223–34.

[268] Li G, Cai F, Yan W, Li C, Wang J. A proteomic analysis of MCLR-induced neurotoxicity: implications for Alzheimer's disease. Toxicol Sci 2012;127:485–95.

[269] de la Monte SM, Neusner A, Chu J, Lawton M. Epidemiological trends strongly suggest exposures as etiologic agents in the pathogenesis of sporadic Alzheimer's disease, diabetes mellitus, and non-alcoholic steatohepatitis. J Alzheimers Dis 2009;17:519–29.

[270] McCutcheon JW. Nitrosamines in bacon: a case study of balancing risks. Public Health Rep 1984;99:360–4.

[271] Liu C, Russell RM. Nutrition and gastric cancer risk: an update. Public Health Rep 2008;66:237–49.

[272] Lijinsky W. *N*-Nitroso compounds in the diet. Mutat Res 1999;443:129–38.

[273] Le Roux J, Gallard H, Croué JP. Chloramination of nitrogenous contaminants (pharmaceuticals and pesticides): NDMA and halogenated DBPs formation. Water Res 2011;45:3164–74.

[274] Oury B, Limasset JC, Protois JC. Assessment of exposure to carcinogenic *N*-nitrosamines in the rubber industry. Int Arch Occup Environ Health 1997;70:261–71.

[275] Bolzan AD, Bianchi MS. Genotoxicity of streptozotocin. Mutat Res 2002;512:121–34.

[276] Ahles TA, Saykin AJ. Candidate mechanisms for chemotherapy-induced cognitive changes. Nat Rev Cancer 2007;7:192–201.

[277] Wefel JS, Kayl AE, Meyers CA. Neuropsychological dysfunction associated with cancer and cancer therapies: a conceptual review of an emerging target. Br J Cancer 2004;90:1691–6.

[278] Miklossy J, Steele JC, Yu S, McCall S, Sandberg G, McGeer EG, et al. Enduring involvement of tau, beta-amyloid, alpha-synuclein, ubiquitin and TDP-43 pathology in the amyotrophic lateral sclerosis/parkinsonism-dementia complex of Guam (ALS/PDC). Acta Neuropathol 2008;116:625–37.

[279] Jiang L, Eriksson J, Lage S, Jonasson S, Shams S, Mehine M, et al. Diatoms: a novel source for the neurotoxin BMAA in aquatic environments. Plos One 2014;9:e84578.

[280] Shankar SK, Yanagihara R, Garruto RM, Grundke-Iqbal I, Kosik KS, Gajdusek DC. Immunocytochemical characterization of neurofibrillary tangles in amyotrophic lateral sclerosis and parkinsonism-dementia of Guam. Ann Neurol 1989;25:146–51.

[281] Schwab C, Steele JC, McGeer PL. Neurofibrillary tangles of Guam Parkinson-dementia are associated with reactive microglial and complement proteins. Brain Res 1996;707:196–205.

[282] de la Monte SM, Tong M. Mechanisms of nitrosamine-mediated neurodegeneration: potential relevance to sporadic Alzheimer's disease. J Alzheimer Dis 2009;17:817–25.

Environmental Exposures and Risks for Parkinson's Disease

Harvey Checkoway, Susan Searles Nielsen, Brad A. Racette

INTRODUCTION

Parkinson's disease (PD), initially described by James Parkinson in 1817 [1], is a debilitating neurodegenerative motor disorder that also can involve cognitive and autonomic nervous system impairment. PD represents the most severe form of the clinical syndrome of parkinsonism (PS). The characteristic motor signs of PD are bradykinesia (slow movement), rigidity, rest tremor, and postural instability [2]. These signs result from the loss of dopaminergic neurons, predominantly in the midbrain substantia nigra. Lewy bodies and neurites, abnormal protein formations that are found in the remaining neurons, are the signature pathological features of PD and can be detected at autopsy [3,4]. Approximately two percent of persons aged 65 and older have PD [5], and its prevalence is expected to increase in many countries as the world's population ages.

Both the prevalence and incidence of PD vary considerably by sex, age, race, and ethnicity [6,7]. A 30–50% greater incidence in men than women is consistently observed in most populations. PD is uncommon before age 50 [7], and incidence and prevalence increase dramatically with age [6,7]. PD incidence is higher among Caucasians and Hispanics than it is among African-Americans and Asians [6,7].

Despite extensive toxicological and epidemiologic research, the causes of PD remain largely unknown. Although genetics clearly play a role in PD development, especially among patients diagnosed at a relatively young age, monogenic forms are rare [8]. Environmental factors, including those in the workplace, in the ambient environment, from various medications, and related to lifestyle, alone or in combination with genetics, are presumed to account for the majority of PD cases [9]. In this chapter, we will focus mainly on epidemiologic evidence for environmental agents as potential risk factors.

ENVIRONMENTAL EXPOSURES

Historically, the environmental agents receiving the most attention as possible risk factors for PD have included metals, pesticides, and solvents, owing to their widespread occurrence and the biological plausibility that they might induce dopaminergic neuron degeneration, given that many of these agents readily cross the blood–brain barrier and are neurotoxicants. Epidemiologic and experimental studies of environmental exposures in general, and metals and pesticides in particular, were spurred by case reports of PS, as early as Couper's 1837 description of five Scottish ore crushers who developed PS following overexposure to manganese [10]. An association between manganese poisoning and signs of PS was subsequently corroborated by Rodier [11] in his survey of Moroccan miners in 1955. We will return later to possible causal links between manganese and PS, including PD, later in this chapter in the discussion of metals as potential risk factors. Similar to the case of metals, interest in pesticides as possible risk factors for PD was enhanced in the 1980s following the identification of a cluster of six cases of irreversible PS in Northern California among intravenous drug abusers who had injected a meperidine analog contaminated with 1-methyl-4-phenyl-1,2,3,6-tetrahydropyridine (MPTP) [12]. Experimental research subsequently established that the active metabolite of MPTP, 1-methyl-4-phenylpyridinium (MPP+), is a potent nigral toxin in rodents and nonhuman primates. Those observations, coupled with recognition of the structural similarity of MPP+ to the herbicide paraquat, prompted extensive toxicologic and epidemiologic research on pesticides [13], which also will be reviewed subsequently.

Although our focus is on occupational and ambient environmental exposures that may increase the risk of PD and other forms of PS, it is important to note that some nongenetic factors appear to reduce the risk of PD—that is, they appear to be protective. It is possible that in the future these associations will shed light on the mechanisms of PD development and the role of putative exposures. Cigarette smoking has been the most prominent and consistently observed "protective" factor. PD risk among ever-smokers is roughly half that of never-smokers, and numerous studies have detected dose–response relationships with duration, intensity, and pack-years smoked [14]. Selective survival to older ages among nonsmokers does not adequately explain this phenomenon [15].

The possibility that a component of tobacco such as nicotine is truly protective is supported by possible inverse associations between PD and both environmental tobacco smoke [16] and consumption of peppers [17], which are in the tobacco plant family. However, the more established inverse association with active tobacco smoking may simply reflect a reduced response to nicotine or a "premorbid personality" type related to risk avoidance among persons who eventually develop PD. Consumption of coffee and other caffeinated beverages [18], as well as some prescription medications, especially nonsteroidal anti-inflammatories [19,20] and lipid-lowering statins [21], have also been associated with reduced PD risk, although not as consistently as the reduction associated with smoking. Multiple studies have also reported inverse associations between PD and other environmental exposures ranging from outdoor work [22,23] to dietary consumption of some types of fat [24].

Characterizing underlying pathogenesis mechanisms remains an important area of toxicological research, and these findings are clearly helpful in guiding environmental epidemiologic research and interpreting observations. Although no single predominant mechanism has been established, strong evidence suggests that nigral system toxicity, which ultimately culminates in PD, results from complex interactions among oxidative stress, inflammation, and abnormal aggregation and removal of various proteins—especially α-synuclein, a major component of Lewy bodies [25] (Figure 12.1).

In the following section, we summarize epidemiologic evidence for associations with environmental exposures. The intention is not to be comprehensive, but rather to highlight important epidemiologic findings, especially those supported by clinical or toxicological evidence.

Pesticides

As mentioned above, the MPTP episode focused considerable attention on pesticides, starting with the herbicide paraquat. Animal models of nigral injury have been established for the combination of paraquat and the manganese-containing fungicide maneb [26], and also the insecticide rotenone [27]. Development of PS after acute exposure to pesticides, including paraquat [28], has been reported in humans. In addition, a study comparing 50 maneb-exposed and 19 nonexposed rural workers observed that the exposed group was significantly more likely to have rigidity than was the nonexposed group [29].

A summary of findings from some of the major epidemiologic investigations of pesticides and PD conducted since 2000 is provided in Table 12.1. Associations with insecticides and herbicides have been examined most frequently, largely in the context of case–control studies [30–40]; some cohort studies have also been conducted [41–43]. These numerous studies have addressed both occupational and nonoccupational (e.g., residential or drift) exposures to pesticides. As can be seen from Table 12.1, the findings are heterogeneous, no doubt due to great variability in the chemical specificity that was permitted by various approaches to the assessment of exposures. The majority of studies have been population-based case–control studies in which pesticide exposures were based on questionnaire self-reports, and were categorized into broad groupings of insecticides, herbicides, or other (e.g., fungicides). Some studies in which chemical-specific exposure data were derived to test toxicologically based hypotheses, however, suggest that elevated risks are

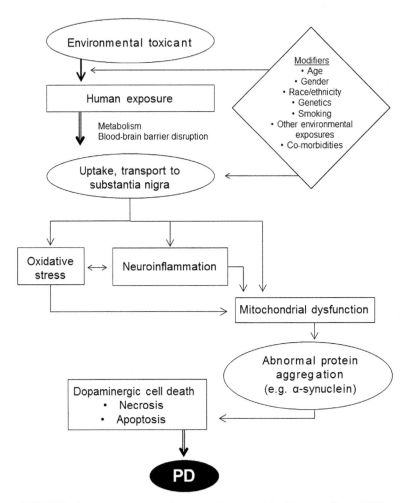

FIGURE 12.1   **Environmental neurotoxicants and Parkinson's disease (PD).**

related to a combination of paraquat and maneb [34], and rotenone [36]. None of these pesticides, nor mancozeb—which like maneb contains manganese—have been among the most commonly used pesticides in the United States as a whole, even within the agricultural sector [44], making epidemiologic studies more challenging. Moreover, for the most part evidence for other specific pesticides is limited, although exposure misclassification may contribute to underestimated associations, while self-reporting of pesticide exposure may tend to create reporting bias toward an association [40].

Several epidemiologic studies of PD have explored gene-pesticide interactions [45–50]. Although gene-environment (including gene-pesticide) interaction studies typically suffer from limited statistical power to detect associations, the results of these studies can complement those focused on environmental main effects, because gene-environment associations are less likely to originate from recall or reporting bias [51] or selection bias [52]. Interactions

TABLE 12.1   Epidemiologic Studies of Parkinson's Disease and Pesticide Exposures

Study	Location	Number of subjects	Exposure assessment method	Relative risk estimate (95% confidence interval)
COHORT STUDIES				
[41]	USA	143,325 including 413 PD cases	Questionnaire	Pesticides: 1.7 (1.2–2.3)
[42]	USA	55,931 including 78 PD cases	Questionnaire for specific pesticides (type, frequency, duration of use)	Cumulative use all pesticides ≥397 days: 2.3 (1.2–4.5)
[43]	Sweden	14,169 (Swedish twin registry) including 293 PD cases	Job/exposure matrix applied to questionnaire data	Highest exposed vs. no exposure: 0.9 (0.5–1.4)
CASE–CONTROL STUDIES				
[30]	5 European countries	767 cases, 1989 controls	Job/exposure matrix applied to questionnaire data	"High" vs. no pesticide exposure 1.4 (1.0–1.9)
[31]	USA	833 cases, 833 controls	Questionnaire	Age <60: Pesticides: 1.8 (1.1–2.9) Herbicides: 2.5 (1.3–4.5)
[32]	USA	319 cases, 296 controls	Questionnaire for specific pesticides (type, frequency, duration of use)	Negative family history of PD: Pesticides cumulative use >179 days 3.3 (1.8–5.7)
[33]	France	224 cases, 557 controls	Questionnaire for specific pesticides (type, frequency, duration of use); expert assessment	Men: Organochlorine insecticides: 2.4 (1.2–5.0)
[34]	USA	368 cases, 341 controls	Geographic information system linkage of addresses to statewide data on agricultural application of pesticides	Age ≤60: Paraquat+maneb: 5.1 (1.8–14.7)
[35]	USA, Canada	511 cases, 511 controls	Questionnaires, Classification of Standard Occupational Job Codes	Pesticides: 1.9 (1.1–3.2) 2,4 Dichlorophenoxyacetic acid: 2.6 (1.0–6.5) Paraquat: 2.8 (0.8–9.7)
[36]	USA	Nested case–control study, 110 cases, 358 controls	Questionnaire for specific pesticides (type, frequency, duration of use)	Rotenone: 2.5 (1.3–4.7) Paraquat: 2.5 (1.4–4.7)
[40]	Canada	403 cases, 405 controls	Questionnaire combined with industrial hygiene review	1.5 (0.9–2.7)
[37]	USA	357 cases, 807 controls	Questionnaire	Any frequent residential pesticide use: 1.5 (1.1–1.9)

(Continued)

II. NEURODEGENERATIVE DISORDERS

TABLE 12.1   Epidemiologic Studies of Parkinson's Disease and Pesticide Exposures—cont'd

Study	Location	Number of subjects	Exposure assessment method	Relative risk estimate (95% confidence interval)
[38]	USA	357 cases, 750 controls	Job/exposure matrix applied to questionnaire data	Men: Highest vs. lowest cumulative lifetime occupational exposure: 2.2 (1.3–3.8)
[39]	USA	357 cases, 752 controls	Geographic information system linkage of addresses to statewide data on agricultural application of organophosphorus insecticides	Residential or occupational ambient exposure: 2.2 (1.6–3.2)

among PD, pesticide exposure, and diet also have been investigated; a recent study suggested that associations among PD, paraquat, and rotenone are modified by dietary fat consumption [24]. Replication of this and the observed pesticide–gene interactions will be required, however, before drawing any firm conclusions about the role of pesticides in PD, both overall and in combination with other factors.

Metals

Some metals, including manganese, lead, and mercury, have widespread occupational and environmental exposures and well-established neurotoxic properties [53]. Consequently, metals have been a common focus of PD etiologic research. Manganese has been of particular concern as a potential PD risk factor, largely prompted by the long-recognized clinical similarities of manganism and PD. Metals have been examined as risk factors in numerous population-based case–control studies, several community surveys, and cohort studies of welders. Complementing these studies are case–control studies of environmental exposure that suggest an increased risk of PD in relation to manganese levels in air emissions or soil [54–56]. In contrast, the associations found for welding and occupational exposure to manganese and other metals, in population-based case–control studies, have been mostly weak or null [57], although a subsequent large case–control study suggests that welders may have a threefold increase in PD risk [58]. Challenges in these case–control studies of occupational welding exposure include the relative infrequency of work as a welder in the general population, and ambiguities of exposure assessment.

Welders have been a target study group in cohort studies because of their common and relatively high exposures to manganese and other metals. The evidence for a causal relationship between welding and PD in these cohort studies has also been largely null (see Table 12.2 for summary) [43,59–66]. Case definitions, ascertainment, and exposure assessment approaches among these welder studies have been heterogeneous, which precludes simple conclusions regarding manganese specifically, or the occupation of welding more broadly, as a PD risk factor. In general, studies in which PD was assessed via records indicated no

TABLE 12.2 Cohort Studies of Parkinsonism and Parkinson's Disease among Welders

Study	Location	Cohort size	Outcome and assessment	Exposure assessment	Relative risk estimate (95% confidence interval)
PS ASSESSED BY NEUROLOGICAL EXAM					
[59]	USA	1423	PS, exam with movement disorders specialist	Welder occupation type	Overall: 7.6 (3.3–17.7) Boilermakers: 10.3 (2.6–40.5) vs. community survey rates in Alabama
[60]	USA	811	PS, exam with disorders specialist	Weighted welding years	16% welders vs. 0% nonwelders; highest vs. lowest exposure among welders: 1.0 (0.6–1.6)
PS/PD ASSESSED THROUGH RECORDS					
[61]	Denmark	6163	PD, hospitalization	Welder occupation	Overall: 0.9 (0.4–1.8) Employed >20 yrs: 0.8 (0.2–2.0) vs. national rates
[62]	Sweden	49,488	"Basal ganglia and movement disorders," hospitalization	Welder occupation	All: 0.9 (0.8–1.0) PD only: 0.9 (0.8–1.0) vs. national rates
[63]	USA	12,595 (nested case–control, 66 PD cases, 660 controls)	PS, including PD and "secondary" PS, medical insurance claims	Welding job title	Overall: 1.0 (0.3–2.3) >30 yrs Welding: 1.0 (0.3–1.6) vs. non-PD controls
[64]	South Korea	38,560	PD, hospitalization billing claims, with neurologist review of medical charts	Welding job title classified by airborne manganese levels	Low exposure: 3.6 (0.7–18.6) High exposure: 2.0 (0.3–12.5) vs. nonexposed white collar workers
[65]	USA	4,252,490	PD, mortality	Welder occupation on death certificate	Men, mortality odds ratio: 0.83 (0.78–0.87) vs. other occupations
[43]	Sweden	14,169 (Swedish twin registry)	PD and "parkinsonian disorders," hospitalization and mortality records	Job/exposure matrix applied to questionnaire data	PD, highest exposed to "welding smoke" vs. no exposure: 0.7 (0.2–2.6)
[66]	Denmark	7602	PD, hospitalization	Occupational history by questionnaire	1.1 (0.7–1.6) vs. national rates

increased PD risk, whereas studies in which PS was defined more broadly and assessed by a clinician indicated marked elevations of PS among welders. These are important distinctions. Although PD and manganism do share some clinical features [60], pathological and neuroimaging studies suggest that these may be different phenotypes [67].

Solvents

Solvents constitute another class of compounds that are commonly used in industry, are widespread workplace and environmental contaminants, and have a range of neurotoxic properties. Here again, findings from most but not all epidemiologic studies have indicated weak or null associations [30,43,68–72] (see Table 12.3 for summary) that often fail to identify exposures to either specific compounds or their classes (e.g., aliphatic vs. aromatic). A notable exception is a study of twins suggesting elevated risks associated with exposures to trichloroethylene and tetrachloroethylene [72]. On balance, the evidence for associations with solvents is inconsistent, as summarized in a recent review of the epidemiologic literature [73].

Other Environmental Exposures

Another relatively widespread class of chemicals that has received attention in epidemiologic PD research is polychlorinated biphenyls (PCBs) [70,74–78]. Some but not all epidemiologic studies suggest increased risk of PD in relation to PCBs, and experimental

TABLE 12.3   Epidemiologic Studies of Solvents and Parkinson's Disease

Study	Location	Number of subjects	Exposure assessment method	Relative risk estimate (95% confidence interval)
COHORT STUDIES				
[68]	UK	Nested case–control 182 cases, 423 controls	Job title	Overall: 1.5 (0.8–2.9) >30 yrs: 3.6 (1.3–10.3)
[69]	USA	2,614,346 including 33,678 with PD	Usual occupation on death certificate translated into a solvent exposure index	Mortality odds ratio for any solvent exposure: 1.07 (1.00–1.13)
[43]	Sweden	14,169 (Swedish twin registry) including 293 with PD	Job/exposure matrix applied to questionnaire data	Highest exposed vs. no exposure: 1.4 (0.6–2.9)
[72]	USA	Nested case–control study (Veteran twins cohort); 99 cases, 99 controls	Questionnaire, classified by industrial hygienists	Trichloroethylene: 6.1 (1.2–33) Perchloroethylene: 10.5 (1.0–113) Toluene: 1.3 (0.5–3.3)
CASE–CONTROL STUDIES				
[30]	5 European countries	767 cases, 1089 controls	Job/exposure matrix applied to questionnaire data	Low: 1.2 (0.9–1.6) High: 0.9 (0.7–1.2)
[70]	Faroe islands/ Denmark	79 cases, 154 controls	Questionnaire	1.7 (0.8–3.5)
[71]	USA	404 cases, 526 controls	Questionnaire, classified by industrial hygienists	Men: 1.0 (0.7–1.3)

studies support this possible association. Experimental studies also implicate other environmental agents such as endotoxin. While there has been little focused examination of this exposure within epidemiologic studies, it is an example of an agent, other than pesticides, that may account for an association between PD and agriculture/farming or farm animals [40,58,79–81]. Nonchemical environmental exposures, ranging from magnetic fields [82] to whole body vibration [83], have been less studied.

DISCUSSION AND RECOMMENDATIONS

To date, no single environmental agent or class of agents has been conclusively demonstrated epidemiologically to increase PD risk. Most emphasis has been on agents (e.g., paraquat) or classes of agents (e.g., mitochondrial inhibitors) for which experimental evidence supports contributions to pathogenesis, or for which there have been striking case reports of PS among heavily exposed workers such as manganese miners.

Investigation of environmental risk factors for PD uses two basic epidemiologic approaches: population- or clinic-based case–control studies conducted in the general population, and cohort studies of persons with characteristically high exposures to suspected nigral system toxicants. Case–control studies have the advantages of accruing large numbers of cases whose diagnoses can be verified clinically or pathologically, and the potential to collect data on nonenvironmental factors such as smoking and family history of PD, which may be confounders or modifiers of associations with environmental exposures. However, case–control studies are often severely limited by nonspecific exposure assessments to environmental toxicants, some of which may be rare in the population at large. Occupational cohorts, by definition, include relatively large numbers of persons exposed to agents of interest, and in some cases enable quantitative exposure assessment. Limitations of occupational cohorts include the small number of PD cases in all but the largest cohorts, and estimation of incidence that is often from mortality data that may be incomplete or clinically inaccurate.

A somewhat idealized epidemiologic study of environmental exposures is depicted in Figure 12.2. Identification of study cohorts with high levels of quantifiable exposures to presumed etiologic agents, comparable nonexposed reference cohorts, and longitudinal follow-up are essential elements of this approach. The exposed cohort could be an occupational group or a cohort defined on the basis of ambient exposures to neurotoxicants, such as residents of communities that experience pesticide contamination in soil and drinking water. At baseline, study subjects undergo clinical exams, preferably with movement disorders specialists, to determine prevalence and severity of the cardinal signs and symptoms of PD. Data on risk factors are determined at this time to address the confounding effects of modification in later statistical analyses. Disease progression, and identification of new PS/PD cases, can be determined from follow-up neurologic assessments, which also include updating of exposure and confounder profiles. Studies constructed in this manner will substantively contribute to our understanding of the relationship between PD and environmental exposures such as pesticides, metals, and solvents.

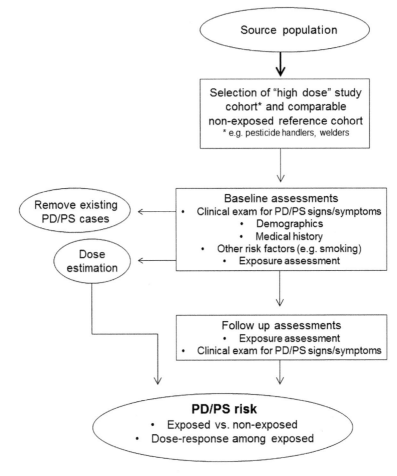

PD = Parkinson Disease, PS = Parkinsonism

FIGURE 12.2 **Idealized epidemiologic study of environmental toxicants and PD/PS.**

References

[1] Parkinson J. An essay on the shaking palsy. London: Sherwood, Neely, and Jones; 1817.

[2] Samii A, Nutt JG, Ransom BR. Parkinson's disease. Lancet 2004;363:1783–93.

[3] Agid Y. Parkinson's disease: pathophysiology. Lancet 1991;337:1321–4.

[4] Hughes AJ, Daniel SE, Kilford L, Lees AJ. Accuracy of clinical diagnosis of idiopathic Parkinson's disease: a clinico-pathological study of 100 cases. J Neurol Neurosurg Psychiatry 1992;55:181–4.

[5] Elbaz A, Bower JH, Maraganore DM, McDonnell SK, Peterson BJ, Ahlskog JE, et al. Risk tables for parkinsonism and Parkinson's disease. J Clin Epidemiol 2002;55:25–31.

[6] Wright Willis A, Evanoff BA, Lian M, Criswell SR, Racette BA. Geographic and ethnic variation in Parkinson disease: a population-based study of us medicare beneficiaries. Neuroepidemiology 2010;34:143–51.

[7] Van Den Eeden SK, Tanner CM, Bernstein AL, Fross RD, Leimpeter A, Bloch DA, et al. Incidence of Parkinson's disease: variation by age, gender, and race/ethnicity. Am J Epidemiol 2003;157:1015–22.

[8] Trinh J, Farrer M. Advances in the genetics of Parkinson disease. Nat Rev Neurol 2013;9:445–54.

[9] Wirdefeldt K, Adami HO, Cole P, Trichopoulos D, Mandel J. Epidemiology and etiology of Parkinson's disease: a review of the evidence. Eur J Epidemiol 2011;26(Suppl. 1):S1–58.

[10] Couper J. On the effects of black oxide of manganese when inhaled into the lungs. Br Ann Med Pharm Vital Stat Gen Sci 1837;1:41–2.

[11] Rodier J. Manganese poisoning in moroccan miners. Br J Ind Med 1955;12:21–35.

[12] Langston JW, Ballard P, Tetrud JW, Irwin I. Chronic parkinsonism in humans due to a product of meperidine-analog synthesis. Science 1983;219:979–80.

[13] Langston JW, Irwin I, Ricaurte GA. Neurotoxins, parkinsonism and Parkinson's disease. Pharmacol Ther 1987;32:19–49.

[14] Ritz B, Ascherio A, Checkoway H, Marder KS, Nelson LM, Rocca WA, et al. Pooled analysis of tobacco use and risk of Parkinson disease. Arch Neurol 2007;64:990–7.

[15] Morens DM, Grandinetti A, Davis JW, Ross GW, White LR, Reed D. Evidence against the operation of selective mortality in explaining the association between cigarette smoking and reduced occurrence of idiopathic Parkinson disease. Am J Epidemiol 1996;144:400–4.

[16] Searles Nielsen S, Gallagher LG, Lundin JI, Longstreth Jr WT, Smith-Weller T, Franklin GM, et al. Environmental tobacco smoke and Parkinson's disease. Mov Disord 2012;27:293–6.

[17] Searles Nielsen S, Franklin GM, Longstreth Jr WT, Swanson PD, Checkoway H. Nicotine from edible *Solanaceae* and risk of Parkinson disease. Ann Neurol 2013;74:472–7.

[18] Qi H, Li S. Dose-response meta-analysis on coffee, tea and caffeine consumption with risk of Parkinson's disease. Geriatr Gerontol Int 2014;14:430–9.

[19] Wahner AD, Bronstein JM, Bordelon YM, Ritz B. Nonsteroidal anti-inflammatory drugs may protect against Parkinson disease. Neurology 2007;69:1836–42.

[20] Manthripragada AD, Schernhammer ES, Qiu J, Friis S, Wermuth L, Olsen JH, et al. Non-steroidal anti-inflammatory drug use and the risk of Parkinson's disease. Neuroepidemiology 2011;36:155–61.

[21] Gao X, Simon KC, Schwarzschild MA, Ascherio A. Prospective study of statin use and risk of Parkinson disease. Arch Neurol 2012;69:380–4.

[22] Kenborg L, Lassen CF, Ritz B, Schernhammer ES, Hansen J, Gatto NM, et al. Outdoor work and risk for Parkinson's disease: a population-based case-control study. Occup Environ Med 2011;68:273–8.

[23] Kwon E, Gallagher LG, Searles Nielsen S, Franklin GM, Littell CT, Longstreth Jr WT, et al. Parkinson's disease and history of outdoor occupation. Park Relat Disord 2013;19:1164–6.

[24] Kamel F, Goldman SM, Umbach DM, Chen H, Richardson G, Barber MR, et al. Dietary fat intake, pesticide use, and Parkinson's disease. Park Relat Disord 2014;20:82–7.

[25] Cannon JR, Greenamyre JT. The role of environmental exposures in neurodegeneration and neurodegenerative diseases. Toxicol Sci 2011;124:225–50.

[26] Thiruchelvam M, Brockel BJ, Richfield EK, Baggs RB, Cory-Slechta DA. Potentiated and preferential effects of combined paraquat and maneb on nigrostriatal dopamine systems: environmental risk factors for Parkinson's disease? Brain Res 2000;873:225–34.

[27] Betarbet R, Sherer TB, MacKenzie G, Garcia-Osuna M, Panov AV, Greenamyre JT. Chronic systemic pesticide exposure reproduces features of Parkinson's disease. Nat Neurosci 2000;3:1301–6.

[28] Sanchez-Ramos JR, Hefti F, Weiner WJ. Paraquat and Parkinson's disease. Neurology 1987;37:728.

[29] Ferraz HB, Bertolucci PH, Pereira JS, Lima JG, Andrade LA. Chronic exposure to the fungicide maneb may produce symptoms and signs of CNS manganese intoxication. Neurology 1988;38:550–3.

[30] Dick FD, De Palma G, Ahmadi A, Scott NW, Prescott GJ, Bennett J, et al. Environmental risk factors for Parkinson's disease and parkinsonism: the Geoparkinson Study. Occup Environ Med 2007;64:666–72.

[31] Brighina L, Frigerio R, Schneider NK, Lesnick TG, de Andrade M, Cunningham JM, et al. Alpha-synuclein, pesticides, and Parkinson disease: a case-control study. Neurology 2008;70:1461–9.

[32] Hancock DB, Martin ER, Mayhew GM, Stajich JM, Jewett R, Stacy MA, et al. Pesticide exposure and risk of Parkinson's disease: a family-based case-control study. BMC Neurol 2008;8:6.

[33] Elbaz A, Clavel J, Rathouz PJ, Moisan F, Galanaud JP, Delemotte B, et al. Professional exposure to pesticides and Parkinson disease. Ann Neurol 2009;66:494–504.

[34] Costello S, Cockburn M, Bronstein J, Zhang X, Ritz B. Parkinson's disease and residential exposure to maneb and paraquat from agricultural applications in the Central Valley of California. Am J Epidemiol 2009;169:919–26.

[35] Tanner CM, Ross GW, Jewell SA, Hauser RA, Jankovic J, Factor SA, et al. Occupation and risk of parkinsonism: a multicenter case-control study. Arch Neurol 2009;66:1106–13.

[36] Tanner CM, Kamel F, Ross GW, Hoppin JA, Goldman SM, Korell M, et al. Rotenone, paraquat, and Parkinson's disease. Environ Health Perspect 2011;119:866–72.

[37] Narayan S, Liew Z, Paul K, Lee PC, Sinsheimer JS, Bronstein JM, et al. Household organophosphorus pesticide use and Parkinson's disease. Int J Epidemiol 2013;42:1476–85.

[38] Liew Z, Wang A, Bronstein J, Ritz B. Job exposure matrix (JEM)-derived estimates of lifetime occupational pesticide exposure and the risk of Parkinson's disease. Arch Environ Occup Health 2014;69:241–51.

[39] Wang A, Cockburn M, Ly TT, Bronstein JM, Ritz B. The association between ambient exposure to organophosphates and Parkinson's disease risk. Occup Environ Med 2014;71:275–81.

[40] Rugbjerg K, Harris MA, Shen H, Marion SA, Tsui JK, Teschke K. Pesticide exposure and risk of Parkinson's disease – a population-based case-control study evaluating the potential for recall bias. Scand J Work Environ Health 2011;37:427–36.

[41] Ascherio A, Chen H, Weisskopf MG, O'Reilly E, McCullough ML, Calle EE, et al. Pesticide exposure and risk for Parkinson's disease. Ann Neurol 2006;60:197–203.

[42] Kamel F, Tanner C, Umbach D, Hoppin J, Alavanja M, Blair A, et al. Pesticide exposure and self-reported Parkinson's disease in the agricultural health study. Am J Epidemiol 2007;165:364–74.

[43] Feldman AL, Johansson AL, Nise G, Gatz M, Pedersen NL, Wirdefeldt K. Occupational exposure in Parkinsonian disorders: a 43-year prospective cohort study in men. Park Relat Disord 2011;17:677–82.

[44] Grube A, Donaldson D, Kiely T, Wu L. Pesticides industry sales and usage: 2006 and 2007 market estimates. EPA 733-R-11–001. Washington, D.C.: US Environmental Protection Agency; 2011 (EPA), page 14.

[45] Rhodes SL, Fitzmaurice AG, Cockburn M, Bronstein JM, Sinsheimer JS, Ritz B. Pesticides that inhibit the ubiquitin-proteasome system: effect measure modification by genetic variation in SKP1 in Parkinsons disease. Environ Res 2013;126:1–8.

[46] Fitzmaurice AG, Rhodes SL, Cockburn M, Ritz B, Bronstein JM. Aldehyde dehydrogenase variation enhances effect of pesticides associated with Parkinson disease. Neurology 2014;82:419–26.

[47] Goldman SM, Kamel F, Ross GW, Bhudhikanok GS, Hoppin JA, Korell M, et al. Genetic modification of the association of paraquat and Parkinson's disease. Mov Disord 2012;27:1652–8.

[48] Gatto NM, Rhodes SL, Manthripragada AD, Bronstein J, Cockburn M, Farrer M, et al. Alpha-synuclein gene may interact with environmental factors in increasing risk of Parkinson's disease. Neuroepidemiology 2010;35:191–5.

[49] Manthripragada AD, Costello S, Cockburn MG, Bronstein JM, Ritz B. Paraoxonase 1, agricultural organophosphate exposure, and Parkinson disease. Epidemiology 2010;21:87–94.

[50] Ritz BR, Manthripragada AD, Costello S, Lincoln SJ, Farrer MJ, Cockburn M, et al. Dopamine transporter genetic variants and pesticides in Parkinson's disease. Environ Health Perspect 2009;117:964–9.

[51] Garcia-Closas M, Rothman N, Lubin J. Misclassification in case-control studies of gene-environment interactions: assessment of bias and sample size. Cancer Epidemiol Biomarkers Prev 1999;8:1043–50.

[52] Morimoto LM, White E, Newcomb PA. Selection bias in the assessment of gene-environment interaction in case-control studies. Am J Epidemiol 2003;158:259–63.

[53] Caudle WM, Guillot TS, Lazo CR, Miller GW. Industrial toxicants and Parkinson's disease. Neurotoxicology 2012;33:178–88.

[54] Lucchini RG, Albini E, Benedetti L, Borghesi S, Coccaglio R, Malara EC, et al. High prevalence of parkinsonian disorders associated to manganese exposure in the vicinities of ferroalloy industries. Am J Ind Med 2007;50:788–800.

[55] Finkelstein MM, Jerrett M. A study of the relationships between Parkinson's disease and markers of traffic-derived and environmental manganese air pollution in two Canadian cities. Environ Res 2007;104:420–32.

[56] Wright Willis A, Evanoff BA, Lian M, Galarza A, Wegrzyn A, Schootman M, et al. Metal emissions and urban incident Parkinson disease: a community health study of medicare beneficiaries by using geographic information systems. Am J Epidemiol 2010;172:1357–63.

[57] Mortimer JA, Borenstein AR, Nelson LM. Associations of welding and manganese exposure with Parkinson disease: review and meta-analysis. Neurology 2012;79:1174–80.

[58] Teschke K, Marion SA, Tsui JK, Shen H, Rugbjerg K, Harris MA. Parkinson's disease and occupation: differences in associations by case identification method suggest referral bias. Am J Ind Med 2014;57:163–71.

[59] Racette BA, Tabbal SD, Jennings D, Good L, Perlmutter JS, Evanoff B. Prevalence of parkinsonism and relationship to exposure in a large sample of Alabama welders. Neurology 2005;64:230–5.

[60] Racette BA, Criswell SR, Lundin JI, Hobson A, Seixas N, Kotzbauer PT, et al. Increased risk of parkinsonism associated with welding exposure. Neurotoxicology 2012;33:1356–61.

[61] Fryzek JP, Hansen J, Cohen S, Bonde JP, Llambias MT, Kolstad HA, et al. A cohort study of Parkinson's disease and other neurodegenerative disorders in Danish welders. J Occup Environ Med 2005;47:466–72.

[62] Fored CM, Fryzek JP, Brandt L, Nise G, Sjogren B, McLaughlin JK, et al. Parkinson's disease and other basal ganglia or movement disorders in a large nationwide cohort of Swedish welders. Occup Environ Med 2006;63:135–40.

[63] Marsh GM, Gula MJ. Employment as a welder and Parkinson disease among heavy equipment manufacturing workers. J Occup Environ Med 2006;48:1031–46.

[64] Park J, Yoo CI, Sim CS, Kim JW, Yi Y, Shin YC, et al. A retrospective cohort study of Parkinson's disease in Korean shipbuilders. Neurotoxicology 2006;27:445–9.

[65] Stampfer MJ. Welding occupations and mortality from Parkinson's disease and other neurodegenerative diseases among United States men, 1985–1999. J Occup Environ Hyg 2009;6:267–72.

[66] Kenborg L, Lassen CF, Hansen J, Olsen JH. Parkinson's disease and other neurodegenerative disorders among welders: a Danish cohort study. Mov Disord 2012;27:1283–9.

[67] Criswell SR, Perlmutter JS, Crippin JS, Videen TO, Moerlein SM, Flores HP, et al. Reduced uptake of FDOPA PET in end-stage liver disease with elevated manganese levels. Arch Neurol 2012;69:394–7.

[68] McDonnell L, Maginnis C, Lewis S, Pickering N, Antoniak M, Hubbard R, et al. Occupational exposure to solvents and metals and Parkinson's disease. Neurology 2003;61:716–7.

[69] Park RM, Schulte PA, Bowman JD, Walker JT, Bondy SC, Yost MG, et al. Potential occupational risks for neurodegenerative diseases. Am J Ind Med 2005;48:63–77.

[70] Petersen MS, Halling J, Bech S, Wermuth L, Weihe P, Nielsen F, et al. Impact of dietary exposure to food contaminants on the risk of Parkinson's disease. Neurotoxicology 2008;29:584–90.

[71] Firestone JA, Lundin JI, Powers KM, Smith-Weller T, Franklin GM, Swanson PD, et al. Occupational factors and risk of Parkinson's disease: a population-based case-control study. Am J Ind Med 2010;53:217–23.

[72] Goldman SM, Quinlan PJ, Ross GW, Marras C, Meng C, Bhudhikanok GS, et al. Solvent exposures and Parkinson disease risk in twins. Ann Neurol 2012;71:776–84.

[73] Lock EA, Zhang J, Checkoway H. Solvents and Parkinson disease: a systematic review of toxicological and epidemiological evidence. Toxicol Appl Pharmacol 2013;266:345–55.

[74] Steenland K, Hein MJ, Cassinelli 2nd RT, Prince MM, Nilsen NB, Whelan EA, et al. Polychlorinated biphenyls and neurodegenerative disease mortality in an occupational cohort. Epidemiology 2006;17:8–13.

[75] Wermuth L, Pakkenberg H, Jeune B. High age-adjusted prevalence of Parkinson's disease among Inuits in Greenland. Neurology 2002;58:1422–5.

[76] Weisskopf MG, Knekt P, O'Reilly EJ, Lyytinen J, Reunanen A, Laden F, et al. Polychlorinated biphenyls in prospectively collected serum and Parkinson's disease risk. Mov Disord 2012;27:1659–65.

[77] Ruder AM, Hein MJ, Hopf NB, Waters MA. Mortality among 24,865 workers exposed to polychlorinated biphenyls (PCBs) in three electrical capacitor manufacturing plants: a ten-year update. Int J Hyg Environ Health 2014;217:176–87.

[78] Hatcher-Martin JM, Gearing M, Steenland K, Levey AI, Miller GW, Pennell KD. Association between polychlorinated biphenyls and Parkinson's disease neuropathology. Neurotoxicology 2012;33:1298–304.

[79] Harris MA, Tsui JK, Marion SA, Shen H, Teschke K. Association of Parkinson's disease with infections and occupational exposure to possible vectors. Mov Disord 2012;27:1111–7.

[80] Goldman SM, Tanner CM, Olanow CW, Watts RL, Field RD, Langston JW. Occupation and parkinsonism in three movement disorders clinics. Neurology 2005;65:1430–5.

[81] Lee E, Burnett CA, Lalich N, Cameron LL, Sestito JP. Proportionate mortality of crop and livestock farmers in the United States, 1984–1993. Am J Ind Med 2002;42:410–20.

[82] Savitz DA, Checkoway H, Loomis DP. Magnetic field exposure and neurodegenerative disease mortality among electric utility workers. Epidemiology 1998;9:398–404.

[83] Harris MA, Marion SA, Spinelli JJ, Tsui JK, Teschke K. Occupational exposure to whole-body vibration and Parkinson's disease: results from a population-based case-control study. Am J Epidemiol 2012;176:299–307.

Parkinson's Disease: Mechanisms, Models, and Biological Plausibility

Eric E. Beier, Jason R. Richardson

DESCRIPTION OF PARKINSON'S DISEASE

Parkinson's disease (PD) is an idiopathic, progressive neurodegenerative disorder of the midbrain, which manifests by tremor, muscular inflexibility, difficulty in initiating motor activity (bradykinesia), and loss of postural stability. It affects a large scope of the population, with over 50,000 Americans diagnosed annually [1]. The future prospects of the PD epidemic are disconcerting, as the incidence of afflicted individuals is projected to massively increase over the next 20 years.

PD was initially described by Dr James Parkinson in 1817 and, since the 1930s, pathologically defined by loss of pigmented cells in the substantia nigra, yet the connection that a corresponding loss of dopamine was occurring in the striatum was only discovered around 1965.

Environmental Factors in Neurodevelopmental and Neurodegenerative Disorders
http://dx.doi.org/10.1016/B978-0-12-800228-5.00013-3

Dopamine is a catecholamine neurotransmitter that is centralized to the basal ganglia. Several central dopaminergic systems in the brain control a variety of tasks, such as reward, motivation, and motor control. The substantia nigra is a part of the basal ganglia in the midbrain, with the majority of dopaminergic neurons concentrated in the pars compacta [2]. These neurons innervate with the neostriatum, subthalamic nucleus, and the globus pallidus, which have been termed the mesolimbic dopamine projections and are known to be involved with voluntary moment. As it relates to PD, the archetypical clinical demonstration that L-dihydroxyphenylalanine, or L-DOPA, given to patients improved the extrapyramidal symptoms of the disease solidified that depleted dopamine levels were involved in deteriorated motor control [3].

PD generally develops in humans over 55, thus age is a factor. However, the pattern of neurodegeneration associated with PD is drastically different than traditional loss of neurons due to aging [4]. Parkinson's is defined biochemically as a dopamine deficiency resulting from degeneration or injury to dopamine neurons of the midbrain in the compact part of substantia nigra [5]. Protein aggregations in the form of Lewy bodies in surviving dopamine neurons or adjacent areas are also a hallmark of PD [6]. Histologically, Lewy bodies are a dense core inclusion encircled by a halo of 10-nm-wide radiating fibrils composed of misfolded α-synuclein. The etiology and pathogenesis of the dopaminergic cell death has remained elusive. Interestingly, symptoms of Parkinson's begin to manifest only when levels drop to 70–80% of striatal dopamine, with the more severe debilitating symptoms occurring with dopamine depletion greater than 90% [7].

The majority of PD cases are idiopathic, and the pathophysiological mechanisms likely involve interaction of genetic and environmental factors. Current research efforts in the field tend to focus on the identification of relevant factors, better characterizations of preclinical models, and the development of new therapeutic strategies to suspend or cease symptomatic progression. The combination of genetic susceptibility and environmental exposures is at the forefront of understanding PD pathogenesis and is the focus of this chapter.

GENETIC ETIOLOGY OF PD

Although significant effort has been applied to searching for genetic linkages in PD, initial twin studies and familial history studies have indicated that a significant number of PD cases were not monogenic forms [8]. However, as genetic techniques improve and various pathways associated with the disease are identified, information on the function of genes interconnected to PD has been revealed. Many of these genes have been reviewed in depth [9,10], but for this chapter, we will focus on those that have been shown to interact with environmental factors (see Tables 13.1 and 13.2 for review) and contribute to nigrostriatal cell death.

Alpha-Synuclein

The protein α-synuclein, encoded by the *SNCA* gene, was the first associated with rare familial cases of the disease in 1997 [11]. The physiological role of α-synuclein is uncertain. It is located in the nucleus, the mitochondria, and at presynaptic terminals, in which it is either free or membrane bound. Mitochondrial expression is region specific. Alpha-synuclein

TABLE 13.1 Characteristics of Genes Associated with Parkinson's Disease

Gene-locus	Expression pattern	Pathology	Phenotype
SNCA (α-synuclein) PARK1/4	Autosomal dominant multiplication	Lewy bodies Decreased dopamine	Parkinsonism, dementia Early-onset parkinsonism
Parkin PARK2	Autosomal recessive	Mitochondrial dysfunction Lewy bodies	Early-onset, slow progression of parkinsonism
PINK1 PARK6	Autosomal recessive	Mitochondrial dysfunction	Early-onset, slow progression of parkinsonism
DJ-1 PARK7	Autosomal recessive	Elevated oxidative stress Mitochondrial dysfunction	Early-onset parkinsonism
LRRK2 PARK8	Autosomal dominant	Lewy bodies Mitochondrial dysfunction	Late-onset parkinsonism

is highly expressed in the olfactory bulb and striatum, and expression is low in the cerebral cortex and cerebellum [12], suggesting that perhaps the olfactory bulb and striatum may be more predisposed to neurodegeneration than the latter regions. Many of its functions are stipulated based on its association with other proteins. For instance, it may act as a molecular chaperone involved in soluble N-ethylmaleimide-sensitive factor attachment protein receptor (SNARE) vesicle fusion due to its C-terminus domain binding of synaptobrevin-2 and N-terminus domain binding of phospholipids [13]. Alpha-synuclein was termed PARK1, as mammalian researchers identified and analyzed several other Parkinson-associated protein products linked to the disease.

Subsequent to the identification of PARK1 as a causative factor for familial PD, researchers identified that fibrils of the protein product were a major constituent of the Lewy body lesions [6]. Lewy bodies are a granular core comprising nitrated, phosphorylated, and ubiquitinated proteins surrounded by a halo of α-synuclein neurofilaments. The exact means of how these develop are uncertain, though research has suggested that they may not initiate the neurodegenerative process, but rather be a consequence of the disease process. Perhaps, a plausible explanation is that they form as aggresomes to facilitate removal of toxic proteins that primarily contribute to the degeneration. However, it should be noted that when the SNCA gene is duplicated or tripled in the human DNA, this leads to a form of PD [14], suggesting that overexpression of the wildtype protein may play a role. In support of this idea, cytoplasmic α-synuclein is known to increase in the aging brain, and transgenic mice overexpressing α-synuclein result in protein aggregates and loss of nigral dopaminergic neurons [15]. Furthermore, gene delivery strategies to substantia nigra compacta (SNc) in both rats and monkeys have demonstrated similar neurodegeneration [16]. Most recently, it was reported that overexpression can also facilitate its interneuronal transfer to axonal projections and spreading of α-synuclein lesions throughout the brain [17]. Finally, gastrointestinal dysfunction is a prominent feature long diagnosed with PD, and these deleterious symptoms may precede motor impairment by decades [18]. Pathologic accumulation of α-synuclein is observed in both gut and brain, suggesting that the two may be linked mechanistically [19,20]. Thus, further investigations have been targeted to establish whether peripheral inflammation may trigger central degeneration in susceptible individuals.

TABLE 13.2 Features of Environmental Neurotoxic Models of Parkinson's Disease

Chemical inducer	Pathology					Mechanisms/limitations
	Nigrostriatal damage				PI	
	SN cell	Str terminal	Str DA	LC		
MPTP						
Nonhuman primates (i.p., intramuscular (i.m.), infusion)	Y	Y	Y	Y	Y	• Metabolism to toxic intermediate MPP$^+$ by MAO-B in astrocytes • Motor deficits, treatable by L-dopa and apomorphine
Mouse						
Acute/subacute (i.p.)	Y	Y	Y	N	N	• Cognitive, dementia-associated behaviors
Chronic (osmotic pump)	Y	Y	Y	Y	N	• Inconsistent protein aggregation
6-OHDA						
Rat (stereotactic injection)	Y	Y	Y	N	N	• Targets dopamine neurons through DAT • Psychiatric and motor impairments • Cannot cross the blood–brain barrier
Paraquat						
Mouse (i.p.)	Y	Inconsistent	N	Y	Y	• Suggested to promote α-synuclein aggregates • Does not interact with DAT
Paraquat+maneb						
Mouse (i.p.)	Y	unk	Y	unk	Y	• Targets cortisone synthesis • Alters dopamine metabolism
Rotenone						
Rat (infusion) (i.p.)	Y Y	Y Y	Y Y	unk unk	Y Y	• Dopaminergic cell death is controversial, but damage to the neurons is known • Restricted use, short environmental duration
PCB/PBDE						
Mouse	Y	Y	Y	N	Y	• Disrupts calcium and dopamine homeostasis • Have current use in manufacturing
Organochlorines						
Mouse	N	Y	Y	N	Y	• Week evidence of dopamine neuron targeting • Cell loss and motor deficits absent
Manganese						
Mouse	N	N	N	N	N	• Locomotor impairment, mitochondrial toxicity • Cell loss in globus pallidus

Abbreviations: PI, protein inclusions; LC, locus coeruleus; SN, substantia nigra; Str, striatal; unk, unknown.

Parkin and PINK1

Phosphatase and tensin homolog (PTEN)-induced putative kinase 1 (PINK1) is a serine/threonine protein kinase proposed to provide security against mitochondrial dysfunction. When released cytosolic, PINK1 is involved with initiating dendritic growth and neuronal development [21]. Specifically, it is a sensor of damaged mitochondria and can subsequently target them for autophagy [22]. It accomplishes this, in part, by accumulating on depolarized mitochondria. Once at a critical threshold, PINK1 will recruit parkin, an E3 ubiquitin ligase, to degrade the dysfunctional mitochondria [23]. Parkin (PARK2) is thus assumed to be involved in mitochondrial regulation and protein degradation, along with UCHL-1 and ATP12A2 (PARK5 and PARK9, respectively). Ubiquitination of proteins as a posttranslational modification is a fundamental mechanism used by the cell to regulate protein stability and activity. One current hypothesis is that parkin plays a neuroprotective role by targeting proteins with deleterious changes that are toxic to dopamine cells. Loss-of-function mutations of parkin and PINK1 have been found to produce an accrual of misfolded proteins in mitochondria and an early onset juvenile form PD [24,25].

DJ-1

The protein DJ-1 was also identified with an autosomal recessive early-onset form of PD [26]. The gene *PARK7* encodes for DJ-1, a cysteine protease. Another function of parkin has been discovered to positively regulate DJ-1 expression though the p53 signaling cascade via X-box-binding protein-1S [27]. DJ-1 has been indicated to function in several different capacities. It has been shown to cooperate with androgen receptor-mediated transcription and has been indicated as a redox-sensitive chaperone. Research has identified DJ-1 to participate in a protective role against the cytotoxicity of reactive metals, such as copper and mercury, by chelation at sites A104T and D149A [28]. These sites are recognized as polymorphisms that contribute to the early onset of PD. Another role for DJ-1 is to regulate mitochondrial structure and tethering, as well as monitor mitochondrial calcium homeostasis [29].

LRRK2

The protein leucine-rich repeat kinase 2 (LRRK2) has been receiving much consideration recently because it has been genetically linked to autosomal-dominant PD in a larger proportion of the population, as it is known to exhibit variable and incomplete penetrance [30,31]. LRRK2 is located in the cytoplasm and outer mitochondrial membrane and interacts with parkin [32]. In that study, it was shown that overexpression led to an increase in ubiquitinated protein aggregates and to enhanced neuronal degradation. Alternatively, mouse models with mutations in LRRK2 result in dysfunction of neurogenesis and dendritic outgrowth through alteration in macroautophagy and misregulation of mitochondrial clearance [33–35].

ENVIRONMENTAL CONTRIBUTORS TO PD

The environmental contributory factors to PD have drawn much consideration over the past years due to a lack of single genes being associated with a significant percentage of PD cases. Studies on familial PARK proteins have found that only between 5 and 10% of all PD cases are related to monogenic forms of PD. Accordingly, environmental susceptibility factors, in isolation or cumulatively, interacting with genetic variation may contribute to an idiopathic form of the disease. To date, no environmental causative agent has been identified as responsible for contributing to the majority of Parkinson's cases. Indeed, PD is probably not caused by a singular compound, but by a mixture of exposures interacting on a background of genetic susceptibility and lifestyle factors. The human experience results in varying duration and quantity of exposures to potentially neurotoxic agents. This presents a scientific challenge in accurately assigning contributing factors. However, a number of natural and industrial chemicals and metals have been found associated with the disease and used to explore potential pathophysiological mechanisms in PD.

CLASSES OF ENVIRONMENTAL TOXICANTS ASSOCIATED WITH PD

The key finding establishing an environmental linkage to PD was the observation of a parkinsonian syndrome that became prevalent among opioid drug users, who were self-administering a synthetic form of meperidine that was contaminated with the compound N-methyl-4-phenyl-1,2,3,6-tetrahydropyridine (MPTP) [36]. MPTP is metabolized by the enzyme monoamine oxidase B (MAO-B) in glial cells to produce the neurotoxin 1-methyl-4-phenylpyridinium (MPP$^+$), which can uncouple mitochondrial oxidative respiration on complex one resulting in adenosine triphosphate (ATP) depletion and cell death. The dopamine reuptake transporter (DAT) is known to readily pass MPP$^+$ into dopaminergic neurons. Autopsies performed on these patients verified that substantial cell death occurred in the substantia nigra. For summary of chemically induced models of PD, see Table 13.3; for structures, see Figure 13.1.

Pesticides

The discovery that MPTP giving rise to an acute parkinsonism in the early 1980s, and that its toxic metabolite MPP$^+$ exhibited close structural similarity to the herbicide paraquat, gave rise to the idea that pesticide exposure may be linked to PD. This spawned many descriptive studies, cohort studies, separate case–control studies, and meta-analysis studies investigating the relationship between various pesticides and PD incidence [37]. The majority of these studies reported an increased odd risk of PD with pesticide exposure, the strongest of which exists with herbicides and insecticides with long exposure durations. With regard to pesticide exposure and PD, exhaustive meta-analyses of random effects, and meta-regression of heterogeneity of investigated sources covering 46 studies found that exposure will elevate risk odds on average 1.6- to 2.5-fold over the baseline population [38].

A detailed case-controlled study connecting PD and pesticides published in 2011 examined exposure to different pesticides and the incidence of PD among farm workers. Two specific pesticides, paraquat and rotenone, were found to significantly increase odd risk by 2% and 3%, respectively [39]. Although speculated to cause PD symptoms, these findings aided to more definitively confirm these two as risk factor neurotoxicants for human PD. These pesticides, in isolation or in combination with the fungicide maneb, have for years been used in various animal models to mimic the neurotoxicity of parkinsonism [40,41]. This gives

TABLE 13.3   Features of Transgenetic Animal Models of Parkinson's Disease

| Genetic manipulation | Pathology | | | | | Mechanisms/limitations |
| | Nigrostriatal damage | | | | | |
	SN cell	Str terminal	Str DA	LC	PI	
Alpha-synuclein						• No significant nigrostriatal degeneration
Overexpression	N	N	Y	Y	Y	
Knockout	N	N	Y	N	Y	• Reduced TH expression in striatum
						• Motor dysfunction and responsiveness to DA treatment
						• Olfactory deficits and gastrointestinal alterations
Parkin						• No substantial dopaminergic or behavioral abnormalities
Knockout	N	N	N	Y	N	
Overexpression	Y	N	Y	N	Y	• Progressive motor deficits
						• Age-dependent nigrostriatal degeneration
PINK1						• Mild mitochondrial deficits
Knockout	N	N	Inconsistent	N	N	• Increased susceptibility to oxidative stress and reactive oxygen species (ROS) production
						• DA levels, motor activity reduced beyond 18 months
DJ-1						• No neuronal abnormalities
Knockout	N	N	N	N	N	• Increased susceptibility to toxins and oxidative stress
LRRK2						• Modulates DA transmission, release and uptake
Overexpression	N	unk	Y	unk	Y	
Knockout	N	N	Y	unk	N	• Age-dependent progressive decrease in motor activity, treatable with L-dopa

Abbreviations: PI, protein inclusions; LC, locus coeruleus; SN, substantia nigra; Str., striatal; unk, unknown.

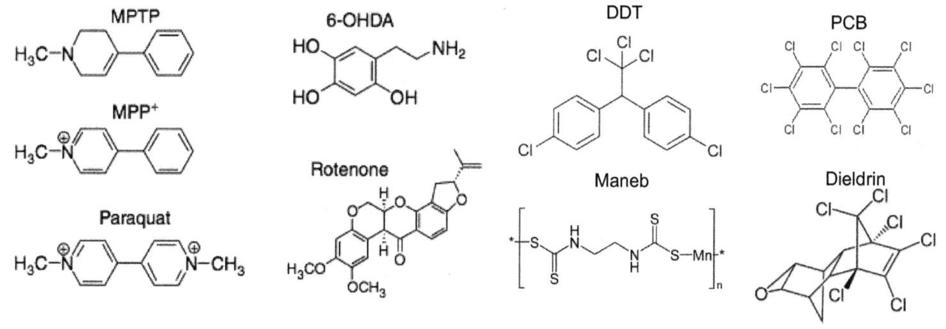

FIGURE 13.1   Chemical structures of neurotoxic molecules presumed to induce dopaminergic neuron damage in animal models of PD.

credence for these exposure models as accurate recapitulations of the human parkinsonian characteristics. Commercially, pesticides are marketed in single preparations or in synergy with other compounds.

Rotenone is a highly lipophilic compound, which is produced naturally by some plants as an insecticide and piscicide. It interferes with the iron–sulfur complexes of mitochondrial complex I to block ion transfer to ubiquinone leading to increased radical formation [42]. This has been demonstrated to induce oxidative damage, ubiquitin accumulation, proteosomal inhibition, and inflammation in neuronal cultures [43,44]. Dopaminergic cell death is uncertain though, as other studies have shown no death of nigral dopamine neuronal bodies and only damage to the dopamine fibers [45]. Rotenone was first identified as a pesticide that could produce parkinsonian lesions experimentally when injected into rats [46]. Specifically, it produced cytoplasmic inclusions redolent of early Lewy bodies and selective dopamine neuron degeneration. The degeneration was perceived in a subset of exposed rats occurring first in the striatum, followed in the substantia nigra pars compacta (SNpc), Since then, low doses of rotenone given by intragastric intubation over 6–12 weeks in mice produced similar lesions, and interestingly no sign of elevated levels in the brain upon necropsy [47]. This provoked epidemiological investigations to test whether the accusation that chronic exposure to rotenone associates with elevated PD risk, most notably the Tanner et al. study provided evidence of this. Notwithstanding the specific damage to dopamine signals, rotenone is probably not a major contributor to PD because of its pharmacokinetics (short environmental half-life and low bioavailability).

Paraquat is a widespread herbicide, which has similar bipyridyl structure to that of MPTP, thus was hypothesized to have similar activity. It ceases photosynthesis in plants, and can redox cycle to continually generate radicals, largely superoxide [48]. It was manufactured and produced as a pesticide in 1961. The ability of paraquat to produce PD is uncertain. The uptake of paraquat to the central nervous system (CNS) was originally rejected by several studies on the bases of its low partition coefficient and low brain deposition in treated animals [49,50]. However, recent studies reject those claims, suggesting that paraquat may penetrate the blood–brain barrier (BBB) by the neutral amino acid transporter [51]. Some researchers suggest little to no neurotoxic effect of paraquat [52]. Others have shown that paraquat exposure causes selective damage to dopamine cells by increasing expression and aggregation of α-synuclein [53], and subsequently cause loss of dopamine neurons in the SNpc that translated to alterations in locomotor activity [54]. Although paraquat is chemically similar to other compounds as mentioned previously, it exerts its neurotoxicity on the dopaminergic axis in a manner quite distinct from MPTP and rotenone [55]. Namely, paraquat does not appear to interact with the DAT, and its ability to activate caspace pathways and oxidize redox reactions differ as well, as it is not presumed to be a complex I inhibitor.

Maneb is a fungicide, which inhibits 11β-hydroxysteroid dehydrogenase type 2, an important enzyme for synthesizing cortisone in the hypothalamic–pituitary–adrenal axis [56]. In regard to PD, it is often coupled with paraquat exposure in laboratory models. Such studies have indicated that these compounds alter striatal dopamine metabolism and cycling in terminal synaptosomes, decreased tyrosine hydroxylase in the striatum, and decreased DAT expression [40,57], potentiating the idea of a two-hit hypothesis. The validity of this co-exposure occurring among the human population is questioned, however, as maneb and paraquat are not applied together, and temporal and environmental-availability differences exist [58].

Organochlorines have been evaluated for their plausible contribution to PD because of their toxicity, biomagnification up food chains, and widespread use on crops. They are banned in the US, but are persistent in the environment and still in use in other countries. Epidemiological studies have linked organochlorine exposure to increased risk of PD and elevated levels of some organochlorine pesticides have been reported in the serum of patients with PD. The primary organochlorines studied for their potential contribution to PD include DDT, DDE, dieldrin, and chlordane. DDT and DDE have little to no effects on the dopamine system in experimental models. Dieldrin was found to be elevated in brain tissue of PD patients [59]. Cyclodienes aldrin and dieldrin ranked second among the highest in-use pesticides in the US in the 1960s and show evidence of neuropathological changes seen in PD. Dieldrin has been shown in experimental models to promote several parkinsonian events: induce fibrillization and aggregation of α-synuclein, deplete intracellular dopamine, activate caspases, depolarize mitochondrial membrane potential, and promote ubiquitin-proteasome system dysfunction [60,61]. In addition, developmental exposure to dieldrin alters expression of key transporters (DAT and vesicular monoamine transporter 2 (VMAT2)), which alters the susceptibility of neurons to toxicants, such as MPTP [62]. One stipulation of these in vitro studies is that the concentrations used are greater than those reported in regions of the PD brain. In addition, the cell loss and locomotor deficits are not readily produced following dieldrin exposure.

Although organochlorines are the most widely used insecticides, much less evidence supports a role for organophosphate insecticides. Organophosphates act primarily on the cholinergic system through inhibition of acetylcholinesterase. However, mice exposed to a high dose of the organophosphate chlorpyrifos (100 mg/kg) showed diminished DAT-dependent ^{3}H-dopamine uptake and corresponding decreased in motor activity [63]. Acute exposure to organophosphate insecticides has been linked epidemiologically to movement defects that share the features of PD [64,65]. However, the development of true PD following organophosphate exposure is uncertain, as the symptoms of the disease lack responsiveness to normal treatments like L-DOPA. Thus, the interaction between dopaminergic and cholinergic signaling holds relevance in PD, and further evidence points to the finding that polymorphisms in acetylcholinesterase or paroxonases (enzymes that hydrolyze organophosphates) were found in higher frequency in PD patients [66], but not in other reports [67].

The epidemiologic and toxicologic studies continue to provide evidence to suggest a biologic basis for various pesticide exposure and elevated presence of parkinsonian outcomes as outlined earlier. However, as of yet, insufficient data exist to conclude a causal relationship with any of these pesticides. Thus, further understanding detailing developmental exposures, gene–environment interactions, or multifaceted exposures have been suggested to help unravel PD etiology [68,69].

Metals

Metals are understood to serve important biological functions as important catalysts in enzyme function. They have the ability to redox cycle and facilitate the transfer of electrons. Because many are bioreactive and participate in oxygenation reactions, cells have developed important systems for monitoring their use, localization, and bioavailability. The systemic toxicity of metals dates back to antiquity, yet the impacts of metals on the CNS, and their impairment of the basal ganglia as it relates to PD, has only been more recently investigated.

Iron (Fe) is an essential metal for life, which is transported across cell membranes by the divalent metal transporter 1 and absorbed by the shuttle protein transferrin and storage protein ferritin. Fe needs to stay bound, as free forms will undergo Fenton reactions to convert hydrogen peroxide to reactive hydroxyl radicals. These species readily damage DNA, lipid membrane, and other cellular proteins. Friedrich's ataxia, a silencing mutation of the frataxin gene, accumulates cytosolic Fe levels and produces motor deficits and dementia. Several studies reported an accumulation of Fe in the SNpc of PD patients in connection with increased expression of Fe transporters in dopaminergic cells and decreased Fe-buffering protein L-ferritin [70,71]. This misappropriation of Fe has been recapitulated in rodent and cell models. Neonatal mice given supplemented iron similar to what is found in baby formula produced a reduction in nigral TH neurons and dopamine levels in aged animals [72]. On the contrary, a nigral neuron-specific overexpression of H-ferritin and pharmacological chelation provided resistance to MPTP-induced death of dopamine neurons [73]. Fe is thought to produce some of these effects presumably through oxidative stress, but additional research has shown that Fe, including other metals such as copper, can accelerate the aggregation of α-synuclein [74]. Whether Fe itself or a reactive oxygen species (ROS) facilitates this conformational change in α-synuclein is uncertain.

Copper (Cu) is an essential metal comparable to iron, and is important for mitochondrial respiration and enzyme activity, such as superoxide dismutase. Cu is transported mainly by the copper membrane transporter 1, is bound by ceruloplasmin and similarly undergoes Fenton reaction. Wilson's disease is a genetic disorder, which produces an accrual of Cu and is known for the neurological deficits in the basal ganglia nuclei and substantia nigra with pathological lesions in dopamine markers. It produces parkinsonian-like movement impairments that is moderately responsive to L-DOPA [75]. Cu poisoning has been assessed for a link to PD, and in a 20-year occupational cohort, Gorell et al. reported a roughly 2.5-fold increased risk [76]. Oxidative damage is likely to be the main contributor to the toxic effect of accumulated free Cu, and one article shows preferential action against the mitochondria [77].

The transition metal manganese (Mn) has also been connected with an atypical form of PD. Mn plays a role for neurotransmitter metabolism and synthesis. Inhalation is especially dangerous as vaporized forms easily bypass the BBB and deposit in the CNS through retrograde transport, although the plausibility of olfactory transport is uncertain in human populations [78]. The most prominent manifestation of Mn poisoning is motor dysfunction, referred to as manganism, which has been common among welders. Some of the cellular systems affected by manganese exposure are inhibition of mitochondrial function, reactive oxygen species (ROS), increased NMDA-mediated neurotoxicity, and reduced glutathione levels [79,80] produced through free-radical generation and dopamine oxidation [81]. The exact contribution of Mn to PD is less certain than the other metals described. Mn insults are largely concentrated to neurons in the globus pallidus, whereas the caudate or SNpc are essentially unaffected with only minor alterations in dopamine levels or function in these areas [82,83]. In addition, Lewy body inclusions do not exist in patients with manganism, and they are nonresponsive to L-DOPA.

Lead (Pb) and mercury (Hg) are unessential metals to the human body, and no systems are available to regulate their bioavailability. Pb will replace other metals and antagonize or alter normal protein functions; therefore, the exact mechanism of Pb-mediated effects is difficult to isolate. Children are known to be very susceptible to the neurotoxic effects of Pb, with increased neuron death and coupled motor-based peripheral ataxia. It has been suggested that PD risk increases two to three-fold as a consequence of lifetime and occupational exposure to Pb [84,85]. In addition to this, rats treated with Pb had demonstrated dysfunction of the nigrostriatal dopamine system with decreased firing rate of dopamine neurons [86]. As it relates to PD, Pb can alter calcium signaling in and out of the mitochondria and dissipate proton gradients to reduce synaptic vesicle effectiveness possibly affecting dopamine sequestration [87]. Data suggest that Hg can interfere with the dopamine nigrostriatal system. Workers occupationally exposed to Hg vapors were found to have a significant dysfunction of DAT [88], which also corroborates with the findings in vitro of depressed dopamine uptake in the neuronal synapses [89]. Along with this, methylmercury exposure reduced dopamine neuron neurites and decreased dopamine levels in the striatum of mice [90]. However, the association of Pb and Hg with PD is much weaker than that of the other metals described previously.

Polychlorinated Biphenyls/Polybrominated Diphenyl Ethers

Polychlorinated biphenyls (PCBs) structurally are composed of two phenyl rings with varying chlorine substitutions about them, resulting in 209 possible conformations and congeners. Before 1977, they were commercially packaged as mixtures and had wide industrial usage as coolants and lubricants in transformers and capacitors, in hydraulic fluids, and in electrical equipment. They are persistent chemicals, which do not readily degrade and do accumulate in the environment. They also are not easily metabolized or secreted from the body, meaning they are retained within lipophilic regions of the body. PCBs perturb several neuronal pathways: generation of oxidative stress, deleterious alterations in calcium homeostasis, and disruption of dopamine synthesis enzymes tyrosine hydroxylase and aromatic acid decarboxylase [91–93]. PCB mixtures also suppress DAT and VMAT2 levels and activity. These depressed dopamine levels were shown in the striatum of PD patients occupationally exposed to high levels of PCBs, who displayed reduced DAT density via single-photon emission computerized tomography (SPECT) imaging [94]. In addition to these molecular pathway alterations, PCB exposure has been associated with the occurrence of PD. A small cohort of occupationally exposed workers in an electrical capacitor plant demonstrated an increased incidence of PD compared to the US population [95].

Polybrominated diphenyl ethers (PBDEs) were introduced to replace PCBs in the late 1970s as polyurethane foam for housewares and flame retardants in manufacturing. Although some mixtures are restricted, PBDEs are still in use and on the rise. One unique property is that they bind to each other and do not chemically incorporate into products, increasing the propensity for leaching. The molecular effects of PBDEs are similar to those described because of PCBs, and at low concentrations (micromolar) and, again, interfere mainly with the dopaminergic system compared to other neurotransmitters. Therefore, these chemicals cause concern as they relate to PD progression, and warrant much interest. Currently, epidemiological data on PBDEs, as they relate to PD risk, are unavailable.

Solvents

Solvents include a wide array of industrialized chemicals; the common ones include trichloroethylene (TCE), toluene, acetone, hexane, and carbon disulfide. Industrial exposure is primary inhalation or dermal, and nonoccupational exposure also occurs through ingestion of contaminated water. Neurological deficits have been reported with these chemicals, largely as movement and cognitive disorders that have parkinsonian-related consistencies. The one solvent, TCE, has been proposed to recapitulate behavioral defects associated with PD in rats and humans [96,97]. Indeed, chronic exposure to TCE has been reported to increase PD risk in several case reports, population-based epidemiology, and twin studies, showing risk elevated six-fold [98]. The action of TCE is primarily as a disruptor of cellular membrane integrity and lipid peroxidation because of free radicals, often associated with nerve demyelination [99]. Metabolism of solvents may play a role as well, as patients with PD have been shown to inadequately clear the hydrocarbon solvent n-hexane [100], which may prolong exposure or result in alternate pathways of metabolism that produce greater toxic intermediates. An example of this is 1-Trichloromethyl-1,2,3,4-tetrahydro-beta-carboline (TaClo), an intermediate of TCE metabolism, which is structurally similar to MPTP, can severely disturb dopamine metabolism and is a potential complex 1 inhibitor of mitochondria [96]. This also resulted in preferential loss of SNpc dopamine neurons and simultaneous reduced levels in the striatum and midbrain, whereas other gamma-aminobutyric acid (GABA)ergic and cholinergic neurons of the SNpc were spared [97].

ANIMAL MODELS RELEVANT TO PD

Although several genes have been identified that are causative of PD as discussed in an earlier section, the creation of transgenic mice harboring these mutations has been disappointing, as most do not produce a parkinsonian phenotype (Table 13.2). Some of the transgenic lines have shown increased nigrostriatal damage when combined with some neurotoxicant models. Mutations in α-synuclein [101], PINK1 [102], and DJ-1 [103] result in increased dopaminergic cell loss when exposed to neurotoxicants, whereas overexpression of these genes have served neuroprotectively against these agents. Yet, overall, the limitations with many of the genetic models alone fail to reliably produce PD pathology and associated behavioral and locomotor deficits. Transgenic rat models of PD mutations have recently been developed and are becoming more widely available [104]. Investigations in the rat may be superior to the mouse models because the neuronal circuitry is more analogous to humans and rats are less prone to anxiety issues that mice encounter, making behavioral testing more reliable.

Based on the failure of many of the transgenic mouse lines to reproduce pathological and behavioral hallmarks of PD, the pursuit of many in the PD field is to examine the various exposures that have been associated with the disease, and come to conclusions relative to the plausibility of such factors contributing to the human disease. Two toxicant models of PD have been used extensively in research to investigate mechanism and follow the neurodegenerative properties of the disease over the years. The first is MPTP, a lipophilic compound that penetrates the BBB and produces a rapid onset of parkinsonian-like symptoms remarkably similar to those of idiopathic PD. Drugs used to ameliorate MPTP-induced PD symptoms, such as L-DOPA or bromocriptine, show improvement. In the laboratory, although doses vary slightly among groups, the general protocol of MPTP administration is by subcutaneous injections of 10–20 mg/kg in mice, 2–4 times at 2 h intervals [105,106]. This dosing paradigm produces robust microglial activation, dopamine depletion resulting from TH nitration, cell damage to SNpc DA neurons, inhibited ATP production, and DNA-damaging peroxynitrate formation within 48 h [107,108]. One caveat of the MPTP model is that it does not effectively recapitulate the Lewy body development, which is seen in PD pathology. Further, it is now known that other brain areas are affected in PD including formation of protein inclusions and degeneration of noradrenergic neurons detected in the locus coeruleus [109]. The locus coeruleus, a nucleus in the pons brain stem, is involved with physiological stress response and is implicated in several neurodegenerative diseases including Alzheimer's disease [110]. However, because of the specific uptake by the DAT, MPTP does not target these areas as extensively as the nigrostriatal system.

Another classic parkinsonism-inducing compound is 6-hydroxydopamine (6-OHDA), which enters the target cells by the DAT and norepinephrine transporter (NET). Because of uptake through the NET, the use of a noradrenaline inhibitor such as desipramine is often employed. Animals are often premedicated with pargyline, which inhibits extrasynaptic breakdown of 6-OHDA [111]. 6-OHDA targets the mitochondria and produces superoxide radicals in these dopaminergic cells resulting in cell death. The nigrostriatal region (DAT expressing) and the locus coeruleus (NET expressing) are highly sensitive to this neurotoxin [112]. The damage caused by 6-OHDA recapitulates the disruption in basal ganglia circuitry and deterioration of the nigrostriatal pathway observed in parkinsonian patients [113]. However, 6-OHDA has to be administered by stereotaxic injection (10 μg 6-OHDA in 0.9% NaCl with 0.02% ascorbic acid) as it cannot cross the BBB [114].

Two pesticides linked to increased PD risk, rotenone and paraquat, are commonly used as environmental toxicants to model parkinsonism. Early studies using acute high doses of rotenone showed pervasive lesions beyond the SNpc [45,115] making it an unappealing model. However, the development of a chronic low-dose protocol via jugular vein cannula infusion attached to a subcutaneous osmotic minipump produced cytoplasmic α-synuclein-positive inclusions and selective nigrostriatal neurodegeneration, highlighting their vulnerability to neurotoxins like rotenone [46]. This also suggested that chronic low doses might be required to produce the Lewy bodies' inclusions. Despite these features, the consistency by which rotenone produces these lesions is highly variable in different rat strains [116,117]. Because, a revised rotenone model using daily intraperitoneal injection (i.p. injection) has been reported to produce consistent loss of SNpc neurons and appearance of α-synuclein aggregation [118].

When injected into mice, paraquat is reported to induce specific dopaminergic neuron death and motor deficits dependent on dose and age [119,120], presumed because of BBB permeability as suggested previously. However, the deleterious effect of paraquat proposed on dopamine neurons has also not been shown to lower striatal dopamine levels, potentially due to a compensatory increase in TH activity that has been reported [54,121]. Therefore, it remains difficult to conclude that the reported changes in motor activity could be caused by a modest reduction in dopamine levels. It has not been shown whether ʟ-dopa can alleviate the motor defects associated with paraquat exposure, and could plausibly be a result of some pulmonary toxicity that affects performance [122,123].

PD pathogenesis is postulated to involve neuroinflammation as many of the described chemicals stimulate an inflammatory response. Whether this is a primary cause or simply a consequence of lesioned dopaminergic cells is uncertain. Elevated inflammatory cytokines and chemokines are believed to promote a destructive cycle leading to cell death. Lipopolysaccharide (LPS) is an endotoxin that stimulates toll-like receptors on microglia to create a reproducible inflammatory response [124]. Various delivery models have been used including stereotactic injection to specific brain regions that produces consistent loss of nigral dopamine neurons at 24 h in rats [125]. Minipump administration (5 ng/h over 2 weeks in rats) activates microglia by three days, produces a sustained response out to eight weeks, and induces subsequent progressive cell loss in the SNpc [126]. Systemic i.p. injection of LPS induces measurable microglial action by three hours but reported only delayed and modest nigral cell loss (47% after 10 months) [127]. LPS has proven to be a useful device in ascertaining whether neuroinflammation is associated with progressive nigral dopaminergic cell death. However, methods to study neuroprotection are difficult in such a chronic model, and protein inclusions or extranigral pathologies have not been reported with LPS.

ALTERNATIVE MODELS

Given the high conservation of pathways from invertebrates to mammals, several different alternative models have been employed to study PD pathogenesis. This is especially prudent given the difficulty of identifying molecular mechanisms underlying parkinsonian pathophysiology. The organisms *Caenorhabditis elegans*, *Drosophila melanogaster*, *Danio rerio*, and *Saccharomyces cerevisiae* are valuable in assessment of gene interactions and screening drug therapies relevant to PD pathogenesis. The nematode roundworm *C. elegans* is an effective model for describing genetic systems, and offers different technical advantages, such as high-throughput approaches and specific neural cell mapping. Of the 302 neurons that compose the nervous system, eight neurons are dopaminergic [128]. The transparency of the animal offers the ability to visualize the neurons with a simple green fluorescent protein (GFP) tag and monitor neurodegeneration. The molecular alterations of α-synuclein and LRRK2 have demonstrated misfolding and aggregation, mitochondrial stress, dopamine cell degeneration, and reduced motor movement [129,130]. In addition, they have provided support of protective mechanisms, such as reduced mitochondrial stress with LRRK2 overexpression [131]. Neurotoxicant treatment with suspected PD implications has been studied in *C. elegans*. 6-OHDA, MPP$^+$, and Mn-caused dopamine neurodegeneration and reduced dopamine levels in various genetic models [132–134]. Paraquat produced elevated oxidative stress in a PINK-1 mutant model to inhibit neurite outgrowth [135]. In all, *C. elegans* is a valuable instrument for evaluating putative pathways and offers immense opportunity for chemical and genetic screens for PD-associated inquires [136].

As pesticides are a large class of chemicals linked to increased PD risk, it is important to understand the biology of sublethal dose effects on insect models, such as the fruit fly *D. melanogaster*. Clusters of dopaminergic neurons are present and metabolic pathways and enzymes are highly conserved between humans and *Drosophila*. Overexpression of the human α-synuclein was the first PD-like *Drosophila* model, which resulted in age-dependent specific loss of dopaminergic neurons, fibrillary inclusions, and gradual loss of climbing ability [137]. Chronic rotenone produced selective loss of dopaminergic neurons in brain clusters and the addition of L-DOPA to feed improved some of the locomotor impairments, as is similar in human PD patients [138]. DJ-1 mutant flies were more susceptible to rotenone, H$_2$O$_2$, and paraquat deficits, supporting a neuroprotective role for the gene [139]. *Drosophila* may also be used to identify new sources of selective dopamine toxicants. The volatile fungal 1-octen-3-ol, popularly known as mushroom alcohol emitted by mold spores, can alter VMAT2 and DAT activity to reduce locomotor activity and cause dopaminergic neuron death [140].

TABLE 13.4 Alternative Models of Parkinson's Disease

Model	Gene/Chemical	Phenotypes
C. elegans (nematode)	Alpha-synuclein	Mitochondrial stress, DA degeneration, irregular DA vesicle dissemination
	Parkin and PINK-1	Protein insolubility and aggregations, proteotoxic stress hypersensitivity
	MPTP/MPP$^+$	DA degeneration, reduced mobility
	Rotenone/PQ	Oxidative stress, mitochondrial stress, reduced DA release/activity
D. melanogaster (fruitfly)	Alpha-synuclein	DA cell loss, progressive climbing defect
	Parkin and PINK-1	Locomotor reduction, abnormal mitochondrial morphology, reduced lifespan
	MPTP/MPP$^+$	N/A
	Rotenone/PQ	DA cell loss, locomotor impairments responsiveness to L-DOPA
D. rerio (zebrafish)	Alpha-synuclein	Protein aggregation
	Parkin and PINK-1	Moderate DA cell loss, decreased complex I activity, increased toxin susceptibility
	MPTP/MPP$^+$	Altered dopaminergic projections, locomotor deficits
	Rotenone/PQ	Reduced DAT expression, swim defects

Protein kinase B (Akt) and Jun N-terminal kinase provided protective benefit to the loss of dopamine activity, elucidating potential signaling pathways relevant to PD pathophysiology. Genetic PD models of *D. melanogaster* have been described for α-synuclein, parkin, and PINK-1 and are summarized in Table 13.4.

Zebrafish, *Danio rerio*, also hold exciting prospects for rapid functional assessments of putative modifier genes and live observation of PD pathogenic mechanisms. Embryos are transparent and develop externally, therefore permitting for fluorescent reporter techniques to monitor organogenesis and phenotypic changes induced by genetic or environmental manipulations. They have high fecundity and short lifespans, which make them ideal for genetic study. Key brain regions between zebrafish and humans are highly conserved. For example, motor movements involving the basal ganglia in mammals are present in zebra fish telencephalon, and TH-positive neurons are present that resemble anatomical comparisons to the striatum [141]. Mutated forms of LRRK2, PINK-1, and parkin zebrafish have been developed over the past five years that may provide important mechanistic data and drug therapy screening for PD [142–144]. Zebrafish treated with MPTP produces loss of dopaminergic neurons in the diencephalic brain region (as zebrafish do not have a mesencephalic region comparable to a substantia nigra in other models), altered neurite projections, and aberrant swim patterns [145]. Paraquat-treated zebrafish exhibited motor-related behavioral defects and reduced DAT expression, whereas nonmotor behaviors, such as social interaction and anxiety challenges, were unaltered [146]. This also allows for testing of novel therapies, such as the "molecular tweezer" CLR01, which encompasses lysine residues and not only prevented α-synuclein accumulation, but untangled existing aggregates, in an α-synuclein transgenic zebrafish model [147].

EMERGING FRONTIERS AND FUTURE DIRECTIONS

The collective benefit the variety of models for Parkinson's research have uncovered at least 15 genes and identified classes of environmental toxicants associated with elevating risk of PD. Epidemiological research has linked exposure to solvents, manganese, and some pesticides and herbicides with elevated Parkinson's risk. Strengthening the association, a recent report suggests that PD patients are 61% more likely to report direct contact with pesticides than unaffected relatives [148], yet many patients report minimal direct exposure with such chemicals [149]. Parkinson's clusters, or a higher occurrence among subsets of the population, have been reported. Such cases may be coincidental, or rather share commonalities in toxicant triggers to disease or similar population genetics that enhanced susceptibility.

To assist in uncovering new pathways and molecular contributors to PD, examination of different genetic variants, mainly of single-nucleotide polymorphisms (SNPs) that associate in case-controlled PD populations has led to new putative targets [150]. These unbiased genome-wide association studies (GWAS) have implicated novel associations with several neurodegenerative diseases; however, many of the implicated genes explain only very small proportions of disease heritability. Thus, focus on biological pathways and networks maybe more valuable in incorporating GWAS and other modalities to improve understanding and clinical treatment. Epigenetic modifications by environmental and stochastic factors have not thoroughly been discussed in this chapter, such as DNA methylation, but maybe play a significant role in PD etiology. Systematic studies of epigenetic lesions in PD individuals is more challenging than GWAS based on subtissue variations in methylation, yet epigenome-wide association studies holds promise in identifying relevant modifications that promote PD etiology [151].

Remote sensing technologies such as geographic information systems, which better tracks pesticide and other environmental exposures, as opposed to relying on records-based data and residential proximity to associated pesticide applications, may provide more statistical power and plausibility of exposure paradigms to epidemiologic and associative studies [152]. Combining several of these approaches, as well as subsequent computational biology and functional studies of identified variants, should improve understanding and diagnostic techniques related to PD. Furthermore, with every novel discovery we may advance understanding of the biochemical and mechanistic basis of gene–environment interactions in PD.

Acknowledgments

This research was supported in part by NIH grants R01ES021800, U01NS079249, P30ES005022, and T32ES007148.

References

[1] Dorsey ER, Constantinescu R, Thompson JP, et al. Projected number of people with Parkinson disease in the most populous nations, 2005 through 2030. Neurology 2007;68(5):384–6.

[2] Bjorklund A, Dunnett SB. Dopamine neuron systems in the brain: an update. Trends Neurosci 2007;30(5):194–202.

[3] Cotzias GC, Papavasiliou PS, Gellene R. L-dopa in Parkinson's syndrome. N Engl J Med 1969;281(5):272.

[4] Kish SJ, Shannak K, Rajput A, Deck JH, Hornykiewicz O. Aging produces a specific pattern of striatal dopamine loss: implications for the etiology of idiopathic Parkinson's disease. J Neurochem 1992;58(2):642–8.

[5] Cannon JR, Greenamyre JT. Neurotoxic in vivo models of Parkinson's disease recent advances. Prog Brain Res 2010;184:17–33.

[6] Spillantini MG, Schmidt ML, Lee VM, Trojanowski JQ, Jakes R, Goedert M. Alpha-synuclein in Lewy bodies. Nature 1997;388(6645):839–40.

[7] Fahn S. Description of Parkinson's disease as a clinical syndrome. Ann NY Acad Sci 2003;991:1–14.

[8] Wirdefeldt K, Gatz M, Schalling M, Pedersen NL. No evidence for heritability of Parkinson disease in Swedish twins. Neurology 2004;63(2):305–11.

[9] Bekris LM, Mata IF, Zabetian CP. The genetics of Parkinson disease. J Geriatr Psychiatry Neurol 2010;23(4):228–42.

[10] Shulman JM, De Jager PL, Feany MB. Parkinson's disease: genetics and pathogenesis. Annu Rev Pathol 2011;6:193–222.

[11] Polymeropoulos MH, Lavedan C, Leroy E, et al. Mutation in the alpha-synuclein gene identified in families with Parkinson's disease. Science 1997;276(5321):2045–7.

[12] Liu G, Zhang C, Yin J, et al. Alpha-synuclein is differentially expressed in mitochondria from different rat brain regions and dose-dependently down-regulates complex I activity. Neurosci Lett 2009;454(3):187–92.

[13] Burre J, Sharma M, Tsetsenis T, Buchman V, Etherton MR, Sudhof TC. Alpha-synuclein promotes SNARE-complex assembly in vivo and in vitro. Science 2010;329(5999):1663–7.

[14] Singleton AB, Farrer M, Johnson J, et al. Alpha-synuclein locus triplication causes Parkinson's disease. Science 2003;302(5646):841.

[15] Neumann M, Kahle PJ, Giasson BI, et al. Misfolded proteinase K-resistant hyperphosphorylated alpha-synuclein in aged transgenic mice with locomotor deterioration and in human alpha-synucleinopathies. J Clin Invest 2002;110(10):1429–39.

[16] Burton EA, Glorioso JC, Fink DJ. Gene therapy progress and prospects: Parkinson's disease. Gene Ther 2003;10(20):1721–7.

[17] Ulusoy A, Rusconi R, Pérez-Revuelta BI, et al. Caudo-rostral brain spreading of alpha-synuclein through vagal connections. EMBO Mol Med 2013;5(7):1051–9.

[18] Pfeiffer RF. Gastrointestinal dysfunction in Parkinson's disease. Lancet Neurol 2003;2(2):107–16.

[19] Braak H, de Vos RA, Bohl J, Del Tredici K. Gastric alpha-synuclein immunoreactive inclusions in Meissner's and Auerbach's plexuses in cases staged for Parkinson's disease-related brain pathology. Neurosci Lett 2006;396(1):67–72.

[20] Natale G, Pasquali L, Ruggieri S, Paparelli A, Fornai F. Parkinson's disease and the gut: a well known clinical association in need of an effective cure and explanation. Neurogastroenterol Motil 2008;20(7):741–9.

[21] Dagda RK, Pien I, Wang R, et al. Beyond the mitochondrion: cytosolic PINK1 remodels dendrites through protein kinase A. J Neurochem 2014;128(6):864–77.

[22] Narendra DP, Jin SM, Tanaka A, et al. PINK1 is selectively stabilized on impaired mitochondria to activate Parkin. PLoS Biol 2010;8(1):e1000298.

[23] Youle RJ, van der Bliek AM. Mitochondrial fission, fusion, and stress. Science 2012;337(6098):1062–5.

[24] Periquet M, Latouche M, Lohmann E, et al. Parkin mutations are frequent in patients with isolated early-onset parkinsonism. Brain 2003;126(Pt 6):1271–8.

[25] Deas E, Wood NW, Plun-Favreau H. Mitophagy and Parkinson's disease: the PINK1-parkin link. Biochim Biophys Acta 2011;1813(4):623–33.

[26] Bonifati V, Rizzu P, van Baren MJ, et al. Mutations in the DJ-1 gene associated with autosomal recessive early-onset parkinsonism. Science 2003;299(5604):256–9.

[27] Duplan E, Giaime E, Viotti J, et al. ER-stress-associated functional link between Parkin and DJ-1 via a transcriptional cascade involving the tumor suppressor p53 and the spliced X-box binding protein XBP-1. J Cell Sci 2013;126(Pt 9):2124–33.

[28] Björkblom B, Adilbayeva A, Maple-Grødem J, et al. Parkinson disease protein DJ-1 binds metals and protects against metal-induced cytotoxicity. J Biol Chem 2013;288(31):22809–20.

[29] Ottolini D, Cali T, Negro A, Brini M. The Parkinson disease-related protein DJ-1 counteracts mitochondrial impairment induced by the tumour suppressor protein p53 by enhancing endoplasmic reticulum-mitochondria tethering. Hum Mol Genet 2013;22(11):2152–68.

[30] Lesage S, Ibanez P, Lohmann E, et al. G2019S LRRK2 mutation in French and North African families with Parkinson's disease. Ann Neurol 2005;58(5):784–7.

[31] Ozelius LJ, Senthil G, Saunders-Pullman R, et al. LRRK2 G2019S as a cause of Parkinson's disease in Ashkenazi Jews. N Engl J Med 2006;354(4):424–5.

[32] Smith WW, Pei Z, Jiang H, et al. Leucine-rich repeat kinase 2 (LRRK2) interacts with parkin, and mutant LRRK2 induces neuronal degeneration. Proc Natl Acad Sci USA 2005;102(51):18676–81.

[33] MacLeod D, Dowman J, Hammond R, Leete T, Inoue K, Abeliovich A. The familial parkinsonism gene LRRK2 regulates neurite process morphology. Neuron 2006;52(4):587–93.

[34] Friedman LG, Lachenmayer ML, Wang J, et al. Disrupted autophagy leads to dopaminergic axon and dendrite degeneration and promotes presynaptic accumulation of alpha-synuclein and LRRK2 in the brain. J Neurosci 2012;32(22):7585–93.

[35] Cherra 3rd SJ, Steer E, Gusdon AM, Kiselyov K, Chu CT. Mutant LRRK2 elicits calcium imbalance and depletion of dendritic mitochondria in neurons. Am J Pathol 2013;182(2):474–84.

[36] Langston JW, Ballard Jr PA. Parkinson's disease in a chemist working with 1-methyl-4-phenyl-1,2,5,6-tetrahydropyridine. N Engl J Med 1983;309(5):310.

[37] Brown TP, Rumsby PC, Capleton AC, Rushton L, Levy LS. Pesticides and Parkinson's disease–is there a link? Environ Health Perspect 2006;114(2):156–64.

[38] van der Mark M, Brouwer M, Kromhout H, Nijssen P, Huss A, Vermeulen R. Is pesticide use related to Parkinson disease? Some clues to heterogeneity in study results. Environ Health Perspect 2012;120(3):340–7.

[39] Tanner CM, Kamel F, Ross GW, et al. Rotenone, paraquat, and Parkinson's disease. Environ Health Perspect 2011;119(6):866–72.

[40] Cory-Slechta DA. Studying toxicants as single chemicals: does this strategy adequately identify neurotoxic risk? Neurotoxicology 2005;26(4):491–510.

[41] Desplats P, Patel P, Kosberg K, et al. Combined exposure to Maneb and Paraquat alters transcriptional regulation of neurogenesis-related genes in mice models of Parkinson's disease. Mol Neurodegener 2012;7:49.

[42] Hayes Jr WJ, Laws Jr ER. Handbook of Pesticide Toxicology, 3 vols. San Diego: Academic Press; 1991.

[43] Sherer TB, Betarbet R, Stout AK, et al. An in vitro model of Parkinson's disease: linking mitochondrial impairment to altered alpha-synuclein metabolism and oxidative damage. J Neurosci 2002;22(16):7006–15.

[44] Liu B, Gao HM, Hong JS. Parkinson's disease and exposure to infectious agents and pesticides and the occurrence of brain injuries: role of neuroinflammation. Environ Health Perspect 2003;111(8):1065–73.

[45] Ferrante RJ, Schulz JB, Kowall NW, Beal MF. Systemic administration of rotenone produces selective damage in the striatum and globus pallidus, but not in the substantia nigra. Brain Res 1997;753(1):157–62.

[46] Betarbet R, Sherer TB, MacKenzie G, Garcia-Osuna M, Panov AV, Greenamyre JT. Chronic systemic pesticide exposure reproduces features of Parkinson's disease. Nat Neurosci 2000;3(12):1301–6.

[47] Pan-Montojo F, Anichtchik O, Dening Y, et al. Progression of Parkinson's disease pathology is reproduced by intragastric administration of rotenone in mice. PLoS One 2010;5(1):e8762.

[48] Bus JS, Gibson JE. Paraquat: model for oxidant-initiated toxicity. Environ Health Perspect 1984;55:37–46.

[49] Koller WC. Paraquat and Parkinson's disease. Neurology 1986;36(8):1147.

[50] Widdowson PS, Farnworth MJ, Simpson MG, Lock EA. Influence of age on the passage of paraquat through the blood–brain barrier in rats: a distribution and pathological examination. Hum Exp Toxicol 1996;15(3):231–6.

[51] Shimizu K, Ohtaki K, Matsubara K, et al. Carrier-mediated processes in blood–brain barrier penetration and neural uptake of paraquat. Brain Res 2001;906(1–2):135–42.

[52] Woolley DE, Gietzen DW, Gee SJ, Magdalou J, Hammock BD. Does paraquat (PQ) mimic MPP+ toxicity? Proc West Pharmacol Soc 1989;32:191–3.

[53] Manning-Bog AB, McCormack AL, Li J, Uversky VN, Fink AL, Di Monte DA. The herbicide paraquat causes up-regulation and aggregation of alpha-synuclein in mice: paraquat and alpha-synuclein. J Biol Chem 2002;277(3):1641–4.

[54] Ossowska K, Wardas J, Smialowska M, et al. A slowly developing dysfunction of dopaminergic nigrostriatal neurons induced by long-term paraquat administration in rats: an animal model of preclinical stages of Parkinson's disease? Eur J Neurosci 2005;22(6):1294–304.

[55] Richardson JR, Quan Y, Sherer TB, Greenamyre JT, Miller GW. Paraquat neurotoxicity is distinct from that of MPTP and rotenone. Toxicol Sci 2005;88(1):193–201.

[56] Atanasov AG, Tam S, Rocken JM, Baker ME, Odermatt A. Inhibition of 11 beta-hydroxysteroid dehydrogenase type 2 by dithiocarbamates. Biochem Biophys Res Commun 2003;308(2):257–62.

[57] Thiruchelvam M, Brockel BJ, Richfield EK, Baggs RB, Cory-Slechta DA. Potentiated and preferential effects of combined paraquat and maneb on nigrostriatal dopamine systems: environmental risk factors for Parkinson's disease? Brain Res 2000;873(2):225–34.

[58] Hatcher JM, Pennell KD, Miller GW. Parkinson's disease and pesticides: a toxicological perspective. Trends Pharmacol Sci 2008;29(6):322–9.

[59] Corrigan FM, Wienburg CL, Shore RF, Daniel SE, Mann D. Organochlorine insecticides in substantia nigra in Parkinson's disease. J Toxicol Environ Health A 2000;59(4):229–34.

[60] Kanthasamy AG, Kitazawa M, Kanthasamy A, Anantharam V. Dieldrin-induced neurotoxicity: relevance to Parkinson's disease pathogenesis. Neurotoxicology 2005;26(4):701–19.

[61] Kitazawa M, Anantharam V, Kanthasamy AG. Dieldrin induces apoptosis by promoting caspase-3-dependent proteolytic cleavage of protein kinase Cdelta in dopaminergic cells: relevance to oxidative stress and dopaminergic degeneration. Neuroscience 2003;119(4):945–64.

[62] Richardson JR, Caudle WM, Wang M, Dean ED, Pennell KD, Miller GW. Developmental exposure to the pesticide dieldrin alters the dopamine system and increases neurotoxicity in an animal model of Parkinson's disease. FASEB J: official Publ Fed Am Soc Exp Biol 2006;20(10):1695–7.

[63] Karen DJ, Li W, Harp PR, Gillette JS, Bloomquist JR. Striatal dopaminergic pathways as a target for the insecticides permethrin and chlorpyrifos. Neurotoxicology 2001;22(6):811–7.

[64] Muller-Vahl KR, Kolbe H, Dengler R. Transient severe parkinsonism after acute organophosphate poisoning. J Neurol Neurosurg Psychiatry 1999;66(2):253–4.

[65] Senanayake N, Sanmuganathan PS. Extrapyramidal manifestations complicating organophosphorus insecticide poisoning. Hum Exp Toxicol 1995;14(7):600–4.

[66] Benmoyal-Segal L, Vander T, Shifman S, et al. Acetylcholinesterase/paraoxonase interactions increase the risk of insecticide-induced Parkinson's disease. FASEB J 2005;19(3):452–4.

[67] Fong CS, Cheng CW, Wu RM. Pesticides exposure and genetic polymorphism of paraoxonase in the susceptibility of Parkinson's disease. Acta Neurol Taiwan 2005;14(2):55–60.

[68] Freire C, Koifman S. Pesticide exposure and Parkinson's disease: epidemiological evidence of association. Neurotoxicology 2012;33(5):947–71.

[69] Moretto A, Colosio C. Biochemical and toxicological evidence of neurological effects of pesticides: the example of Parkinson's disease. Neurotoxicology 2011;32(4):383–91.

[70] Sofic E, Riederer P, Heinsen H, et al. Increased iron (III) and total iron content in post mortem substantia nigra of parkinsonian brain. J Neural Transm 1988;74(3):199–205.

[71] Friedman A, Arosio P, Finazzi D, Koziorowski D, Galazka-Friedman J. Ferritin as an important player in neurodegeneration. Parkinsonism Relat Disord 2011;17(6):423–30.

[72] Kaur D, Peng J, Chinta SJ, et al. Increased murine neonatal iron intake results in Parkinson-like neurodegeneration with age. Neurobiol Aging 2007;28(6):907–13.

[73] Kaur D, Yantiri F, Rajagopalan S, et al. Genetic or pharmacological iron chelation prevents MPTP-induced neurotoxicity in vivo: a novel therapy for Parkinson's disease. Neuron 2003;37(6):899–909.

[74] Uversky VN, Li J, Fink AL. Metal-triggered structural transformations, aggregation, and fibrillation of human alpha-synuclein. A possible molecular NK between Parkinson's disease and heavy metal exposure. J Biol Chem 2001;276(47):44284–96.

[75] Oder W, Prayer L, Grimm G, et al. Wilson's disease: evidence of subgroups derived from clinical findings and brain lesions. Neurology 1993;43(1):120–4.

[76] Gorell JM, Johnson CC, Rybicki BA, et al. Occupational exposure to manganese, copper, lead, iron, mercury and zinc and the risk of Parkinson's disease. Neurotoxicology 1999;20(2–3):239–47.

[77] Paris I, Perez-Pastene C, Couve E, Caviedes P, Ledoux S, Segura-Aguilar J. Copper dopamine complex induces mitochondrial autophagy preceding caspase-independent apoptotic cell death. J Biol Chem 2009;284(20):13306–15.

[78] Aschner M, Erikson KM, Dorman DC. Manganese dosimetry: species differences and implications for neurotoxicity. Crit Rev Toxicol 2005;35(1):1–32.

[79] Brouillet EP, Shinobu L, McGarvey U, Hochberg F, Beal MF. Manganese injection into the rat striatum produces excitotoxic lesions by impairing energy metabolism. Exp Neurol 1993;120(1):89–94.

[80] Gavin CE, Gunter KK, Gunter TE. Manganese and calcium transport in mitochondria: implications for manganese toxicity. Neurotoxicology 1999;20(2–3):445–53.

[81] Halliwell B. Manganese ions, oxidation reactions and the superoxide radical. Neurotoxicology 1984;5(1):113–7.

[82] Wright AK, Atherton JF, Norrie L, Arbuthnott GW. Death of dopaminergic neurones in the rat substantia nigra can be induced by damage to globus pallidus. Eur J Neurosci 2004;20(7):1737–44.

[83] Guilarte TR, Chen MK, McGlothan JL, et al. Nigrostriatal dopamine system dysfunction and subtle motor deficits in manganese-exposed non-human primates. Exp Neurol 2006;202(2):381–90.

[84] Coon S, Stark A, Peterson E, et al. Whole-body lifetime occupational lead exposure and risk of Parkinson's disease. Environ Health Perspect 2006;114(12):1872–6.

[85] Weisskopf MG, Weuve J, Nie H, et al. Association of cumulative lead exposure with Parkinson's disease. Environ Health Perspect 2010;118(11):1609–13.

[86] Tavakoli-Nezhad M, Barron AJ, Pitts DK. Postnatal inorganic lead exposure decreases the number of spontaneously active midbrain dopamine neurons in the rat. Neurotoxicology 2001;22(2):259–69.

[87] Borisova T, Krisanova N, Sivko R, et al. Presynaptic malfunction: the neurotoxic effects of cadmium and lead on the proton gradient of synaptic vesicles and glutamate transport. Neurochem Int 2011;59(2):272–9.

[88] Lin CY, Liou SH, Hsiech CM, Ku MC, Tsai SY. Dose-response relationship between cumulative mercury exposure index and specific uptake ratio in the striatum on Tc-99m TRODAT SPECT. Clin Nucl Med 2011;36(8):689–93.

[89] Dreiem A, Shan M, Okoniewski RJ, Sanchez-Morrissey S, Seegal RF. Methylmercury inhibits dopaminergic function in rat pup synaptosomes in an age-dependent manner. Neurotoxicol Teratol 2009;31(5):312–7.

[90] Bourdineaud JP, Fujimura M, Laclau M, Sawada M, Yasutake A. Deleterious effects in mice of fish-associated methylmercury contained in a diet mimicking the western populations' average fish consumption. Environ Int 2011;37(2):303–13.

[91] Kodavanti PR, Derr-Yellin EC, Mundy WR, et al. Repeated exposure of adult rats to Aroclor 1254 causes brain region-specific changes in intracellular Ca2+ buffering and protein kinase C activity in the absence of changes in tyrosine hydroxylase. Toxicol Appl Pharmacol 1998;153(2):186–98.

[92] Richardson JR, Miller GW. Acute exposure to aroclor 1016 or 1260 differentially affects dopamine transporter and vesicular monoamine transporter 2 levels. Toxicol Lett 2004;148(1–2):29–40.

[93] Lee DW, Notter SA, Thiruchelvam M, et al. Subchronic polychlorinated biphenyl (Aroclor 1254) exposure produces oxidative damage and neuronal death of ventral midbrain dopaminergic systems. Toxicol Sci 2012;125(2):496–508.

[94] Caudle WM, Richardson JR, Delea KC, et al. Polychlorinated biphenyl-induced reduction of dopamine transporter expression as a precursor to Parkinson's disease-associated dopamine toxicity. Toxicol Sci 2006;92(2):490–9.

[95] Steenland K, Hein MJ, Cassinelli 2nd RT, et al. Polychlorinated biphenyls and neurodegenerative disease mortality in an occupational cohort. Epidemiology 2006;17(1):8–13.

[96] Gash DM, Rutland K, Hudson NL, et al. Trichloroethylene: parkinsonism and complex 1 mitochondrial neurotoxicity. Ann Neurol 2008;63(2):184–92.

[97] Liu M, Choi DY, Hunter RL, et al. Trichloroethylene induces dopaminergic neurodegeneration in Fisher 344 rats. J Neurochem 2010;112(3):773–83.

[98] Goldman SM, Quinlan PJ, Ross GW, et al. Solvent exposures and Parkinson disease risk in twins. Ann Neurol 2012;71(6):776–84.

[99] Feldman RG, Mayer RM, Taub A. Evidence for peripheral neurotoxic effect of trichloroethylene. Neurology 1970;20(6):599–606.

[100] Canesi M, Perbellini L, Maestri L, et al. Poor metabolization of n-hexane in Parkinson's disease. J Neurol 2003;250(5):556–60.

[101] Lam HA, Wu N, Cely I, et al. Elevated tonic extracellular dopamine concentration and altered dopamine modulation of synaptic activity precede dopamine loss in the striatum of mice overexpressing human alpha-synuclein. J Neurosci Res 2011;89(7):1091–102.

[102] Gautier CA, Kitada T, Shen J. Loss of PINK1 causes mitochondrial functional defects and increased sensitivity to oxidative stress. Proc Natl Acad Sci USA 2008;105(32):11364–9.

[103] Kim RH, Smith PD, Aleyasin H, et al. Hypersensitivity of DJ-1-deficient mice to 1-methyl-4-phenyl-1,2,3,6-tetrahydropyrindine (MPTP) and oxidative stress. Proc Natl Acad Sci USA 2005;102(14):5215–20.

[104] Welchko RM, Leveque XT, Dunbar GL. Genetic rat models of Parkinson's disease. Parkinsons Dis 2012;2012:128356.

[105] Richardson JR, Caudle WM, Guillot TS, et al. Obligatory role for complex I inhibition in the dopaminergic neurotoxicity of 1-methyl-4-phenyl-1,2,3,6-tetrahydropyridine (MPTP). Toxicol Sci 2007;95(1):196–204.

[106] Morgan WW, Richardson AG, Nelson JF. Dietary restriction does not protect the nigrostriatal dopaminergic pathway of older animals from low-dose MPTP-induced neurotoxicity. J Gerontol A Biol Sci Med Sci 2003;58(5):B394–9.

[107] Przedborski S, Jackson-Lewis V, Djaldetti R, et al. The parkinsonian toxin MPTP: action and mechanism. Restor Neurol Neurosci 2000;16(2):135–42.

[108] Yasuda Y, Shimoda T, Uno K, et al. The effects of MPTP on the activation of microglia/astrocytes and cytokine/chemokine levels in different mice strains. J Neuroimmunol 2008;204(1–2):43–51.

[109] Fornai F, Schluter OM, Lenzi P, et al. Parkinson-like syndrome induced by continuous MPTP infusion: convergent roles of the ubiquitin-proteasome system and alpha-synuclein. Proc Natl Acad Sci USA 2005;102(9):3413–8.

[110] Heneka MT, Nadrigny F, Regen T, et al. Locus ceruleus controls Alzheimer's disease pathology by modulating microglial functions through norepinephrine. Proc Natl Acad Sci USA 2010;107(13):6058–63.

[111] Henry B, Crossman AR, Brotchie JM. Characterization of a rodent model in which to investigate the molecular and cellular mechanisms underlying the pathophysiology of L-dopa-induced dyskinesia. Adv Neurol 1998;78:53–61.

[112] Jacobsson SE, Jonsson S, Lindberg C, Svensson LA. Determination of terbutaline in plasma by gas chromatography chemical ionization mass spectrometry. Biomed Mass Spectrom 1980;7(6):265–8.

[113] Alvarez-Fischer D, Henze C, Strenzke C, et al. Characterization of the striatal 6-OHDA model of Parkinson's disease in wild type and alpha-synuclein-deleted mice. Exp Neurol 2008;210(1):182–93.

[114] da Conceicao FS, Ngo-Abdalla S, Houzel JC, Rehen SK. Murine model for Parkinson's disease: from 6-OH dopamine lesion to behavioral test. J Vis Exp 2010;(35).

[115] Heikkila RE, Nicklas WJ, Vyas I, Duvoisin RC. Dopaminergic toxicity of rotenone and the 1-methyl-4-phenylpyridinium ion after their stereotaxic administration to rats: implication for the mechanism of 1-methyl-4-phenyl-1,2,3,6-tetrahydropyridine toxicity. Neurosci Lett 1985;62(3):389–94.

[116] Zhu C, Vourc'h P, Fernagut PO, et al. Variable effects of chronic subcutaneous administration of rotenone on striatal histology. J Comp Neurol 2004;478(4):418–26.

[117] Lapointe N, St-Hilaire M, Martinoli MG, et al. Rotenone induces non-specific central nervous system and systemic toxicity. FASEB J 2004;18(6):717–9.

[118] Cannon JR, Tapias V, Na HM, Honick AS, Drolet RE, Greenamyre JT. A highly reproducible rotenone model of Parkinson's disease. Neurobiol Dis 2009;34(2):279–90.

[119] McCormack AL, Thiruchelvam M, Manning-Bog AB, et al. Environmental risk factors and Parkinson's disease: selective degeneration of nigral dopaminergic neurons caused by the herbicide paraquat. Neurobiol Dis 2002;10(2):119–27.

[120] Thiruchelvam M, McCormack A, Richfield EK, et al. Age-related irreversible progressive nigrostriatal dopaminergic neurotoxicity in the paraquat and maneb model of the Parkinson's disease phenotype. Eur J Neurosci 2003;18(3):589–600.

[121] Thiruchelvam M, Richfield EK, Baggs RB, Tank AW, Cory-Slechta DA. The nigrostriatal dopaminergic system as a preferential target of repeated exposures to combined paraquat and maneb: implications for Parkinson's disease. J Neurosci 2000;20(24):9207–14.

[122] Smith P, Heath D. Paraquat. CRC Crit Rev Toxicol 1976;4(4):411–45.

[123] Cicchetti F, Lapointe N, Roberge-Tremblay A, et al. Systemic exposure to paraquat and maneb models early Parkinson's disease in young adult rats. Neurobiol Dis 2005;20(2):360–71.

[124] Herrera AJ, Castano A, Venero JL, Cano J, Machado A. The single intranigral injection of LPS as a new model for studying the selective effects of inflammatory reactions on dopaminergic system. Neurobiol Dis 2000;7(4):429–47.

[125] Iravani MM, Leung CC, Sadeghian M, Haddon CO, Rose S, Jenner P. The acute and the long-term effects of nigral lipopolysaccharide administration on dopaminergic dysfunction and glial cell activation. Eur J Neurosci 2005;22(2):317–30.

[126] Gao HM, Jiang J, Wilson B, Zhang W, Hong JS, Liu B. Microglial activation-mediated delayed and progressive degeneration of rat nigral dopaminergic neurons: relevance to Parkinson's disease. J Neurochem 2002;81(6):1285–97.

[127] Qin L, Wu X, Block ML, et al. Systemic LPS causes chronic neuroinflammation and progressive neurodegeneration. Glia 2007;55(5):453–62.

[128] Bargmann CI. Neurobiology of the Caenorhabditis elegans genome. Science 1998;282(5396):2028–33.

[129] Hamamichi S, Rivas RN, Knight AL, Cao S, Caldwell KA, Caldwell GA. Hypothesis-based RNAi screening identifies neuroprotective genes in a Parkinson's disease model. Proc Natl Acad Sci USA 2008;105(2):728–33.

[130] Lakso M, Vartiainen S, Moilanen AM, et al. Dopaminergic neuronal loss and motor deficits in Caenorhabditis elegans overexpressing human alpha-synuclein. J Neurochem 2003;86(1):165–72.

[131] Saha S, Guillily MD, Ferree A, et al. LRRK2 modulates vulnerability to mitochondrial dysfunction in Caenorhabditis elegans. J Neurosci 2009;29(29):9210–8.

[132] Nass R, Hall DH, Miller 3rd DM, Blakely RD. Neurotoxin-induced degeneration of dopamine neurons in Caenorhabditis elegans. Proc Natl Acad Sci USA 2002;99(5):3264–9.

[133] Settivari R, Levora J, Nass R. The divalent metal transporter homologues SMF-1/2 mediate dopamine neuron sensitivity in Caenorhabditis elegans models of manganism and Parkinson disease. J Biol Chem 2009;284(51):35758–68.

[134] Braungart E, Gerlach M, Riederer P, Baumeister R, Hoener MC. Caenorhabditis elegans MPP+ model of Parkinson's disease for high-throughput drug screenings. Neurodegener Dis 2004;1(4–5):175–83.

[135] Samann J, Hegermann J, von Gromoff E, Eimer S, Baumeister R, Schmidt E. Caenorhabditits elegans LRK-1 and PINK-1 act antagonistically in stress response and neurite outgrowth. J Biol Chem 2009;284(24):16482–91.

[136] Kamath RS, Ahringer J. Genome-wide RNAi screening in Caenorhabditis elegans. Methods 2003;30(4):313–21.

[137] Feany MB, Bender WW. A Drosophila model of Parkinson's disease. Nature 2000;404(6776):394–8.

[138] Coulom H, Birman S. Chronic exposure to rotenone models sporadic Parkinson's disease in Drosophila melanogaster. J Neurosci 2004;24(48):10993–8.

[139] Hirth F. Drosophila melanogaster in the study of human neurodegeneration. CNS Neurol Disord Drug Targets 2010;9(4):504–23.

[140] Inamdar AA, Masurekar P, Hossain M, Richardson JR, Bennett JW. Signaling pathways involved in 1-octen-3-ol-mediated neurotoxicity in Drosophila melanogaster: implication in Parkinson's disease. Neurotox Res 2014;25(2):183–91.

[141] Blandini F, Armentero MT. Animal models of Parkinson's disease. FEBS J 2012;279(7):1156–66.

[142] Sheng D, Qu D, Kwok KH, et al. Deletion of the WD40 domain of LRRK2 in Zebrafish causes Parkinsonism-like loss of neurons and locomotive defect. PLoS Genet 2010;6(4):e1000914.

[143] Anichtchik O, Diekmann H, Fleming A, Roach A, Goldsmith P, Rubinsztein DC. Loss of PINK1 function affects development and results in neurodegeneration in zebrafish. J Neurosci 2008;28(33):8199–207.

[144] Flinn L, Mortiboys H, Volkmann K, Koster RW, Ingham PW, Bandmann O. Complex I deficiency and dopaminergic neuronal cell loss in parkin-deficient zebrafish (Danio rerio). Brain 2009;132(Pt 6):1613–23.

[145] Panula P, Sallinen V, Sundvik M, et al. Modulatory neurotransmitter systems and behavior: towards zebrafish models of neurodegenerative diseases. Zebrafish 2006;3(2):235–47.

[146] Bortolotto JW, Cognato GP, Christoff RR, et al. Long-term exposure to paraquat alters behavioral parameters and dopamine levels in adult zebrafish (danio rerio). Zebrafish 2014;11(2):142–53.

[147] Prabhudesai S, Sinha S, Attar A, et al. A novel "molecular tweezer" inhibitor of alpha-synuclein neurotoxicity in vitro and in vivo. Neurotherapeutics 2012;9(2):464–76.

[148] Hancock DB, Martin ER, Mayhew GM, et al. Pesticide exposure and risk of Parkinson's disease: a family-based case-control study. BMC Neurol 2008;8:6.

[149] Rugbjerg K, Harris MA, Shen H, Marion SA, Tsui JK, Teschke K. Pesticide exposure and risk of Parkinson's disease–a population-based case-control study evaluating the potential for recall bias. Scand J Work Environ Health 2011;37(5):427–36.

[150] Ramanan VK, Saykin AJ. Pathways to neurodegeneration: mechanistic insights from GWAS in Alzheimer's disease, Parkinson's disease, and related disorders. Am J Neurodegener Dis 2013;2(3):145–75.

[151] Rakyan VK, Down TA, Balding DJ, Beck S. Epigenome-wide association studies for common human diseases. Nat Rev Genet 2011;12(8):529–41.

[152] Allpress JL, Curry RJ, Hanchette CL, Phillips MJ, Wilcosky TC. A GIS-based method for household recruitment in a prospective pesticide exposure study. Int J Health Geogr 2008;7:18.

Genetic Models of Parkinson's Disease: Behavior, Signaling, and Pathological Features

Amy M. Palubinsky, Britney N. Lizama-Manibusan, Dana Miller, BethAnn McLaughlin

INTRODUCTION

Parkinson's disease (PD) is the most common neurodegenerative movement disorder with a global prevalence of approximately 1% for individuals 65 years old and 5% in those over 85 [1,2]. The majority of PD cases are sporadic in nature, meaning their cause is unknown,

Environmental Factors in Neurodevelopmental and Neurodegenerative Disorders
http://dx.doi.org/10.1016/B978-0-12-800228-5.00014-5

whereas 10–20% of all cases are associated with genetic mutations [3,4]. Although the average onset of the disease is 70 years old, approximately 4% of patients develop PD well before the age of 50 [5].

Although a clear etiology is unknown, analysis of genetic forms of the disease in combination with epidemiological work increasingly suggests that a combination of genes and environmental factors contribute to PD. The goal of this work is to provide a brief overview of the pathological features of clinical PD and discuss how well the more than a dozen genetic mutants of PD-related genes capture these features. The chapter starts with a description of the tools used to introduce genetic mutations. Although we emphasize mouse models, we also discuss a few nonmammalian models that may prove particularly helpful for drug and environmental stress screening.

Clinical Features of PD

The main pathological hallmarks of PD are the progressive and selective degeneration of dopaminergic neurons in the substantia nigra pars compacta (SNpc), the formation of eosinophilic inclusions known as Lewy bodies and the presence of Lewy neurites, which are enriched in ubiquitin and α-synuclein [3,6–8]. Ultimately, profound molecular and cellular changes result in the onset of the cardinal clinical motor symptoms associated with PD, including resting tremor, bradykinesia, postural instability, and rigidity, as well as nonmotor symptoms, such as autonomic dysfunction, anxiety, depression, cognitive decline, and dementia. The neurochemical dysfunction associated with dopaminergic cell death is thought to lead to compensation within the basal ganglia (Figure 14.1) [3,6,9]. Behavioral models suggest that significant changes not only occur in the basal ganglia, but pathological and biochemical changes occur throughout the entire nervous system including the anterior olfactory nucleus, cortical areas, and mesocortex [2,7,10–13]. Recent reviews of the clinical features of PD have underscored the largely overlooked heterogeneity of cognitive decline and motor impairment within the clinical population [14]. The nonmotor symptoms, such as depression, cognitive impairment, sleep disturbances, gastrointenstinal distress, and orthostatic hypotension are often poorly addressed by current medical therapy, yet result in significant disability in patients [11].

Dopamine supplementation by means of levodopa (L-DOPA) remains the current gold standard treatment used to alleviate the major motor symptoms in PD, yet its effectiveness decreases with the progression of the disease and its continuous use has been associated with the development of debilitating motor complications including dyskinesias [3,9]. When medical therapies fail to control symptoms, the use of deep brain stimulation has become an increasingly popular tool that also shows promise as an early interventional technique in patients [15].

Animal models that recapitulate the pathological and symptomatic changes in PD are critical for identifying factors, such as environmental agents, that either cause PD or hasten its onset. Moreover, these models offer an opportunity to identify small-molecule therapies that can attenuate the progression of the disease or increase the efficacy of drugs and device-based therapies. Animal models also present a unique opportunity to evaluate presymptomatic carriers of gene mutations. Understanding how the course of symptom onset and endogenous adaptations occur prior to behavioral and cognitive manifestations of the disease are essential

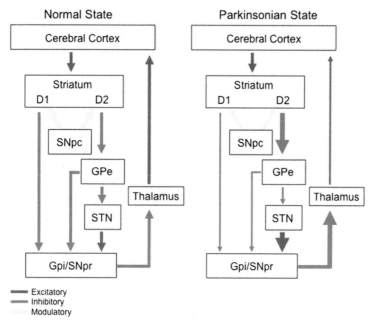

Normal State

Parkinsonian State

FIGURE 14.1 **Schematic representation of the basal ganglia and its associated pathological changes in Parkinson's disease (PD).** This schematic indicates the nuclei involved in the proper function of the basal ganglia circuitry, highlighting the excitatory (red), inhibitory (blue), or modulatory (yellow) effects of each nucleus in humans without PD (Normal state). Loss of dopaminergic neurons results in changes of both the indirect and direct pathways of the basal ganglia circuitry, resulting in the overinhibition of the thalamus and a subsequent reduction of excitatory stimulation of the cortex (Parkinsonian state). These changes ultimately result in the cardinal clinical motor features in patients diagnosed with PD.

to developing biomarkers for PD progression and drug therapies that promote neuronal stabilization and survival. Similarly, identifying individuals who are not only at high risk due to genetic predisposition, but who are also considered high risk due to occupational exposures to pesticides, heavy metals, traumatic brain injury, and other factors, will greatly improve our ability to diagnose patients earlier and develop the best combination of therapies to treat their individual disease [16,17].

Parkinson's Disease Genetics

In the last 30 years, mitochondrial dysfunction, redox imbalance, and changes in energetic tone have been recognized as critical factors in PD. The work of Drs Beal, Greenamyre, and Langston fueled research on a group of stressors that recapitulate many of the behavioral and pathological features of PD [18]. In the last decade, over a dozen genes and hundreds of unique mutations have been shown to cause PD [1,2] (Table 14.1).

Monogenic forms of Parkinsonism with a clear pattern of Mendelian inheritance have been the most accessible in terms of the identification of genetic causes of primary Parkinsonism and/or PD. These mutations occur in genes that are autosomal dominant in inheritance including

TABLE 14.1   Known genetic mutations resulting in Parkinson's disease or Parkinsonism

Gene	Locus	Position	Inheritance
SNCA	PARK1/4	4q21-q23	Autosomal dominant
Parkin	PARK2	6q25.2-q27	Autosomal recessive
Unknown	PARK3	2p13	Autosomal dominant
UCHL1	PARK5	4p13	Autosomal dominant
PINK1	PARK6	1p35-p36	Autosomal recessive
DJ-1	PARK7	1p36	Autosomal recessive
LRRK2 (Darda)	PARK8	12q12	Autosomal dominant
ATP13A2	PARK9	1p36	Autosomal recessive
Unknown	PARK10	1p32	Risk factor
GIGYF2	PARK11	2q36-q37	Autosomal dominant
Unknown	PARK12	Xq21-q25	Risk factor
Omi/HTRA2	PARK13	2p12	Autosomal dominant or risk factor
PLA2G6	PARK14	18q11	Autosomal recessive
FBX07	PARK15	22q12-q13	Autosomal recessive
Unknown	PARK16	1q32	Risk factor
Unknown	PARK17	16q11.2	Autosomal dominant
EIF4G1	PARK18	3q27.1	Autosomal dominant

α-synuclein and *leucine-rich repeat kinase 2 (LRRK2)* and those that are autosomal recessive such as *Parkin, DJ-1*, and *phosphatase and tensin homolog (PTEN)-inducible putative kinase 1(PINK1)* [1,3,6].

In addition to these (Table 14.1), other mutations in genes such as *β-glucocerebrosidase (GBA)* can change one's susceptibility to developing PD [19–22]. The *GBA* susceptibility variant population is particularly intriguing as patients with these mutations have also been found to have Gaucher's disease, the most common recessively inherited glycolipid storage disorder [20–26].

DESIRABLE CHARACTERISTICS OF AN ANIMAL MODEL FOR PD

One of the main advantages that mouse models have over other model systems is that they allow for the study of genetic mutations in the context of a fully functional neuronal network [3]. However, as the prevalence of PD increases with age, the ideal model system would have an age-dependent, progressive loss of dopaminergic neurons in the SNpc. The lifespan of a mouse limits the ability to capture the decades of accumulated biochemical signaling that contributes to PD onset that typically occurs in midlife [27].

Capturing the neuropathological features of PD has been a major obstacle in developing models of this disease. To have predictive validity as a screening tool, the desirable pathological features of PD models would include late-in-life loss of dopaminergic neurons, formation of Lewy bodies and Lewy neurites, motor deficits, and symptom alleviation through dopamine supplementation. Although we discuss neurochemical features of introducing unique genetic mutations, none of the animals generated thus far fully capture the substantial, selective, and progressive loss of dopaminergic neurons of the SNpc [3,8,28].

METHODS TO GENERATE GENETIC MODELS OF DISEASE

Knock-Out and Knock-In Mouse Models

The generation of knock-out mice involves inserting a gene (synthesized in the lab) into embryonic stem cells which disrupts the allele(s) of interest [29,30]. From a technical perspective, knock-out mice are most commonly generated by replacing an essential coding sequence with a drug-selective marker, such as the neomycin resistant (neo^r) gene [29]. This selection marker disrupts the synthesis of the protein encoded by the targeted gene when the antibiotic is present.

An alternate perturbation of recombinant technology is to "knock in" a new gene, placing it in the native gene site [29]. The gene of interest can be either a human ortholog or a lab-generated genetic mutant. This technique provides the additional biological assurance as the gene remains under the control of its normal transcriptional promoters [31–34].

Unfortunately, a high level of lethality often occurs when knock-out and knock-in mice are generated [30,32], underscoring the biological stress placed on the host organism.

Lox and Flox

To minimize off-target effects of genetic manipulation, conditional knock-out and knock-in mouse models have been developed. With this technology, a gene of interest can be deleted or mutated in a subset of tissues or cells rather than altering expression of the gene of interest throughout the entire body. An additional advantage of this technology is that it offers the potential to limit expression to certain developmental periods [29,30,32,34,35].

The most widely used systems for generating conditional knock-out mice is Cre recombinase/*loxP* technology [30,35] wherein two 34 base pair sequences, known as *loxP* sites, are placed around an exon of the gene of interest. The mouse that arises from this manipulation is referred to as "floxed" and a floxed mouse carries the gene of interest that can be manipulated when bred with a Cre mouse (Figure 14.3).

Cre animals have been bred to express a gene encoding the DNA recombinase Cre [30]. Matings of a "Cre" and "floxed" animal therefore results in progeny that have excised the exon flanked by the *loxP* sites [30]. This technology can be further refined using specific promoters that allow the expression of Cre recombinase in a defined subset of tissues or cell types [29,30,35] (Figure 14.3). The expression of the gene mutation can also be controlled temporally using a tamoxifen-inducible Cre system, whereby transcription of Cre recombinase is activated only when tamoxifen is administered [29,30].

Generation of conditional knock-out mice is generally more successful than conditional knock in [32]. The goal of knock-in mice is to place a new gene under the control of an endogenous promoter to assure that the mutated gene is expressed in the same tissue and at the same time as the wild-type allele [32].

Three strategies have been used to generate conditional knock-in mice. The first strategy, known as the "heterozygous" method, involves flanking the wild-type allele by *loxP* sites, whereas the mutated sequence is located on the second allele not containing *loxP* sites. In the presence of Cre recombinase, the floxed wild-type allele is excised allowing only the mutated sequence of the other allele to be expressed. The disadvantage of this strategy is that the resulting offspring have a mixed genetic phenotype. One allele is expressed in regions where Cre recombinase is promoted, but both alleles are expressed in tissues lacking Cre recombinase. Further, having only one functional allele in Cre-positive regions likely leads to substantially less protein being expressed than would normally occur in an animal with both functional alleles [34].

In conditions in which a regional specificity is needed, a "minigene" approach allows insertion of a floxed mini-DNA cassette into the gene of interest assuring the expression of exons upstream and under the control of the endogenous promoter. Upon recombination with Cre, the minigene cassette is excised, allowing for the expression of the mutant sequence by means of the endogenous promoter in a highly tissue-specific manner [34].

Finally, a newer perturbation of knock-in technology is the use of "inversion" in which the sequence and orientation of the *loxP* sites are modified allowing for the insertion of the mutated gene antiparallel to the wild-type gene. In the absence of Cre recombinase, the wild-type gene will be expressed, whereas in the presence of Cre recombinase there would be transcriptional drive only for the mutated gene [34].

Transgenic Mouse Models

In contrast to using knock-out mice for the study of autosomal recessive genetic mutations, autosomal dominant mutations have predominately been studied using transgenic mice. This technology drives expression of a protein at levels that likely would never occur even under pathophysiological conditions. However, one advantage of this approach is that it presents a biological stress that can be a useful model to promote neurodegeneration. By presenting the central nervous system (CNS) with a new genetic load, researchers can identify phenotypes that often do not develop for decades in humans. From a technical standpoint, it is not possible to control the chromosomal location where the introduced transgene will integrate or to control the absolute number of transgenes [35]. As a result, aberrant and unexpected phenotypes can occur in response to random placement and/or multiple copy-number integrations of the transgene [35].

To control for spatial and temporal expression of transgenes, tissue-specific promoters can be used to drive the expression of a transgene in a region of interest [35,36].

Although several different classes of eukaryotic promoters have been developed, of predominant interest for the study of PD is the 9-kb tyrosine hydroxylase (TH) promoter, as this promoter drives the expression of a transgene specifically in catecholaminergic neurons [37,38].

From a strategic perspective, transgenic mice are generated by means of pronuclear microinjections [36]. The gene of interest is generally placed downstream of a promoter of choice, which will govern the temporal and spatial expression patterns [36]. Other structural components of the transgene include a Kozak sequence for ribosomal recognition of the ATG start codon, as well as introns which are placed at the 5′ and 3′ end of the transgene, thereby supporting gene splicing and hence stabilization of mature mRNA [36]. To assure transgene expression, a termination sequence along with a poly (A) tail is placed at the 3′ end of the transgene [36].

Like knock-out mice, the expression of a transgene from embryonic stages on can result in the initiation of physiological compensatory mechanisms that can mask the effect of the overexpression of the transgene [35]. To prevent physiological adaptation, tetracycline-controlled inducible-gene expression systems (Tet) were developed in the early 1990s by Gossen and colleagues [39–41] to generate conditional transgenic mice.

Tet-Inducible Systems

Tet can be used to spatially and temporally control gene expression in a reversible manner [39,42]. The main elements required for the function of a Tet system are the tetracycline-controlled transactivator (tTA) and the tet-responsive promoter (tetP). The tTA gene is generated by fusing the transposon Tn10 from *Escherichia coli*, which encodes the tetracycline repressor (tetR), to the activating domain of the virion protein 16 of the herpes simplex virus [39].

tTA binds to promoters with tetracycline operator (tetO) sequences and fusion ultimately results in transcriptional activation of the transgene located downstream of tetO. For tet-OFF systems, even low concentrations of tetracycline or a derivative (doxycycline) inhibits binding of tTA to tetO, thereby preventing activation and transgene expression [39]. In contrast, tetracycline drives tet-ON systems [41] (Figure 14.2). Tet systems are useful in studying PD as combination with prion protein (PrP) and/or Ca^{2+}/calmodulin-dependent protein kinase II (CamKII) promoters result in regionally selective CNS mouse lines [43]. For example, the CamKII promoter can drive expression of the transgene specifically in forebrain neurons and the PrP promoter in both neurons and glia of the brain stem and cerebellum [43].

BACs and YACs

One obstacle in generating transgenic mice is the failure rate associated with attempting to insert larger DNA constructs into the mouse genome [44,45]. To overcome this limitation, bacterial artificial chromosome (BAC), as well as yeast artificial chromosome (YAC) vectors have been developed to drive the expression of 100–1000 kb vectors [44,45]. Given that the LRRK2 gene is so large, these systems have been especially helpful in designing relevant constructs [45].

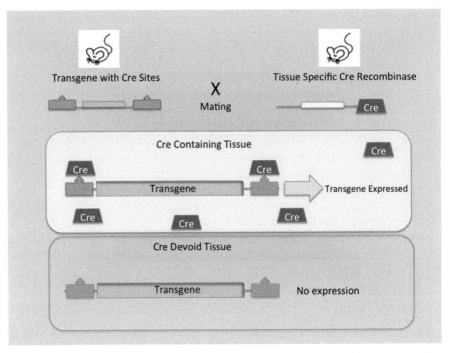

FIGURE 14.2 **Schematic depicting the generation of conditional knock-out mice.** Mice carrying the gene encoding the enzyme Cre recombinase (Cre) under the control of a tissue-specific promoter are crossed to mice in which the target gene has been flanked by *loxP* sites. The target gene will be removed in a tissue-specific and time-specific manner in offspring that express both the floxed target gene as well as Cre recombinase, thereby resulting in the generation of a conditional knock-out mouse.

GENETIC MOUSE MODELS FOR PD

Autosomal Dominant Mutations

The first gene that was linked to familial PD was the *SNCA* gene, which encodes the protein α-synuclein. Polymeropolous and colleagues identified this gene in 1996 when studying a large Italian family in which 60 individuals over five generations were diagnosed with PD [1,3,46,47]. The mean age of onset in this cohort was 46 years old [48], and mutations were found by linkage studies in α-synuclein located on chromosome 4q21-q23, within a genetic locus referred to as PARK1 [6,47–49] (Table 14.1).

Alpha-synuclein is a highly conserved, 140 amino acid (14 kDa) phosphoprotein expressed throughout the adult mammalian CNS. Expression is noted particularly in brain regions that exhibit extensive synaptic plasticity, where it is found predominantly at presynaptic terminals and associated with membranous structures, including synaptic vesicles through its amino-terminal repeats [1,49,50].

Alpha-synuclein plays a role in the assembly of the *N*-ethylmaleimide-sensitive factor attachment protein receptor (SNARE) in which it is thought to form a complex with the

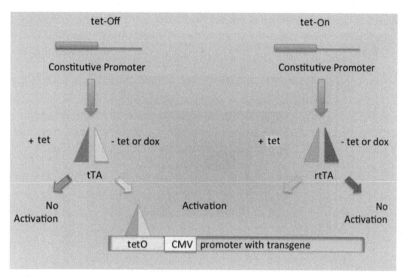

FIGURE 14.3 **Schematic representing the tet-ON/OFF system.** In the tet-OFF system, expression of tTA results in its binding to pTet, thereby driving the expression of the gene of interest. Addition of doxycycline (Dox) inhibits the expression of the gene of interest. In contrast, in the tet-ON system, binding of rtTA only results to the initiation of gene expression in the presence of Dox.

SNARE-protein synaptobrevin-2/vesicle-associated membrane protein [51]. Chandra and colleagues found that deletion of any of the three α-synuclein isoforms that exist induces age-dependent neurological impairments and decreases assembly of functional SNARE complexes [51–53]. Taken together, these results support a role for α-synuclein in synaptic vesicle maturation and stability during aging [51,52].

Three discrete point mutations in the *SNCA* gene have been linked to PD. Outright mutations in the α-synuclein gene are rare, but when present, they are autosomal dominant in inheritance and completely penetrant [1,3,6,49]. The first missense mutation identified in α-synuclein was the A53T mutation, followed by the discovery of the A30P and the E46K mutations [1,3,47,54,55]. Simply having duplication of even a wild-type α-synuclein gene increases the risk an individual will develop PD [27,56]. Screens of individuals who have been diagnosed with PD but do not have a family history of the disease have failed to identify any *de novo* mutations that have arisen in the α-synuclein gene [49]. Both missense mutations in α-synuclein as well as multiplications of the SNCA gene result in toxic gain-of-function scenarios [57]. This suggests an intricate relationship exists between gene, protein, and disease [1,3].

Aberrant posttranslational modifications of α-synuclein including phosphorylation and nitration have been noted in postmortem analyses of the CNS in PD patients [57]. Given α-synuclein's function as a phosphoprotein, studies aimed at identifying the phosphorylation sites demonstrate that serine 129 (Ser129) is predominantly phosphorylated in patients with PD [57–59]. To date, kinases, including the G-protein-coupled receptor kinase 2 (Gprk2), as well as casein kinase (CK) 1 and CK 2 [59,60], have all been shown to phosphorylate α-synuclein; however, the precise role of phosphorylation at Ser129 on α-synuclein remains controversial [60].

Analyses in a transgenic *Drosophila melanogaster* model harboring missense mutations in α-synuclein result in increases in α-synuclein phosphorylation levels [57]. Currently, a leading hypothesis regarding this model is that phosphorylation results in the maintenance of α-synuclein in its soluble state [57]. Patients with point mutations in *SNCA* present with earlier motor symptoms, severe autonomic dysfunction, and subsequent dementia and psychosis [1,6,59–61]. These patients also have pronounced dopaminergic degeneration and Lewy body formation [48]. The progression of the disease is generally accelerated in patients with *SNCA* duplications, and even faster progression and severity is noted in patients with triplications of the *SNCA* gene [6]. Moreover, postmortem analyses demonstrate much broader brain pathology in patients with α-synuclein multiplications [6].

Alpha-synuclein natively exists as an elongated and unstructured protein in solution, yet it undergoes conformational alterations into an alpha helix-rich protein upon membranous structure association [1,6,49]. When mutated or expressed at high levels, α-synuclein is highly prone to form β-sheet-rich insoluble structures such as oligomers, protofibrils, fibrils, and filaments. Protofibrils and fibrils are currently believed to be the most neurotoxic forms of α-synuclein [1,3,49].

In addition to mutations and multiplications of α-synuclein, other factors including oxidative and nitrosylative stress have been associated with the aggregation of α-synuclein [1,3]. Increased levels of oxidative stress as well as tyrosine nitration of α-synuclein have been reported in postmortem tissue samples of PD patients [1,62] suggesting that increased levels of reactive oxygen and nitrogen species can damage α-synuclein and further promote its aggregation. Moreover, decreased proteasome activity in PD patients as well as in numerous model systems of the disease [63] also suggest exacerbation of the accumulation of both mutated and posttranslationally modified α-synuclein within neurons, ultimately leading to abnormal mitochondrial function and subsequent neuronal cell death [63].

Failure of the lysosomal system is also implicated in the progression of PD [64,65]. Meredith and colleagues describe an intricate relationship between lysosomal dysfunction and α-synuclein aggregation [66]. The authors postulate that the accumulation of aggregated α-synuclein in dopaminergic neurons of mice treated with the neurotoxin 1-methyl-4-phenyl-1,2,3,6-tetrahydropyridine (MPTP) may not only be due to the buildup of insoluble proteins, but also to the dysfunction of protein degradation machineries [66].

Additionally, inhibiting the lysosome results in the aggregation of insoluble α-synuclein [67], and studies have demonstrated that mutations in the lysosomal membrane protein ATP13A2 (PARK9) increase both α-synuclein accumulation and neuronal toxicity [68,69]. Taken together, these data support the hypothesis that cellular inability to degrade proteins results in dopaminergic degeneration in PD and that therapeutics that can increase the function of both the proteasome and the lysosome may be neuroprotective [65].

Desplats and colleagues performed an intricate study demonstrating that α-synuclein is transmitted to neighboring neurons by means of endocytosis [70]. This neuron-to-neuron transmission of α-synuclein was further supported by experiments in which normal cells grafted into PD brains develop Lewy pathology [70]. These data are a sobering reminder of the limitations of stem cell therapeutics—suggesting that they may have limited long-term utility in individuals in which pathology is still progressing.

Alpha-synuclein fibrils are the main component found in Lewy bodies and Lewy neurites [1]. Lewy bodies and Lewy neurites have been detected predominantly in the soma and the neurites of both dopaminergic and nondopaminergic neurons of brain regions including the cortex and the brain stem [1]. The aggregation of α-synuclein into Lewy bodies and Lewy neurites in other neurodegenerative diseases (commonly referred to as α-synucleinopathies), suggest a prominent role of α-synuclein and its associated mutations in a large number of disorders [1,3].

Although neither *D. melanogaster* nor *C. elegans* express α-synuclein, mutations in these genes have been introduced into these model organisms resulting in the progressive loss of dopaminergic neurons, the formation of Lewy body-like structures, and motor deficits [3].

Alpha-synuclein knock-out mice are viable, develop normally, and have no loss of dopaminergic neurons [50,53]. Although these animals lack changes in short- and long-term synaptic plasticity, synaptic vesicle recycling, and neurotransmitter release [3,53], other groups have reported changes in dopaminergic tone evidenced by lower levels of striatal dopamine following exposure to multiple electrical pulse stimuli as compared to wild-type mice [50]. αβγ-Synuclein triple knock out, however, is sufficient to alter synaptic structure and function, and to induce neuronal dysfunction and cell death [71].

To date, several different promoters have been used to drive the expression of the human α-synuclein transgene in different regions of the brain [72]. The first report of the generation of an α-synuclein transgenic mouse was in 2000 by Masliah and colleagues [73]. In these mice, the α-synuclein transgene is under the control of the human platelet-derived growth factor-β (PDGF-β) promoter [72,73]. Histopathological analysis demonstrates the presence of α-synuclein-positive inclusion bodies in several brain regions, including midbrain, neocortex, and olfactory bulb. However, the morphology of the inclusions found in these mice is not comparable to that of Lewy bodies found in PD patients [72,73]. Moreover, high expression levels of α-synuclein under control of the PDGF-β promoter results in reduced levels and activity of TH in the striatum of these animals [73]. Unfortunately, no changes in TH levels were present in the substantia nigra of these mice [73]. From a behavioral perspective, these transgenic α-synuclein mice have impaired rotorod performance as older adults (12 months) [72,73].

In addition to the generation of transgenic mice, which overexpress human wild-type α-synuclein [73], transgenic mice overexpressing mutant α-synuclein have also been created. Placing α-synuclein under the control of the murine Thy1 promoter, researchers report a robust decline in motor function [74]. Neuropathological changes, including α-synuclein- and ubiquitin-positive inclusion bodies, are more prominent in spinal cord than midbrain [74], whereas degeneration of dopaminergic neurons of the substantia nigra is absent [72,74]. Further biochemical analyses of these transgenic mice reveal relatively low expression levels of the A53T α-synuclein protein in the ventral midbrain, suggesting that the lack of nigral pathology may be due to inadequate amounts of A53T α-synuclein overexpression in this region critical to maintaining dopaminergic tone [74]. Transgenic mice expressing either wild-type α-synuclein or the A30P α-synuclein mutant under the control of the Thy1 promoter do not have motor dysfunctions, despite the fact that α-synuclein accumulates in multiple brain regions including the substantia nigra, neocortex, cerebellum, and brain stem [72], implying that the A53T finding was unique in being able to promote formation of cytoplasmic filamentous inclusion bodies [75].

Another transgenic mouse line generated early on following the identification of α-synuclein's role in the onset of PD is a line in which the transgene has been placed under control of the TH promoter [76]. Given that TH is the rate-limiting enzyme in the synthesis of catecholamines, it was believed that this promoter would selectively drive the expression of the transgene in catecholamergic neurons, including the dopaminergic neurons of the SNpc [76]. Unfortunately, a robust effect of α-synuclein overexpression on the selective loss of dopaminergic neurons is only detected in mice in which either part of the α-synuclein protein is expressed (amino acids 1–130) or in which the α-synuclein protein has multiple mutations [76–78]. Although a progressive loss of dopaminergic neurons occurs in mice when α-synuclein is multiply mutated, this is not the typical progression in PD patients [76]. Mice that overexpress amino acids 1–130 of α-synuclein do not have dopaminergic cell death but rather have decreased levels of striatal dopamine as well as its metabolite homovanillic acid [76].

Using the rat TH promoter to drive α-synuclein expression results in a robust increase in wild-type as well as mutated forms of α-synuclein in the dopaminergic neurons of the substantia nigra without dopaminergic degeneration [76]. Although such outcomes are disappointing, they do suggest that the sole overexpression of α-synuclein is not sufficient to induce the cardinal pathological and symptomatic changes observed in PD patients.

Transgenic mouse lines used to study the role of α-synuclein in the onset and progression of PD that have been the most successful in recapitulating the symptoms and neuropathological hallmarks of the disease in patients are those in which human wild-type or mutated A53T α-synuclein are under the control of the mouse prion promoter [3,72,75,79]. This promoter drives the expression of the transgene at high levels in most neurons of the CNS [75,79]. Mice overexpressing A53T mutant, but not wild-type α-synuclein, present with complex and severe motor impairments at 8 months of age, without loss of dopaminergic neurons [75]. These motor features eventually give way to paralysis and death [75]. Accumulation of α-synuclein- and ubiquitin-positive inclusion bodies in neuronal cytosol are comparable to those found in PD patients and occur as these animals age [75]. Overexpression of wild-type α-synuclein under the control of the mouse prion promoter does not, however, cause any cell death, further supporting that the point mutation, rather than the genetic manipulation of the animals, results in degeneration [75].

As all transgenic animals with A53T mutations in α-synuclein form filamentous inclusion bodies, these animals are highly valuable models to study α-synuclein neurotoxicity [79]. Indeed, this model has been incredibly useful in clarifying the normal role of α-synuclein— demonstrating its importance in the assembly of a functional SNARE complex, and thereby assuring synaptic integrity and health during aging [51–53,80].

A second gene associated with autosomal dominant PD was identified in 2002 by Funayama and colleagues during a genome-wide linkage analysis of a large Japanese family [81]. The average age of onset of the disease in this cohort was 51 years [81]. The locus of the gene was mapped to the 13.6 cM region of chromosome 12p11.23-q13.11 and termed *PARK8* [81]. Postmortem analyses of brain tissue samples of these patients revealed degeneration of dopaminergic neurons in the SNpc, without Lewy bodies and Lewy neurites [81]. Further neuropathological analyses of postmortem brain samples from other cohorts with *PARK8* mutations from England, Spain, and western Nebraska demonstrated a wide range of pathological changes including the formation and accumulation of Lewy bodies [82]. These data suggest that not only are the pathological changes associated with the *PARK8* gene dependent on the population under investigation, but also that different mutations may result in varying pathological outcomes.

In 2004, a missense mutation that segregated with *PARK8* in families with PD in England and Spain was published [82]. The protein encoded by *PARK8* was termed dardarin, which is derived from the Basque word *dardara*, meaning tremor [82]. Dardarin is distributed throughout the body, including the brain; however, the highest expression levels are in peripheral tissues [82]. Structural analyses found that dardarin contains a kinase domain, a leucine-rich repeat domain, an RAS domain, and a WD40 domain suggesting that it may function as a cytoplasmic kinase resembling LRRK1 [82].

In the same year, Zimprich and colleagues identified another protein encoded by the *PARK8*. A 2527 amino acid (285 kDa), multifunctional protein [5,83,84] termed leucine-rich repeat kinase 2 or LRRK2 [83]. The authors further identified six disease-segregating mutations in LRRK2 including Y1699C, R1441C, I1122V, and I2020T [83]. LRRK2 is a large, multidomain protein composed of several functional domains including an ankyrin repeat region, a leucine-rich repeat domain, a Roc GTPase domain, a C-terminal of Ras domain, a kinase domain, and a WD40 domain [85–87]. LRRK2 is widely expressed in the CNS, including neurons of the substantia nigra, cortex, striatum, hippocampus, and cerebellum, in which it associates with membranes and vesicular structures [86–88]. The physiological role of LRRK2 remains largely unknown [87,89,90]. To date, over 40 LRRK2 mutations have been identified, most of which are due to single amino acid substitutions [89]. The most prevalent LRRK2 mutation, which has been linked to the majority of sporadic and familial cases of PD, is an amino acid substitution whereby a highly conserved glycine (Gly^{2019}) residue, located within the kinase domain, is replaced by a serine residue [87,89,91]. This G2019S mutation is associated with an increase in kinase activity [86,91].

Patients with LRRK2 mutations present late in life with symptoms that are indistinguishable from sporadic PD [6]. These patients have also been found to be L-DOPA resistant [6]. Postmortem analyses demonstrate pathological hallmarks including loss of dopaminergic neurons in the SNpc, as well as the accumulation of ubiquitin- and α-synuclein-positive Lewy bodies and Lewy neurites [6]. However, mutations in LRRK2 are not always associated with Lewy bodies [6].

Several model systems have been developed to study LRRK2. LRRK2 over expression in *D. melanogaster* or *C. elegans* results in loss of dopaminergic neurons as well as modified dopamine (DA)-associated behaviors; however, the α-synuclein gene is not inherently expressed in these model systems, therefore formation of Lewy bodies and Lewy neurites is not noted. Transgenic mice that express mutant forms of LRKK2, exhibit progressive degeneration of dopaminergic neurons of the SNpc and motor dysfunctions that are parkinsonian in nature [92,93]. A recent study by Ramonet and colleagues describes mice expressing wild-type LRRK2, as well as G2019S, and R1441C mutants under the control of the cytomegalovirus-enhanced human platelet-derived growth factor β-chain (CMVE–PDGFβ) promoter [93]. The advantage of this promoter is that it selectively drives the expression of the transgene in neurons throughout the brain, including dopaminergic neurons of the substantia nigra. LRRK2 expression increases 3–5-fold globally and 2.7-fold in TH-positive neurons of the SNpc [93]. In contrast to the majority of all other LRRK2 transgenic mice, the G2019S mutant mice described by Ramonet and colleagues have age-dependent degeneration of the dopaminergic neurons in the SNpc with associated changes in mitochondrial function and autophagy [93].

Recently, Chen and colleagues published an elegant study wherein LRRK2 transgenic mice display progressive degeneration of dopaminergic neurons, as well as other parkinsonian-like phenotypes [92]. The authors placed a Human influenza hemagglutinin (HA)-tagged transgene, either wild-type (WT) or G2019S, under the transcriptional control of a CMV-enhanced/PDGF-β promoter. Immunohistochemical analyses revealed a significant decrease in TH-positive neurons in 12-month-old G2019S mice when compared to age-matched WT animals. Because no degeneration is observed in 8–9-month-old mice, the degeneration of dopaminergic neurons was clearly progressive in nature and more akin to that found in PD patients [92]. Unfortunately, further analyses demonstrated that these mice did not form Lewy bodies or Lewy neurites, even when aged up to 16 months [92]. The authors did, however, observe progressive behavioral deficits in G2019S LRRK2 mice compared to WT animals, and administration of dopamine supplementation ameliorated hypoactivity [92].

LRRK2 transgenic mice have also been generated using BAC technology [3]. The main advantage of this technique being that BACs can hold up a transgene of a size range from 100 to 300 kb, giving them the capacity to introduce even large genes such as LRRK2 into the mouse genome [45]. BAC technology has therefore been used to generate WT, G2019S, and R1441G LRRK2 transgenic mice [3,45,94]. BAC-LRRK2 WT transgenic mice demonstrate transgene expression throughout the entire brain, but with higher expression levels compared to those of endogenous murine LRRK2 [45]. These mice have no behavioral dysfunction or dopaminergic cell death [3,45]. However, in a recent study by Li and colleagues, it was found that BAC-LRRK2-WT mice had elevated striatal dopamine release, whereas overall dopamine uptake and tissue content remained unaltered, suggesting a role for LRRK2 in striatal dopamine transmission [95]. BAC-LRRK2 G2019S transgenic mice have a decrease in evoked dopamine release in aged animals, without locomotor deficits or loss of dopaminergic neurons [45,95]. Finally, BAC-LRRK2 R1441G mice express 5–10-fold higher transgene than endogenous LRRK2. Progressive locomotor deficits are reported in older mice developing to near immobility at 10–12 months, symptoms that are ameliorated with L-DOPA [45,95]. Neuropathologically, transgenic BAC-LRRK2 R1441G mice demonstrate axonal pathology as well as impaired dopamine release [45,95]. Although the number of dopaminergic neurons in the SNpc remain the same in aged BAC-LRRK2 R1441G as compared to control mice, the overall average cell body size and number of TH-positive neurites is markedly reduced [45,95].

Conditional LRRK2 transgenic mice have been generated wherein the expression of human WT or mutant G2019S LRRK2 is regulated by tetracycline [94,96]. Transgenic lines from these animals are viable, develop normally, and have an 8–16-fold increase in transgene expression levels [84,96]. LRRK2 is modestly increased in the dopaminergic neurons of the substantia nigra of these animals using a calcium/calmodulin-dependent protein kinase II (CamKII) promoter, however, no PD-associated pathology is reported [84,96]. The LRRK2 G2019S animals have progressive and age-dependent locomotor impairment [94]. These motor deficits can be ameliorated upon dopamine supplementation, demonstrating that conditional LRRK2 transgenic mice may, in part, recapitulate PD [94].

Lin and colleagues demonstrate that the co-expression of WT or G2019S LRRK2 with an A53T α-synuclein mutant robustly enhances and accelerates the progression of α-synuclein-mediated neurotoxicity, and that this neurotoxicity can be inhibited when LRRK2 expression is reduced [96]. These results imply a synergistic effect of LRRK2 and α-synuclein on the neuropathology of PD and suggest that inhibition of LRRK2 expression may present a potential avenue for the development of neurotherapeutics for PD [96]. Moreover, these data suggest that the interaction of multiple factors may be required for the development of this disease, and therapeutics will likely have to target several pathways to be effective long term.

In keeping with the theory that LRRK2 and α-synuclein interact in a common pathway that ultimately leads to the degeneration of dopaminergic neurons of the substantia nigra, Tong and colleagues published an intricate study demonstrating robust accumulation and aggregation of α-synuclein, as well as other ubiquitinated proteins, in the kidneys but not the brain of LRRK2 knock-out mice [97]. Additionally, the authors found an impairment of the lysosomal system, suggesting that LRRK2 may play a regulatory role in protein homeostasis and that loss of LRRK2 may lead to α-synuclein aggregation and ultimately loss of dopaminergic neurons in the SNpc [97]. LRRK2 knock-out mice are viable, develop normally, and live into adulthood, that LRRK2 does not play a major role in development [3,94,98]. Moreover, susceptibility to MPTP-induced neurotoxicity does not differ between LRRK2 knock-out and WT mice, indicating that this kinase may only play a minor role in governing dopaminergic survival [98].

In addition to knock-out mice, knock-in mice carrying the R1441C mutation have also been generated [99]. In these mice, no apparent phenotypic changes or degeneration of dopaminergic neurons in the SNpc are observed. These mice, however, do have reduced amphetamine-stimulated locomotion and impaired D2 receptor function [99]. Although these mice do not present with the classical parkinsonian pathological and phonotypical changes, they do represent a valuable tool to better understand the pathogenic mechanism associated with R1441C mutations, such as impairments in dopamine neurotransmission, as well as dopamine receptor function [99].

It remains to be determined why LRRK2 transgenic mice do not have consistent neuropathological and phenotypic changes [3]. Conditional LRRK2 transgenic mice, whereby the expression of the transgene can be prevented throughout development by means of the administration of doxycycline and then turned on during the animal's adulthood, may prove to be a beneficial alternative to study the effects of LRRK2 overexpression on dopaminergic survival.

AUTOSOMAL RECESSIVE MUTATIONS

Parkin Gene

The first genes associated with autosomal-recessive mutations were identified by Kitada and colleagues in 1998. Screening families in Japan, this group identified mutations in the *parkin* or *PARK2* gene associated with a juvenile-onset variant of PD [100] was mapped to region 6q25.2-q27 on the long arm of chromosome 6 spans 1.3 Mb, making it one of the largest genes in the genome. The *parkin* gene encodes a 465 amino acid protein, which is widely expressed in human tissue, including the brain, heart, skeletal muscle, and testis [100].

Patients with mutations in *parkin* present with the classical motor symptoms of PD, which can be ameliorated by dopamine supplementation therapy [100]. The progression of the disease is slow and notable in that it lacks cognitive decline [101]. Postmortem analyses of PD patients harboring *parkin* mutations demonstrate the characteristic loss of dopaminergic neurons of the SNpc, as well as those in the locus coeruleus [100]. Interestingly, autosomal recessive juvenile PD (AR-JP) patients lack Lewy bodies or Lewy neurites [100].

Structural analyses of the Parkin protein reveal the presence of a ubiquitin-like domain at the amino-terminal as well as a cysteine-rich motif composed of two really interesting new gene (RING)-finger domains surrounding an in-between RING (IBR) domain [102,103]. These characteristics place Parkin in a family of proteins known as RING-finger containing E3 ubiquitin ligases. Such proteins are involved in the tagging of damaged, misfolded, or oxidized proteins with ubiquitin, marking them for subsequent proteasomal degradation [104].

The *parkin* gene is highly prone to chromosomal rearrangements. Patients diagnosed with autosomal recessive PD often demonstrate deletion of one or multiple exons, leading to loss of Parkin's E3 ubiquitin ligase activity [100,102,104]. In addition to exon rearrangements, numerous missense mutations have also been identified, all of which abolish Parkin's ability to ubiquitinate client proteins [102,105–107]. Due to this high propensity for mutations, it is now suggested that *parkin* is indeed one of the major causes of autosomal recessive PD [108].

The loss of Parkin's E3 ligase function implies that AR-JP patients would accumulate damaged client proteins as they cannot be removed from the cellular milieu by means of proteasomal degradation [109].

Ko and colleagues demonstrated that parkin also plays an important role in governing cell death by ubiquitinating the stress kinase, p38 [106]. In this study, the authors show that p38 is upregulated in both young and old *parkin*-deficient mice, and that increased levels of p38 occur in postmortem brain tissue samples from patients diagnosed with AR-JP, idiopathic PD, and Lewy body disease. Knock out of the *parkin* gene in *D. melanogaster* results in decreased lifespan, impaired motor activity when assessed by the flies' ability to fly and climb, as well as male sterility [3,105,110]. Despite the observed loss of dopaminergic neurons in aged parkin knock-out flies, locomotor deficits are likely due to muscle degeneration [3,105,110].

The majority of parkin knock-out mice lack the characteristic progressive and selective loss of SNpc dopaminergic neurons [111,112]. Mice demonstrate motor and cognitive deficits [111] as well as reduced synaptic excitability [112]. Itier and colleagues detected increased levels of the antioxidant glutathione in mesencephalic neurons of fetal parkin knock-out mice that may contribute to the lack of dopaminergic cell death observed in these animals [111]. Goldberg and colleagues found that steady-state levels of known parkin client substrates such as CDCrel-1, synphilin-1, and α-synuclein are not altered in parkin knock-out mice [112]. This suggests that parkin is either not sufficient in conjugating ubiquitin onto these substrates or other E3 ligases are not fully compensating for lack of parkin in this system.

In an effort to overcome potential compensatory mechanisms, Shin and colleagues developed conditional parkin knock-out mice, which allowed the authors to not only observe a selective and progressive loss of dopaminergic neurons of the SNpc, but also to identify a novel parkin-interacting substrate (PARIS) [113]. The authors reported that overexpression of PARIS leads to enhanced selective degeneration of dopaminergic neurons, an effect that can be rescued by co-expression of either parkin or the peroxisome coactivator-1α (PGC-1α) [113]. Parkin tightly controls PARIS expression levels by ubiquitination. These authors also noted increased PARIS levels in the cingulate cortex of individuals with parkin mutations [113].

Periquet and colleagues performed proteomic analyses comparing the CNS from parkin knock-out and WT mice [114]. This study found that proteins associated with energy metabolism, the cellular antioxidant system, the ubiquitin proteasome pathway, as well as stress-related molecular chaperones were all compromised in parkin-deficient mice even though the animals were not anatomically impaired [114]. In order to generate a model that more fully recapitulates the pathological characteristics of PD, Lu and colleagues generated a BAC parkin transgenic mouse in which the parkin mutant Q311X was overexpressed [115]. This mutation was initially identified in a Turkish patient who presented with an early-onset form of PD. Q311X mutation results in the premature truncation of the final 155 amino acids of the parkin protein [115]. BAC-parkin Q311X transgenic mice exhibit age-dependent locomotor deficits along with a substantial loss of dopaminergic neurons of the SNpc and concomitant loss of dopamine neuron terminals in the striatum [115]. Dopamine depletion in the striatum of these animals may contribute to motor deficits in these mice [115]. Neuropathologically, these mice present with age-dependent accumulation of α-synuclein increased levels of the oxidative stress marker 3-nitortyrosine in dopaminergic neurons, suggesting that a Q311X mutation in *parkin* may have a dominant-negative effect on the survival of dopaminergic neurons [115].

PARK Mutations

Within the entire population of patients with PARK mutations, the autosomal recessive, early-onset form of PD due to *PARK6* gene mutations was noted in a very small cluster of patients. PARK6 was identified in 2001 in a screen of a large Sicilian family in which four members presented with early-onset PD [116]. Analysis of the precise location of the *PARK6* gene by means of a genome-wide screen found that the *PARK6* gene maps to the 12.5 cM interval of chromosome 1p35-p36 [116].

Members of the Marsala kindred present with a slow progression of the disease as well as a sustained response to levodopa treatment [116]. However, levodopa-induced complications can occur in this subpopulation of patients [116]. Neuropathologically, a selective loss of dopaminergic neurons of the nigrostriatal system, as well as the formation and accumulation of Lewy bodies and Lewy neurites, is reported [117].

PINK1 Genes

In 2004, Valente and colleagues identified homozygous mutations in the *PTEN-inducible putative kinase 1 (PINK1)* gene in PD patients of three consanguineous families [116]. Further analyses of the *PINK1* revealed eight exons, which span approximately 1.8 kb of the human genome [116]. The PINK1 protein is composed of 581 amino acids, a molecular weight of 63 kDa. Structural analyses of the protein kinase domain of PINK1 revealed significant similarities to the serine/threonine kinases belonging to the Ca^{2+}/calmodulin protein kinase family [116]. PINK1 protein is ubiquitously expressed throughout the human brain including the substantia nigra, caudate, putamen, frontal cortex, and hippocampus, as well as the periphery with the highest expression levels found in heart, skeletal muscle, and testes [116,117]. PINK1 localizes to mitochondria by means of an N-terminal mitochondrial targeting sequence. Once located at the mitochondria, PINK1's C-terminal kinase domain faces the cytoplasm in which it can interact with substrates [116–118] (see also [119]). Under neuropathological conditions, PINK1 collocates with Lewy bodies even in individuals without PINK1 mutations [117].

Disease-causing mutations in *PINK1* are often found in the kinase domain, resulting in loss of kinase activity [116]. The first two mutations identified were a missense mutation (G309D) in a Spanish kindred, and a truncation mutation (W437X) in an Italian family [116,117]. Since the identification of the first two disease segregating mutations in 2004, a number of other mutations spanning the *PINK1* gene have been identified [120]. Loss of PINK1 kinase function increases susceptibility to cellular stressors, including misfolded proteins and oxidative stress, and underscores the role of mitochondrial dysfunction in dopaminergic degeneration [116,121].

D. melanogaster transgenic models have revealed that the phenotype resulting from deletion of the PINK1 homologue CG4523 in flies looks remarkably similar to that of parkin-deficient mammalian models. They have locomotor defects, muscle fiber degeneration, male sterility, abnormal mitochondrial morphology, increased sensitivity to numerous stressors [121,122]. PINK1 overexpression does not ameliorate the phenotype induced by parkin deficiency [3,121–123], suggesting that PINK1 and parkin function in the same molecular pathway with PINK1 acting upstream of parkin [121,122].

PINK1 knock-out mice do not have SNpc degeneration or alterations in dopamine levels or receptors in the striatum [124,125]. Additionally, no accumulation of α-synuclein-, and ubiquitin-positive inclusion bodies are observed in PINK1 knock-out mice [125]. However, modest changes in nigrostriatal dopamine release suggests that PINK1 may contribute to striatal synaptic plasticity [124]. In keeping with PINK1's role in governing mitochondrial dynamics, PINK1 knock-out mice have deficits in mitochondrial function and dynamics [126,127]. Overt changes in mitochondrial number or ultrastructure do not occur in PINK1 knock-out mice, yet an overall increase in the number of enlarged and swollen mitochondria is reported [127,128]. Biochemical analyses of these mitochondria suggest that swelling may be due to impaired mitochondrial respiration in these animals. Like fly models, PINK1 knock-out mice also have increased sensitivity to biological stress [126–128], making these models valuable tools for understanding gene/environment interactions.

DJ-1 Genetic Mutations

In 2003, the *PARK7* gene was identified in a Dutch and Spanish cohort affected with autosomal recessive, early-onset PD [129]. The *PARK7* gene was mapped to a 20 cM interval on chromosome 1p36 and is composed of eight exons encoding a 189-amino acid protein known as DJ-1 [129–133].

Patients harboring mutations in the *DJ-1* gene present with the same cardinal symptoms as those with mutations in *parkin* or *PINK1*, including slow progression of the disease accompanied by characteristic locomotor deficits and a positive response to dopamine supplementation therapy [134,135]. No neuropathological examinations of patients with DJ-1 mutations have been published as of yet, therefore it remains to be determined if Lewy bodies are present within this population [135].

DJ-1 belongs to the DJ-1/ThiJ/Pfpl protein superfamily of molecular chaperones. The protein is ubiquitously expressed in the human brain [3,129,136–138] and is present in some peripheral tissues as well. The single-domain, DJ-1 protein exists as a homodimer localized in the nucleus, cytosol, and the intermembrane space of the mitochondria [3,129,130,136–139]. Functional analysis demonstrates a role for DJ-1 in response to oxidative stress. Loss of DJ-1 increases sensitivity to oxidative stress whereas overexpression is cytoprotective [128,136,139–141]. DJ-1 also functions as an RNA-binding protein in a redox-sensitive manner [3,138,142].

The first two mutations identified in the *DJ-1* gene were noted in Dutch and Italian families in which a large genomic deletion resulted in a 14 kb deletion of the gene's promoter region as well as exons 1–5, and a homozygous missense mutation (L166P), leading to loss of DJ-1 stability, localization, and expression [129,130,134,136,139]. Introduction of the L166P mutation in model systems results in a protein that is unable to homodimerize and is quickly subject to proteasomal degradation [130,134]. Since the identification of the initial mutations, several others have been noted, including a M26I mutation and D194A and A104T mutations [130,136]. The precise effect of these mutations on neuronal survival, however, remains to be elucidated [130,136].

DJ-1 flies are unique in that they contain two separate DJ-1 orthologs, DJ-1α and DJ-1β [122,132,143]. Both genes exhibit approximately 50% amino acid identity and a 69% similarity to the human DJ-1 gene [132,143,144]. The homozygous L166P mutation found in PD patients was also noted in flies at positions L192 and L181 in the DJ-1α and DJ-1β genes, respectively, further supporting the importance of this residue for DJ-1 function [143]. To investigate the contribution of loss of DJ-1 function on the pathogenesis of PD, flies in which both the DJ-1α and DJ-1β genes were knocked out were generated [132]. These flies are viable, fertile, have a normal number of dopaminergic neurons, and present with no difference in lifespan when compared to WT flies. They do, however, exhibit increased sensitivity to oxidative stress-inducing agents including rotenone and paraquat, which have been linked to PD [132]. DJ-1β homozygous mutant flies have impaired locomotor activity in the absence of dopaminergic degeneration [143]. Acute knockdown of the *Drosophila* DJ-1 ortholog DJ-1α by means of RNA interference results in flies with high levels of reactive oxygen species (ROS) and an increased sensitivity to oxidative stress, as well as selective loss of dopaminergic and photoreceptor neurons [145].

DJ-1 genetic deletion in mice results in modest changes in the nigrostriatal system, motor behavior deficits, and increased sensitivity to MPTP. The mice do not, however, have dopaminergic cell death or changes in the levels of dopamine or the dopamine transporter in the striatum [3,57,141,142,145]. Also absent are α-synuclein- and ubiquitin-positive Lewy bodies [146]. However, DJ-1 deficiency is associated with loss of mitochondrial aconitase activity [142].

MitoPark Mouse

Although animal models associated with the study of mutations in PD-linked genes have not captured the slow, progressive, and selective loss of dopaminergic neurons, with a robust parkinsonian phenotype, a recently characterized mouse model may meet these criteria. Ekstrand and colleagues generated a conditional knock-out mouse in which the gene encoding mitochondrial transcription factor A (*Tfam*) was disrupted specifically in dopaminergic neurons via a DAT promoter [147].

Tfam knock-out mice demonstrate decreased expression of mitochondrial DNA along with deficits at the levels of the mitochondrial electron transport chain in dopaminergic neurons of the SNpc [147]. MitoPark mice have a slow progression of parkinsonian motor deficits including decreased locomotion, limb rigidity, tremor, and twitching [147]. Notably, dopamine supplementation therapy with levodopa ameliorates the locomotor deficits in these mice and much like PD patients, the levodopa responsiveness decreases as mice age [147]. MitoPark mice also exhibit progressive loss of dopaminergic neurons of the SNpc along with accumulation of inclusion bodies within midbrain dopamine neurons. Inclusion bodies do not contain α-synuclein and are therefore not considered to be Lewy bodies [147]. However, the authors noted a close physical association of these intraneuronal inclusion bodies with mitochondria, suggesting that dysfunctional mitochondria may play a role in their formation [147].

Although the MitoPark mouse does not harbor any of the known mutations associated with PD, it provides strong support for the hypothesis originally developed in toxicant-induced cell death—that mitochondrial dysfunction contributes to PD [147]. At this time, it remains to be determined if the pathological changes in MitoPark transgenic mice are specific to dopaminergic neurons of the SNpc [13].

CONCLUSIONS

Over the past 30 years, numerous important discoveries have shaped our general understanding of the pathology underlying PD. Although it was initially thought that environmental factors and aging were the main causes of PD, the discovery of several PD-linked genes has significantly broadened our perception regarding the factors that can result in the onset and progression of this devastating disease [1]. Genetic and technological advancements have not only allowed researchers to identify, but also to elucidate which are associated with particular symptoms, the time of onset, susceptibility, progression, and prognosis of disease. These discoveries have further fueled the development of novel model systems in which the expression of genes can be altered and the behavioral and neuroanatomical changes tracked in real time. Moreover, researchers have begun developing increasingly powerful screens for biomarkers of dysfunction and new means to interrogate the CNS.

When it comes to understanding environmental influences on neuropathological cascades, any number of these models have the potential to generate a rich pool of data for genetic "priming," in which cell death can be induced by addition of external triggers. The utility of conditional transgenic mouse models also presents researchers with systems in which to address the spatial and temporal factors that influence cell survival and begin to further understand the roles of PD-related genes in development.

Several genetic PD models demonstrate selective and progressive loss of dopaminergic neurons in the SNpc, with a smaller subset also demonstrating behavioral deficits related to PD [92–93,113]. Although no model system is perfect, it would be shortsighted to discount the value of those that do not meet all criteria.

Acknowledgments

This work was supported by grants to BAM, AMP, and BLM from The Suzanne and Walter Scott Foundation, The Marino Foundation, The American Heart Association, and NIH 1 UH2 NS080701. The authors sincerely thank Dr Jeannette Stankowski who conceptualized this entire chapter and generated many initial drafts but declined authorship.

References

[1] Martin I, Dawson VL, Dawson TM. The impact of genetic research on our understanding of Parkinson's disease. Prog Brain Res 2010;183:21–41.

[2] Shulman JM, De Jager PL, Feany MB. Parkinson's disease: genetics and pathogenesis. Annu Rev Pathol 2011;6:193–222.

[3] Dawson TM, Ko HS, Dawson VL. Genetic animal models of Parkinson's disease. Neuron 2010;66(5):646–61.

[4] Emborg ME. Evaluation of animal models of Parkinson's disease for neuroprotective strategies. J Neurosci Methods 2004;139(2):121–43.

[5] Farrer MJ. Genetics of Parkinson disease: paradigm shifts and future prospects. Nat Rev Genet 2006;7(4):306–18.

[6] Savitt JM, Dawson VL, Dawson TM. Diagnosis and treatment of Parkinson disease: molecules to medicine. J Clin Invest 2006;116(7):1744–54.

[7] Gerlach M, Riederer P. Animal models of Parkinson's disease: an empirical comparison with the phenomenology of the disease in man. J Neural Transm 1996;103(8–9):987–1041.

[8] Beal MF. Experimental models of Parkinson's disease. Nat Rev Neurosci 2001;2(5):325–34.

[9] Fahn S. Description of Parkinson's disease as a clinical syndrome. Ann N Y Acad Sci 2003;991:1–14.

[10] Melzer TR, Watts R, MacAskill MR, Pitcher TL, Livingston L, Keenan RJ, et al. Grey matter atrophy in cognitively impaired Parkinson's disease. J Neurol Neurosurg Psychiatry 2012;83(2):188–94.

[11] Chaudhuri KR, Healy DG, Schapira AHV. Non-motor symptoms of Parkinson's disease: diagnosis and management. Lancet Neurol 2006;5(3):235–45.

[12] Churchyard A, Lees AJ. The relationship between dementia and direct involvement of the hippocampus and amygdala in Parkinson's disease. Neurology 1997;49(6):1570–6.

[13] Braak H, Del Tredici K, Rub U, de Vos RA, Jansen Steur EN, Braak E. Staging of brain pathology related to sporadic Parkinson's disease. Neurobiol Aging 2003;24(2):197–211.

[14] Kehagia AA, Barker RA, Robbins TW. Neuropsychological and clinical heterogeneity of cognitive impairment and dementia in patients with Parkinson's disease. Lancet Neurol 2010;9(12):1200–13.

[15] Phillips L, Litcofsky KA, Pelster M, Gelfand M, Ullman MT, Charles PD. Subthalamic nucleus deep brain stimulation impacts language in early Parkinson's disease. PLoS ONE 2012;7(8):e42829.

[16] Stanwood GD, Leitch DB, Savchenko V, Wu J, Fitsanakis VA, Anderson DJ, et al. Manganese exposure is cytotoxic and alters dopaminergic and GABAergic neurons within the basal ganglia. J Neurochem 2009;110(1):378–89.

[17] Langston JW, Ballard P, Tetrud JW, Irwin I. Chronic Parkinsonism in humans due to a product of meperidine-analog synthesis. Science 1983;219(4587):979–80.

[18] Langston JW. Progress in understanding the mechanisms of neuronal dysfunction and degeneration in Parkinson's disease. In: Uversky V, Fink A, editors. Protein misfolding, aggregation, and conformational diseases. Protein reviews, vol. 6. US: Springer; 2007. pp. 49–59.

[19] Clark LN, Ross BM, Wang Y, Mejia-Santana H, Harris J, Louis ED, et al. Mutations in the glucocerebrosidase gene are associated with early-onset Parkinson disease. Neurology 2007;69(12):1270–7.

[20] Neudorfer O, Giladi N, Elstein D, Abrahamov A, Turezkite T, Aghai E, et al. Occurrence of Parkinson's syndrome in type I Gaucher disease. QJM 1996;89(9):691–4.

[21] Mazzulli JR, Xu YH, Sun Y, Knight AL, McLean PJ, Caldwell GA, et al. Gaucher disease glucocerebrosidase and alpha-synuclein form a bidirectional pathogenic loop in synucleinopathies. Cell 2011;146(1):37–52.

[22] Tayebi N, Callahan M, Madike V, Stubblefield BK, Orvisky E, Krasnewich D, et al. Gaucher disease and parkinsonism: a phenotypic and genotypic characterization. Mol Genet Metab 2001;73(4):313–21.

[23] Neumann J, Bras J, Deas E, O'Sullivan SS, Parkkinen L, Lachmann RH, et al. Glucocerebrosidase mutations in clinical and pathologically proven Parkinson's disease. Brain 2009;132(Pt 7):1783–94.

[24] Aharon-Peretz J, Rosenbaum H, Gershoni-Baruch R. Mutations in the glucocerebrosidase gene and Parkinson's disease in Ashkenazi Jews. N Engl J Med 2004;351(19):1972–7.

[25] Farfel-Becker T, Vitner EB, Futerman AH. Animal models for Gaucher disease research. Dis Model Mech 2011;4(6):746–52.

[26] Manning-Bog AB, Schule B, Langston JW. Alpha-synuclein-glucocerebrosidase interactions in pharmacological Gaucher models: a biological link between Gaucher disease and parkinsonism. Neurotoxicology 2009;30(6):1127–32.

[27] Beal MF. Parkinson's disease: a model dilemma. Nature 2010;466(7310):S8–10.

[28] Chesselet MF, Fleming S, Mortazavi F, Meurers B. Strengths and limitations of genetic mouse models of Parkinson's disease. Park Relat Disord 2008;14(Suppl. 2):S84–7.

[29] Hall B, Limaye A, Kulkarni AB. Overview: generation of gene knockout mice. Curr Protoc Cell Biol 2009. Chapter 19:Unit 19 2 2 1-7.

[30] Friedel RH, Wurst W, Wefers B, Kuhn R. Generating conditional knockout mice. Methods Mol Biol 2011;693:205–31.

[31] Chen J, Roop DR. Mouse models in preclinical studies for pachyonychia congenita. J Investig Dermatol Symp Proc 2005;10(1):37–46.

[32] Skvorak K, Vissel B, Homanics GE. Production of conditional point mutant knockin mice. Genesis 2006;44(7):345–53.

[33] Sacca R, Engle SJ, Qin W, Stock JL, McNeish JD. Genetically engineered mouse models in drug discovery research. Methods Mol Biol 2010;602:37–54.

[34] Bayascas JR, Sakamoto K, Armit L, Arthur JS, Alessi DR. Evaluation of approaches to generation of tissue-specific knock-in mice. J Biol Chem 2006;281(39):28772–81.

[35] Rudmann DG, Durham SK. Utilization of genetically altered animals in the pharmaceutical industry. Toxicol Pathol 1999;27(1):111–4.

[36] Gama Sosa MA, De Gasperi R, Elder GA. Modeling human neurodegenerative diseases in transgenic systems. Hum Genet 2012;131(4):535–63.

[37] Rockenstein E, Crews L, Masliah E. Transgenic animal models of neurodegenerative diseases and their application to treatment development. Adv Drug Deliv Rev 2007;59(11):1093–102.

[38] Chesselet MF, Richter F, Zhu C, Magen I, Watson MB, Subramaniam SR. A progressive mouse model of Parkinson's disease: the Thy1-aSyn ("Line 61") mice. Neurotherapeutics 2012;9(2):297–314.

[39] Gossen M, Bujard H. Tight control of gene expression in mammalian cells by tetracycline-responsive promoters. Proc Natl Acad Sci USA 1992;89(12):5547–51.

[40] Schonig K, Bujard H. Generating conditional mouse mutants via tetracycline-controlled gene expression. Methods Mol Biol 2003;209:69–104.

[41] Sprengel R, Hasan MT. Tetracycline-controlled genetic switches. Handb Exp Pharmacol 2007;178:49–72.

[42] Schonig K, Schwenk F, Rajewsky K, Bujard H. Stringent doxycycline dependent control of CRE recombinase in vivo. Nucleic Acids Res 2002;30(23):e134.

[43] Odeh F, Leergaard TB, Boy J, Schmidt T, Riess O, Bjaalie JG. Atlas of transgenic Tet-Off Ca2+/calmodulin-dependent protein kinase II and prion protein promoter activity in the mouse brain. Neuroimage 2011;54(4):2603–11.

[44] Schedl A, Larin Z, Montoliu L, Thies E, Kelsey G, Lehrach H, et al. A method for the generation of YAC transgenic mice by pronuclear microinjection. Nucleic Acids Res 1993;21(20):4783–7.

[45] Johnson SJ, Wade-Martins R. A BACwards glance at neurodegeneration: molecular insights into disease from LRRK2, SNCA and MAPT BAC-transgenic mice. Biochem Soc Trans 2011;39(4):862–7.

[46] Golbe LI, Di Iorio G, Sanges G, Lazzarini AM, La Sala S, Bonavita V, et al. Clinical genetic analysis of Parkinson's disease in the Contursi kindred. Ann Neurol 1996;40(5):767–75.

[47] Polymeropoulos MH, Higgins JJ, Golbe LI, Johnson WG, Ide SE, Di Iorio G, et al. Mapping of a gene for Parkinson's disease to chromosome 4q21-q23. Science 1996;274(5290):1197–9.

[48] Warner TT, Schapira AH. Genetic and environmental factors in the cause of Parkinson's disease. Ann Neurol 2003;53(Suppl. 3):S16–23. Discussion S-5.

[49] Dauer W, Przedborski S. Parkinson's disease: mechanisms and models. Neuron 2003;39(6):889–909.

[50] Abeliovich A, Schmitz Y, Farinas I, Choi-Lundberg D, Ho WH, Castillo PE, et al. Mice lacking alpha-synuclein display functional deficits in the nigrostriatal dopamine system. Neuron 2000;25(1):239–52.

[51] Burre J, Sharma M, Tsetsenis T, Buchman V, Etherton MR, Sudhof TC. Alpha-synuclein promotes SNARE-complex assembly in vivo and in vitro. Science 2010;329(5999):1663–7.

[52] Chandra S, Gallardo G, Fernandez-Chacon R, Schluter OM, Sudhof TC. Alpha-synuclein cooperates with CSPalpha in preventing neurodegeneration. Cell 2005;123(3):383–96.

[53] Chandra S, Fornai F, Kwon HB, Yazdani U, Atasoy D, Liu X, et al. Double-knockout mice for alpha- and beta-synucleins: effect on synaptic functions. Proc Natl Acad Sci USA 2004;101(41):14966–71.

[54] Kruger R, Kuhn W, Muller T, Woitalla D, Graeber M, Kosel S, et al. Ala30Pro mutation in the gene encoding alpha-synuclein in Parkinson's disease. Nat Genet 1998;18(2):106–8.

[55] Zarranz JJ, Alegre J, Gomez-Esteban JC, Lezcano E, Ros R, Ampuero I, et al. The new mutation, E46K, of alpha-synuclein causes Parkinson and Lewy body dementia. Ann Neurol 2004;55(2):164–73.

[56] Singleton AB, Farrer M, Johnson J, Singleton A, Hague S, Kachergus J, et al. Alpha-synuclein locus triplication causes Parkinson's disease. Science 2003;302(5646):841.

[57] Chen L, Cagniard B, Mathews T, Jones S, Koh HC, Ding Y, et al. Age-dependent motor deficits and dopaminergic dysfunction in DJ-1 null mice. J Biol Chem 2005;280(22):21418–26.

[58] Wang Y, Shi M, Chung KA, Zabetian CP, Leverenz JB, Berg D, et al. Phosphorylated alpha-synuclein in Parkinson's disease. Sci Transl Med 2012;4(121):121ra20.

[59] Okochi M, Walter J, Koyama A, Nakajo S, Baba M, Iwatsubo T, et al. Constitutive phosphorylation of the Parkinson's disease associated alpha-synuclein. J Biol Chem 2000;275(1):390–7.

[60] Iwatsubo T. Aggregation of alpha-synuclein in the pathogenesis of Parkinson's disease. J Neurol 2003;250(Suppl. 3):III 11–4.

[61] Kaufmann H, Hague K, Perl D. Accumulation of alpha-synuclein in autonomic nerves in pure autonomic failure. Neurology 2001;56(7):980–1.

[62] Halliwell B. Reactive oxygen species and the central nervous system. J Neurochem 1992;59(5):1609–23.

[63] Tanaka Y, Engelender S, Igarashi S, Rao RK, Wanner T, Tanzi RE, et al. Inducible expression of mutant alpha-synuclein decreases proteasome activity and increases sensitivity to mitochondria-dependent apoptosis. Hum Mol Genet 2001;10(9):919–26.

[64] Schneider L, Zhang J. Lysosomal function in macromolecular homeostasis and bioenergetics in Parkinson's disease. Mol Neurodegener 2010;5:14.

[65] Chu Y, Dodiya H, Aebischer P, Olanow CW, Kordower JH. Alterations in lysosomal and proteasomal markers in Parkinson's disease: relationship to alpha-synuclein inclusions. Neurobiol Dis 2009;35(3):385–98.

[66] Meredith GE, Totterdell S, Petroske E, Santa Cruz K, Callison Jr RC, Lau YS. Lysosomal malfunction accompanies alpha-synuclein aggregation in a progressive mouse model of Parkinson's disease. Brain Res 2002;956(1):156–65.

[67] Lee HJ, Khoshaghideh F, Patel S, Lee SJ. Clearance of alpha-synuclein oligomeric intermediates via the lysosomal degradation pathway. J Neurosci 2004;24(8):1888–96.

[68] Usenovic M, Krainc D. Lysosomal dysfunction in neurodegeneration: the role of ATP13A2/PARK9. Autophagy 2012;8(6):987–8.

[69] Usenovic M, Tresse E, Mazzulli JR, Taylor JP, Krainc D. Deficiency of ATP13A2 leads to lysosomal dysfunction, alpha-synuclein accumulation, and neurotoxicity. J Neurosci 2012;32(12):4240–6.

[70] Desplats P, Lee HJ, Bae EJ, Patrick C, Rockenstein E, Crews L, et al. Inclusion formation and neuronal cell death through neuron-to-neuron transmission of alpha-synuclein. Proc Natl Acad Sci USA 2009;106(31):13010–5.

[71] Greten-Harrison B, Polydoro M, Morimoto-Tomita M, Diao L, Williams AM, Nie EH, et al. Alphabeta-gamma-synuclein triple knockout mice reveal age-dependent neuronal dysfunction. Proc Natl Acad Sci USA 2010;107(45):19573–8.

[72] Fernagut PO, Chesselet MF. Alpha-synuclein and transgenic mouse models. Neurobiol Dis 2004;17(2):123–30.

[73] Masliah E, Rockenstein E, Veinbergs I, Mallory M, Hashimoto M, Takeda A, et al. Dopaminergic loss and inclusion body formation in alpha-synuclein mice: implications for neurodegenerative disorders. Science 2000;287(5456):1265–9.

[74] van der Putten H, Wiederhold KH, Probst A, Barbieri S, Mistl C, Danner S, et al. Neuropathology in mice expressing human alpha-synuclein. J Neurosci 2000;20(16):6021–9.

[75] Giasson BI, Duda JE, Quinn SM, Zhang B, Trojanowski JQ, Lee VM. Neuronal alpha-synucleinopathy with severe movement disorder in mice expressing A53T human alpha-synuclein. Neuron 2002;34(4):521–33.

[76] Magen I, Chesselet MF. Genetic mouse models of Parkinson's disease: the state of the art. Prog Brain Res 2010;184:53–87.

[77] Wakamatsu M, Ishii A, Iwata S, Sakagami J, Ukai Y, Ono M, et al. Selective loss of nigral dopamine neurons induced by overexpression of truncated human alpha-synuclein in mice. Neurobiol Aging 2008;29(4):574–85.

[78] Thiruchelvam MJ, Powers JM, Cory-Slechta DA, Richfield EK. Risk factors for dopaminergic neuron loss in human alpha-synuclein transgenic mice. Eur J Neurosci 2004;19(4):845–54.

[79] Lee MK, Stirling W, Xu Y, Xu X, Qui D, Mandir AS, et al. Human alpha-synuclein-harboring familial Parkinson's disease-linked Ala-53--> Thr mutation causes neurodegenerative disease with alpha-synuclein aggregation in transgenic mice. Proc Natl Acad Sci USA 2002;99(13):8968–73.

[80] Sudhof TC, Rizo J. Synaptic vesicle exocytosis. Cold Spring Harb Perspect Biol 2011;3(12).

[81] Funayama M, Hasegawa K, Kowa H, Saito M, Tsuji S, Obata F. A new locus for Parkinson's disease (PARK8) maps to chromosome 12p11.2-q13.1. Ann Neurol 2002;51(3):296–301.

[82] Paisan-Ruiz C, Jain S, Evans EW, Gilks WP, Simon J, van der Brug M, et al. Cloning of the gene containing mutations that cause PARK8-linked Parkinson's disease. Neuron 2004;44(4):595–600.

[83] Zimprich A, Biskup S, Leitner P, Lichtner P, Farrer M, Lincoln S, et al. Mutations in LRRK2 cause autosomal-dominant parkinsonism with pleomorphic pathology. Neuron 2004;44(4):601–7.

[84] Yue Z, Lachenmayer ML. Genetic LRRK2 models of Parkinson's disease: Dissecting the pathogenic pathway and exploring clinical applications. Mov Disord 2011;26(8):1386–97.

[85] Mata IF, Wedemeyer WJ, Farrer MJ, Taylor JP, Gallo KA. LRRK2 in Parkinson's disease: protein domains and functional insights. Trends Neurosci 2006;29(5):286–93.

[86] Biskup S, Moore DJ, Celsi F, Higashi S, West AB, Andrabi SA, et al. Localization of LRRK2 to membranous and vesicular structures in mammalian brain. Ann Neurol 2006;60(5):557–69.

[87] Cookson MR. The role of leucine-rich repeat kinase 2 (LRRK2) in Parkinson's disease. Nat Rev Neurosci 2010;11(12):791–7.

[88] Biskup S, West AB. Zeroing in on LRRK2-linked pathogenic mechanisms in Parkinson's disease. Biochim Biophys Acta 2009;1792(7):625–33.

[89] Nichols RJ, Dzamko N, Hutti JE, Cantley LC, Deak M, Moran J, et al. Substrate specificity and inhibitors of LRRK2, a protein kinase mutated in Parkinson's disease. Biochem J 2009;424(1):47–60.

[90] Lee BD, Shin JH, VanKampen J, Petrucelli L, West AB, Ko HS, et al. Inhibitors of leucine-rich repeat kinase-2 protect against models of Parkinson's disease. Nat Med 2010;16(9):998–1000.

[91] Mata IF, Ross OA, Kachergus J, Huerta C, Ribacoba R, Moris G, et al. LRRK2 mutations are a common cause of Parkinson's disease in Spain. Eur J Neurol 2006;13(4):391–4.

[92] Chen CY, Weng YH, Chien KY, Lin KJ, Yeh TH, Cheng YP, et al. (G2019S) LRRK2 activates MKK4-JNK pathway and causes degeneration of SN dopaminergic neurons in a transgenic mouse model of PD. Cell Death Differ 2012;19(10):1623–33.

[93] Ramonet D, Daher JP, Lin BM, Stafa K, Kim J, Banerjee R, et al. Dopaminergic neuronal loss, reduced neurite complexity and autophagic abnormalities in transgenic mice expressing G2019S mutant LRRK2. PLoS ONE 2011;6(4):e18568.

[94] Hisahara S, Shimohama S. Toxin-induced and genetic animal models of Parkinson's disease. Parkinson's Dis 2010;2011:951709.

[95] Li X, Patel JC, Wang J, Avshalumov MV, Nicholson C, Buxbaum JD, et al. Enhanced striatal dopamine transmission and motor performance with LRRK2 overexpression in mice is eliminated by familial Parkinson's disease mutation G2019S. J Neurosci 2010;30(5):1788–97.

[96] Lin X, Parisiadou L, Gu XL, Wang L, Shim H, Sun L, et al. Leucine-rich repeat kinase 2 regulates the progression of neuropathology induced by Parkinson's-disease-related mutant alpha-synuclein. Neuron 2009;64(6):807–27.

[97] Tong Y, Yamaguchi H, Giaime E, Boyle S, Kopan R, Kelleher 3rd RJ, et al. Loss of leucine-rich repeat kinase 2 causes impairment of protein degradation pathways, accumulation of alpha-synuclein, and apoptotic cell death in aged mice. Proc Natl Acad Sci USA 2010;107(21):9879–84.

[98] Andres-Mateos E, Mejias R, Sasaki M, Li X, Lin BM, Biskup S, et al. Unexpected lack of hypersensitivity in LRRK2 knock-out mice to MPTP (1-methyl-4-phenyl-1,2,3,6-tetrahydropyridine). J Neurosci 2009;29(50):15846–50.

[99] Tong Y, Pisani A, Martella G, Karouani M, Yamaguchi H, Pothos EN, et al. R1441C mutation in LRRK2 impairs dopaminergic neurotransmission in mice. Proc Natl Acad Sci USA 2009;106(34):14622–7.

[100] Kitada T, Asakawa S, Hattori N, Matsumine H, Yamamura Y, Minoshima S, et al. Mutations in the parkin gene cause autosomal recessive juvenile parkinsonism. Nature 1998;392(6676):605–8.

[101] Bonifati V, Breedveld GJ, Squitieri F, Vanacore N, Brustenghi P, Harhangi BS, et al. Localization of autosomal recessive early-onset parkinsonism to chromosome 1p36 (PARK7) in an independent dataset. Ann Neurol 2002;51(2):253–6.

[102] Zhang Y, Gao J, Chung KK, Huang H, Dawson VL, Dawson TM. Parkin functions as an E2-dependent ubiquitin-protein ligase and promotes the degradation of the synaptic vesicle-associated protein, CDCrel-1. Proc Natl Acad Sci USA 2000;97(24):13354–9.

[103] Imai Y, Soda M, Takahashi R. Parkin suppresses unfolded protein stress-induced cell death through its E3 ubiquitin-protein ligase activity. J Biol Chem 2000;275(46):35661–4.

[104] Shimura H, Hattori N, Kubo S, Mizuno Y, Asakawa S, Minoshima S, et al. Familial Parkinson disease gene product, parkin, is a ubiquitin-protein ligase. Nat Genet 2000;25(3):302–5.

[105] Mata IF, Lockhart PJ, Farrer MJ. Parkin genetics: one model for Parkinson's disease. Hum Mol Genet 2004;13(1):R127–33.

[106] Ko HS, von Coelln R, Sriram SR, Kim SW, Chung KK, Pletnikova O, et al. Accumulation of the authentic parkin substrate aminoacyl-tRNA synthetase cofactor, p38/JTV-1, leads to catecholaminergic cell death. J Neurosci 2005;25(35):7968–78.

[107] Wang C, Ko HS, Thomas B, Tsang F, Chew KC, Tay SP, et al. Stress-induced alterations in parkin solubility promote parkin aggregation and compromise parkin's protective function. Hum Mol Genet 2005;14(24):3885–97.

[108] Dawson TM, Dawson VL. Molecular pathways of neurodegeneration in Parkinson's disease. Science 2003;302(5646):819–22.

[109] Di Napoli M, McLaughlin B. The ubiquitin-proteasome system as a drug target in cerebrovascular disease: therapeutic potential of proteasome inhibitors. Curr Opin Investig Drugs 2005;6(7):686–99.

[110] Greene JC, Whitworth AJ, Kuo I, Andrews LA, Feany MB, Pallanck LJ. Mitochondrial pathology and apoptotic muscle degeneration in Drosophila parkin mutants. Proc Natl Acad Sci USA 2003;100(7):4078–83.

[111] Itier JM, Ibanez P, Mena MA, Abbas N, Cohen-Salmon C, Bohme GA, et al. Parkin gene inactivation alters behaviour and dopamine neurotransmission in the mouse. Hum Mol Genet 2003;12(18):2277–91.

[112] Goldberg MS, Fleming SM, Palacino JJ, Cepeda C, Lam HA, Bhatnagar A, et al. Parkin-deficient mice exhibit nigrostriatal deficits but not loss of dopaminergic neurons. J Biol Chem 2003;278(44):43628–35.

[113] Shin JH, Ko HS, Kang H, Lee Y, Lee YI, Pletnikova O, et al. PARIS (ZNF746) repression of PGC-1alpha contributes to neurodegeneration in Parkinson's disease. Cell 2011;144(5):689–702.

[114] Periquet M, Corti O, Jacquier S, Brice A. Proteomic analysis of parkin knockout mice: alterations in energy metabolism, protein handling and synaptic function. J Neurochem 2005;95(5):1259–76.

[115] Lu XH, Fleming SM, Meurers B, Ackerson LC, Mortazavi F, Lo V, et al. Bacterial artificial chromosome transgenic mice expressing a truncated mutant parkin exhibit age-dependent hypokinetic motor deficits, dopaminergic neuron degeneration, and accumulation of proteinase K-resistant alpha-synuclein. J Neurosci 2009;29(7):1962–76.

[116] Valente EM, Abou-Sleiman PM, Caputo V, Muqit MM, Harvey K, Gispert S, et al. Hereditary early-onset Parkinson's disease caused by mutations in PINK1. Science 2004;304(5674):1158–60.

[117] Gandhi S, Muqit MM, Stanyer L, Healy DG, Abou-Sleiman PM, Hargreaves I, et al. PINK1 protein in normal human brain and Parkinson's disease. Brain 2006;129(Pt 7):1720–31.

[118] Zhou C, Huang Y, Shao Y, May J, Prou D, Perier C, et al. The kinase domain of mitochondrial PINK1 faces the cytoplasm. Proc Natl Acad Sci USA 2008;105(33):12022–7.

[119] Haque ME, Thomas KJ, D'Souza C, Callaghan S, Kitada T, Slack RS, et al. Cytoplasmic Pink1 activity protects neurons from dopaminergic neurotoxin MPTP. Proc Natl Acad Sci USA 2008;105(5):1716–21.

[120] Brooks J, Ding J, Simon-Sanchez J, Paisan-Ruiz C, Singleton AB, Scholz SW. Parkin and PINK1 mutations in early-onset Parkinson's disease: comprehensive screening in publicly available cases and control. J Med Genet 2009;46(6):375–81.

[121] Clark IE, Dodson MW, Jiang C, Cao JH, Huh JR, Seol JH, et al. Drosophila pink1 is required for mitochondrial function and interacts genetically with parkin. Nature 2006;441(7097):1162–6.

[122] Park J, Lee SB, Lee S, Kim Y, Song S, Kim S, et al. Mitochondrial dysfunction in Drosophila PINK1 mutants is complemented by parkin. Nature 2006;441(7097):1157–61.

[123] Deng H, Dodson MW, Huang H, Guo M. The Parkinson's disease genes pink1 and parkin promote mitochondrial fission and/or inhibit fusion in Drosophila. Proc Natl Acad Sci USA 2008;105(38):14503–8.

[124] Kitada T, Pisani A, Karouani M, Haburcak M, Martella G, Tscherter A, et al. Impaired dopamine release and synaptic plasticity in the striatum of parkin-/- mice. J Neurochem 2009;110(2):613–21.

[125] Gispert S, Ricciardi F, Kurz A, Azizov M, Hoepken HH, Becker D, et al. Parkinson phenotype in aged PINK1-deficient mice is accompanied by progressive mitochondrial dysfunction in absence of neurodegeneration. PLoS ONE 2009;4(6):e5777.

[126] Yang Y, Ouyang Y, Yang L, Beal MF, McQuibban A, Vogel H, et al. Pink1 regulates mitochondrial dynamics through interaction with the fission/fusion machinery. Proc Natl Acad Sci USA 2008;105(19):7070–5.

[127] Gautier CA, Kitada T, Shen J. Loss of PINK1 causes mitochondrial functional defects and increased sensitivity to oxidative stress. Proc Natl Acad Sci USA 2008;105(32):11364–9.

[128] Poole AC, Thomas RE, Andrews LA, McBride HM, Whitworth AJ, Pallanck LJ. The PINK1/Parkin pathway regulates mitochondrial morphology. Proc Natl Acad Sci USA 2008;105(5):1638–43.

[129] Bonifati V, Rizzu P, van Baren MJ, Schaap O, Breedveld GJ, Krieger E, et al. Mutations in the DJ-1 gene associated with autosomal recessive early-onset parkinsonism. Science 2003;299(5604):256–9.

[130] Moore DJ, Dawson VL, Dawson TM. Genetics of Parkinson's disease: what do mutations in DJ-1 tell us? Ann Neurol 2003;54(3):281–2.

[131] Lev N, Roncevic D, Ickowicz D, Melamed E, Offen D. Role of DJ-1 in Parkinson's disease. J Mol Neurosci 2006;29(3):215–25.

[132] Meulener M, Whitworth AJ, Armstrong-Gold CE, Rizzu P, Heutink P, Wes PD, et al. *Drosophila* DJ-1 mutants are selectively sensitive to environmental toxins associated with Parkinson's disease. Curr Biol 2005;15(17):1572–7.

[133] Meulener MC, Graves CL, Sampathu DM, Armstrong-Gold CE, Bonini NM, Giasson BI. DJ-1 is present in a large molecular complex in human brain tissue and interacts with alpha-synuclein. J Neurochem 2005;93(6):1524–32.

[134] Abou-Sleiman PM, Healy DG, Wood NW. Causes of Parkinson's disease: genetics of DJ-1. Cell Tissue Res 2004;318(1):185–8.

[135] Hardy J, Lewis P, Revesz T, Lees A, Paisan-Ruiz C. The genetics of Parkinson's syndromes: a critical review. Curr Opin Genet Dev 2009;19(3):254–65.

[136] Zhang L, Shimoji M, Thomas B, Moore DJ, Yu SW, Marupudi NI, et al. Mitochondrial localization of the Parkinson's disease related protein DJ-1: implications for pathogenesis. Hum Mol Genet 2005;14(14):2063–73.

[137] Macedo MG, Anar B, Bronner IF, Cannella M, Squitieri F, Bonifati V, et al. The DJ-1L166P mutant protein associated with early onset Parkinson's disease is unstable and forms higher-order protein complexes. Hum Mol Genet 2003;12(21):2807–16.

[138] van der Brug MP, Blackinton J, Chandran J, Hao LY, Lal A, Mazan-Mamczarz K, et al. RNA binding activity of the recessive parkinsonism protein DJ-1 supports involvement in multiple cellular pathways. Proc Natl Acad Sci USA 2008;105(29):10244–9.

[139] Bonifati V, Rizzu P, Squitieri F, Krieger E, Vanacore N, van Swieten JC, et al. DJ-1(PARK7), a novel gene for autosomal recessive, early onset parkinsonism. Neurol Sci 2003;24(3):159–60.

[140] Martinat C, Shendelman S, Jonason A, Leete T, Beal MF, Yang L, et al. Sensitivity to oxidative stress in DJ-1-deficient dopamine neurons: an ES-derived cell model of primary Parkinsonism. PLoS Biol 2004;2(11):e327.

[141] Kim RH, Smith PD, Aleyasin H, Hayley S, Mount MP, Pownall S, et al. Hypersensitivity of DJ-1-deficient mice to 1-methyl-4-phenyl-1,2,3,6-tetrahydropyridine (MPTP) and oxidative stress. Proc Natl Acad Sci USA 2005;102(14):5215–20.

[142] Andres-Mateos E, Perier C, Zhang L, Blanchard-Fillion B, Greco TM, Thomas B, et al. DJ-1 gene deletion reveals that DJ-1 is an atypical peroxiredoxin-like peroxidase. Proc Natl Acad Sci USA 2007;104(37):14807–12.

[143] Park J, Kim SY, Cha GH, Lee SB, Kim S, Chung J. *Drosophila* DJ-1 mutants show oxidative stress-sensitive locomotive dysfunction. Gene 2005;361:133–9.

[144] Meulener MC, Xu K, Thomson L, Ischiropoulos H, Bonini NM. Mutational analysis of DJ-1 in *Drosophila* implicates functional inactivation by oxidative damage and aging. Proc Natl Acad Sci USA 2006;103(33):12517–22.

[145] Yang Y, Gehrke S, Haque ME, Imai Y, Kosek J, Yang L, et al. Inactivation of *Drosophila* DJ-1 leads to impairments of oxidative stress response and phosphatidylinositol 3-kinase/Akt signaling. Proc Natl Acad Sci USA 2005;102(38):13670–5.

[146] Goldberg MS, Pisani A, Haburcak M, Vortherms TA, Kitada T, Costa C, et al. Nigrostriatal dopaminergic deficits and hypokinesia caused by inactivation of the familial Parkinsonism-linked gene DJ-1. Neuron 2005;45(4):489–96.

[147] Ekstrand MI, Terzioglu M, Galter D, Zhu S, Hofstetter C, Lindqvist E, et al. Progressive parkinsonism in mice with respiratory-chain-deficient dopamine neurons. Proc Natl Acad Sci USA 2007;104(4):1325–30.

15

Alzheimer's Disease and the Search for Environmental Risk Factors

Walter A. Kukull

INTRODUCTION

In this chapter we will discuss the context of research for environmental risk factors of Alzheimer's disease (AD). While AD is only one cause of cognitive impairment and dementia, it is the most frequent one. However, AD cannot be assumed to be the only pathology active in those diagnosed with clinical AD dementia. Neuropathologists have known for some time that more than one neurodegenerative disease and one vascular disease often coexist in patients diagnosed clinically with AD dementia. Because of this, one must keep in mind that an observed risk/protective factor or gene might have its main association with one of the coexisting, though perhaps unobservable, pathologies. We will concentrate more on methodological issues and advances in knowledge of the disease, which influence how we evaluate risk factor associations, than on creating a catalog of all past studies of environmental risk factors. Because of advances in methodology, as well as our understanding of AD and ability to detect it, many past studies of environmental factors could have provided questionable results. As a result, the aim is to provide the reader with better information to evaluate existing studies of environmental risk factors.

HISTORY AND BIOLOGICAL CONTEXT

In order to study risk factors for AD effectively, one must also be aware of some of the history and biology of the disease, as well as its clinical and pathologic contexts. While AD has only come crashing into the collective consciousness within the last 30 years or so, AD was actually pathologically characterized near the beginning of the twentieth century. Alois Alzheimer was treating a 51-year-old woman (Auguste D.) for several years prior to her death in 1906. The woman showed symptoms of memory loss, paranoia as well as other severe cognitive deficits. Dr Alzheimer conducted an autopsy following her death and thereby described the pathologic features found in her brain. Those features are the amyloid plaques and neurofibrillary tangles that still form the pathologic basis of diagnosis of the disease [1]. However, because of the patient's age, the disease was regarded by the medical community as a very unusual or rare condition—even though presumably similar severe cognitive deficits occurred frequently in the very old and were then ascribed to senility, hardening of the arteries, or chronic brain syndrome. It was not until almost 70 years later that the medical research community began to understand that both the unusual, early onset form of the disease seen by Alzheimer, and the more common occurrence of cognitive deficits among the very old, have essentially the same pathologic basis [2,3]. Then in 1984, Glenner and Wong characterized the amyloid-beta protein that made up the amyloid plaques first noted pathologically by Dr Alzheimer [4]. Similarly, advances were made in the characterization of the neurofibrillary tangles [5–7]. Soon afterward, Braak and Braak published their famous paper describing the progression and proliferation of neurofibrillary tangles in the brain, thus characterizing the six "stages" of tangle formation commonly used today as a measure of AD pathology [8].

Once the primary pathologic changes and proteins associated with AD were characterized, there was added impetus for geneticists to seek the genes that might make those proteins [9]. Goldgaber et al. were among the groups who established that the amyloid precursor protein (APP) was located on chromosome 21. When cleaved in an abnormal fashion, the APP yielded the pathologic amyloid-beta protein segment that ultimately would aggregate into the amyloid plaques of AD. This led to greatly increased work in the genetics of AD, especially to explain the genetic autosomal-dominant forms of AD caused by mutations on the APP gene or other mutations in presenilin 1 or 2 [10–12]. Determination of the structure and the importance of the A-beta peptide led to the preeminence of what is called the "amyloid cascade hypothesis" that describes the seminal role of amyloid in the pathogenesis of AD [13,14].

In about 1994, the most substantial genetic "risk factor" (i.e., not strictly causal, as autosomal-dominant gene mutations were) was discovered. The apolipoprotein E (APOE) epsilon 4 allele, one of three normally occurring alleles of APOE, increased the risk of AD but could be shown to be neither necessary nor sufficient to cause the disease itself [15–17]. One copy of the epsilon 4 (e4) allele was shown to increase the risk of AD by approximately four-fold, whereas two copies increased the risk by a range of 6- to 10-fold [18]. APOE e4, remains one of the strongest risk factors for AD. Although the exact mechanism by which APOE may promote AD symptoms or pathology remains unknown, many roles for APOE have been hypothesized; among them is the hypothesis that APOE acts in a way to enhance the supposed "amyloid cascade" [19]. For researchers studying the effects of environmental exposure on AD occurrence, failure to account for the APOE genotype in study design and analysis could lead to spurious results. This is because unequal distribution of such a strong risk factor among comparison groups formed to study environmental exposures could cause the unsuspected APOE effect to show an association between AD and the environmental factor being studied (due to confounding). In other words, the APOE genotype must be accounted for in the analysis of literally any putative environmental risk factors that would be studied. Further, because APOE could also be the basis of a gene by environmental interaction, the possibility of "effect modification" by APOE must also be investigated when studying environmental risk factors. Work to identify many other genetic associations has continued and expanded with dramatic progress in the technology. Many completed studies have observed the overwhelming effect of APOE on AD, but other observed associations, while statistically significant, are quite modest in comparative magnitude [20–22] and at present do not carry the same weight for environmental risk factor research as does the APOE genotype.

CLINICAL AND STUDY DESIGN CONSIDERATIONS

The pathologic description of AD is most directly available only at autopsy, and thus risk factor studies have generally had to rely on presumptive clinical diagnoses, or face the substantial delay, as well as the likely selection and recall biases, that would accompany a risk factor study based only on an autopsy series. In order to sharpen the clinical diagnosis of AD and improve concurrence between clinical and neuropathologic diagnoses, clinical diagnostic criteria were developed. Most of the early AD risk factor studies specified that an AD case must meet full clinical diagnostic criteria for dementia and AD [23,24], and relied on the symptoms and degree of cognitive impairment present to form a probabilistic diagnosis of AD. This symptomatic diagnosis of AD served well for concurrence with the pathologic diagnosis within autopsy cases for those that also showed symptoms; however, it was less than effective at delineating which of the "control" subjects (i.e., with normal cognition) might also have active and progressive AD pathology in their brains. Compounding the matter further, the NIA–Reagan consensus criteria for the neuropathologic definition of AD required, in addition to pathologic features, relatively frequent plaques and tangles in the brain, and patient symptoms consistent with clinical criteria for dementia [25,26]. This posed a major complication for risk factor studies, since case definition was limited to only those with clinical dementia symptoms present, regardless of the degree of AD pathology in the brain. For example, suppose that the nidus of AD pathology would begin perhaps 20 years prior to any cognitive symptoms. If researchers could observe that pathology in the brain near its inception, they may be able to also correctly determine which environmental exposures could have occurred prior to that time, then the probability of a valid association between exposure and the initial disease onset could be potentially observed. Whereas, if a substantial proportion of the "normal cognition" control group actually carried AD pathology in their brains but had no clinical symptoms, then other things being equal, the observed measure of association between exposure and disease could be biased away from the null (i.e., providing a spurious overestimate of the effect of a risk factor).

EARLY DISEASE RECOGNITION AND BIOMARKERS INFLUENCE STUDY DESIGN

Since about 2000, the AD research community has become more actively involved in characterizing the early symptomatic and/or asymptomatic disease process for AD. This interest has spurred the broader application of the concept of mild cognitive impairment (MCI) [27], and also increased interest in imaging and other biomarkers. With the development of new technology for in vivo imaging of amyloid deposits [28,29] and techniques for measurement and quantification of amyloid-beta, as well as other potential biomarkers in cerebrospinal fluid (CSF) [30–34], this important area of research remains a major focus of AD research today. Quite recently, methods are being developed to provide in vivo imaging of tau protein, the major component of the neurofibrillary tangles of AD, also found in neuritic plaques [35,36]. Thus, there is now the potential, through neuroimaging, to directly observe the primary pathologies of AD [37]. As this "biomarker" research began to accumulate and be validated, Jack et al. synthesized a unifying hypothetical model for the occurrence of AD pathologies and clinical symptoms [38] that showed a time-ordered development of amyloid deposits and tau protein, as well as the development of cognitive and functional deficits of clinical AD. While still considered a "hypothesis," its predictions seem to hold quite well, and it has become a widely accepted working model to characterize the progression of AD through biomarkers and clinical symptoms.

Because of its timeliness and apparent validity, the Jack model has contributed to the revision of both clinical and neuropathologic diagnostic criteria for AD. The clinical criteria now allow for a "preclinical" asymptomatic phase [39] as well as an MCI phase [27,40] in addition to a revised statement of the original AD dementia criteria [41]. Notably, the clinical criteria no longer cast AD as a so-called "diagnosis of exclusion," nor require that AD be the only diagnosis present. The clinical criteria also implicitly and explicitly recognize AD as a progressive neurodegenerative brain disease that leads to increasing cognitive impairment, dementia, and ultimately death. Similarly, the criteria for neuropathologic diagnosis of AD also have been significantly revised. The new neuropathologic criteria provide for an AD diagnosis that is based strictly on the neuropathologic features in the brain, regardless of clinical diagnosis [42,43]. Thus, the neuropathologic diagnosis of "asymptomatic" or preclinical disease, which was not possible under the old criteria, is now available to researchers. For researchers interested in exploring the effects of environmental factors, being able to include asymptomatic AD patients as "cases" is important, because it opens up new possibilities to reduce threats to study validity.

The evolution of knowledge and development of broader knowledge concerning the pathophysiological processes of AD will have profound effects on the investigation of etiologic risk factors for the disease. As Figure 15.1 shows, the temporal relationship between an exposure and the preclinical recognition of disease may allow us to find actual risk factors for the pathogenesis, if we are able to observe the incidence of amyloid or tau changes through positron emission tomography (PET) imaging or CSF biomarker assays. Large cohort studies with substantial, sequential neuroimaging may become the primary means to observe pathology as it begins and progresses, once those methods are well validated and standardized. Determining the time window in which past exposures could have affected the beginning of AD pathology is still a substantial challenge. However, it will be not nearly as confusing as the current situation, where a clinical dementia diagnosis is required for cases, while the critical period for an effective exposure might have ended 20 years earlier, concurrent with when the pathology actually began.

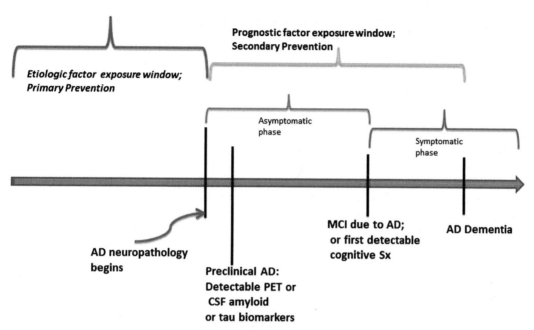

FIGURE 15.1　Time line showing typical temporal relationships for Alzheimer's disease (AD). The time line include the start of neuropathology, preclinical recognition, asymptomatic and symptomatic phases, and primary and secondary prevention time windows. Following primary prevention efforts and the start of AD neuropathology, the secondary prevention approach attempts to identify those factors associated with the onset or progression of observable cognitive decline from AD. Temporality of exposure relative to the inception of disease pathology often relies on a case definition based on dementia diagnosis, while it can be seen that for AD dementia as the defined outcome, initiation of pathology may have begun more than 20 years earlier.

A "secondary prevention" approach (e.g., preventing symptom onset among those who have AD brain pathology) is now frequently adopted because of the logistic difficulty of a "primary prevention" model in determining the incident occurrence of AD pathology in the brains of normal individuals [44]. Specifically, a secondary prevention approach would focus on those factors that may be associated with onset or progression of observable cognitive decline [45], as also shown in Figure 15.1. In other words, similar to initial risk factor studies where AD dementia was the categorical end point, now an alternative end point could be diagnosis of MCI due to AD. Or at an even earlier stage, one might choose the first noticeable symptoms thought to represent a change from customary and usual cognitive abilities as the categorical end point. Thus, the "risk factors" examined would actually be "prognostic" factors for the progression of the clinical/symptomatic disease. The problem of having a proportion of "controls" that harbor the AD pathology asymptomatically, however, remains to some extent. Other methods of comparison would require longitudinal measurement of cognitive and functional decline so that trajectories or slopes of the decline could then be compared. These sorts of comparisons may shed light on those factors that could delay the onset of cognitive symptoms, even among those who may have disease pathology as well as the progression of that pathology [30]. They could also be important, if eventually sanctioned for use by the FDA, in determining whether specific treatments slow the progression of symptoms.

The "A4 trial" [46] is an example of one such secondary prevention clinical trial [47]; this and several other trials used antiamyloid drugs to treat preclinical AD (i.e., amyloid positive biomarkers, but asymptomatic) with the hope of delaying or preventing cognitive symptoms. Similarly, in the context of strictly familial, autosomal-dominant AD, the DIAN study and the Alzheimer's Prevention Initiative [48] also sought to identify factors (and test medications) that would accelerate or delay the expected symptomatic onset, within mutation carriers, as compared with unexposed mutation carriers [49,50]. It was based on work from the DIAN study, estimating the interval between onset of observable brain pathology and cognitive symptom onset as 10–25 years [49,51]. Admittedly, these are genetically determined patients with family histories of young dementia onset, who inevitably will experience symptoms if they live long enough. The interval was estimated based on the usual age of symptomatic onset in the family. It is open to some discussion and conjecture whether a similar interval might be expected for nonautosomal-dominant (so-called "sporadic") cases; however, many in the AD research community subscribe to the view that it does apply.

CONSIDERING RISK FACTORS

Environmental risk factors must be viewed in context with genetic risks and host characteristics. Environmental factors are also potentially modified by genetics, and may act to increase or decrease risk of disease. It is often said that "age is the greatest risk factor for AD and dementia." While in one sense the statement is true, because the incidence of AD and dementia increase essentially exponentially with age, the increase is likely not due to age itself, but to the cellular effects of aging coupled with the age-related penetrance of genetic factors. Age itself is not a modifiable risk factor, unless an age limit were to be placed on lifespan by the government—and that is only the stuff of draconian science fiction. Examining factors that could modify the molecular and cellular effects of aging is a more laudable and researchable goal.

Historically, most early studies on environmental risk factors for AD were of rather rudimentary case–control design. AD cases with dementia formed the case groups and cognitively intact persons, usually of similar age and gender, constituted the control group. Exposure histories, prior to onset of the disease (or a similar time for controls), were then obtained in some way from each subject, and the exposure frequencies between groups were compared. While this sounds simple and straightforward, it was fraught with pitfalls. Potential biases from case or control selection methods, and whether selection might in some way be associated with exposure, often caused great and unrecognized difficulties. The differential recollection of past exposures by cases and controls led to recall bias in the measures of effects. Exposure history was often inadequately or incompletely determined. Poor assessment of temporality between risk factors and disease onset—for example, failure to account for whether putative risk factors could be actual effects of the disease syndrome (e.g., depression)—often led to so-called "reverse causation." Figure 15.1 shows clearly the difficulty in determining correctly the temporality of exposure relative to the inception of disease pathology when the case definition rests on dementia diagnosis while the initiation of pathology may have begun 20 years earlier—causing the etiologic exposure window to include only exposures prior to that time. Because of these and other methodological flaws, most of the case–control studies of environmental risk factors occurring between 1975 and 1990 should be viewed through the prism of skepticism, even though investigators of the time were attempting to conduct state-of-the-art investigations.

By about 1990, researchers realized that more valid answers to risk factor questions could be obtained using "cohort" study designs; that is, enrolling cognitively normal persons and examining them periodically over time to determine whether their cognition had worsened, or worsened to the point of dementia [52–54]. These designs eliminated much of the potential selection bias seen in case–control studies and also had the advantage of obtaining exposure histories directly from cognitively intact subjects, since despite imperfect recall the subjects did not know if they would later develop symptomatic disease. The results of environmental risk factor studies based on cohort designs are generally more likely to be valid than those previously obtained from case–control studies. Still, case–control studies that are very well done can provide valid estimates. Thus, it is the reader's task to examine each such study critically in order to make a judgment about the potential value and validity of its results.

EXAMPLES OF ENVIRONMENTAL RISK FACTORS

One example of contrasting results obtained from the two types of designs is seen in the study of smoking as a risk factor for AD. Initial case–control studies conducted by a number of distinguished investigators were summarized in a detailed review [55]. The case–control studies almost uniformly showed a "protective" effect of smoking on AD. Worried about potential biases, one investigator proposed scenarios by which subject selection factors could potentially account for the observed beneficial effect of smoking seen in these case–control studies [56]. These explanatory scenarios revolved around whether AD cases who were smokers would be less available to enroll in research studies because of smoking-related illnesses or earlier death, as compared with AD patients who were nonsmokers. Of course, some studies may have excluded subjects with any indications of vascular disease, unwittingly also eliminating smokers from the case series. The hypothetical scenarios were, in a sense, validated as cohort studies began to accumulate and were evaluated. The effect of smoking on AD is now generally recognized as *increasing* the risk of AD rather than decreasing it [54,57]. The increased risk is thought to be due to smoking's influence on vascular disease.

Perhaps the largest effective advance in thinking about risk factors for AD and other neurodegenerative diseases has come from the realization that vascular disease, once excluded or held separate from the study of AD, may act synergistically to promote AD pathology as well as symptomatic progression [58–61]. This is a vitally important shift, because it opens the door to potential existing and conventional treatments that could delay the onset of cognitive symptoms. It is also important because vascular disease at some level exists in practically all elderly brains [58,62,63]. Vascular disease, including hypertension, hyperlipidemia, atherosclerosis, and similar conditions may "promote subclinical brain injury" (p. 1, [58]), and could also be the basis for previously observed associations of AD with physical activity, statins, homocysteine, diabetes/metabolic syndrome, and even smoking. Further, most of these vascular risk factors are modifiable with existing treatments, and therefore could impact the occurrence of dementia generally [64]. The key of course, is to start intervention on these vascular-related factors at midlife or even earlier if present. Effective treatment could act to reduce the incidence of dementia due to AD and other neurodegenerative and vascular causes [65].

Greater years of education are said to contribute to "cognitive reserve," or the ability of individuals to cope with and delay the expression of cognitive impairment, given the presence of an underlying brain pathology that would ultimately cause such impairment [65]. However, the effect of education on the expression of cognitive symptoms appears to be trumped by the presence of more severe pathology [66]. One complication of studying the effect of education is that in recruitment for these studies, higher education is strongly related to participation, especially for those without cognitive impairment. Therefore, especially among case–control studies, this effective selection bias would generally make education appear to be protective. Among cohort studies the problem is less severe, but because the range of educational attainment may be restricted in initial recruitment, inferences concerning education may be biased. In settings where a fair sampling of the population educational range occurs, the initial cohort could more directly be analyzed for the unbiased effect of education. As with vascular factors, however, additional education attempted in later life, after cognitive decline has begun, is unlikely to result in a modified course of cognitive decline.

The potential association between nonsteroidal anti-inflammatory drugs and AD has been reported for many years [67–69]. However, it too may have been the victim of inappropriate timing of exposures relative to disease onset. As shown above in Figure 15.1, if the outcome measure is AD dementia, the actual pathology for AD may begin many years earlier (perhaps 10–25 years earlier [49]). Typically, case–control and some cohort studies have counted exposures occurring up to a few years prior to onset of cognitive impairment or diagnosis of dementia. While those exposures fall in the time window for secondary prevention, they would not likely be accurate for the earlier period necessary for identifying them as etiologic risk factors. Further, the use of anti-inflammatory drugs could be prompted by early non-specific effects of the disease, adding to this complicated scenario. The ADAPT trial was a randomized placebo-controlled trial of nonsteroidal anti-inflammatory drugs administered at the onset of dementia. Subjects entering the trial were relatively older in age, in order to maximize the potential to observe sufficient dementia cases [70]. If it were done today, care would be taken to conduct amyloid-PET imaging or CSF biomarker assays to at least stratify the presence of AD pathology indicated by those biomarkers. However, as with most risk factor studies, those means were not available at the time. The ADAPT trial was stopped early due to adverse events associated with the treatments, and no association with dementia was seen. Neuroinflammation is still considered an important potential promoter or by-product of neurodegenerative disease, and its presence may serve as an appropriate drug target to potentially slow the disease process [71].

In a number of early studies, hormone therapy in the form of estrogen replacement (estradiol with or without progestin) was observed to be a potential protective factor against AD dementia [72–77]. However, a large randomized trial of hormone therapy, the Women's Health Initiative Memory Study [78], showed a two-fold increased risk of dementia in the treated groups. Randomized trials are usually thought to carry more weight than observational studies, and many concluded that the observational studies had simply been biased in some way, especially because of the lack of protective effect for cardiovascular disease, which was also thought to be a reason for their prescription. Nonetheless, the controversy continues about the type and direction of effects, as well as for whom and at what age, that may be result from estrogen therapy [79].

The effects of head trauma on AD have enjoyed a rather checkered and equivocal past [80–85], leaving it difficult interpret its actual effect on AD. Head injury, of course, can result in immediate and long-lasting cognitive impairment. As shown more recently [86], chronic repetitive head injury, perhaps without loss of consciousness, can also result in chronic traumatic encephalopathy (CTE), as seen among football players and boxers [87]—but CTE is not AD. Falls, motor vehicle accidents, and violence account for the vast majority of head injuries. The recent study by Mielke [80] found no significant association between the presence of PET-PiB detected amyloid (a biomarker for AD) and reports of past head injury among cognitively normal individuals. However, when the same analysis was run among persons with MCI, an association appeared. These conflicting results raise questions concerning whether a reporting bias may have been active. The study is nonetheless an excellent example of the use of new technology to define the earliest stages of AD and to attempt to examine etiologic risk factors. With millions of Americans now having served in Iraq and Afghanistan, and many having suffered traumatic brain injury or postconcussive blast injuries, the potential exists for an epidemic of dementia and/or AD, as well as a variety of other behavioral and neurological conditions due to those injuries [88–90].

CONCLUSIONS

Attempting to find a few good, modifiable environmental risk factors for AD (or even dementia generally) seemed like a very achievable goal when most risk factor studies began in the early 1980s. However, it has proved to be quite an elusive goal. Changes in how we understand and define the disease seem to compound the difficulties in obtaining valid and accurate exposure histories. In risk factor studies as in comedy, timing is everything. If we cannot accurately place the exposure prior to the onset of disease, we simply cannot draw the conclusion that an observed association is causal for that onset. As pointed out, however, we now have a significant opportunity to identify those factors that may be associated with symptomatic progression of the disease among those asymptomatic persons who have imaging-observed pathology. Further, because of the common occurrence of vascular factors and their effect on cognitive processes, we may be able to delay or prevent cognitive decline significantly by providing known and accepted treatments for hypertension and other cardiovascular risk factors. It would not hurt to promote exercise, good diet, avoidance of tobacco, and other everyday actions that promote good health. The problem with this is that it all must occur in midlife (e.g., 30–50 years of age); it is not likely to be effective at all in 75-year-old subjects.

Our lack of success at finding stable and replicable environmental risk factors specifically for AD should be remedied, as we are more able to directly observe the very initial, asymptomatic amyloid and tau deposits characteristic of AD through new imaging techniques; there is also the possibility of even better and simpler biomarkers. We must be careful to follow rigorous study-design tenets and draw from the information included in Figure 15.1, if we are to eventually succeed in this effort.

References

[1] Alzheimer A. On a distinctive disease of the cerebral cortex. Z Psychiatry 1907;64:146–8.
[2] Tyler HR. Alzheimer's disease: senile dementia and related disorders. In: Katzman R, Terry RD, Bick KL, editors. New York, NY: Raven Press; 1978. pp. 614, $49.00. Wiley Online Library; 1979.
[3] Terry RD, Katzman RK. Senile dementia of the Alzheimer type. Ann Neurol 1983;14(5):497–506.

[4] Glenner GG, Wong CW. Alzheimer's disease: initial report of the purification and characterization of a novel cerebrovascular amyloid protein. Biochem Biophys Res Commun 1984;120(3):885–90.

[5] Wisniewski H, Wen G. Substructures of paired helical filaments from Alzheimer's disease neurofibrillary tangles. Acta Neuropathol 1985;66(2):173–6.

[6] Masters CL, Multhaup G, Simms G, Pottgiesser J, Martins R, Beyreuther K. Neuronal origin of a cerebral amyloid: neurofibrillary tangles of Alzheimer's disease contain the same protein as the amyloid of plaque cores and blood vessels. EMBO J 1985;4(11):2757.

[7] Iqbal K, Zaidi T, Wen G, Grundke-Iqbal I, Merz P, Shaikh S, et al. Defective brain microtubule assembly in Alzheimer's disease. Lancet 1986;328(8504):421–6.

[8] Braak H, Braak E. Staging of Alzheimer's disease-related neurofibrillary changes. Neurobiol Aging 1995;16(3): 271–8.

[9] Goldgaber D, Lerman M, McBride O, Saffiotti U, Gajdusek D. Characterization and chromosomal localization of a cDNA encoding brain amyloid of Alzheimer's disease. Alzheimer Dis Assoc Disord 1988;2(2):135.

[10] St George-Hyslop PH. Molecular genetics of Alzheimer's disease. Biol Psychiatry 2000;47(3):183–99.

[11] Van Broeckhoven C, Backhovens H, Cruts M, De Winter G, Bruyland M, Cras P, et al. Mapping of a gene predisposing to early–onset Alzheimer's disease to chromosome 14q24. 3. Nat Genet 1992;2(4):335–9.

[12] Schellenberg GD, Payami H, Wijsman EM, Orr HT, Goddard KA, Anderson L, et al. Chromosome 14 and late-onset familial Alzheimer disease (FAD). Am J Hum Genet 1993;53(3):619–28.

[13] Selkoe DJ. Alzheimer's disease: a central role for amyloid. J Neuropathol Exp Neurol 1994;53(5):438–47.

[14] Hardy J, Selkoe DJ. The amyloid hypothesis of Alzheimer's disease: progress and problems on the road to therapeutics. Science 2002;297(5580):353–6.

[15] Saunders AM, Strittmatter WJ, Schmechel D, George-Hyslop PH, Pericak-Vance MA, Joo SH, et al. Association of apolipoprotein E allele epsilon 4 with late-onset familial and sporadic Alzheimer's disease. [see comments] Neurology 1993;43(8):1467–72.

[16] Roses AD, Saunders AM. APOE is a major susceptibility gene for Alzheimer's disease. Curr Opin Biotechnol 1994;5(6):663–7.

[17] Corder EH, Saunders AM, Strittmatter WJ, Schmechel DE, Gaskell PC, Small GW, et al. Gene dose of apolipoprotein E type 4 allele and the risk of Alzheimer's disease in late onset families. [see comments] Science 1993;261(5123):921–3.

[18] Farrer LA, Cupples LA, Haines JL, Hyman B, Kukull WA, Mayeux R, et al. Effects of age, sex, and ethnicity on the association between apolipoprotein E genotype and Alzheimer disease. A meta-analysis. APOE and Alzheimer Disease Meta Analysis Consortium. [see comments] JAMA: J Am Med Assoc 1997;278(16):1349–56.

[19] Potter H, Wisniewski T, Apolipoprotein E. Essential catalyst of the Alzheimer amyloid cascade. Int J Alzheimer's Dis 2012;2012.

[20] Lambert J-C, Ibrahim-Verbaas CA, Harold D, Naj AC, Sims R, Bellenguez C, et al. Meta-analysis of 74,046 individuals identifies 11 new susceptibility loci for Alzheimer's disease. Nat Genet 2013;45:1452–8.

[21] Guerreiro R, Wojtas A, Bras J, Carrasquillo M, Rogaeva E, Majounie E, et al. TREM2 variants in Alzheimer's disease. N Engl J Med 2013;368(2):117–27.

[22] Naj AC, Jun G, Beecham GW, Wang LS, Vardarajan BN, Buros J, et al. Common variants at MS4A4/MS4A6E, CD2AP, CD33 and EPHA1 are associated with late-onset Alzheimer's disease. Nat Genet 2011;43(5):436–41.

[23] McKhann G, Drachman D, Folstein M, Katzman R, Price D, Stadlan EM. Clinical diagnosis of Alzheimer's disease: report of the NINCDS-ADRDA Work Group under the auspices of Department of Health and Human Services Task Force on Alzheimer's Disease. Neurology 1984;34(7):939–44.

[24] American Psychiatric Association A. Diagnostic criteria from DSM-IV. Washington, DC: American Psychiatric Association; 1994.

[25] Phelps C, Khachaturian Z, Trojanowski J. NIA/Reagan Institute Working Group on diagnostic criteria for the neuropathological assessment of Alzheimer's disease: consensus recommendations for the post-mortem diagnosis of Alzheimer's disease. Neurobiol Aging 1997;81:S1–2.

[26] Hyman BT, Trojanowski JQ. Editorial on consensus recommendations for the postmortem diagnosis of Alzheimer disease from the National Institute on Aging and the Reagan Institute Working Group on diagnostic criteria for the neuropathological assessment of Alzheimer disease. J Neuropathol Exp Neurol 1997;56(10):1095–7.

[27] Petersen RC. Mild cognitive impairment as a diagnostic entity. J Intern Med 2004;256(3):183–94.

[28] Klunk WE, Engler H, Nordberg A, Bacskai BJ, Wang Y, Price JC, et al. Imaging the pathology of Alzheimer's disease: amyloid-imaging with positron emission tomography. Neuroimaging Clin N Am 2003;13(4):781–9, ix.

[29] Mintun MA, Larossa GN, Sheline YI, Dence CS, Lee SY, Mach RH, et al. [11C]PIB in a nondemented population: potential antecedent marker of Alzheimer disease. Neurology 2006;67(3):446–52.

[30] Fagan AM, Chhatwal J, Morris JC, Rossor M, Bateman R, Benzinger T. Longitudinal change in csf biomarkers in dian. Alzheimer's Dementia: J Alzheimer's Assoc 2014;10(4):P127–8.

[31] Fagan AM, Roe CM, Xiong C, Mintun MA, Morris JC, Holtzman DM. Cerebrospinal fluid tau/beta-amyloid(42) ratio as a prediction of cognitive decline in nondemented older adults. Arch Neurol 2007;64(3):343–9.

[32] Da X, Toledo JB, Zee J, Wolk DA, Xie SX, Ou Y, et al. Integration and relative value of biomarkers for prediction of MCI to AD progression: spatial patterns of brain atrophy, cognitive scores, APOE genotype and CSF biomarkers. NeuroImage Clin 2014;4:164–73.

[33] Alves G, Lange J, Blennow K, Zetterberg H, Andreasson U, Forland MG, et al. CSF Abeta42 predicts early-onset dementia in Parkinson disease. Neurology 2014;82(20):1784–90.

[34] Palmqvist S, Zetterberg H, Blennow K, Vestberg S, Andreasson U, Brooks DJ, et al. Accuracy of brain amyloid detection in clinical practice using cerebrospinal fluid beta-amyloid 42: a cross-validation study against amyloid positron emission tomography. JAMA Neurol 2014;71(10):1282–9.

[35] Xia C-F, Arteaga J, Chen G, Gangadharmath U, Gomez LF, Kasi D, et al. [18F] T807, a novel tau positron emission tomography imaging agent for Alzheimer's disease. Alzheimer's Dementia 2013;9(6):666–76.

[36] Chien DT, Szardenings AK, Bahri S, Walsh JC, Mu F, Xia C, et al. Early clinical PET imaging results with the novel PHF-tau radioligand [F18]-T808. J Alzheimer's Dis 2014;38(1):171–84.

[37] Trojanowski JQ, Hampel H. Neurodegenerative disease biomarkers: guideposts for disease prevention through early diagnosis and intervention. Prog Neurobiol 2011;95(4):491.

[38] Jack Jr CR, Knopman DS, Jagust WJ, Petersen RC, Weiner MW, Aisen PS, et al. Tracking pathophysiological processes in Alzheimer's disease: an updated hypothetical model of dynamic biomarkers. Lancet Neurol 2013;12(2):207–16.

[39] Sperling RA, Aisen PS, Beckett LA, Bennett DA, Craft S, Fagan AM, et al. Toward defining the preclinical stages of Alzheimer's disease: recommendations from the National Institute on Aging-Alzheimer's Association workgroups on diagnostic guidelines for Alzheimer's disease. Alzheimer's Dementia: J Alzheimer's Assoc 2011;7(3):280–92.

[40] Albert MS, DeKosky ST, Dickson D, Dubois B, Feldman HH, Fox NC, et al. The diagnosis of mild cognitive impairment due to Alzheimer's disease: recommendations from the National Institute on Aging-Alzheimer's Association workgroups on diagnostic guidelines for Alzheimer's disease. Alzheimer's Dementia: J Alzheimer's Assoc 2011;7(3):270–9.

[41] McKhann GM, Knopman DS, Chertkow H, Hyman BT, Jack Jr CR, Kawas CH, et al. The diagnosis of dementia due to Alzheimer's disease: recommendations from the National Institute on Aging-Alzheimer's Association workgroups on diagnostic guidelines for Alzheimer's disease. Alzheimer's Dementia: J Alzheimer's Assoc 2011;7(3):263–9.

[42] Hyman BT, Phelps CH, Beach TG, Bigio EH, Cairns NJ, Carrillo MC, et al. National Institute on Aging-Alzheimer's Association guidelines for the neuropathologic assessment of Alzheimer's disease. Alzheimer's Dementia: J Alzheimer's Assoc 2012;8(1):1–13.

[43] Montine TJ, Phelps CH, Beach TG, Bigio EH, Cairns NJ, Dickson DW, et al. National Institute on Aging–Alzheimer's Association guidelines for the neuropathologic assessment of Alzheimer's disease: a practical approach. Acta Neuropathol 2012;123(1):1–11.

[44] Bateman RJ, Aisen PS, De Strooper B, Fox NC, Lemere CA, Ringman JM, et al. Autosomal-dominant Alzheimer's disease: a review and proposal for the prevention of Alzheimer's disease. Alzheimers Res Ther 2011;3(1):1.

[45] Fratiglioni L, Qiu C. Prevention of cognitive decline in ageing: dementia as the target, delayed onset as the goal. Lancet Neurol 2011;10(9):778–9.

[46] Sperling R, Donohue M, Aisen P. The A4 trial: anti-amyloid treatment of asymptomatic Alzheimer's disease. Alzheimer's Dementia 2012;8(4):P425–6.

[47] Selkoe DJ. Preventing Alzheimer's disease. Science 2012;337(6101):1488–92.

[48] Reiman E, Lopera F, Langbaum J, Fleisher A, Ayutyanont N, Quiroz Y, et al. The Alzheimer's prevention initiative. Alzheimer's Dementia 2012;8(4):P426.

[49] Bateman RJ, Xiong C, Benzinger TL, Fagan AM, Goate A, Fox NC, et al. Clinical and biomarker changes in dominantly inherited Alzheimer's disease. N Engl J Med 2012;367(9):795–804.

[50] Mills S, Mallmann J, Santacruz A, Fuqua A, Carril M, Aisen P, et al. Preclinical trials in autosomal dominant AD: implementation of the DIAN-TU trial. Rev Neurol 2013;169(10):737–43.

[51] Bateman R, Benzinger T, Cairns N, Fagan A, Goate A, Marcus D, et al. Presymptomatic Alzheimer's disease in the dominantly inherited Alzheimer's network (DIAN). Alzheimer's Dementia 2012;8(4):P4.

[52] Tangney CC, Kwasny MJ, Li H, Wilson RS, Evans DA, Morris MC. Adherence to a mediterranean-type dietary pattern and cognitive decline in a community population. Am J Clin Nutr 2011;93(3):601–7.

[53] Negash S, Wilson RS, Leurgans SE, Wolk DA, Schneider JA, Buchman AS, et al. Resilient brain aging: characterization of discordance between Alzheimer's disease pathology and cognition. Curr Alzheimer Res 2013;10(8):844–51.

[54] Launer LJ, Andersen K, Dewey ME, Letenneur L, Ott A, Amaducci LA, et al. Rates and risk factors for dementia and Alzheimer's disease: results from EURODEM pooled analyses. EURODEM Incidence Research Group and Work Groups. European studies of dementia. Neurology 1999;52(1):78–84.

[55] Lee PN. Smoking and Alzheimer's disease: a review of the epidemiological evidence. Neuroepidemiology 1994;13(4):131–44.

[56] Kukull WA. The association between smoking and Alzheimer's disease: effects of study design and bias. Biol Psychiatry 2001;49(3):194–9.

[57] Breteler MM, Bots ML, Ott A, Hofman A. Risk factors for vascular disease and dementia. Haemostasis 1998;28(3–4):167–73.

[58] Chui HC, Zheng L, Reed BR, Vinters HV, Mack WJ. Vascular risk factors and Alzheimer's disease: are these risk factors for plaques and tangles or for concomitant vascular pathology that increases the likelihood of dementia? An evidence-based review. Alzheimer's Res Ther 2012;3(6):36.

[59] Kalaria RN, Akinyemi R, Ihara M. Does vascular pathology contribute to Alzheimer changes? J Neurol Sci 2012;322(1):141–7.

[60] Kling MA, Trojanowski JQ, Wolk DA, Lee VM, Arnold SE. Vascular disease and dementias: paradigm shifts to drive research in new directions. Alzheimer's Dementia 2013;9(1):76–92.

[61] DeCarli C, Mungas D, Carmichael OT, Villeneuve S, Fletcher E, Singh B, et al. Vascular risk factors impact cognition independent of pib pet and mri measures of ad and vascular brain injury. Alzheimer's Dementia: J Alzheimer's Assoc 2014;10(4):P2–3.

[62] Gorelick PB, Scuteri A, Black SE, DeCarli C, Greenberg SM, Iadecola C, et al. Vascular contributions to cognitive impairment and dementia a statement for healthcare professionals from the American Heart Association/ American Stroke Association. Stroke; A J Cereb Circ 2011;42(9):2672–713.

[63] Toledo JB, Arnold SE, Raible K, Brettschneider J, Xie SX, Grossman M, et al. Contribution of cerebrovascular disease in autopsy confirmed neurodegenerative disease cases in the National Alzheimer's Coordinating Centre. Brain 2013;136(9):2697–706.

[64] Norton S, Matthews FE, Barnes DE, Yaffe K, Brayne C. Potential for primary prevention of Alzheimer's disease: an analysis of population-based data. Lancet Neurol 2014;13(8):788–94.

[65] Imtiaz B, Tolppanen A-M, Kivipelto M, Soininen H. Future directions in Alzheimer's disease from risk factors to prevention. Biochem Pharmacol 2014;88(4):661–70.

[66] Koepsell TD, Kurland BF, Harel O, Johnson EA, Zhou XH, Kukull WA. Education, cognitive function, and severity of neuropathology in Alzheimer disease. Neurology 2007;70(19 Pt 2):1732–9.

[67] Breitner JC. Inflammatory processes and antiinflammatory drugs in Alzheimer's disease: a current appraisal. Neurobiol Aging 1996;17(5):789–94.

[68] in t' Veld BA, Ruitenberg A, Hofman A, Launer LJ, van Duijn CM, Stijnen T, et al. Nonsteroidal antiinflammatory drugs and the risk of Alzheimer's disease. N Engl J Med 2001;345(21):1515–21.

[69] Szekely CA, Thorne JE, Zandi PP, Ek M, Messias E, Breitner JC, et al. Nonsteroidal anti-inflammatory drugs for the prevention of Alzheimer's disease: a systematic review. Neuroepidemiology 2004;23(4):159–69.

[70] Meinert CL, McCaffrey LD, Breitner JC. Alzheimer's disease anti-inflammatory prevention trial: design, methods, and baseline results. Alzheimer's Dementia: J Alzheimer's Assoc 2009;5(2):93–104.

[71] Cudaback E, Jorstad NL, Yang Y, Montine TJ, Keene CD. Therapeutic implications of the prostaglandin pathway in Alzheimer's disease. Biochem Pharmacol 2014;88(4):565–72.

[72] Birge SJ. Hormones and the aging brain. Geriatrics 1998;53(Suppl. 1):S28–30.

[73] Henderson VW. Estrogen, cognition, and a woman's risk of Alzheimer's disease. Am J Med 1997;103(3A):11S–8S.

[74] Henderson VW. The epidemiology of estrogen replacement therapy and Alzheimer's disease. Neurology 1997;48(5 Suppl. 7):S27–35.

[75] Henderson VW, Paganini-Hill A, Miller BL, Elble RJ, Reyes PF, Shoupe D, et al. Estrogen for Alzheimer's disease in women: randomized, double-blind, placebo-controlled trial. Neurology 2000;54(2):295–301.

[76] Slooter AJ, Bronzova J, Witteman JC, Van Broeckhoven C, Hofman A, van Duijn CM. Estrogen use and early onset Alzheimer's disease: a population-based study. J Neurol Neurosurg Psychiatry 1999;67(6):779–81.

[77] Yaffe K, Sawaya G, Lieberburg I, Grady D. Estrogen therapy in postmenopausal women: effects on cognitive function and dementia. JAMA: J Am Med Assoc 1998;279(9):688–95.

[78] Shumaker SA, Legault C, Rapp SR, Thal L, Wallace RB, Ockene JK, et al. Estrogen plus progestin and the incidence of dementia and mild cognitive impairment in postmenopausal women: the women's health initiative memory study: a randomized controlled trial. JAMA: J Am Med Assoc 2003;289(20):2651–62.

[79] Henderson VW, Brinton RD. Menopause and mitochondria: windows into estrogen effects on Alzheimer's disease risk and therapy. Prog Brain Res 2010;182:77–96.

[80] Mielke MM, Savica R, Wiste HJ, Weigand SD, Vemuri P, Knopman DS, et al. Head trauma and in vivo measures of amyloid and neurodegeneration in a population-based study. Neurology 2014;82(1):70–6.

[81] Molgaard CA, Stanford EP, Morton DJ, Ryden LA, Schubert KR, Golbeck AL. Epidemiology of head trauma and neurocognitive impairment in a multi-ethnic population. Neuroepidemiology 1990;9(5):233–42.

[82] Brayne C. The EURODEM collaborative re-analysis of case-control studies of Alzheimer's disease: implications for public health. Int J Epidemiol 1991;20(Suppl. 2):S68–71.

[83] Heyman A, Wilkinson WE, Stafford JA, Helms MJ, Sigmon AH, Weinberg T. Alzheimer's disease: a study of epidemiological aspects. Ann Neurol 1984;15(4):335–41.

[84] O'Meara ES, Kukull WA, Sheppard L, Bowen JD, McCormick WC, Teri L, et al. Head injury and risk of Alzheimer's disease by apolipoprotein E genotype. Am J Epidemiol 1997;146(5):373–84.

[85] Tang MX, Maestre G, Tsai WY, Liu XH, Feng L, Chung WY, et al. Effect of age, ethnicity, and head injury on the association between APOE genotypes and Alzheimer's disease. Ann N Y Acad Sci 1996;802:6–15.

[86] McKee AC, Stein TD, Nowinski CJ, Stern RA, Daneshvar DH, Alvarez VE, et al. The spectrum of disease in chronic traumatic encephalopathy. Brain 2013;136(Pt 1):43–64.

[87] Vent J, Koenig J, Hellmich M, Huettenbrink K-B, Damm M. Impact of recurrent head trauma on olfactory function in boxers: a matched pairs analysis. Brain Res 2010;1320:1–6.

[88] Brenner LA, Ivins BJ, Schwab K, Warden D, Nelson LA, Jaffee M, et al. Traumatic brain injury, posttraumatic stress disorder, and postconcussive symptom reporting among troops returning from Iraq. J Head Trauma Rehabil 2010;25(5):307–12.

[89] Bogdanova Y, Verfaellie M. Cognitive sequelae of blast-induced traumatic brain injury: recovery and rehabilitation. Neuropsychol Rev 2012;22(1):4–20.

[90] Eskridge SL, Macera CA, Galarneau MR, Holbrook TL, Woodruff SI, MacGregor AJ, et al. Influence of combat blast-related mild traumatic brain injury acute symptoms on mental health and service discharge outcomes. J Neurotrauma 2013;30(16):1391–7.

Environmental Factors and Amyotrophic Lateral Sclerosis: What Do We Know?

Pam Factor-Livak

INTRODUCTION

Amyotrophic lateral sclerosis (ALS) is a devastating neurodegenerative disease of motor neurons that results in worsening weakness of voluntary muscles, progressive paralysis, and death, usually from respiratory failure [1]. The median survival time from symptom onset is approximately 3 years, although survival can range from several months to over 20 years. Recent developments in molecular genetics have increased the understanding of many genetic/familial forms of ALS and other motor neuron diseases (MNDs) [2–6]; nevertheless, most ALS cases are not attributable to genes alone, and there is much to be learned from studying environmental factors in the etiology and progression of ALS and other MNDs. This review first discusses the basic epidemiology of ALS and the challenges of studying it. Second, because many identified risk factors for ALS are related to oxidative stress, it reviews the key studies suggesting such a relationship with disease etiology and progression. Third, it reviews other selected environmental risk factors related to antioxidants. The chapter ends with recommendations for future work. The chapter focuses on sporadic ALS, unless otherwise noted.

EPIDEMIOLOGY AND CHALLENGES TO THE STUDY OF AMYOTROPHIC LATERAL SCLEROSIS

Classification, diagnostic criteria, and case ascertainment: Three major forms of ALS have been identified: (a) sporadic; (b) familial or heredity; and (c) associated with clusters in the Western Pacific (Mariana Islands). The last of these was first identified among the Chamorro population of Guam, and is characterized by symptoms similar to those of Parkinson's disease (PD) and Alzheimer's disease in addition to ALS characteristics [7,8]. Several other indigenous populations in the Pacific, notably areas of Western New Guinea and the Kii Peninsula of Japan, also demonstrate an excess of this form of ALS [9,10], suggesting either a genetic variant common among these people or a common environmental factor; unfortunately, the causes of these clusters remain unknown.

A recent PubMed search identified over 1000 papers with the key words ALS and epidemiology. These studies have provided information regarding sociodemographic characteristics, disease phenotype, and geographic variation of ALS cases, and case–control studies have examined risk factors such as lead exposure, smoking, and pesticide exposure. Several prospective cohort studies have also identified risk and protective factors.

Challenges: Despite the increase in the size of the research base, as Armon points out [11], there are inherent limitations to the epidemiologic investigations of rare, and on average rapidly fatal, diseases such as ALS (Table 16.1).

First, case ascertainment and follow-up pose specific problems that lead to selection bias. Adequate and complete case ascertainment is crucial to the study of environmental risk factors for ALS. Case–control studies recruit patients from ALS specialty clinics, resulting in a nonrepresentative case group in comparison with the general ALS population, many of whom receive care in the community. Patients who are older, more debilitated or of lower socioeconomic circumstances are likely to be underrepresented in these clinics. Yoshida et al. [12] found that patients recruited into studies from specialty clinics are more likely to be younger and have longer survival times than patients recruited from the general community. Further, case–control

TABLE 16.1   Challenges to the Study of Environmental Factors and ALS

- Case ascertainment (selection bias)
  - Cases with longer survival time are more likely to join studies
  - Recruitment from specialty clinics
  - Willingness to participate in nontreatment studies
  - Loss to follow-up in large prospective cohort studies
- Poor definition of phenotype
- Data collection
  - Queries regarding past exposures
  - Biomarkers collected after disease onset

studies likely ascertain cases with longer survival times; these do not reflect the larger pool of ALS cases. A second factor leading to selection bias is refusal to participate in nontreatment studies, as ALS patients perceive no direct benefit. Third, although large cohort studies ensure that exposure occurred prior to disease onset, there are some inherent bases in these studies as well. For example, loss to follow-up of subjects leads to bias if the proportion lost is differential in its incidence of the disease, compared with the entire study group. Further, large cohort studies focusing on special populations; for example, the Nurses' Health Study or the Health Professionals Health Study, may not be representative of the general population. The prospective Cancer Prevention Study II cohort (CPS-II) has been used to explore associations between a variety of risk factors and ALS (e.g., [13]); this is an important step forward, as the CPS-II recruited over one million US adults who were enrolled in 1982. In 1989, death certificates for members of the cohort were used to identify 621 deaths from ALS. Data on possible risk factors, ascertained at cohort recruitment, have been and are being used to identify risk factors in this large cohort. Wang et al. also combine several large prospective cohorts, the Nurses' Health Study, the Health Professionals Follow-Up Study, the Cancer Prevention Study II Nutrition (CPS-IIN), the Multiethnic Cohort, and the NIH–American Association of Retired Persons Diet and Health Study (AARP-DH), for an even larger sample size and number of cases, in order to evaluate risk factors for ALS [15]. Several large population-based cohorts are also available in Europe. For example, Chio et al. [15] used the Piemonte and Valle d'Aosta Register for ALS (PARALS), which monitored cases of ALS in two regions of northwestern Italy (over four million persons) for 10 years. Overall, 1347 ALS cases were identified, most of them (1260) categorized as definite or probable. These prospective studies, if data are available, have the most promise for identifying environmental etiologic factors for ALS in larger populations.

A second limitation in epidemiologic studies of ALS is poor definition of the phenotype. The lack of precise diagnoses and measures of disease severity are likely to introduce measurement error. Correct diagnosis and classification of ALS is crucial for etiologic and progression studies. ALS is primarily a clinical diagnosis, as no biomarkers have been identified. Two major consensus conferences have formalized diagnostic criteria, known as the El Escorial World Federation of Neurology Criteria with the Airlie House revised criteria, which include clinical, electrophysiological, and/or neuropathological criteria. Patients are classified as definite, probable, possible, and suspected ALS under these criteria. Studies undertaken in large research consortia widely use the El Escorial ALS criteria [16] to characterize their patient populations, while older smaller studies do not.

The third limitation is one common to many epidemiologic studies, and concerns data collection methods. Many epidemiologic studies rely, at least in part, on questionnaires administered to subjects. In case–control studies, recall may differ between cases and controls, as cases may search their memories in order to explain the cause of their disease. This is less of a problem in cohort designs. More recent studies have used biomarkers of exposure to circumvent such bias. If biospecimens are available before disease onset in cohort studies, then their analysis provides evidence of exposure preceding disease onset. However, it is likely that the initiation of the disease process occurred many years prior to disease diagnosis. Thus, analysis of biospecimens in cohort as well as in case–control studies requires more caution in interpretation, as disease processes may change the accumulation of exposures or metabolism of exposures. Further, using generally accepted data collection instruments, including questionnaires, allows for comparisons across studies and combining of data across studies.

It is likely that associations between environmental risk factors and ALS will be small, with odds ratios (ORs) on the order of 1.5–2. Moreover, because many risk factors are relatively ubiquitous—for example, electromagnetic field exposure—a large number of cases is required. In order to study these exposures fully, one option is to use nested case–control studies; i.e., cases embedded in the large cohorts (e.g., CPS-II), and selected controls without ALS from the same cohorts. This is an efficient design option that may be more representative of ALS cases than those from referral centers.

Overview of ALS descriptive epidemiology: In the United States, excluding cases in the Pacific cluster regions, the incidence rate of sporadic ALS is approximately two per 100,000 per year. Internationally, the incidence ranges from 0.86 to 2.4 per 100,000 per year (reviewed in [17]).

In a pilot study of mortality from ALS in Havana, Cuba, the incidence rate was estimated as 0.83/100,000 [18], a rate consistent with that of US Hispanics, estimated from a systematic review to be 0.9/100,000 (95% confidence interval (CI)=0.8–1.1/100,000) [19]. In this study, case recruitment is likely free of ascertainment bias due to equal access to health care, with the caveat that ALS may be underdiagnosed, especially in rural areas, due to decreased availability of neurological services.

Age-adjusted incidence rates are generally higher among men than among women, although this difference is less pronounced in studies in the United States. In general, incidence rates are higher among those living in northern latitudes than among those in southern latitudes. A variety of studies suggest that rates are higher in White than in nonWhite (including Hispanic) populations (reviewed in [17]); however, there could be ascertainment bias among minorities, who may not present to specialty clinics for adequate diagnosis.

ENVIRONMENTAL RISK FACTORS FOR AMYOTROPHIC LATERAL SCLEROSIS

The definition of environmental risk factors can be narrow—that is, encompassing contaminants in the external environment such as lead, pesticides, or electromagnetic fields—or broad, including nutritional variables, physical activity, and lifestyle characteristics. Here we take a broad approach and argue that the commonality among these environmental risk factors is their relationship to oxidative stress (Table 16.2). In 2003, evidence-based methods to evaluate the quality of ALS epidemiological studies were proposed [11]. These included assessment of (1) general methodological criteria (selection of control group, high response rate, blinding, recall bias, quantification of exposure, accounting for confounding and bias, and appropriate analytical approach); (2) the diagnostic certainty of ALS based on established criteria; and (3) temporal precedence of exposure and disease onset. Further, more stringent criteria and a new method to qualify exposure accuracy and reproducibility have subsequently been developed based on (1) self-reported exposures, (2) assigned exposure by an expert or computer-based assessment of exposure levels, and (3) evidence of external exposure, internal exposure (e.g., metals), or a biomarker of exposure (usually a metabolite) to suspected agents (e.g., pesticides) [20]. When possible, the review below will follow these criteria. A realistic appraisal of these multiple risk factors must also account for their combination, as individuals are usually not exposed to only one risk factor. Few methods are developed to handle such multiple exposures, but several are currently under development. A schematic for creation of an "oxidative stress index" is shown in Figure 16.1.

TABLE 16.2   Sources of Oxidative Stress

- Occupation and hobbies
  - Agricultural chemicals
  - Lead
  - Electromagnetic fields
  - Exposures in the military
- Physical activity
- Diet
  - Oxidants (e.g., high fat intake, trans fats, refined sugars, and excessive caloric intake)
  - Antioxidants (e.g., vitamins A and E, carotenoids, selenium, and omega-3 fatty acids)
- Smoking
- Psychological variables
  - Psychological stress
  - Distress
  - Social support as a protective mechanism

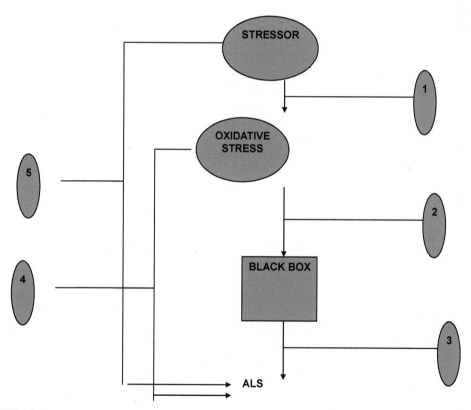

FIGURE 16.1   **Conceptual model relating stressors to ALS.** Step 1 entails relating the stressor to a biological marker of oxidative stress. Steps 2 and 3 require basic science (in cell cultures, animal models, or stored biospecimens) to determine the mechanisms relating oxidative stress to ALS. Step 4 entails associations between the oxidative stress biomarker and ALS. Step 5 entails associations between the stressor and ALS.

Agricultural chemicals: Until the recent ban, organophosphate pesticides were among the most widely used synthetic chemicals for control of domestic and agricultural pests. Several epidemiological studies suggest that exposure to organophosphate pesticides is associated with ALS [13,21–31]. Based on evidence-based criteria, Armon reviewed two population-based, one case–control, and one mortality surveillance study [21,27,30,31]. Only two studies meet the more stringent exposure assessment criteria [21,23], and both of those find that increased risk for ALS is associated with pesticide exposure.

The first [21] is a population-based case–control study of 174 cases and 348 age-and-sex-matched controls recruited in Washington state between 1990 and 1994. After adjustment for age and education, exposure to agricultural chemicals was found to be associated with ALS (OR = 2.0, 95% CI = 1.1–3.5) in men (OR = 2.4, 95% CI = 1.2–4.8), but not in women (OR = 0.9, 95% CI = 0.2–3.8). The second study [23] is a mortality surveillance of populations over two million in 22 states during 1992–1998, which analyzed the association of mortality from neurodegenerative diseases including MND, with occupations having probable pesticide exposure: (1) all farming occupations, (2) farming occupations with likely pesticide exposure (farm workers, horticultural specialty workers, etc.), and (3) farmers alone [28]. Risk for MND was strongest for farmers (OR = 1.23, 95% CI = 1.03–1.46). However, in other occupations where pesticide use might be expected (e.g., farm workers, horticultural specialty workers), significantly elevated ORs were not observed [23].

Although not fulfilling the strict exposure assessment criteria [20], two case–control studies [22,24] originating from Europe [22] and Oceania [24] found positive associations between pesticide exposure and ALS. The first study found an increase in the number of observed to expected ALS cases (observed = 22; 95% Poisson CI = 13.8–32.3; expected = 6) for occupational exposure to agricultural work [22]. A case–control study in Oceania found an elevated OR for ALS for exposure to industrial herbicides, pesticides, or both (OR = 4.18, 95% CI = 1.79–9.74, *P* = 0.006), but not for farm-based herbicides or pesticides [24].

Two European studies, a population-based case–control study [26] and a case–control study [25], found very similar ORs for pesticide exposure (OR = 3.6; 95% CI = 1.2–10.5, and OR = 3.04; 95% CI = 1.1–7.7, respectively). Moreover, the American Cancer Society's recent CPS-II, with nearly 1.2 million participants [32], evaluated the relationship between regular exposure to 11 different chemical classes and ALS mortality [13]. Follow-up from 1989 through 2004 identified 617 deaths from ALS among men and 539 deaths among women, but found no statistically increased risk of ALS mortality from self-reported exposure to pesticides, herbicides, or both (OR = 1.07; 95% CI = 0.79–1.44). Regarding the duration of exposure, ORs were 0.62 (95% CI = 0.09–4.45) for those reporting <4 years of exposure, 1.92 (95% CI = 0.71–5.19) for 4–10 years of exposure, and 1.48 (95% CI = 0.82–2.67) for >10 years of exposure [13].

Therefore, epidemiological evidence of an association between exposure to agricultural chemicals and ALS is present but modest, and thus is suggestive at best. If the association between agricultural pesticides and ALS is small to modest, and if a specific genotype modifies the association, then even large-scale case–control and cohort studies may not be effective in finding consistent associations between organophosphate-induced oxidative stress and ALS.

Most agricultural organophosphates can be clearly linked to oxidative stress [33–36]. Among these, chlorpyrifos, parathion, and malathion have been shown to impair glutathione (GSH) homeostasis by decreasing the level of GSH [37]. GSH is a tripeptide

(L-γ-glutamyl-L-cysteinyl-glycine) with multiple functions in living organisms [38] and can be directly oxidized by hydroxyl radical (HO•) and peroxynitrite (ONOO−), or indirectly during GSH-dependent peroxidase-catalyzed reactions [39]. Increased oxidative stress biomarkers (urinary 8-oxodG, malondialdehyde), reduced GSH levels, and DNA damage (COMET assay) are observed in organophosphate-treated lymphocyte cell cultures [35,40].

Although the molecular mechanism underlying organophosphate induction of oxidative stress remains to be fully clarified [41], active metabolites (various oxon compounds) that are enzymatically converted from organophosphates and capable of deleting intrinsic antioxidants have been identified in human liver, lung, and brain tissues [42–44]. Paraquat, a common herbicide, can induce GSH depletion in the central nervous system (CNS) [45]. In a neuron–microglia culture system, paraquat was toxic to neurons even at low concentrations [46]. Paraquat activates microglial NADPH oxidase–generating superoxide, which is well studied in the microglia [46,47]. Moreover, organophosphate uptake into the mitochondria across the mitochondrial inner membrane leads to reduction by complex I, resulting in redox cycling and mitochondrial toxicity [48]. Paraquat not only weakly inhibits complex I, but impairs complex III function as well, and respiratory chain disruption generates additional oxidants [47,49].

Lead exposure: Many case reports link a variety of metals to ALS [50], and indeed some investigators find higher concentrations of specific metals in the blood, bone, cerebrospinal fluid, urine, or spinal cords of patients with ALS, compared with controls (reviewed in [51]). Metal exposure has also been evaluated in several epidemiological studies [13,23,52–56]. Despite this, the role of these metals in the pathogenesis of ALS, if any, remains unclear (reviewed in [20]). Kamel et al. [54] longitudinally assessed lead exposure in ALS. In a case–control study, they recruited 109 ALS cases from two major referral ALS centers in New England from 1993 to 1996, along with 256 controls (matched to cases by age, sex, and region of residence) identified via random-digit dialing. A structured interview obtained information on occupational, residential, and recreational exposure to lead. In a subset of cases and controls, blood and bone lead concentrations (blood was collected from 107 cases and 39 controls; bone lead was measured at two sites, the midtibial shaft and the patella, in 104 cases and 41 controls). Self-reported occupational exposure to lead was associated with ALS (OR = 1.9; 95% CI = 1.1–3.3) and in the case and control subsets, elevated blood and bone lead were also associated with increased risk; OR was 1.9 (95% CI = 1.4–2.6) for each μg/dL increase in blood lead, 3.6 (95% CI = 0.6–20.6) for each unit (μg/g) increase in log-transformed patella lead, and 2.3 (95% CI = 0.4–14.5) for each unit (μg/g) increase in log-transformed tibia lead.

Further, the association between lead and ALS appeared to be modified by known single-nucleotide polymorphisms (SNPs) in the δ-aminolevulinic acid dehydratase (*ALAD*) gene [55]. In these analyses, the *ALAD* 2 allele (177G to C; K59N) was associated with decreased lead concentration in both the patella and tibia, but not in blood. The association between SNP and ALS risk was increased (OR = 1.9; 95% CI = 0.60–6.3), although the CI was imprecise and the association was not statistically significant. A previously unreported *ALAD* SNP at an *Msp1* site in intron 2 (IVS2 + 299G > A) was associated with decreased bone lead concentration and with a nonsignificant trend toward decreased ALS risk (OR = 0.35; 95% CI = 0.10–1.2), although the latter was imprecise. The ALAD enzyme is the principal lead-binding site in erythrocytes, and the ALAD 2 protein binds lead more tightly than does the ALAD 1 protein [57]. This change in intron 2 alters lead toxicokinetics and may modify the risk associated with lead exposure [58].

More recently, Kamel et al. [56] obtained follow-up data for 91% of the patients who participated in the original case–control study [54]. Surprisingly, lead exposure was associated with increased survival time (OR = 0.5; 95% CI = 0.2–1.0), although the estimate was imprecise.

Environmental lead exposure markedly reduces GSH levels and induces caspase 3 and prostaglandin E2 activation in neuroblastoma cells [59]. In astroglia, lead also causes neuronal oxidative injury by interfering with copper homeostasis [60] and by inducing glial heme oxygenase synthesis [61]. Collectively, this evidence suggests that lead alone, or in combination with other biochemical stressors, leads to oxidative injury [62,63]. Further laboratory studies find that astrocytes can sequester and buffer lead in primary coculture with neurons, modulating its diffusion [64]. In particular, astrocytes induce neuronal cytoprotective and antioxidant gene expression in response to lead exposure [65]. In immortalized human fetal astrocytes, this expression results in increased vascular endothelial growth factor (VEGF) [61] and suggests that VEGF protects motor neurons in culture and in organotypic culture [66–68]. However, further studies are clearly needed.

Military deployment: Although controversial, two studies suggest an approximately 2-fold increase in the risk of ALS among veterans of the 1991 Gulf War [69,70]. A study involving only Gulf War veterans under the age of 45 also found an elevated risk of ALS [71]. Subsequently, secondary analyses of these data found that the increased risk of ALS was limited to the decade following the war and also to deployed military personnel only, suggesting that exposure to potential neurotoxic agents occurred over a relatively brief period of time [72]. However, previous case–control [73] studies, as well as prospective [74] studies, found no association between deployment and ALS among Gulf War veterans.

In 2005, Weisskopf et al. [75] prospectively assessed the relationship between military deployment and ALS mortality among participants in the American Cancer Society's CPS-II, a cohort that included more than 500,000 men [32]. Cause-specific mortality between 281,874 deployed military personnel, excluding Gulf War veterans (military service was ascertained in 1982), and 126,414 men who did not serve in the military was compared. Of 280 deaths due to ALS, 217 occurred in the military group and 63 in the group who had not served, indicating that those who served in the military had an increased death rate from ALS (OR = 1.53; 95% CI = 1.12–2.09). Further, when the relationships between ALS mortality and the different branches of military service were analyzed, mortality was greater for those who served in the Army or National Guard (OR = 1.54; 95% CI = 1.09–2.17), the Navy (OR = 1.87; 95% CI = 1.28–2.74), Air Force (OR = 1.54; 95% CI = 0.99–2.39), and Coast Guard (OR = 2.24; 95% CI = 0.70–7.18), but not the Marines. The increased risk of ALS mortality was similar among those who served in World War II (OR = 1.60; 95% CI = 1.12–2.30), Korea (OR = 1.54; 95% CI = 0.92–2.60), or Vietnam (OR = 1.44; 95% CI = 0.47–4.47) compared with those who had never served.

A more recent study of more than 1000 men enrolled in the National Registry of Veterans with ALS [76] examined a variety of factors (demographics, medical history, major traumatic injury, and smoking status) and aspects of military service (cumulative time of military service, primary branch of service (Air Force, Army, Coast Guard, Marines, or Navy), and deployment (in Korea, Vietnam, or the Persian Gulf regions)) and their effect on ALS survival time. Among enrollees in the registry, 14% reported past deployment to Korea, 21% to Vietnam, 5% to the Persian Gulf, and 4% to more than one location. Shorter survival was associated with deployment to Vietnam, compared with other locations (hazard ratio (HR) = 1.73; 95% CI = 1.36–2.19), suggesting the possibility that those deployed in Vietnam shared some common exposure (e.g., virus, toxin) that shortened ALS survival. There are several limitations to this study, including that only veterans enrolled in the registry were studied (there may be other cases not included in the registry) and that only veterans alive in 2003 and beyond were included.

A 2012 case–control study of 184 veterans with ALS and 194 control subjects from the National Registry of US Veterans [77] found that blood lead concentration was higher among ALS cases, compared with controls ($P<0.0001$, adjusted for age), and a doubling of blood lead was associated with a 1.9-fold increased risk of ALS (95% CI = 1.3–2.7) [78]. However, it is possible that the wasting associated with ALS leads to increased bone turnover and lead leaching from bone into the blood. Thus, higher blood lead levels may be possible in ALS without being causal.

Examination of 2000 medical records of US veterans from April 2003–September 2007 [77] included data on 241 incident ALS cases and 597 controls. Associations were found between head injuries occurring more than 15 years before the reference date (the date of ALS diagnosis for cases, and the date of interview for controls) and adjusted (age, sex, race/ethnicity and education) were associated with ALS (OR = 2.33, 95% CI = 1.18–4.61) [79]. Using these data, the investigators performed a gene–environment interaction analysis (GxE) of 221 cases and 476 controls for cigarette smoking (never vs. ever), head injury (reported by 79 cases and 183 controls at least one time), and apolipoprotein E (*ApoE*) genotype (homozygous *ApoE-3* carriers vs. *ApoE-2* carriers vs. *ApoE-4* carriers). They found weak evidence for GxE; the association between ALS and head injury was stronger in *ApoE-4* carriers (likelihood ratio test of 2.7, 1 df, $P=0.10$) than in *ApoE-2* carriers (0.2; 1 df, $P=0.68$) [153], likely due to the small sample sizes.

Exposure to military service encompasses a variety of exposures, many of which are highly correlated. Inferences from these studies are thus challenging, as an association with one factor may reflect the underlying association with a different factor that is correlated with the first. Biostatistical methods are just being developed to address these issues.

Trauma: Although an association between head trauma and ALS is tenuous [30,80–89], the report of US veterans discussed above suggests that the potential relationship to brain injury may deserve further exploration. The finding of a very high incidence of ALS among Italian professional footballers [90–92] and American football players [93] has led to speculation over whether head injury caused by collision is a risk factor for ALS. A rigorous population-based study of ALS in western Washington state found no associations with previous fractures, head injuries, or hospitalization [30]. In a recent cohort study, the adjusted rate ratio for ALS after head injury was 1.5 (95% CI = 1.1–2.1), but this elevated risk was found only during the first year after injury, possibly suggesting reverse causality—that is, that very early subclinical weakness caused by ALS can increase the risk of head injury due to falls and fractures [89]. A recent review evaluated the epidemiological literature regarding the association between head trauma and ALS published between 1980 and October 2010 [94], using the American Academy of Neurology's evidence-based classification of evidence for inferring causality and assigning level of conclusion [95]. Twelve articles (one with Class II evidence, three with Class III evidence, and eight with Class IV evidence) met the inclusion criteria. The authors concluded that the relationship between a single instance of head trauma or head injury, and the occurrence of ALS, is likely unproven or unsupported (Level U conclusion) [94].

During a traumatic brain injury, the brain and spinal cord undergo shear deformation that transiently elongates axons. Traumatic axonal injury also perturbs the cytoskeleton, causing microtubule and neurofilament dissolution, and pathologic reorganization of neurofilament proteins [96]. Trauma to the CNS triggers stress responses, including oxidative stress due to reactive oxygen species (ROS) generation [97]. Necrotic events due to cellular disintegration resulting from acute injury are associated with the release of cellular contents that elicit inflammatory response and oxygen radical production.

Repetitive head injury is associated with the development of chronic traumatic encephalopathy (CTE), a tauopathy characterized by neurofibrillary tangles throughout the brain in the relative absence of β-amyloid deposits. McKee et al. [98] described 12 cases of CTE, 10 of which had widespread TAR DNA–binding protein-43 (TDP-43) inclusions (TDP-43 proteinopathy). Among these cases, three were athletes who developed progressive MND that clinically resembled an ALS-like syndrome. TDP-43 seems to be critical in mediating the response of the neuronal cytoskeleton to axonal injury [99]. It is intrinsically prone to aggregation, and TDP-43 expression is upregulated after experimental axotomy in mouse spinal motor neurons [100]. Traumatic axonal injury may also accelerate TDP-43 accumulation, aggregation, and translocation to the cytoplasm, thereby enhancing its neurotoxicity. Even though the number of cases was small, the coexistence of MND phenotype and CTE neuropathology has resulted in increased interest in the ALS community.

The importance of TDP-43 in ALS quickly became evident when researchers discovered TDP-43 inclusions in ALS and frontotemporal dementia (FTD) [101,102]. TDP-43 inclusions are now known to be present in the majority of ALS cases, including ALS with FTD, sporadic ALS, non–superoxide dismutase 1 (SOD1) familial ALS, and the ALS–PD complex in Guam [103–106]. Recent studies suggest that oxidative stress may have a close relationship to TDP-43 and its pathologic alteration. Cohen et al. [107] found that oxidative stress promotes TDP-43 crosslinking via cysteine oxidation and disulphide bond formation, leading to decreased TDP-43 solubility. Prostaglandin metabolites, which are major players in oxidative stress and inflammation, can affect TDP-43 proteolysis, solubility, and subcellular localization [108]. Chronic exposure of cultured neuron-like (SH-SY5Y and NSC34) cells to ROS and reactive nitrogen species (RNS) via paraquat and ethacrynic acid administration revealed TDP-43 C-terminal phosphorylation, insolubility, C-terminal fragmentation, and translocation of nuclear TDP-43 to the cytosol [109,110]. Therefore, it appears that oxidative stress may play an important role in TDP-43 proteinopathy, further suggesting an intriguing relationship to ALS disease mechanisms.

Strenuous exercise and physical exertion: Strenuous exercise diminishes lymphocyte GSH content, resulting in oxidative stress-induced apoptosis, and it also increases urinary excretion of 8-Oxo-2'-deoxyguanosine (8-oxo-dG), a product of DNA oxidation, whereas moderate exercise attenuates lymphocyte apoptosis, possibly by improving antioxidative capacity [111]. Strenuous exercise generates a host of complex biological changes; however, there is evidence to suggest that it increases free radical production that leads to oxidative stress [112].

A study of aerobic exercise (incremental workload on a cycloergometer) assessed oxidative stress by measuring blood lipoperoxides [113]. Values were significantly higher in patients with ALS ($P < 0.05$) than in control subjects ($n = 5$ with various chronic neuropathies) at rest, during exercise, and 30 min afterward. Moreover, the increase in lipoperoxide concentration during exercise strongly and positively correlated with lactate accumulation ($r = 0.94$, $P < 0.01$), a finding suggesting impaired mitochondrial function [113].

Epidemiological studies investigating an association between strenuous physical activity and ALS are inconclusive. In a case–control study of 279 patients with ALS and 152 with other neurological diseases, a greater number of subjects with ALS reported they had been varsity athletes (OR = 1.7, 95% CI = 1.04–2.76) and had always been slim (OR = 2.21, 95% CI = 1.40–3.47) compared with controls [114]. A recent matched nested case–control study in a large cohort of more than 680,000 Swedish male subjects born between 1951 and 1965 [115] reported that weight-adjusted physical fitness was associated with ALS (OR = 1.98, 95% CI = 1.32–2.97). In that study, data were collected on body weight and height, physical fitness, resting heart rate, and isometric strength at the time of military conscription (age 18–19 years); there were 85 ALS deaths during the period of follow-up. In Italian professional soccer players, a higher risk for ALS was found among midfielders (standardized morbidity ratio of 12.2, 95% CI = 3.3–31.2), a role characterized by great aerobic effort and a higher proportion of lean body mass [90]. But an absence of ALS cases in professional road cyclists and basketball players, who also have a high level of aerobic fitness and higher proportion of lean body mass [116], suggests that physical exertion or fitness may not explain a causal relationship with ALS. Further studies are clearly needed [91,117].

Diet: The roles of macro- and micronutrients in the etiology of ALS have not been fully studied. Similar to the case of military exposures, dietary constituents are highly correlated, and finding associations between a single nutrient and ALS does not necessarily reflect a causal interpretation. Further, while the diet contains constituents that are pro-oxidants, it also contains those that are antioxidants, making the analysis even more complicated. Nevertheless, as reviewed below, there have been several studies of the role of diet in ALS.

A population-based, case–control study in western Washington state from 1990 to 1994 found that increased dietary fat intake (OR = 2.7, 95% Cl = 0.9–8.0) and glutamate intake (OR = 3.2, 95% Cl = 1.2–8.0) were associated with increased risk of ALS, whereas dietary fiber intake was associated with a decreased risk (OR = 0.3, 95% Cl = 0.1–0.7) [118]. In 2008, Morozova et al. [119] examined the relationship between diet and risk of ALS among participants in the American Cancer Society CPS-II cohort [32]. No associations were found. In a more recent case–control study, 153 ALS cases, and 306 gender- and age-matched controls randomly selected from the general population in the Tokai area of Japan, completed a self-administered food frequency questionnaire to estimate pre-illness intake of food groups and nutrients. A high intake of carbohydrates was significantly associated with an increased risk of ALS (OR = 2.14, 95% CI = 1.05–4.36) [120].

One micronutrient, vitamin E, deserves brief discussion as an extrinsic antioxidant. Vitamin E is a fat-soluble hydrocarbon and an important component of cell membranes and lipoproteins [121]. Its antioxidant properties are attributed to the inhibition of lipid peroxidation and free radical formation, because vitamin E can attack the peroxyl radical more rapidly than polyunsaturated fatty acids can, donating its phenolic hydrogen atom to the radical and converting it to a more stable product [122]. Because of these antioxidant properties, Vitamin E may have a potential protective role in preventing the development of ALS [197] while also slowing down disease progression [123]. A mortality surveillance study conducted in a well-defined population (participants in the American Cancer Society CPS-II [32]) indicated that regular use of vitamin E supplements for 10 years or more is associated with a lower risk of mortality from ALS (relative risk = 0.38, 95% CI = 0.16–0.92, based on trend analysis, $P = 0.004$) [123]. Further, a recent pooled analysis of mortality data from five different prospective cohort studies (among 1,055,546 participants, 805 developed ALS) revealed a positive trend ($P = 0.01$) for a decline in ALS progression within patients (231 cases) with longer duration of vitamin E supplementation, but found no overall protective role for vitamin E [124].

Smoking: Several population-based and case-controlled studies have established a possible relationship between ALS and cigarette smoking [14,79,125–133], although at times results are inconsistent [134,135]. Methodological differences and heterogeneity in studied populations could account in part for these disparities [11,136]. Cigarette smoking is associated with oxidative stress, inducing protein oxidation, DNA damage, and cell death [137,138]. It is hypothesized that cigarette smoking causes lipid peroxidation via formaldehyde formation [139], and increases cadmium and lead levels in seminal fluid, plasma, and blood, all of which may increase oxidative damage [140].

A recent pooled analysis prospectively examined the relationship between smoking and ALS in five well-established large cohorts, with study-specific follow-up ranging from 7 to 28 years [14]. Among 562,804 men and 556,276 women, 832 ALS cases were documented. Smokers had a higher risk of ALS than those who had never smoked, with age- and sex-adjusted relative risks of 1.44 (95% CI = 1.23–1.68) for former smokers and 1.42 (95% CI = 1.07–1.88) for current smokers [14].

Combined effects of multiple stressors: As mentioned above, human subjects are exposed to a multitude of agents having both oxidative and antioxidative properties. The development of analytic strategies to address this issue has only recently been initiated [141–145]. However, there are some animal data that examine concurrent exposure to environmental agents. In experimental models, combined exposure to glutamate and lead, as compared with either agent alone, increased neuronal cell death via mechanisms involving an elevation in the production of oxygen radicals and a decrease in intracellular GSH defenses against oxidative stress [146]. Mouse spinal cord and dorsal root ganglion on embryonic day 12 were exposed to low-level paraquat, glutamate, and thermal stress, singly and in combination [147], and it was found that paraquat administered alone was associated with cell death in a dose-dependent manner, an effect that was multiplied by heat shock. This model is particularly intriguing, because a combination of low-level oxidative stress factors was associated with motor neuron vulnerability to environmental stressors. Such low-level cumulative stress may in part explain mid- to late-life symptom onset in both familial and sporadic ALS patients. Other investigations [148] found various combinations of stressors differentially affected enzymatic and nonenzymatic antioxidant defense systems, protein oxidation, and lipid peroxidation in the brain.

These experiments are important, because they attempt to mimic the human situation in which multiple stressors are likely to be present, resulting in additive or multiplicative interactions that contribute to abnormal oxidative stress. Thus, multidisciplinary epidemiological studies using appropriate analytical methods are needed that consider joint associations between demographic, biological, and genetic risk factors to better understand the disease mechanisms and eventually identify markers of susceptibility in patients with ALS.

ANTIOXIDANTS AND AMYOTROPHIC LATERAL SCLEROSIS

Little is known about the role of intrinsic antioxidants in ALS. Serum total antioxidant status, a measure of peroxyl adduct-scavenging capacity, has been reported to be significantly ($P < 0.05$) higher in ALS patients ($n = 28$) compared with that of healthy controls ($n = 20$), but did not correlate with ALS onset phenotype, disease duration, or clinical state [149]. More recently, the thiobarbituric acid–reactive substances (TBARS) concentration, and antioxidants such as SOD1, catalase, GSH peroxidase, GSH reductase, and glucose-6-phosphate dehydrogenase were assessed in the erythrocytes of 20 ALS patients and 20 controls (221). In ALS, the TBARS concentration was significantly increased in comparison with that of controls ($P < 0.001$); moreover, TBARS started to increase, and antioxidant molecule concentration began to decrease, as ALS progressed over 6–24 months, suggesting a correlation between these variables and the duration of the disease [150]. In a second study, antioxidative stress markers were examined in 31 patients with ALS, 24 patients with PD, and 30 healthy subjects. Superoxide dismutase (SOD) activity was significantly decreased in ALS ($P = 0.001$), whereas GSH peroxidase activity was decreased in PD ($P = 0.001$), suggesting that the affected antioxidant system might differ between ALS and PD [151].

SOD1: SOD1, or Cu/Zn-SOD, was the first identified and comprehensively studied antioxidant enzyme of the SOD family; mitochondrial Mn-SOD2 and extracellular Cu/Zn-SOD3 also exist (for review see [152,153]). SOD1 catalyzes dismutation of O_2^- to H_2O_2 and O_2 in a two-step redox reaction involving the reduction and reoxidization of the copper ion in the active site of the dimeric protein.

The role of SOD1 in ALS pathogenesis has received great attention since the first description of mutations in the gene encoding SOD1 [154]. Since then, more than 160 mutations have been reported. They account for approximately 15%–20% of familial ALS cases, and thus SOD1 mutations are the second most frequent known cause of familial ALS after C9ORF72 [155–157]. How mutant SOD1 leads to the death of motor neurons has not been established, although a number of hypotheses have been presented [152,158–161]. Importantly, the cause of motoneuron degeneration in *SOD1* mutation carriers is not the result of a loss of antioxidative function of the SOD1 enzyme, but instead a gain of function appears to be pathogenic. We refer here to several excellent papers, since this topic is beyond the scope of this chapter [152,161,162].

Although *SOD1* mutations can cause familial ALS, wild-type SOD1 protein might also have a role in sporadic ALS pathogenesis. First, oxidized wild-type SOD1 occurs in lymphocytes of ALS patients, resulting in mitochondrial dysfunction [163]. Second, misfolded SOD1 is found not only in ALS motor neurons [164,165], but also in the motor neurons of FUS- or TDP43-linked familial ALS cases—these are cases with no associated *SOD1* mutation [166]. The mechanisms underlying how wild-type SOD1 is misfolded in these ALS cases remain to be clarified [161,167]. On the other hand, if this process is a frequent phenomenon in ALS or other types of familial ALS, SOD1 misfolding might explain a common molecular event between familial ALS and ALS [166].

Paraoxonase: Paraoxonase (PON) is a potential intrinsic antioxidative enzyme [168,169]. PON hydrolyzes paraoxon, a metabolite of parathion and a highly toxic organophosphate; exposure can lead to acute cholinergic crisis and subacute or chronic neurotoxicity. PON is a family of three independent enzymes: PON1, PON2, and PON3. The genes are located on chromosome 7 [170]. PON1 is synthesized in the liver, and a portion is secreted into the plasma as a component of HDL [171].

Despite its name, PON1 is not very efficient in hydrolyzing paraoxon, which accounts in part for the latter's toxicity. The mechanisms of organophosphate hydrolysis by PON1 have been well studied in laboratory animals, and oxidative stress occurs because organophosphate detoxification consumes GSH [172]. On the other hand, oxidative stress may reduce PON1 activity [173]. In fact, PON1 is highly susceptible to hydroxyl radicals produced from metal (Cu^{2+} or Fe^{2+})-catalyzed oxidation, and the modification of some histidine residues results in the decrease of its antioxidant activity [173].

PON1 has been extensively studied in cardiovascular disease because of its ability to protect against atherosclerosis [174]. Certain SNPs in *PON1* have been associated with differential susceptibility to organophosphate exposure [175]: two SNPs, *Q192R (rs662) and L55M (rs854560)* [176], have the greatest paraoxon hydrolytic activity, but protect less effectively against low-density lipoprotein (LDL) oxidation [177,178]. Decreased PON1 activity is associated with organophosphate toxicity [179]. In laboratory animals, PON1 decreases oxidation of LDL by hydrolyzing hydroperoxides into less reactive hydroxides [174]. Moreover, PON1 has hydrolytic activities on aromatic and long-chain aliphatic lactones, including arachidonate and lipid mediators generated from the oxidation of polyunsaturated fatty acids and exogenous lactone-containing drugs such as prulifloxacin. Statin lactones (simvastatin, lovastatin) and the diuretic spironolactone, previously reported to be hydrolyzed by PON1, are metabolized by PON3 [180]. PON2 and PON3 lack the ability to hydrolyze organophosphates, but have lactonase or peroxidase-like activity, or both [181,182]. Like PON1, PON3 is also found in high-density lipoprotein (HDL) and serum. PON2 is expressed in a number of tissues, including the brain [183,184]. A human cell line overexpressing PON2 had significantly less intracellular oxidative stress following treatment with H_2O_2 [183].

Serum PON1 activity in a given human population can vary by 40-fold. Although a portion of this variation is explained by extensive genetic variation in the PON1-coding regions (more than 160 SNPs have been described), activity may be largely influenced by environmental and perhaps epigenetic factors [184,185]. Age plays the most relevant role, at least early in life; for humans, PON1 activity is very low before birth, and gradually increases during the one to 2 years following [186]. PON1 activity may also decline with aging, possibly because oxidative stress increases with age [187]. A sex effect has also been proposed, as females have higher PON1 activity [185]. A variety of other factors (e.g., environmental chemicals, drugs, smoking, alcohol, diet, and disease conditions) are suspected to reduce PON1 activity [185]. On the other hand, statins increase PON1 expression and activity in in vitro studies [188,189]. PON1 activity increased by ~13% in patients who took low-dose acetylsalicylic acid [190].

The data regarding *PON1* SNPs in ALS is conflicting [191–200]. A meta-analysis of 10 published studies and one unpublished study on *PON1* SNPs found no significant association with ALS [201]. A recent case–control study involving 1160 ALS patients and 1240 control subjects of Dutch descent, which examined *PON1* and *PON3* genetic variants, showed that SNP frequencies did not differ between the two groups [200]. Thus, work to date indicates that *PON1* SNPs that contribute to ALS are not easily detectable. Indeed, if SNP variation is important for ALS, it may affect a susceptible minority of patients. Thus, different approaches to study gene–environment interactions are necessary [202].

FUTURE RESEARCH CONSIDERATIONS

Although the evidence for oxidative damage in ALS pathogenesis is extensive and possibly related to environmental factors, it is not yet clear whether such damage occurs prior to the onset of the disease process. Prospective studies have shown that smoking, a cause of oxidative stress, is associated with ALS, whereas the evidence for other factors is at best equivocal. Smoking is relatively common in the general population, but other factors such as agricultural or heavy metal exposure are less common; thus, statistically significant relationships are more difficult to find. The results may depend heavily on the populations and locations studied. Many epidemiological studies have failed to demonstrate strong relationships, in part due to methodological shortcomings [20,203,204]. However, it is difficult to find strong associations between rare events and a rare multifactorial disease such as ALS. Although a prospective epidemiological study to search for a positive causative factor in ALS is desirable, such a study would involve following a large population for an extended period of time. Such studies are expensive and not feasible.

Another approach may be feasible. In a previous paper [205], our group has proposed that oxidative stress may be an underlying biological risk factor for ALS, and that oxidative stress may accumulate via a variety of environmental and lifestyle factors (Figure 16.2). Five steps are required to test these hypotheses. Throughout the disease course, oxidative stress may further accumulate due to malnutrition, respiratory failure, and psychological stress. Highly neurotoxic lipid peroxidase products such as 4-hydroxy-2-nonenal (HNE) and isoprostanes (IsoPs) may be generated both in a systematic manner and in the periphery. This may result in a feedback loop in that the production of oxidative stress further hastens the disease process by affecting

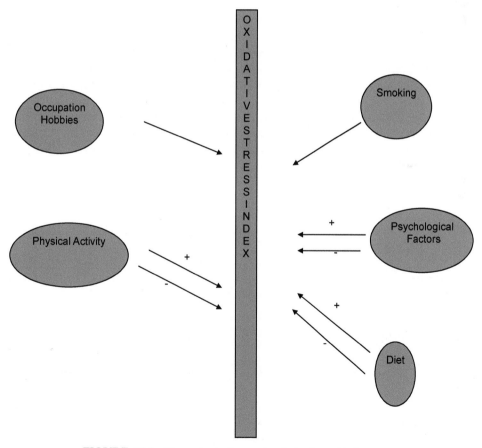

FIGURE 16.2 Conceptual scheme of an index for oxidative stress.

motor neurons and damaging the cerebral microvascular endothelial cells of the blood–brain barrier [206]. Almost certainly, the real picture is far more complex than what is described here, because individual genetic susceptibility and epigenetic mechanisms, which further modify gene–environment interactions, are likely to influence disease onset, disease progression, and even disease phenotype. Studies evaluating these factors are only in their infancy. Nevertheless, our group argues that molecular epidemiological studies focusing on oxidative stress as the biological process may be a more effective approach to identify potential risk factors for ALS.

Answering these simple but fundamental questions requires well-planned, well-designed epidemiological studies that recruit patients from multiple centers to achieve adequate sample sizes. Moreover, a multidisciplinary approach is essential, because both molecular and environmental risk factors are involved. Clinical investigators who define the disease, epidemiologists with different specialties (general, nutritional, occupational, and genetic epidemiology), laboratory scientists who are experts in oxidative stress biology, molecular geneticists, experts in epigenetics, clinical psychologists, and biostatisticians must participate in such studies, because their combined expertise is essential to uncover the etiology of a complex disease like ALS. In addition, the appropriate collection and processing of biospecimens is critical for success.

This approach has a good chance to provide answers to many of the important questions regarding ALS etiology and progression. If oxidative stress is a key factor in the pathogenicity of ALS, innovative clinical trials to investigate antioxidant treatment strategies will be necessary. Although several past clinical trials intended to counteract presumed oxidative stress failed to demonstrate clinical benefit ([207–211]), these failures do not mean that the hypothesis is wrong, as those medications had a limited ability to cross the blood–brain barrier. With a more thorough understanding of these mechanisms of cellular oxidative stress, the treatment and management of ALS can be biologically based, and proactive and appropriate treatments can be developed that slow down or stop the progression of this devastating neurodegenerative disease.

References

[1] Williams DB, Windeband AJ. Motor neuron disease (amyotrophic lateral sclerosis). Mayo Clin Proc 1991;66:54–82.

[2] Hand CK, Rouleau GA. Familial amyotrophic lateral sclerosis. Muscle Nerve 2002;25:135–59.

[3] Majoor-Krakauer D, Willems PJ, Hofman A. Genetic epidemiology of amyotrophic lateral sclerosis. Clin Genet 2003;63:83–101.

[4] Corcia P, Mayeux-Portas V, Khoris J, de Toffol B, Autret A, Muh JP, French ALS research group, et al. Amyotrophic lateral sclerosis. Abnormal gene copy number is a susceptibility factor for amyotrophic lateral sclerosis. Ann Neurol 2002;51:243–6.

[5] Hosler BA, Siddique T, Sapp PC, Sailor W, Huang MC, Hossain A, et al. Linkage of familial amyotrophic lateral sclerosis with frontal temporal dementia to chromosome 9q21–q22. JAMA 2000;284:1664–9.

[6] Keller MF, Ferrucci L, Singleton AB, Tienari PJ, Laaksovirta H, Restagno G, et al. Genome-wide analysis of the heritability of amyotrophic lateral sclerosis. JAMA Neurol 2014;71(9):1123–34. http://dx.doi.org/10.1001/jamaneurol.2014.1184.

[7] Hirano A, Kurland LT, Krooth RS, Lessell S. Parkinsonism-dementia complex, an endemic disease on the island of Guam. I. Clinical features. Brain 1961;84:642–61.

[8] Plato CC, Garruto RM, Galasko D, Craig UK, Plato M, Gamst A, et al. Amyotrophic lateral sclerosis and parkinsonism-dementia complex of Guam: changing incidence rates during the past sixty years. Am J Epidemiol 2003;157:149–57.

[9] Gajdusek DC, Salazar AM. Amyotrophic lateral sclerosis and parkinsonian syndrome in high incidence among the Auyu and Jakai people of West New Guinea. Neurology 1982;32:107–26.

[10] Kihira T, Yoshida S, Kondo T, Iwai K, Wada S, Morinaga S, et al. An increase in ALS incidence on the Kii Peninsula, 1060-2009: a possible link to change in water source. Amyotroph Lateral Scler 2012;13:347–50.

[11] Armon C. An evidence-based medicine approach to the evaluation of the role of exogenous risk factors in sporadic amyotrophic lateral sclerosis. Neuroepidemiology 2003;22:217–28.

[12] Yoshida S, Mulder DW, Kurland LT, Chu CP, Okazaki H. Follow-up study on amyotrophic lateral sclerosis in Rochester, Minnesota, 1925–1984. Neuroepidemiology 1986;5:61–70.

[13] Weisskopf MG, Morozova N, O'Reilly EJ, McCullough ML, Calle EE, Thun MJ, et al. Prospective study of chemical exposures and amyotrophic lateral sclerosis. J Neurol Neurosurg Psychiatry 2009;80:558–61.

[14] Wang H, O'Reilly ÉJ, Weisskopf MG, Logroscino G, McCullough ML, Thun MJ, et al. Smoking and risk of amyotrophic lateral sclerosis: a pooled analysis of 5 prospective cohorts. Arch Neurol 2011;68:207–13.

[15] Chio A, Mora G, Calvo A, Mazzini L, Bottacchi E, Mutani R, PAALS. Epidemiology of ALS in Italy: a 10-year prospective population-based study. Neurology 2009;72:725–31.

[16] Brooks BR. El Escorial World Federation of Neurology criteria for the diagnosis of amyotrophic lateral sclerosis. Subcommittee on Motor Neuron Diseases/Amyotrophic Lateral Sclerosis of the World Federation of Neurology Research Group on Neuromuscular Diseases and the El Escorial "Clinical limits of amyotrophic lateral sclerosis" workshop contributors. J Neurol Sci 1994;124(Suppl.):96–107.

[17] McGuire V, Nelson LM. Epidemiology of ALS. In: Mitsumoto H, Przedborski S, Gordon PH, editors. Amyotrophic lateral sclerosis. New York: Taylor and Francis Group; 2006. p. 17–41.

[18] Zaldivar T, Gutierrez J, Lara G, Carbonara M, Logroscine G, Hardiman O. Reduced frequency of ALS in an ethnically mixed population: a population-based mortality study. Neurology 2009;72:1640–5.

[19] Cronin S, Hardiman O, Traynor B. Ethnic variation in the incidence of ALS: a systematic review. Neurology 2007;68:1002–7.

[20] Sutedja NA, Veldink JH, Fischer K, Kromhout H, Heederik D, Huisman MH, et al. Exposure to chemicals and metals and risk of amyotrophic lateral sclerosis: a systematic review. Amyotroph Lateral Scler 2009;10:302–9.

[21] McGuire V, Longstreth Jr WT, Nelson LM, Koepsell TD, Checkoway H, Morgan MS, et al. Occupational exposures and amyotrophic lateral sclerosis. A population-based case-control study. Am J Epidemiol 1997;145:1076–88.

[22] Govoni V, Granieri E, Fallica E, Casetta I. Amyotrophic lateral sclerosis, rural environment and agricultural work in the local health district of Ferrara, Italy, in the years 1964–1998. J Neurol 2005;252:1322–7.

[23] Park RM, Schulte PA, Bowman JD, Walker JT, Bondy SC, Yost MG, et al. Potential occupational risks for neurodegenerative diseases. Am J Ind Med 2005;48:63–77.

[24] Morahan JM, Pamphlett R. Amyotrophic lateral sclerosis and exposure to environmental toxins: an Australian case-control study. Neuroepidemiology 2006;27:130–5.

[25] Furby A, Beauvais K, Kolev I, Rivain JG, Sebille V. Rural environment and risk factors of amyotrophic lateral sclerosis: a case-control study. J Neurol 2010;257:792–8.

[26] Bonvicini F, Marcello N, Mandrioli J, Pietrini V, Vinceti M. Exposure to pesticides and risk of amyotrophic lateral sclerosis: a population-based case-control study. Ann Ist Super Sanita 2010;46:284–7.

[27] Chancellor AM, Slattery JM, Fraser H, Warlow CP. Risk factors for motor neuron disease: a case-control study based on patients from the Scottish Motor Neuron Disease Register. J Neurol Neurosurg Psychiatry 1993;56:1200–6.

[28] Schulte PA, Burnett CA, Boeniger MF. Neurodegenerative diseases: occupational occurrence and potential risk factors, 1982 through 1991. Am J Public Health 1996;86:1281–8.

[29] Bergomi M, Vinceti M, Nacci G, Pietrini V, Bratter P, Alber D, et al. Environmental exposure to trace elements and risk of amyotrophic lateral sclerosis: a population-based case-control study. Environ Res 2002;89:116–23.

[30] Cruz DC, Nelson LM, McGuire V, Longstreth Jr WT. Physical trauma and family history of neurodegenerative diseases in amyotrophic lateral sclerosis: a population-based case-control study. Neuroepidemiology 1999;18:101–10.

[31] Gunnarsson LG, Lindberg G, Soderfeldt B, Axelson O. Amyotrophic lateral sclerosis in Sweden in relation to occupation. Acta Neurol Scand 1991;83:394–8.

[32] Thun MJ, Calle EE, Rodriguez C, Wingo PA. Epidemiological research at the American Cancer Society. Cancer Epidem Biomar 2000;9:861–8.

[33] Shadnia S, Azizi E, Hosseini R, Khoei S, Fouladdel S, Pajoumand A, et al. Evaluation of oxidative stress and genotoxicity in organophosphorus insecticide formulators. Hum Exp Toxicol 2005;24:439–45.

[34] Franco R, Sanchez-Olea R, Reyes-Reyes EM, Panayiotidis MI. Environmental toxicity, oxidative stress and apoptosis: menage a trois. Mutat Res 2009;674:3–22.

[35] Soltaninejad K, Abdollahi M. Current opinion on the science of organophosphate pesticides and toxic stress: a systematic review. Med Sci Monit 2009;15:RA75–90.

[36] Schmuck G, Rohrdanz E, Tran-Thi QH, Kahl R, Schluter G. Oxidative stress in rat cortical neurons and astrocytes induced by paraquat in vitro. Neurotox Res 2002;4:1–13.

[37] Costa LG. Current issues in organophosphate toxicology. Clin Chim Acta 2006;366:1–13.

[38] Lushchak VI. Glutathione homeostasis and functions: potential targets for medical interventions. J Amino Acids 2012;2012:736837. http://dx.doi.org/10.1155/2012/736837.

[39] Griffith OW, Meister A. Glutathione: interorgan translocation, turnover, and metabolism. Proc Natl Acad Sci USA 1979;76:5606–10.

[40] Muniz JF, McCauley L, Scherer J, Lasarev M, Koshy M, Kow YW, et al. Biomarkers of oxidative stress and DNA damage in agricultural workers: a pilot study. Toxicol Appl Pharmacol 2008;227:97–107.

[41] Franco R, Li S, Rodriguez-Rocha H, Burns M, Panayiotidis MI. Molecular mechanisms of pesticide-induced neurotoxicity: relevance to Parkinson's disease. Chem Biol Interact 2010;188:289–300.

[42] Suntres ZE. Role of antioxidants in paraquat toxicity. Toxicology 2002;180:65–77.

[43] Dutheil F, Beaune P, Loriot MA. Xenobiotic metabolizing enzymes in the central nervous system: contribution of cytochrome P450 enzymes in normal and pathological human brain. Biochimie 2008;90:426–36.

[44] Day BJ, Patel M, Calavetta L, Chang LY, Stamler JS. A mechanism of paraquat toxicity involving nitric oxide synthase. Proc Natl Acad Sci USA 1999;96:12760–5.

[45] Purisai MG, McCormack AL, Cumine S, Li J, Isla MZ, Di Monte DA. Microglial activation as a priming event leading to paraquat-induced dopaminergic cell degeneration. Neurobiol Dis 2007;25:392–400.

[46] Wu XF, Block ML, Zhang W, Qin L, Wilson B, Zhang WQ, et al. The role of microglia in paraquat-induced dopaminergic neurotoxicity. Antioxid Redox Signal 2005;7:654–61.

[47] Kim S, Hwang J, Lee WH, Hwang DY, Suk K. Role of protein kinase Cdelta in paraquat-induced glial cell death. J Neurosci Res 2008;86:2062–70.

[48] Cocheme HM, Murphy MP. Complex I is the major site of mitochondrial superoxide production by paraquat. J Biological Chem 2008;283:1786–98.

[49] Choi WS, Kruse SE, Palmiter RD, Xia ZG. Mitochondrial complex I inhibition is not required for dopaminergic neuron death induced by rotenone, MPP(+), or paraquat. Proc Natl Acad Sci USA 2008;105:15136–41.

[50] Mitsumoto H, Hanson MR, Chad DA. Amyotrophic lateral sclerosis – recent advances in pathogenesis and therapeutic trials. Arch Neurol 1988;45:189–202.

[51] Callaghan B, Feldman D, Gruis K, Feldman E. The association of exposure to lead, mercury, and selenium and the development of amyotrophic lateral sclerosis and the epigenetic implications. Neurodegener Dis 2011;8:1–8.

[52] Vinceti M, Nacci G, Rocchi E, Cassinadri T, Vivoli R, Marchesi C, et al. Mortality in a population with long-term exposure to inorganic selenium via drinking water. J Clin Epidemiol 2000;53:1062–8.

[53] Gait R, Maginnis C, Lewis S, Pickering N, Antoniak M, Hubbard R, et al. Occupational exposure to metals and solvents and the risk of motor neuron disease. A case-control study. Neuroepidemiology 2003;22:353–6.

[54] Kamel F, Umbach DM, Munsat TL, Shefner JM, Hu H, Sandler DP. Lead exposure and amyotrophic lateral sclerosis. Epidemiology 2002;13:311–9.

[55] Kamel F, Umbach DM, Lehman TA, Park LP, Munsat TL, Shefner JM, et al. Amyotrophic lateral sclerosis, lead, and genetic susceptibility: polymorphisms in the delta-aminolevulinic acid dehydratase and vitamin D receptor genes. Environ Health Perspect 2003;111:1335–9.

[56] Kamel F, Umbach DM, Stallone L, Richards M, Hu H, Sandler DP. Association of lead exposure with survival in amyotrophic lateral sclerosis. Environ Health Perspect 2008;116:943–7.

[57] Bergdahl IA, Gerhardsson L, Schutz A, Desnick RJ, Wetmur JG, Skerfving S. Delta-aminolevulinic acid dehydratase polymorphism: influence on lead levels and kidney function in humans. Arch Environ Health 1997;52:91–6.

[58] Kelada SN, Shelton E, Kaufmann RB, Khoury MJ. Delta-aminolevulinic acid dehydratase genotype and lead toxicity: a HuGE review. Am J Epidemiol 2001;154:1–13.

[59] Chetty CS, Vemuri MC, Campbell K, Suresh C. Lead-induced cell death of human neuroblastoma cells involves GSH deprivation. Cell Mol Biol Lett 2005;10:413–23.

[60] Qian Y, Zheng Y, Ramos KS, Tiffany-Castiglioni E. The involvement of copper transporter in lead-induced oxidative stress in astroglia. Neurochem Res 2005;30:429–38.

[61] Cabell L, Ferguson C, Luginbill D, Kern M, Weingart A, Audesirk G. Differential induction of heme oxygenase and other stress proteins in cultured hippocampal astrocytes and neurons by inorganic lead. Toxicol Appl Pharmacol 2004;198:49–60.

[62] Fowler BA, Whittaker MH, Lipsky M, Wang G, Chen XQ. Oxidative stress induced by lead, cadmium and arsenic mixtures: 30-day, 90-day, and 180-day drinking water studies in rats: an overview. Biometals 2004;17:567–8.

[63] Prasanthi RPJ, Devi CB, Basha DC, Reddy NS, Reddy GR. Calcium and zinc supplementation protects lead (Pb)-induced perturbations in antioxidant enzymes and lipid peroxidation in developing mouse brain. Int J Dev Neurosci 2010;28:161–7.

[64] Tiffany-Castiglioni E. Cell culture models for lead toxicity in neuronal and glial cells. Neurotoxicology 1993;14:513–36.

[65] Barbeito AG, Martinez-Palma L, Vargas MR, Pehar M, Manay N, Beckman JS, et al. Lead exposure stimulates VEGF expression in the spinal cord and extends survival in a mouse model of ALS. Neurobiol Dis 2010;37:574–80.

[66] Bogaert E, Van Damme P, Poesen K, Dhondt J, Hersmus N, Kiraly D, et al. VEGF protects motor neurons against excitotoxicity by upregulation of GluR2. Neurobiol Aging 2010;31:2185–91.

[67] Tovar YRLB, Tapia R. VEGF protects spinal motor neurons against chronic excitotoxic degeneration in vivo by activation of PI3-K pathway and inhibition of p38MAPK. J Neurochem 2010;115:1090–101.

[68] Lambrechts D, Storkebaum E, Morimoto M, Del-Favero J, Desmet F, Marklund SL, et al. VEGF is a modifier of amyotrophic lateral sclerosis in mice and humans and protects motoneurons against ischemic death. Nat Genet 2003;34:383–94.

[69] Coffman CJ, Horner RD, Grambow SC, Lindquist J, Studi IV. Estimating the occurrence of amyotrophic lateral sclerosis among Gulf War (1990–1991) Veterans using capture-recapture methods - an assessment of case ascertainment bias. Neuroepidemiology 2005;24:141–50.

[70] Horner RD, Kamins KG, Feussner JR, Grambow SC, Hoff-Lindquist J, Harati Y, et al. Occurrence of amyotrophic lateral sclerosis among Gulf War veterans. Neurology 2003;61:742–9.

[71] Haley RW. Excess incidence of ALS in young Gulf War veterans. Neurology 2003;61:750–6.

[72] Horner RD, Grambow SC, Coffman CJ, Lindquist JH, Oddone EZ, Allen KD, et al. Amyotrophic lateral sclerosis among 1991 Gulf War veterans: evidence for a time-limited outbreak. Neuroepidemiology 2008;31:28–32.

[73] Sharief MK, Priddin J, Delamont RS, Unwin C, Rose MR, David A, et al. Neurophysiologic analysis of neuromuscular symptoms in UK Gulf War veterans: a controlled study. Neurology 2002;59:1518–25.

[74] Kang HK, Bullman TA. Mortality among US veterans of the Persian Gulf War: 7-year follow-up. Am J Epidemiol 2001;154:399–405.

[75] Weisskopf MG, O'Reilly EJ, McCullough ML, Calle EE, Thun MJ, Cudkowicz M, et al. Prospective study of military service and mortality from ALS. Neurology 2005;64:32–7.

[76] Pastula DM, Coffman CJ, Allen KD, Oddone EZ, Kasarskis EJ, Lindquist JH, et al. Factors associated with survival in the National Registry of Veterans with ALS. Amyotroph Lateral Scler 2009;10:332–8.

[77] Allen KD, Kasarskis EJ, Bedlack RS, Rozear MP, Morgenlander JC, Sabet A, et al. The national registry of veterans with amyotrophic lateral sclerosis. Neuroepidemiology 2008;30:180–90.

[78] Fang F, Kwee LC, Allen KD, Umbach DM, Ye WM, Watson M, et al. Association between blood Lead and the risk of amyotrophic lateral sclerosis. Am J Epidemiol 2010;171:1126–33.

[79] Schmidt S, Kwee LC, Allen KD, Oddone EZ. Association of ALS with head injury, cigarette smoking and APOE genotypes. J Neurol Sci 2010;291:22–9.

[80] Kurtzke JF, Beebe GW. Epidemiology of amyotrophic lateral sclerosis: 1. A case-control comparison based on ALS deaths. Neurology 1980;30:453–62.

[81] Kondo K, Tsubaki T. Case-control studies of motor neuron disease: association with mechanical injuries. Arch Neurol 1981;38:220–6.

[82] Deapen DM, Henderson BE. A case-control study of amyotrophic lateral sclerosis. Am J Epidemiol 1986;123:790–9.

[83] Granieri E, Carreras M, Tola R, Paolino E, Tralli G, Eleopra R, et al. Motor neuron disease in the province of Ferrara, Italy, in 1964-1982. Neurology 1988;38:1604–8.

[84] Chio A, Meineri P, Tribolo A, Schiffer D. Risk factors in motor neuron disease: a case-control study. Neuroepidemiology 1991;10:174–84.

[85] Gresham LS, Molgaard CA, Golbeck AL, Smith R. Amyotrophic lateral sclerosis and history of skeletal fracture: a case-control study. Neurology 1987;37:717–9.

[86] Gallagher JP, Sanders M. Trauma and amyotrophic lateral sclerosis: a report of 78 patients. Acta Neurol Scand 1987;75:145–50.

[87] Williams DB, Annegers JF, Kokmen E, O'Brien PC, Kurland LT. Brain injury and neurologic sequelae: a cohort study of dementia, parkinsonism, and amyotrophic lateral sclerosis. Neurology 1991;41:1554–7.

[88] Beghi E, Logroscino G, Chio A, Hardiman O, Millul A, Mitchell D, et al. Amyotrophic lateral sclerosis, physical exercise, trauma and sports: results of a population-based pilot case-control study. Amyotroph Lateral Scler 2010;11:289–92.

[89] Turner MR, Abisgold J, Yeates DGR, Talbot K, Goldacre MJ. Head and other physical trauma requiring hospitalisation is not a significant risk factor in the development of ALS. J Neurol Sci 2010;288:45–8.

[90] Chio A, Benzi G, Dossena M, Mutani R, Mora G. Severely increased risk of amyotrophic lateral sclerosis among Italian professional football players. Brain 2005;128(Pt. 3):472–6.

[91] Chio A, Calvo A, Dossena M, Ghiglione P, Mutani R, Mora G. ALS in Italian professional soccer players: the risk is still present and could be soccer-specific. Amyotroph Lateral Scler 2009;10:205–9.

[92] Belli S, Vanacore N. Proportionate mortality of Italian soccer players: is amyotrophic lateral sclerosis an occupational disease? Eur J Epidemiol 2005;20:237–42.

[93] Abel EL. Football increases the risk for Lou Gehrig's disease, amyotrophic lateral sclerosis. Percept Mot Ski 2007;104(3 Pt 2):1251–4.

[94] Armon C, Nelson LM. Is head trauma a risk factor for amyotrophic lateral sclerosis? an evidence based review. Amyotroph Lateral Scler 2012;13:351–6.

[95] Harden CL, Meador KJ, Pennell PB, Hauser WA, Gronseth GS, French JA, et al. Practice parameter update: management issues for women with epilepsy–focus on pregnancy (an evidence-based review): teratogenesis and perinatal outcomes: report of the Quality Standards Subcommittee and Therapeutics and Technology Assessment Subcommittee of the American Academy of Neurology and American Epilepsy Society. Neurology 2009;73:133–41.

[96] Serbest G, Burkhardt MF, Siman R, Raghupathi R, Saatman KE. Temporal profiles of cytoskeletal protein loss following traumatic axonal injury in mice. Neurochem Res 2007;32:2006–14.

[97] Cole K, Perez-Polo JR. Neuronal trauma model: in search of Thanatos. Int J Dev Neurosci 2004;22:485–96.

[98] McKee AC, Gavett BE, Stern RA, Nowinski CJ, Cantu RC, Kowall NW, et al. TDP-43 proteinopathy and motor neuron disease in chronic traumatic encephalopathy. J Neuropathol Exp Neurol. 69, 918–929.

[99] Strong MJ, Volkening K, Hammond R, Yang WC, Strong W, Leystra-Lantz C, et al. TDP43 is a human low molecular weight neurofilament (hNFL) mRNA-binding protein. Mol Cell Neurosci 2007;35:320–7.

[100] Moisse K, Volkening K, Leystra-Lantz C, Welch I, Hill T, Strong MJ. Divergent patterns of cytosolic TDP-43 and neuronal progranulin expression following axotomy: implications for TDP-43 in the physiological response to neuronal injury. Brain Res 2009;1249:202–11.

[101] Arai T, Hasegawa M, Akiyama H, Ikeda K, Nonaka T, Mori H, et al. TDP-43 is a component of ubiquitin-positive tau-negative inclusions in frontotemporal lobar degeneration and amyotrophic lateral sclerosis. Biochem Biophys Res Commun 2006;351:602–11.

[102] Neumann M, Sampathu DM, Kwong LK, Truax AC, Micsenyi MC, Chou TT, et al. Ubiquitinated TDP-43 in frontotemporal lobar degeneration and amyotrophic lateral sclerosis. Science 2006;314:130–3.

[103] Ferrari R, Kapogiannis D, Huey ED, Momeni P. FTD and ALS: a tale of two diseases. Curr Alzheimer Res 2011;8:273–94.

[104] Yang W, Strong MJ. Widespread neuronal and glial hyperphosphorylated tau deposition in ALS with cognitive impairment. Amyotroph Lateral Scler 2012;13:178–93.

[105] Maekawa S, Leigh PN, King A, Jones E, Steele JC, Bodi I, et al. TDP-43 is consistently co-localized with ubiquitinated inclusions in sporadic and Guam amyotrophic lateral sclerosis but not in familial amyotrophic lateral sclerosis with and without SOD1 mutations. Neuropathology 2009;29:672–83.

[106] Murray ME, DeJesus-Hernandez M, Rutherford NJ, Baker M, Duara R, Graff-Radford NR, et al. Clinical and neuropathologic heterogeneity of c9FTD/ALS associated with hexanucleotide repeat expansion in C9ORF72. Acta Neuropathol 2011;122:673–90.

[107] Cohen TJ, Hwang AW, Unger T, Trojanowski JQ, Lee VM. Redox signalling directly regulates TDP-43 via cysteine oxidation and disulphide cross-linking. Embo J 2012;31:1241–52.

[108] Zhang HX, Tanji K, Yoshida H, Hayakari M, Shibata T, Mori F, et al. Alteration of biochemical and pathological properties of TDP-43 protein by a lipid mediator, 15-deoxy-delta(12,14)-prostaglandin J(2). Exp Neurol 2010;222:296–303.

[109] Meyerowitz J, Parker SJ, Vella LJ, Ng D, Price KA, Liddell JR, et al. C-Jun N-terminal kinase controls TDP-43 accumulation in stress granules induced by oxidative stress. Mol Neurodegener 2011;6:57.

[110] Iguchi Y, Katsuno M, Takagi S, Ishigaki S, Niwa J, Hasegawa M, et al. Oxidative stress induced by glutathione depletion reproduces pathological modifications of TDP-43 linked to TDP-43 proteinopathies. Neurobiol Dis 2012;45:862–70.

[111] Wang JS, Huang YH. Effects of exercise intensity on lymphocyte apoptosis induced by oxidative stress in men. Eur J Appl Physiol 2005;95:290–7.

[112] Reid MB. Free radicals and muscle fatigue: of ROS, canaries, and the IOC. Free Radic Biol Med 2008;44:169–79.

[113] Siciliano G, D'Avino C, Del Corona A, Barsacchi R, Kusmic C, Rocchi A, et al. Impaired oxidative metabolism and lipid peroxidation in exercising muscle from ALS patients. Amyotroph Lateral Scler Other Motor Neuron Disord 2002;3:57–62.

[114] Scarmeas N, Shih T, Stern Y, Ottman R, Rowland LP. Premorbid weight, body mass, and varsity athletics in ALS. Neurology 2002;59:773–5.

[115] Mattsson P, Lonnstedt I, Nygren I, Askmark H. Physical fitness, but not muscle strength, is a risk factor for death in amyotrophic lateral sclerosis at an early age. J Neurol Neurosur Psychiat 2012;83:390–4.

[116] Kamble PDV, Baji PS. Study of anthropological parameters, body composition, strength & endurance in basketball players. Int J Biol Med Res 2012;3:1404–6.

[117] Chiò A, Mora G. Physical fitness and amyotrophic lateral sclerosis: dangerous liaisons or common genetic pathways? J Neurol Neurosurg Psychiatry 2012;83:389.

[118] Nelson LM, Matkin C, Longstreth Jr WT, McGuire V. Population-based case-control study of amyotrophic lateral sclerosis in western Washington State. II. Diet. Am J Epidemiol 2000;151:164–73.

[119] Morozova N, Weisskopf MG, McCullough ML, Munger KL, Calle EE, Thun MJ, et al. Diet and amyotrophic lateral sclerosis. Epidemiology 2008;19:324–37.

[120] Okamoto K, Kihira T, Kondo T, Kobashi G, Washio M, Sasaki S, et al. Nutritional status and risk of amyotrophic lateral sclerosis in Japan. Amyotroph Lateral Scler 2007;8:300–4.

[121] Wang X, Quinn PJ. Vitamin E and its function in membranes. Prog Lipid Res 1999;38:309–36.

[122] Jialal I, Grundy SM. Effect of dietary supplementation with alpha-tocopherol on the oxidative modification of low density lipoprotein. J Lipid Res 1992;33:899–906.

[123] Ascherio A, Weisskopf MG, O'Reilly EJ, Jacobs EJ, McCullough ML, Calle EE, et al. Vitamin E intake and risk of amyotrophic lateral sclerosis. Ann Neurology 2005;57:104–10.

[124] Wang H, O'Reilly EJ, Weisskopf MG, Logroscino G, McCullough ML, Schatzkin A, et al. Vitamin E intake and risk of amyotrophic lateral sclerosis: a pooled analysis of data from 5 prospective cohort studies. Am J Epidemiol 2011;173:595–602.

[125] Weisskopf MG, McCullough ML, Calle EE, Thun MJ, Cudkowicz M, Ascherio A. Prospective study of cigarette smoking and amyotrophic lateral sclerosis. Am J Epidemiol 2004;160:26–33.

[126] Nelson LM, McGuire V, Longstreth Jr WT, Matkin C. Population-based case-control study of amyotrophic lateral sclerosis in western Washington State. I. Cigarette smoking and alcohol consumption. Am J Epidemiol 2000;151:156–63.

[127] Kamel F, Umbach DM, Munsat TL, Shefner JM, Sandler DP. Association of cigarette smoking with amyotrophic lateral sclerosis. Neuroepidemiology 1999;18:194–202.

[128] Gallo V, Bueno-De-Mesquita HB, Vermeulen R, Andersen PM, Kyrozis A, Linseisen J, et al. Smoking and risk for amyotrophic lateral sclerosis: analysis of the EPIC cohort. Ann Neurol 2009;65:378–85.

[129] Sutedja NA, Veldink JH, Fischer K, Kromhout H, Wokke JHJ, Huisman MHB, et al. Lifetime occupation, education, smoking, and risk of ALS. Neurology 2007;69:1508–14.

[130] Alonso A, Logroscino G, Jick SS, Hernan MA. Association of smoking with amyotrophic lateral sclerosis risk and survival in men and women: a prospective study. BMC Neurol 2010;10:6.

[131] Okamoto K, Kihira T, Kondo T, Kobashi G, Washio M, Sasaki S, et al. Lifestyle factors and risk of amyotrophic lateral sclerosis: a case-control study in Japan. Ann Epidemiol 2009;19:359–64.

[132] Fang F, Bellocco R, Hernan MA, Ye W. Smoking, snuff dipping and the risk of amyotrophic lateral sclerosis–a prospective cohort study. Neuroepidemiology 2006;27:217–21.

[133] de Jong SW, Huisman MH, Sutedja NA, van der Kooi AJ, de Visser M, Schelhaas HJ, et al. Smoking, alcohol consumption, and the risk of amyotrophic lateral sclerosis: a population-based study. Am J Epidemiol 2012;176:233–9.

[134] Alonso A, Logroscino G, Hernán MA. Smoking and the risk of amyotrophic lateral sclerosis: a systematic review and meta-analysis. J Neurol Neurosurg Psychiatry 2010;81:1249–52.

[135] Pamphlett R, Cochran Ward E. Smoking is not a risk factor for sporadic amyotrophic lateral sclerosis in an Australian population. Neuroepidemiology 2012;38:106–13.

[136] Armon C. Smoking may be considered an established risk factor for sporadic ALS. Neurology 2009;73:1693–8.

[137] Vayssier M, Banzet N, Francois D, Bellmann K, Polla BS. Tobacco smoke induces both apoptosis and necrosis in mammalian cells: differential effects of HSP70. Am J Physiol 1998;275:L771–9.

[138] Bertram KM, Baglole CJ, Phipps RP, Libby RT. Molecular regulation of cigarette smoke induced-oxidative stress in human retinal pigment epithelial cells: implications for age-related macular degeneration. Am J Physiol Cell Physiol 2009;297:C1200–10.

[139] Yanbaeva DG, Dentener MA, Creutzberg EC, Wesseling G, Wouters EF. Systemic effects of smoking. Chest 2007;131:1557–66.

[140] Kiziler AR, Aydemir B, Onaran I, Alici B, Ozkara H, Gulyasar T, et al. High levels of cadmium and lead in seminal fluid and blood of smoking men are associated with high oxidative stress and damage in infertile subjects. Biol Trace Elem Res 2007;120:82–91.

[141] Wild CP. Complementing the genome with an "exposome": the outstanding challenge of environmental exposure measurement in molecular epidemiology. Cancer Epidemiol Biomarkers Prev 2005;14:1847–50.

[142] Rappaport SM. Implications of the exposome for exposure science. J Expo Sci Environ Epidemiol 2011;21:5–9.

[143] Wild CP. The exposome: from concept to utility. Int J Epidemiol 2012;41:24–32.

[144] Buck Louis GM, Yeung E, Sundaram R, Laughon SK, Zhang C. The exposome–exciting opportunities for discoveries in reproductive and perinatal epidemiology. Paediatr Perinat Epidemiol 2013;27:229–36.

[145] Gennings C, Carrico C, Factor-Litvak P, Krigbaum N, Cirillo PM, Cohn BA. A cohort study evaluation of maternal PCB exposure related to time to pregnancy in daughters. Environ Health 2013;12:66.

[146] Loikkanen J, Naarala J, Vahakangas KH, Savolainen KM. Glutamate increases toxicity of inorganic lead in GT1-7 neurons: partial protection induced by flunarizine. Arch Toxicol 2003;77:663–71.

[147] Kriscenski-Perry E, Durham HD, Sheu SS, Figlewicz DA. Synergistic effects of low level stressors in an oxidative damage model of spinal motor neuron degeneration. Amyotroph Lateral Scler Other Mot Neuron Disord 2002;3:151–7.

[148] Sahin E, Gumuslu S. Alterations in brain antioxidant status, protein oxidation and lipid peroxidation in response to different stress models. Behav Brain Res 2004;155:241–8.

[149] Ilzecka J. Total antioxidant status is increased in the serum of amyotrophic lateral sclerosis patients. Scand J Clin Lab Inv 2003;63:297–302.

[150] Babu GN, Kumar A, Chandra R, Puri SK, Singh RL, Kalita J, et al. Oxidant-antioxidant imbalance in the erythrocytes of sporadic amyotrophic lateral sclerosis patients correlates with the progression of disease. Neurochem Int 2008;52:1284–9.

[151] Baillet A, Chanteperdrix V, Trocme C, Casez P, Garrel C, Besson G. The role of oxidative stress in amyotrophic lateral sclerosis and Parkinson's disease. Neurochem Res 2010;35:1530–7.

[152] Beckman JS, Estevez A. Superoxide dismutase, oxidative stress and ALS. In: Mitsumoto H, Przedborski S, Gordon PH, editors. Amyotrophic lateral sclerosis. New York: Taylor and Francis Group; 2006. p. 339–54.

[153] Maier CM, Chan PH. Role of superoxide dismutases in oxidative damage and neurodegenerative disorders. Neuroscientist 2002;8:323–34.

[154] Rosen DR, Siddique T, Patterson D, Figlewicz DA, Sapp P, Hentati A, et al. Mutations in Cu/Zn superoxide dismutase gene are associated with familial amyotrophic lateral sclerosis. Nature 1993;362:59–62.

[155] DeJesus-Hernandez M, Mackenzie IR, Boeve BF, Boxer AL, Baker M, Rutherford NJ, et al. Expanded GGGGCC hexanucleotide repeat in noncoding region of C9ORF72 causes chromosome 9p-linked FTD and ALS. Neuron 2011;72:245–56.

[156] Renton AE, Majounie E, Waite A, Simon-Sanchez J, Rollinson S, Gibbs JR, et al. A hexanucleotide repeat expansion in C9ORF72 is the cause of chromosome 9p21-linked ALS-FTD. Neuron 2011;72:257–68.

[157] Synofzik M, Ronchi D, Keskin I, Basak AN, Wilhelm C, Gobbi C, et al. Mutant superoxide dismutase-1 indistinguishable from wild-type causes ALS. Hum Mol Genet 2012;21:3568–74.

[158] Shibata N, Nagai R, Miyata S, Jono T, Horiuchi S, Hirano A, et al. Nonoxidative protein glycation is implicated in familial amyotrophic lateral sclerosis with superoxide dismutase-1 mutation. Acta Neuropathol 2000;100:275–84.

[159] Shibata N, Hirano A, Yamamoto T, Kato Y, Kobayashi M. Superoxide dismutase-1 mutation-related neurotoxicity in familial amyotrophic lateral sclerosis. Amyotroph Lateral Scler Other Mot Neuron Disord 2000;1:143–61.

[160] Crow JP, Sampson JB, Zhuang Y, Thompson JA, Beckman JS. Decreased zinc affinity of amyotrophic lateral sclerosis-associated superoxide dismutase mutants leads to enhanced catalysis of tyrosine nitration by peroxynitrite. J Neurochem 1997;69:1936–44.

[161] Kabashi E, Valdmanis PN, Dion P, Rouleau GA. Oxidized/misfolded superoxide dismutase-1: the cause of all amyotrophic lateral sclerosis? Ann Neurol 2007;62:553–9.

[162] Trumbull KA, Beckman JS. A role for copper in the toxicity of zinc-deficient superoxide dismutase to motor neurons in amyotrophic lateral sclerosis. Antioxid Redox Signal 2009;11:1627–39.

[163] Guareschi S, Cova E, Cereda C, Ceroni M, Donetti E, Bosco DA, et al. An over-oxidized form of superoxide dismutase found in sporadic amyotrophic lateral sclerosis with bulbar onset shares a toxic mechanism with mutant SOD1. Proc Natl Acad Sci USA 2012;109:5074–9.

[164] Bosco DA, Morfini G, Karabacak NM, Song Y, Gros-Louis F, Pasinelli P, et al. Wild-type and mutant SOD1 share an aberrant conformation and a common pathogenic pathway in ALS. Nat Neurosci 2010;13:1396–403.

[165] Forsberg K, Jonsson PA, Andersen PM, Bergemalm D, Graffmo KS, Hultdin M, et al. Novel antibodies reveal inclusions containing non-native SOD1 in sporadic ALS patients. PLoS One 2010;5:e11552.

[166] Pokrishevsky E, Grad LI, Yousefi M, Wang J, Mackenzie IR, Cashman NR. Aberrant localization of FUS and TDP43 is associated with misfolding of SOD1 in amyotrophic lateral sclerosis. PLoS One 2012;7:e35050.

[167] Liu HN, Sanelli T, Horne P, Pioro EP, Strong MJ, Rogaeva E, et al. Lack of evidence of monomer/misfolded superoxide dismutase-1 in sporadic amyotrophic lateral sclerosis. Ann Neurol 2009;66:75–80.

[168] Aviram M, Rosenblat M, Bisgaier CL, Newton RS, Primo-Parmo SL, La Du BN. Paraoxonase inhibits high-density lipoprotein oxidation and preserves its functions. A possible peroxidative role for paraoxonase. J Clin Invest 1998;101:1581–90.

[169] Costa LG, Cole TB, Furlong CE. Paraoxonase (PON1): from toxicology to cardiovascular medicine. Acta Biomed 2005;76(Suppl. 2):50–7.

[170] Primo-Parmo SL, Sorenson RC, Teiber J, La Du BN. The human serum paraoxonase/arylesterase gene (PON1) is one member of a multigene family. Genomics 1996;33:498–507.

[171] Gaidukov L, Tawfik DS. The development of human sera tests for HDL-bound serum PON1 and its lipolactonase activity. J Lipid Res 2007;48:1637–46.

[172] Chambers JE. PON1 multitasks to protect health. Proc Natl Acad Sci U S A 2008;105:12639–40.

[173] Nguyen SD, Sok DE. Oxidative inactivation of paraoxonase1, an antioxidant protein and its effect on antioxidant action. Free Radic Res 2003;37:1319–30.

[174] Watson AD, Berliner JA, Hama SY, La Du BN, Faull KF, Fogelman AM, et al. Protective effect of high density lipoprotein associated paraoxonase. Inhibition of the biological activity of minimally oxidized low density lipoprotein. J Clin Invest 1995;96:2882–91.

[175] Furlong CE, Richter RJ, Seidel SL, Motulsky AG. Role of genetic polymorphism of human plasma paraoxonase/arylesterase in hydrolysis of the insecticide metabolites chlorpyrifos oxon and paraoxon. Am J Hum Genet 1988;43:230–8.

[176] Li HL, Liu DP, Liang CC. Paraoxonase gene polymorphisms, oxidative stress, and diseases. J Mol Med 2003;81:766–79.

[177] Watson CE, Draganov DI, Billecke SS, Bisgaier CL, La Du BN. Rabbits possess a serum paraoxonase polymorphism similar to the human Q192R. Pharmacogenetics 2001;11:123–34.

[178] Zhao Y, Ma YS, Fang Y, Liu LL, Wu SD, Fu D, et al. Association between PON1 activity and coronary heart disease risk: a meta-analysis based on 43 studies. Mol Genet Metab 2012;105:141–8.

[179] Singh S, Kumar V, Thakur S, Banerjee BD, Rautela RS, Grover SS, et al. Paraoxonase-1 genetic polymorphisms and susceptibility to DNA damage in workers occupationally exposed to organophosphate pesticides. Toxicol Appl Pharmacol 2011;252:130–7.

[180] Draganov DI, La Du BN. Pharmacogenetics of paraoxonases: a brief review. Naunyn Schmiedebergs Arch Pharmacol 2004;369:78–88.

[181] Zurich MG, Honegger P, Schilter B, Costa LG, Monnet-Tschudi F. Involvement of glial cells in the neurotoxicity of parathion and chlorpyrifos. Toxicol Appl Pharmacol 2004;201:97–104.

[182] Precourt LP, Amre D, Denis MC, Lavoie JC, Delvin E, Seidman E, et al. The three-gene paraoxonase family: physiologic roles, actions and regulation. Atherosclerosis 2011;214:20–36.

[183] Ng CJ, Wadleigh DJ, Gangopadhyay A, Hama S, Grijalva VR, Navab M, et al. Paraoxonase-2 is a ubiquitously expressed protein with antioxidant properties and is capable of preventing cell-mediated oxidative modification of low density lipoprotein. J Biol Chem 2001;276:44444–9.

[184] Reddy ST, Wadleigh DJ, Grijalva V, Ng C, Hama S, Gangopadhyay A, et al. Human paraoxonase-3 is an HDL-associated enzyme with biological activity similar to paraoxonase-1 protein but is not regulated by oxidized lipids. Arterioscler Thromb Vasc Biol 2001;21:542–7.

[185] Costa LG, Vitalone A, Cole TB, Furlong CE. Modulation of paraoxonase (PON1) activity. Biochem Pharmacol 2005;69:541–50.

[186] Richter RJ, Furlong CE. Determination of paraoxonase (PON1) status requires more than genotyping. Pharmacogenetics 1999;9:745–53.

[187] Ferre N, Camps J, Fernandez-Ballart J, Arija V, Murphy MM, Ceruelo S, et al. Regulation of serum paraoxonase activity by genetic, nutritional, and lifestyle factors in the general population. Clin Chem 2003;49:1491–7.

[188] Aviram M, Rosenblat M, Bisgaier CL, Newton RS. Atorvastatin and gemfibrozil metabolites, but not the parent drugs, are potent antioxidants against lipoprotein oxidation. Atherosclerosis 1998;138:271–80.

[189] Deakin S, Guernier S, James RW. Pharmacogenetic interaction between paraoxonase-1 gene promoter polymorphism C-107T and statin. Pharmacogenet Genom 2007;17:451–7.

[190] Blatter-Garin MC, Kalix B, De Pree S, James RW. Aspirin use is associated with higher serum concentrations of the anti-oxidant enzyme, paraoxonase-1. Diabetologia 2003;46:593–4.

[191] Slowik A, Tomik B, Wolkow PP, Partyka D, Turaj W, Malecki MT, et al. Paraoxonase gene polymorphisms and sporadic ALS. Neurology 2006;67:766–70.

[192] Saeed M, Siddique N, Hung WY, Usacheva E, Liu E, Sufit RL, et al. Paraoxonase cluster polymorphisms are associated with sporadic ALS. Neurology 2006;67:771–6.

[193] Valdmanis PN, Kabashi E, Dyck A, Hince P, Lee J, Dion P, et al. Association of paraoxonase gene cluster polymorphisms with ALS in France, Quebec, and Sweden. Neurology 2008;71:514–20.

[194] Cronin S, Greenway MJ, Prehn JH, Hardiman O. Paraoxonase promoter and intronic variants modify risk of sporadic amyotrophic lateral sclerosis. J Neurol Neurosurg Psychiatry 2007;78:984–6.

[195] Morahan JM, Yu B, Trent RJ, Pamphlett R. A gene-environment study of the paraoxonase 1 gene and pesticides in amyotrophic lateral sclerosis. Neurotoxicology 2007;28:532–40.

[196] Landers JE, Shi L, Cho TJ, Glass JD, Shaw CE, Leigh PN, et al. A common haplotype within the PON1 promoter region is associated with sporadic ALS. Amyotroph Lateral Scler 2008;9:306–14.

[197] Zawislak D, Ostrowska M, Golenia A, Marona M, Tomik B, Wolkow P, et al. The -A162G polymorphism of the PON1 gene and the risk of sporadic amyotrophic lateral sclerosis. Neurol Neurochir Pol 2010;44:246–50.

[198] Ricci C, Battistini S, Cozzi L, Benigni M, Origone P, Verriello L, et al. Lack of association of PON polymorphisms with sporadic ALS in an Italian population. Neurobiol Aging 2011;32(552):e7–13.

[199] Ticozzi N, LeClerc AL, Keagle PJ, Glass JD, Wills AM, van Blitterswijk M, et al. Paraoxonase gene mutations in amyotrophic lateral sclerosis. Ann Neurol 2010;68:102–7.

[200] van Blitterswijk M, Blokhuis A, van Es MA, van Vught PW, Rowicka PA, Schelhaas HJ, et al. Rare and common paraoxonase gene variants in amyotrophic lateral sclerosis patients. Neurobiol Aging 2012;33(1845):e1–3.

[201] Wills AM, Cronin S, Slowik A, Kasperaviciute D, Van Es MA, Morahan JM, et al. A large-scale international meta-analysis of paraoxonase gene polymorphisms in sporadic ALS. Neurology 2009;73:16–24.

[202] Costa LG, Giordano G, Furlong CE. Pharmacological and dietary modulators of paraoxonase 1 (PON1) activity and expression: the hunt goes on. Biochem Pharmacol 2011;81:337–44.

[203] Chio A. Risk factors in the early diagnosis of ALS: European epidemiological studies. Amyotroph Lateral Scler Other Mot Neuron Disord 2000;1(Suppl. 1):S13–8.

[204] Armon C. Environmental risk factors for amyotrophic lateral sclerosis. Neuroepidemiology 2001;20:2–6.

[205] D'Amico E, Factor-Litvak P, Santella RM, Mitsumoto H. Clinical perspective on oxidative stress in sporadic amyotrophic lateral sclerosis. Free Radic Biol Med 2013;65:509–27.

[206] Eckl PM. Genotoxicity of HNE. Mol Asp Med 2003;24:161–5.

[207] Louwerse ES, Weverling GJ, Bossuyt PM, Meyjes FE, de Jong JM. Randomized, double-blind, controlled trial of acetylcysteine in amyotrophic lateral sclerosis. Arch Neurol 1995;52:559–64.

[208] Lange DJ, Murphy PL, Diamond B, Appel V, Lai EC, Younger DS, et al. Selegiline is ineffective in a collaborative double-blind, placebo-controlled trial for treatment of amyotrophic lateral sclerosis. Arch Neurol-Chicago 1998;55:93–6.

[209] Desnuelle C, Dib M, Garrel C, Favier A, Ar-tS G. A double-blind, placebo-controlled randomized clinical trial of alpha-tocopherol (vitamin E) in the treatment of amyotrophic lateral sclerosis. Amyotroph Lateral Scler Other Motor Neuron Disord 2001;2:9–18.

[210] Graf M, Ecker D, Horowski R, Kramer B, Riederer P, Gerlach M, et al. High dose vitamin E therapy in amyotrophic lateral sclerosis as add-on therapy to riluzole: results of a placebo-controlled double-blind study. J Neural Transm 2005;112:649–60.

[211] Kaufmann P, Thompson JLP, Levy G, Buchsbaum R, Shefner J, Krivickas LS, et al. Phase II trial of CoQ10 for ALS finds insufficient evidence to justify phase III. Ann Neurol 2009;66:235–44.

Gene–Environment Interactions in Huntington's Disease

Terry Jo Bichell, Michael Uhouse, Emma Bradley, Aaron B. Bowman

Referring to the debate about nature vs. nurture, neuropsychologist Dr. Donald Hebb is said to have asked, "Which contributes more to the area of a rectangle, its length or its width?"

HUNTINGTON'S DISEASE: ONSET, GENETICS, PROTEIN, AND MODELS

The identification of the genetic cause of Huntington's disease (HD) began in the 1980s with a large, isolated Venezuelan pedigree that contained a high incidence of HD. Researchers traced the genetic cause to an increase in CAG repeats in exon 1 of the huntingtin gene (*HTT*),

Environmental Factors in Neurodevelopmental and Neurodegenerative Disorders
http://dx.doi.org/10.1016/B978-0-12-800228-5.00017-0

355

which leads to an expansion of the polyglutamine tract in the huntingtin protein [1]. The inverse relationship between the number of CAG repeats and age at symptom onset explains only 60% of the variability in the age of onset and disease progression [2,3] (Figure 17.1). Sibling studies have established that both genetic and environmental factors have additional influence on age of onset, with modifier genes explaining only an additional 13% of onset variability [4–8]. While the median age of onset is 39, surprising cases of endophenotypes and full disease onset in patients with alleles ranging in length from 27 to 35 repeats suggest that genetic repeat length is not alone in its influence on disease onset [3,9,10]. Moreover, cases of identical twins diverging in onset and symptom manifestation emphasize the influence of the environment on phenotypic variability [4,11,12].

A wide variety of environmental factors may influence HD pathobiology in either a positive (e.g., delayed onset) or negative (earlier onset) manner. Cognitive activity, nutritional factors such as environmental enrichment, increased exercise, and healthy diets have been shown in animal models of HD to delay disease progression via the increase of endogenous brain-derived neurotrophic factor (BDNF) levels [13–16] (Figure 17.2). In contrast, correlations between the earlier age of onset and measures of passive or apathetic lifestyles have been observed in premanifest HD patients [17,18]. Essential heavy metals, such as copper (Cu), zinc (Zn), iron (Fe), and manganese (Mn), have been shown to influence pathologies related to HD, including altered vesicular transport, protein aggregation, mitochondrial dysfunction, and induction of oxidative stress in many neurodegenerative disorders [19,20]. For example, genetic studies of *PARK9* (a gene that has been indicated to play a role in divalent metal sequestration) in *Saccharomyces cervisiae* have demonstrated functional relationships between manganese toxicity, zinc dyshomeostasis, and mitochondrial dysfunction [21–24].

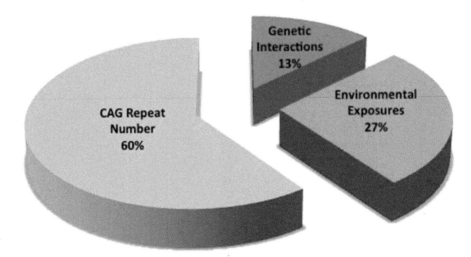

FIGURE 17.1 **Genetics do not explain all of the variability in the age of onset of Huntington's disease.** The number of CAG repeats is the strongest predictor of the age of onset of Huntington's disease, explaining 60% of the variability, while other modifier genes explain another 13%, leaving 27% of variability that is attributed to environmental interactions.

Neurodegenerative processes may be influenced by an amalgam of concurrent mechanisms such as diminished neurotrophic support (i.e., BDNF) and unstable metal ion homeostasis in vivo.

HD is classified as an autosomal-dominant neurodegenerative disorder, coded for by a gene that resides on the short arm of chromosome 4 [25] Interestingly, even though the *HTT* gene is highly conserved across many phyla, there is extreme diversity in the number of CAG repeats in the normal gene of different species [26–28]. Within the context of its autosomal-dominant inheritance, researchers have sought to determine whether HD symptoms are caused by a (1) loss of function of normal *HTT* (e.g., haploinsufficiency); (2) toxic gain of function in mutant *HTT* [29]; or (3) combination of these genetic mechanisms [30]. There are mouse models of HD (e.g., transgenic models) that develop HD-like pathologies while retaining two functional wild-type *Htt* alleles, suggesting that the presence of a mutant CAG-expanded allele of *Htt* is sufficient to cause neurodegeneration [31]. In contrast, a study of homozygous human patients exhibited exacerbated disease progression when compared with heterozygotes and controlled for CAG repeat number on both alleles [32]. Therefore, it remains possible that the loss of wild-type *HTT* function retains an important role in disease pathology.

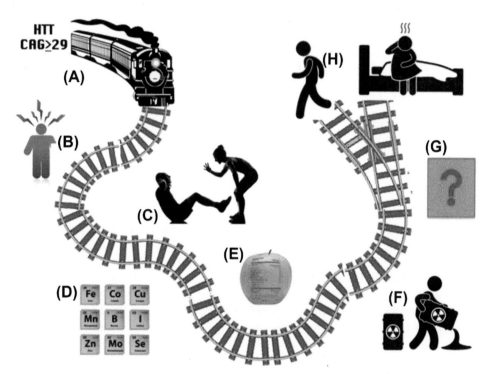

FIGURE 17.2 **Environmental influences affect the timing of onset of Huntington's disease.** Excess CAG repeats in the HTT gene (A) will cause Huntington's disease, but the age of onset is hastened by factors such as stress (B), and delayed by other factors such as exercise (C) and diet (E). Metals (D) accumulate abnormally in Huntington's disease, and other unknown exposures, such as those that increase oxidative stress (F)(G), may alter prognosis (H).

The potential for mutant huntingtin protein aggregates to be influenced by environmental factors, or to alter their vulnerability to environmental factors, remains relatively unexplored. The huntingtin protein becomes much less soluble when it is extended by excess polyglutamine sequences, and this causes a pathological cascade that is set off by misfolding of the extended polyglutamine protein into a β-sheet structure leading to abnormal posttranslational modifications [3,33,34]. The presence of excess copper, added to mutant huntingtin in vitro, causes it to form aggregates, and when copper is prevented from entering neurons, fewer aggregates form in HD models in vivo [35,36]. Another metal, iron, has also been linked to increased aggregates of mutant huntingtin [37]. Mutant huntingtin itself has widespread cellular effects such as altering gene transcription by aggregating transcription factors, as well as decreasing histone acetylation and increasing histone methylation and ubiquitination of chromatin [38–41]. Additionally, mutant huntingtin is associated with the production of abnormal metabolites and oxidative stress [3,42]. Both in vitro and in vivo, mutant huntingtin has been shown to recruit wild-type huntingtin into insoluble aggregates, suggesting a potential mechanism whereby pathology could derive from a deficit in the action of wild-type huntingtin [43–45]. Thus, mutant huntingtin alone has pathological effects, but in combination with environmental agents such as the metals iron and copper, it may become even more toxic to the fragile striatal neurons that degenerate in HD.

NATURAL HISTORY OF HUNTINGTON'S DISEASE AND RELATED DISORDERS

The clinical progression of HD can be described in terms of three main phases: (1) premanifest, in which patients with the gene mutation are symptomatically invariable from controls approximately 10–15 years before disease onset, (2) prodromal, in which subtle cognitive and emotional manifestations occur with little loss of function, and (3) manifest, which involves varying severities of motor, cognitive, and psychiatric deterioration [3,46]. Actual striatal deterioration begins up to 15 years before phenoconversion, and aggressive atrophy continues throughout disease progression. Even presymptomatic HD-model transgenic mice display significantly decreased numbers of dendritic spines, as well as proximal dendrite thickening in both the striatum and the cortex [47]. While other brain regions such as the thalamus, hippocampus, and globus pallidus undergo cellular atrophy, the magnitude of striatal volume reduction is most extensive [48]. The most prominent pathological changes in white matter structures occur in the areas around the striatum and within the corpus callosum and posterior white matter tracts [18]. In terms of loss of cortical volume, HD predominantly affects the large pyramidal neurons in layers III, V, and VI [49]. Prodromal loss of cortical mass followed by loss of striatal GABAergic medium spiny neurons suggests that disease manifestation in the cortex may induce excitotoxic postsynaptic deterioration of striatal medium spiny neurons [50]. In an HD mouse model, the removal of the cortex actually alleviates HD symptoms [51].

Including HD, there are currently at least nine known neurodegenerative diseases that derive from excess CAG repeats. These CAG repeat expansion diseases are characterized by severe chromosomal instability, specifically when inheritance is paternal, as expansions are frequent in the male germline [10]. The neuronal loss of each of these diseases is restricted to specific brain regions, with similar intracellular manifestations such as modified metal processing, inclusions, aggregates, and protein misfolding [52]. These molecular findings are also present in non-CAG repeat neurodegenerative diseases such as Alzheimer's and Parkinson's, both of which have greater environmental than genetic attributions [53]. In each of these neurodegenerative diseases, environmental enrichment, exercise, diet, and xenobiotic exposures have been shown to either exacerbate or alleviate disease [54,55].

ANIMAL MODELS OF HUNTINGTON'S DISEASE AND THE STUDY OF ENVIRONMENTAL FACTORS

To date there are over 20 different mouse models of HD, including both chemically induced lesion models and genetically modified models [56]. These mouse models are divisible into three main categories of construction: (1) N-terminal transgenic mice with the 5′ region of human *HTT*, (2) full-length transgenic models that carry and express the entire *HTT* sequence including the expanded polyglutamine region, and (3) knock-in models in which CAG repeats of different lengths are directly inserted into the mouse *Htt* locus [57]. In many ways, mice are an excellent model for HD because they develop subtle signs in early adulthood and more severe signs at advanced ages, mimicking the human condition, but their symptoms are much less severe than those seen in humans, even with extreme CAG repeat length [58]. Furthermore, as seen in the cases of patients with lower numbers of disease-related CAG repeats, symptoms seem to be greatly affected by environmental conditions such as exercise, diet, and enrichment, such that standardization of these conditions is now seen as crucial for preclinical studies. Following is a brief description of some of the more widely used animal models of HD and their potential for studying the role of environmental factors in HD pathobiology.

The R6/1 and R6/2 mice fall into the N-terminal transgenics category, as they contain a transgene fragment of human exon 1 of *HTT*, 113 and 144 CAG repeats from the 5′ end. Symptoms manifest pronouncedly in R6/2 mice in the form of weight loss, aggregation, brain atrophy, and motor deterioration by 12 weeks, but vary depending on repeat length [59]. These mice have proven to have a high degree of phenotypic variation, most likely because symptomology varies with CAG repeat size, and the randomly integrated pronuclearly injected CAG-repeat sequences are highly unstable in both mouse germlines and somatic cells [59–63].

The YAC128 transgene mouse expresses the full-length human *HTT* gene with 128 repeats, inserted via a yeast artificial chromosome [64–66]. The large size of the full-length transgenic-model construct allows *HTT* mRNA and protein expression levels to correspond better to the number of transgene copies inserted [66]. Although the YAC128 mouse survives for a normal lifespan, it has increased weight; develops motor deterioration; has increased N-methyl-ᴅ-aspartate (NMDA), AMPA, and metabotropic glutamate receptor (mGluR) binding; and reduced striatal and cortical volumes [67]. The bacterial artificial chromosome transgenic mouse (BACHD) carries the full-length mouse *Htt* gene and demonstrates fewer motor symptoms than the YAC128 mouse, though it eventually develops aggregates and brain atrophy at an advanced age [66]. Because the polyglutamine tract consists of both CAG and CAA repeats in this model, these mice are more resistant to both germline and somatic instability [66]. Finally, while mouse models that have knock-in CAG repeat sequences inserted into the mouse *Htt* genes may retain more expression fidelity, as well as the potential to create an allelic series, they do not develop as robust a phenotype as those of transgenic models, especially at earlier ages, although they do display some neuronal defects and motor abnormalities [58,68–72]. Because environmental influences, even minor enhancements to mouse habitats, have been shown to affect the HD course, it is crucial that mouse husbandry is uniform across experiments, especially in preclinical trials.

In addition to HD mouse models, other more simple model organisms can be important tools in identifying disease-modifying genes and investigating HD mechanisms. For example, *Drosophila* models of polyglutamate expansion diseases display the same cell death and aggregation phenomena that humans do [73–76]. *D. melanogaster* models of HD utilize the UAS–GAL4 system to express N-terminal fragments of mutant *HTT* in targeted cells [77]. Fly models exhibit progressive degeneration, motor abnormalities, and reduced survival [77]. *D. melanogaster* models have been constructed to express human huntingtin cDNA encoding both those pathogenic proteins that result in progressive motor deterioration and decreased survival rates, and also nonpathogenic proteins that have no perceptible effect on behavior [78]. Such fly models have become useful in demonstrating the kinetics of protein aggregation in axonal transport disruption and in testing therapeutic approaches such as histone deacetylation inhibitors, and protein abundance modulators such as NUB1 [78–80]. In testing environmental exposures, the simpler organisms provide an efficient way to study toxicities and mechanistic results of gene–environment interactions, because they retain molecular pathways that recapitulate those of higher mammals.

A transgenic *C. elegans* model of HD expresses amino terminal fragments of huntingtin in target neurons [81,82]. The translucent body of the *C. elegans* model allows in vivo visualization of neuronal processes, which is particularly useful in the study of gene–environment interaction effects with metal on neurodegeneration [19]. Even though *C. elegans* lacks an *HTT* ortholog, transgenic expression of mutant *HTT* yields age-dependent mechanosensory defects, neurodegeneration, and neuronal dysfunction [81,82]. Via insertion of a polyglutamine repeat of 40 CAG repeats in length, researchers have used *C. elegans* models to examine the effects of diverse genetic backgrounds on HD pathology, as well as for preclinical work on therapeutic targets such as the copper-chelating drug, PBT2 [83,84]. A human clinical trial of PBT2 as a treatment for HD has shown that the drug is safe and tolerated well, but a study of its effectiveness in the amelioration of symptoms is still underway [85].

Large-animal models such as primates and sheep are particularly useful for studies in which it is vital to replicate the clinical phenotypes of human HD as closely as possible. A transgenic primate model has been developed, but has not been well studied, and at present there is only one living transgenic HD rhesus monkey. This animal expresses exon 1 of human *HTT* with only 29 CAG repeats, which should be in the normal range, but it displays dystonia, chorea, decreased hippocampal and striatal volumes, and behavioral changes such as difficulties with cognitive ability, impulsivity, and spatial recognition [86]. None of these symptoms developed until the model's later years, a progression that mirrors human HD motor degeneration [86]. Other primate models have been produced by lesioning the striatal region

[87] or by exposing the animals to medications that degenerate the striatal neurons, such as 3-nitropropionic acid (3NP) [88]. Such nonhuman primate models have already become relevant in the study of HD gene × environment effects, in a study that shows increases in the spontaneous locomotor activity of both calorically restricted and supplemented primates, as well as increased spatial memory abilities of calorically supplemented animals [89]. Additionally, after chronic manganese exposure, nonhuman wild-type primate models have demonstrated damaged visuospatial associative learning, neurodegenerative aggregation activity, increases in microglia, and general neurodegeneration characteristic of disorders such as HD [90]. These long-term studies of primates demonstrate that mutant human *HTT* may be unique in its mechanisms and its ability to cause neurodegeneration, and that environmental exposures alone may be sufficient to cause an HD-like phenotype, even without the presence of the HD mutation.

Researchers have also utilized sheep in the investigation of environmental effects such as sleep disruption on disease onset and progression of HD. Specifically, even young HD-model sheep that expresses juvenile expanded full-length transgenic human *HTT* display circadian abnormalities in behavior before any other HD symptoms appear, but environment is a crucial element in this finding; the mutant sheep who were housed with other mutants exhibited aberrant behaviors, but those housed with wild-type sheep did not [91]. All of these animal models of HD reveal the overlap between gene and environment in the cause and treatment of the disorder, but the molecular link between the huntingtin protein and the environmental agents that affect symptoms has yet to be elucidated.

BIOLOGICAL FUNCTIONS OF THE HUNTINGTIN PROTEIN

The polyglutamine tract of huntingtin acts as a flexible domain that is essential for appropriate intramolecular proximity, dynamics, and substrate specificity, and is altered by the extended CAG repeats [92]. The large (348-kDa) huntingtin protein is expressed ubiquitously, and has more than 50 binding partners confirmed to date [29]. The crucial polyglutamine tract that lies at the N-terminal end of exon 1 of *HTT* appears to adopt numerous different structures (e.g., helix, coil, and loop) that render it a vital regulator of binding interactions [93,94]. Posttranslation, huntingtin is modified by phosphorylation, ubiquitination, palmitoylation, sumoylation, and acetylation, and is cleaved by caspases and calpain [95–102]. This cleavage of the N-terminal fragment containing the glutamine repeats can accumulate in neuronal nuclei [103] and is a major cause of the toxicity, aggregate formation, and mitochondrial dysfunction of HD pathology [104,105]. This toxic property of the polyglutamine fragment is revealed in the R6/2 transgenic mouse model that carries only the first exon of human *HTT*. Without its full-length tail, this portion of the gene yields an excess of the N-terminal cleavage fragment, leading to an accelerated HD-like pathology [38].

While many of the pleiotropic functions of wild-type *HTT* remain unknown, *HTT* is known to be involved in essential neuronal pathways such as apoptosis, axonal transport, transcriptional regulation, and scaffolding of protein–protein interactions [46]. Wild-type *HTT* is essential in embryological development, as *Htt* knockout mice do not survive past embryonic day seven [106,107], but its specific developmental role is ambiguous. Neurodegeneration caused by inactivation of wild-type *HTT* appears in the brains of mice on postnatal day five, indicating that the protein has a crucial role in postembryonic development [108]. This developmental role may derive from the importance of the huntingtin protein in mitochondrial structure, demonstrated by metabolic aberrations in *Htt*–/– mouse embryonic stem cells [109]. However, the mitochondrial connection is only one function of the huntingtin protein.

Wild-type huntingtin is generally localized in the cytoplasm, and associated with the clathrin scaffold of vesicle membranes via a linking protein known as huntingtin-interacting protein-1 (HIP-1), as well as with microtubules [110,111]. When the huntingtin protein carries an extended polyglutamine sequence, it interferes with vesicular trafficking along axons [112]. It is also likely that huntingtin plays a role in trafficking between the nucleus and the cytoplasm, via its C-terminus active nuclear export signal and an N-terminus interaction with the translocated promoter region that plays a role in nuclear export [113,114]. Additionally, it is possible that the huntingtin protein travels from one cell to another in the amyloid fibril form, proliferating its effects as a prion does [115]. Even though further research is required to determine the full functional spectrum of huntingtin, it is undeniable that the huntingtin protein plays diverse and vital roles in many different cellular processes, thereby leaving it vulnerable to environmental toxicants in multiple intracellular compartments and organelles.

ENVIRONMENTAL INFLUENCES IN HUNTINGTON'S DISEASE: LIFESTYLE EFFECTS

Cognitive stimulation: Research mice are conventionally housed in small boxes with nothing but bedding, food, and water. Under these standardized conditions, mice expressing mutant huntingtin develop motor and cognitive disease [116]. However, when allowed access to an enriched environment including exercise wheels, natural bedding materials, stimulating toys, and novel objects, the onset of symptoms is delayed [117–119]. In response to concerns that clinical treatments tested in HD mouse models were not successful in human clinical trials because of differing animal husbandry conditions, as well as differences in mouse models, the Jackson Laboratory developed a comprehensive guide to standardizing colony maintenance [57] based on evidence that even a minimal change in environmental conditions alters survivability in HD mouse models [117–120]. In some studies, exercise alone prolongs the premanifest phase in rodent models, as does environmental enrichment alone [121,122]. However, in one carefully crafted study in a rat model, exercise actually accelerated symptom onset [123]. Other presumably benign environmental conditions, such as the housing of mutants without wild types, can unmask progressive circadian rhythm abnormalities, as described above for HD model sheep [91]. Stress hormones accelerate the progression of symptoms in HD model mice [124], pointing to a negative role for stress on age of onset as well. Environmental conditions appear to be crucial to the rate of progression of HD pathology, and thus may also be important to the underlying mechanisms of the disease.

The most apparent molecular connector between environmental enrichment and HD progression is BDNF. Both exercise and environmental enrichment have been shown to increase the levels of striatal BDNF in wild-type mice [125], and serum BDNF in healthy young men [126,127], as well as in patients with Parkinson's disease [128]. Environmental enrichment and exercise (and treatment with BDNF) also increase neurogenesis [129]. This addition to the reserve pool of healthy neurons may explain the protective effects of environmental enrichment observed in neurodegenerative diseases in general [130]. With HD in particular, BDNF gene transcription [131], and serum and cortical BDNF levels, are reduced in patients compared with healthy controls [13,16]. However, BDNF protein levels are increased via environmental enrichment even in the HD murine model [132]. An increase in BDNF alone, caused by environmental conditions, could have a large impact on the underlying disease pathology of HD.

The primary function of BDNF is the promotion of neurogenesis and neuronal survival through binding to the tyrosine kinase B receptor (TrkB), thereby phosphorylating and activating neuroprotective pathways [133,134]. Striatal neurons express TrkB to receive BDNF transported from the cortex or substantia nigra, but have substantially less BDNF than found in other brain regions [135]. One of the recognized roles of the huntingtin protein is interaction with huntingtin-associated protein 1 (HAP1) to facilitate transport of BDNF along cortical axons to synapses on medium spiny neurons in the striatum [14]. In addition, wild-type *HTT* (but not mutant *HTT*) regulates transcription of BDNF by suppressing a key site in the BDNF promoter [136,137]. Reductions in BDNF levels correlate directly with impairments in motor coordination in the R6/2 mouse model [138]. Emerging evidence has shown that increasing the amount of BDNF protein protects postsynaptic medium spiny neurons, even in the presence of mutant *HTT* [139]. Furthermore, increasing the transcription of cortical BDNF ameliorates symptoms in HD model mice [140] and protects mitochondria [141]. Increasing the level of BDNF in the brain, either directly or indirectly, has been suggested as a way to improve the symptoms observed in HD and other neurodegenerative diseases such as Alzheimer's disease, Parkinson's disease, and amyotrophic lateral sclerosis (ALS) [31]. Interference with the crucial role of BDNF in the prevention of neuronal death, via mutations found on specific proteins associated with each of these neurodegenerative diseases, may explain the regional specificity noted in these disorders as well as delays in their onset.

Diet: Diet also has a role in delaying or expediting the inevitable genetic destiny of HD. Carbohydrate metabolism is altered in HD, with early weight gain followed by hyperglycemia and severe weight loss [142–144]. Hypometabolism in critical brain regions precedes symptoms in HD [145], and a recent study showed that increased copy number of SLC2A3, a gene that regulates glucose metabolism in neurons, delays age of onset of HD [146]. Baseline leptin levels are normal in premanifest human patients [147,148], but as BMI increases, leptin levels do not increase appropriately [144,149], and in mouse models, the level of leptin is higher in mutants than it is in wild types [15,150]. The R6/2 mouse develops metabolic and motor symptoms similar to those observed in human patients with HD [151], and these mice respond to dietary supplements of essential fatty acids (linoleic and α-linoleic acids) with reductions in motor signs, such as foot clasping and locomotor deficits, but these fatty acids did not correct weight loss or the reduction of dopamine receptors [152]. A clinical trial with a randomized placebo-controlled double-blind model showed that supplementation of fatty acids in patients with HD also significantly improved symptoms, compared with placebo [153]. Interestingly, restriction of α-linoleic acid reduces BDNF in a striatal-specific manner in wild-type mice [154]. Other dietary manipulations, such as dietary restriction often implemented through fasting on alternate days, have been shown to be neuroprotective in wild-type animals [155], resulting in delayed locomotor dysfunction, reduced oxidative stress, restored BDNF levels and glucose metabolism, and increased lifespans in mice [156].

The dietary restriction model has also been shown to increase longevity in *C. elegans* [19,157]. Ironically, both dietary restriction and fatty acid supplementation increase BDNF, which may be the protective mechanism of dietary manipulation in HD.

Oxidative Stress: In addition to the beneficial factors that prolong the presymptomatic premanifest phase, HD pathology is hastened by injurious factors such as oxidative stress and mitochondrial insults [158], from genetic and/or environmental causes. There are severe reductions in mitochondrial complexes II–IV in postmortem striatal tissue, but with no effects seen in the blood [42]. Furthermore, abnormalities in striatal energy metabolism have been revealed by PET studies prior to striatal neurodegeneration [159]. Systemic treatment with the complex II inhibitor, 3NP, causes abnormal HD-like motor behavior [160], striatal-specific neurodegeneration [88], and reduced phosphorylation of DARPP-32 [161] (a protein encoded by a modifier gene reported to affect HD onset) [162]. Interestingly, pretreatment with BDNF protects neurons from the effects of 3NP [141]. Environmental agents that increase oxidative stress, especially in environmental conditions that cause decreased BDNF, may be especially toxic to neurons made vulnerable by the presence of the *HTT* mutation.

The *HTT* mutation also causes mitochondria to be abnormal in HD models, with alterations in enzymatic complexes [163] and calcium (Ca^{2+}) kinetics [159]. In the YAC128 mouse, mutant huntingtin interacts with the NR2B subunit of the NMDA receptor, enhancing Ca^{2+} influx that increases excitotoxicity in striatal medium spiny neurons, which carry the NR2B subunit longer in adulthood than most other neurons [163]. Mitochondria are both a source and target of reactive oxygen species (ROS) [164], and oxidative stress further hastens pathology, increasing apoptosis and aggregation in cultured cells expressing mutant *HTT* [165]. Overexpression of the mitochondrial enzyme superoxide dismutase 1 (SOD1), which binds copper and zinc, reverses oxidative stress in cultured murine cells [165]. Systemic supplementation with mitochondrial components such as creatine and coenzyme Q10 (also known as CoQ10 or ubiquinone) demonstrated improvements in HD symptomology in HD animal models, and human clinical trials of creatine and CoQ10 demonstrated improvements in HD symptoms [166,167]. Perhaps diets highly enriched with creatine and CoQ10 may delay the onset of HD, while exposure to toxins that target or accumulate in mitochondria may hasten the age of onset.

ENVIRONMENTAL INFLUENCES IN HUNTINGTON'S DISEASE: METALS

Pollutants such as heavy metals can be defined by many criteria, and conventionally included in this grouping are transition metals that pose environmental concern. These transition metals tend to ionize, and are involved in catalyzing many redox reactions within the cells of the body [168]. Copper (Cu), iron (Fe), zinc (Zn), magnesium (Mg), and manganese (Mn) are common transition metals that serve as essential nutrients, but can also be neurodegenerative disease–causing toxicants, sometimes in conditions of overexposure, and at other times in conditions of depletion [163,169,170]. Metal homeostasis is integral to neurobiology, and the brain exhibits region-specific vulnerability as well as specific metal-accumulation patterns. The medium spiny neurons of the striatum appear to be especially vulnerable to mitochondrial toxins [171]. Mutant *HTT* may decrease the resiliency of mitochondrial membranes, principally in the striatum, in response to excitotoxic insults [172,173]. Additionally, mitochondrial biometal homeostasis must remain highly regulated to prevent oxidative damage; current research on the dysregulation of metals in mitochondria is exploiting this as a potential therapeutic target for HD [174].

Heavy metals selectively accumulate in specific brain regions. Specifically, Mn^{2+} accumulates in the thalamus and substantia nigra, Fe^{2+} in the globus pallidus, and Cu^{2+} in the striatum and thalamus, following excessive exposures in rats [175] and also as a result of normal aging in humans [176]. Emerging studies of metals in HD have demonstrated accumulation of Cu^{2+} and Fe^{2+20}, and decreased serum ferritin, in the striatum of HD [177]. Copper interacts with wild-type huntingtin and decreases its solubility [178], and copper has been found to bind directly to regions of the huntingtin protein encoded by exon 1 of *HTT*, encouraging fibrillation and impeding clearance [36]. Furthermore, nuclear inclusions of fragments of the mutant polyglutamine protein are associated with iron-dependent oxidation [37]. The cause of this derangement of metals in HD is unknown, but may very well stem from an underlying function of the huntingtin protein that is altered in the presence of the CAG repeat.

Although not all functions of the huntingtin protein are fully elucidated, huntingtin is known to be involved with both endocytic and microtubule-mediated vesicle transports, mechanisms for transporting both BDNF and various metals into cells and organelles. The huntingtin protein is closely associated with vesicles, endosomes [179–182], and microtubules [111,180], and is directly associated with the plasma membrane [183]. Via HAP1, huntingtin interacts with an integral member of the microtubule transport system, dynactin subunit p150Glued [184,185], and cofractionates with the transferrin receptor (TfR) [186,187]. In zebra fish, knockdown of *HTT* expression causes increased TfR 1 transcription in the presence of hypochromic blood. Surprisingly, this phenotype is reversed upon administration of bioavailable iron, demonstrating that normal huntingtin has a role in making endocytosed iron accessible [188]. The authors theorize that the function of normal huntingtin must be related to the release of iron from endocytic vesicles. The tight regulation of metals in neurons and within their organelles, which is necessary for normal biomolecular functions, makes them both potential culprits and therapeutic targets for neurodegenerative disorders. In HD, manganese may be especially important, not necessarily as a toxicant but as a crucial cofactor for neuronal enzymes.

Manganese: Manganese is an essential nutrient for many cellular processes, but in excess it is also a neurotoxin. Overexposure to manganese can lead to neurodegeneration and result in a parkinsonian-like condition known as manganism [189,190]. Similar in some ways to HD, manganism targets the motor pathways of the basal ganglia that are also primarily affected in both HD and Parkinson's disease. However, there is a loss of nigrostriatal dopaminergic pathways in Parkinson's disease, whereas deterioration occurs in striatal GABAergic medium spiny neurons in HD; as well, motor symptoms differ [191] between the three conditions. When exposed to elevated levels of manganese, both humans [192–194] and rats [195] develop selective accumulation in the mitochondria of the brain and liver [193], especially in the basal ganglia. Manganese exposure also causes apoptosis from mitochondrial cytochrome c release [196,197] in a caspase-dependent pathway [198]. Recently, the prion protein was shown to bind manganese as well as copper, giving the metal a role in prion-related disorders [199].

Moreover, cells expressing abnormal prion proteins are also resistant to Mn toxicity [200]. Subtoxic manganese exposure causes greater susceptibility to 1-methyl-4-phenylpyridine (MPP+)-related apoptosis, a common Parkinson's disease model, and yet this vulnerability to manganese can be reversed by *n*-acetyl creatine [201]. Manganese plays an undefined role in these diverse neurodegenerative conditions, demonstrating that it must be so tightly regulated in neuronal cells that its increase or decrease causes pathological changes.

Manganese is especially interesting with regard to HD, because of the HD-like symptomology of manganism caused by overexposure, but surprisingly, a screen of potential metal toxicants revealed that Mn levels were reduced rather than increased in a murine cell culture model of striatal neurons bearing a mutation in *HTT* with 111 CAG repeats [202]. In parallel, a neuroprotective response to manganese intoxication has been demonstrated in both the knock-in mouse models of HD and the immortalized striatal cell models of HD, compared with wild type [202–204]. One hypothesis to explain this effect is that manganese transport dynamics in HD cells are altered, leading to decreased manganese uptake and storage capabilities following manganese exposure [205]. It is possible that mutant huntingtin interacts with constituents of the neuronal manganese transport system and dysregulates manganese kinetics. This gene–environment interaction between mutant *HTT* and manganese may serve to explain how xenobiotics influence genetic functions. A reduction of manganese in neurons could possibly alter the normal neuronal and glial functions of proteins that cannot function without sufficient available manganese, and thus downstream products would be reduced in HD. A review of studies on manganoproteins shows that all known manganese-dependent enzymes are reduced, either directly or indirectly, in the presence of mutant *HTT*.

Manganoproteins: Manganese is required for many cellular enzymatic processes (Figure 17.3). The best characterized Mn-dependent enzymes include glutamine synthetase (GLN), superoxide dismutase 2 (SOD2), arginase 1 and 2 (ARG1 and ARG2), pyruvate carboxylase (PC), ataxia telangiectasia mutated (ATM), and selected serine/threonine phosphatases (PPMs) [194]. In addition, two proteins intimately involved in DNA repair are also manganoproteins: meiotic recombination 11 homolog A (also known as MRE11) and DNA polymerase iota. Importantly, there is either direct or indirect evidence of reduced activity of all of these manganoproteins in HD models.

GLN: Glutamine synthetase (GLN) works primarily in astrocytes to convert glutamate and ammonia to glutamine in order to shield neurons from excitotoxicity. A reduction in the activity of GLN in the brain could potentially explain the neurodegeneration seen in striatal medium spiny neurons [194,206]. Typically manganese is selectively deposited in astrocytes [207], and astrocyte-related glial dysfunction is one of the major sources of the pathology of HD [208]. In HD, the glutamate–glutamine cycle is known to be dysregulated [209] and there is a reduction of GLN in HD animal models and patients [210,211]. Insufficient manganese, leading to a reduction in GLN, could be one of the causes of excess glutamate and excitotoxicity seen in HD.

SOD2: Superoxide dismutase 2 (SOD2) is an antioxidant metalloenzyme positioned in the mitochondrial matrix providing resistance to oxidative stress. This is accomplished by converting the free-radical superoxides produced as a byproduct of the mitochondrial electron transport chain into diatomic oxygen and hydrogen peroxide [212]. The structure of SOD2 shares a highly conserved homology with the other members of the superoxide dismutase family, and is that of a tetrameric apoprotein, with the ability to bind one manganese ion per

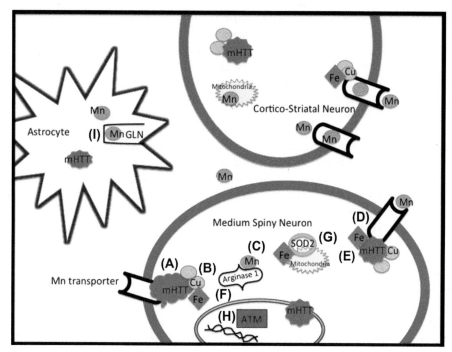

FIGURE 17.3 **Abnormal metal accumulation in Huntington's disease affects crucial cellular processes.** Mutant huntingtin (mHTT) forms aggregates (A), especially in the presence of copper (B), and copper accumulates in excess in Huntington's disease (HD) models, whereas the amount of striatal manganese (C) is reduced in HD. Metal transporters (D) often transport iron (E) and copper, as well as manganese, and may be affected by mutant huntingtin. Manganese-dependent enzymes such as arginase 1 (F), superoxide dismutase 2 (SOD2) (G), ataxia telangiectasia mutated (ATM) (H), and glutamine synthetase (GLN) (I) may be reduced in HD due to the lack of striatal manganese.

subunit. Of particular interest is the high affinity of SOD2 binding site for iron as well as for manganese, and SOD2 is only catalytically active if bound to manganese, and the introduction of iron inactivates the complex [213]. It is theorized that the cell compensates for this bipolar reaction through specialized compartmentalization of the two metal ions to prevent active competition for the site, as iron is found in much higher levels than manganese intracellularly [214,215]. In HD, a reduction in SOD2 worsens the severity of HD symptoms as confirmed in HD brain tissue [94].

ARG1 and ARG2: Both arginase 1 (ARG1) and arginase 2 (ARG2) are manganese-dependent enzyme isoforms that convert L-arginine to L-ornithine and urea, but each has unique tissue and subcellular distribution and localization. ARG1 is highly concentrated in the cytosol of the liver and serves as an integral part of the urea cycle, whereas ARG2 is broadly expressed across the body and is located in the mitochondria. Both isoforms require two Mn^{2+} ions to form the activated binuclear manganese complex, and perform at maximal velocity between pH 9.0–9.5 [216]. ARG2 is implicated in the neuronal metabolism of arginine, where it competes with nitric oxide synthase for substrate, thereby regulating nitric oxide production [217]. Metabolic pathways downstream of arginase are known to be disrupted in HD models. For example, arginine is metabolized by either nitric oxide synthase or arginase, so a lack of arginase could push the balance toward the nitric oxide synthase pathway, and alterations in nitric oxide signaling and nitric oxide synthase have been identified in HD models [218–220]. The urea cycle is deficient in HD and has been linked to increased blood citrulline levels [221]. Patients with HD have an abnormal growth hormone response to arginine infusion [222], and a HD mouse model fed with diets high in arginine has an earlier disease onset [223], suggesting that arginine is not being metabolized properly. The only study that directly measured arginase in HD models focused on circadian gene transcription, but not in striatum. Interestingly, ARG1 is transcribed in a circadian manner by the liver, but in R6/2 model mice, the gene transcription becomes arrhythmic [224]. ARG1 also has a neuroprotective role; it has been found to prevent neuronal death in trophic factor-deprived cell cultures [225]. All of these indirect links between arginase levels and HD point to arginase as a potential regional indicator for manganese.

PC: Pyruvate carboxylase (PC) catalyzes the conversion of pyruvate to oxaloacetate and is implicated as an intermediary in many pathways, including gluconeogenesis and the TCA cycle, earning the classification of an anaplerotic enzyme [226–228]. PC relies on either manganese or magnesium as cofactors, and dysregulated activity of the enzyme has wide-ranging implications. Deficiencies of PC activity can cause acidosis through a build up of lactate in the blood [229], a serious medical condition. Minor reductions in PC can reduce the availability of TCA cycle precursors, which has been hypothesized to contribute to the reduced adenosine triphosphate (ATP) levels seen in HD models [230–233]. In patients, two downstream metabolites have been identified as potential biomarkers for HD; N-acetyl aspartate (a metabolite of oxaloacetate), and lactate. The concentrations of both N-acetyl aspartate and lactate together have been identified as potential biomarkers for HD, as the concentration of both increases in correlation with the duration since symptom onset [234,235]. PC deficiency has also been linked to dysfunction of the urea cycle [236], as has arginase. A region-specific deficiency of manganese could reduce the availability of PC in the fragile medium spiny neurons of the striatum, leading to derangements in these basic energetic cycles, which may explain some of the degeneration of neurons in this region.

PPM: Metalloprotein phosphatases (PPMs) are a subset of serine/threonine protein phosphatases that require both Mn^{2+} and Mg^{2+} ions to perform the vital function of dephosphorylation at the serine and threonine sites, which is a crucial regulatory step for myriad neuronal proteins [237]. One such manganese-dependent enzyme, protein phosphatase 1 (PP1) dephosphorylates serine in AKT and P53, two signaling pathways implicated in HD [238,239]. In addition, some of PPM/PP2C family members have been identified as mediators in a g-protein-coupled receptor dopamine signaling pathway, through the striatal dopamine D2 receptor (D2R) [240]. D2R activation decreases dephosphorylation of huntingtin on its AKT site through the PPM/PP2C family, independent of B-arrestin, and this can then downregulate AKT activation and reduce AKT's phosphorylation of huntingtin at Serine-421, known to be necessary fro proper anterograde vesicular transport [240], This connection between the manganese-dependent PPM and the AKT pathway through huntingtin could have far ranging downstream effects, and AKT is already known to have reduced phosphorylation in HD models [203]. Exposure to 3NP that causes neurodegeneration similar to HD, also results in reduced striatal PPM gene expression [241]. Evidence that PPMs are reduced in HD is lacking, but pathways that would be affected by such a reduction in dephosphatase activity have shown derangements that could stem from such a deficiency.

ATM: Ataxia telangiectasia mutated (ATM) is another manganese-dependent, serine/threonine protein kinase that is involved in repairing double-stranded DNA breaks [242,243]. ATM phosphorylates many targets that are involved in signaling pathways implicated in HD pathology, such as P53, H2AX, and CHK2 [244,245]. Dysfunctional signaling and repair of double-strand DNA breaks have been observed in HD [246], and interestingly, the cellular machinery that is responsible for DNA repair contains several proteins that are dependent on manganese, including MRE11 and DNA polymerase iota described below.

MRE11: Meiotic recombination 11 homolog A (Mre11A) is a manganese-binding nuclear enzyme that is an integral part of the protein complex of Rad50- Mre11A- Nbs1 known as the MRN complex [247]. This MRN complex is another part of the DNA damage repair mechanism, correcting double-stranded breaks through the homologous recombination pathway [248]. The Mre11A protein functions as a nuclease by cleaving DNA with both exonuclease and endonuclease activity, required for both DNA repair and genomic stability [249,250]. As part of the MRN complex, Mre11A also assists in DNA damage response signaling by activating ATM kinase [250,251]. Research has shown that Mre11A can also be activated by magnesium, but nuclease activity appears to be highly variable and limited in regards to temperature, direction of nuclease activity, and substrate length [247–249,252]. In response to irradiation, Mre11 expression is higher in HD fibroblasts, and takes a longer time to return to baseline [246]. It appears that manganese is crucial to Mre11a's activity, and decreased manganese availability in HD may result in the reduced ability to repair DNA double-strand breaks, and thus lead to increased genomic instability.

POLI: DNA polymerase iota is yet another manganese-dependent enzyme involved with translesion synthesis [253]. When a DNA lesion is present, translesion synthesis is necessary for a cell to continue DNA replication. POLI marks the site by inserting a nucleotide directly across from the DNA lesion [253]. Although historically it was thought that DNA polymerases utilized the divalent cation of magnesium, it is now apparent that certain polymerases such as DNA polymerase iota may preferentially utilize manganese, as shown by greater activation in Mn^{2+}-containing rather than Mg^{2+}-containing solutions [254]. Additionally, when Mn was substituted for Mg, the efficiency of nucleotide incorporation in DNA translesion synthesis was enhanced [255]. This increased kinetic activity may result from the stronger stabilizing effect of Mn^{2+}, allowing an enhanced rate of conformational change [255]. In theory, HD models may alter translesion synthesis through deficiencies in available manganese. If reduced manganese leads to inefficiency in DNA repair, these lesions can lead to senescence or apoptosis in affected cells.

Although it is not yet clear which of these known manganoproteins are most affected by manganese deficiency, striatal-specific reductions in manganese, such as those found in HD models, clearly could have widespread repercussions for cellular processes. A proteomics study of gene–environment interactions between manganese and mutant *HTT* in the YAC128 model mouse showed that manganese exposure caused disruptions in proteins involved in glycolysis, excitotoxicity, and cytoskeletal dynamics, but that the enzymes listed above were not among the affected proteins that were found [256]. The proteins that were most likely to represent markers of the mutant *HTT*/manganese interaction were UBQLN1 (a ubiquilin), ENO1 (enolase), and SAE1 (a sumoylation factor). Beyond the proteins that are directly affected by changes in the availability of manganese in HD, it is important to understand how manganese is transported and handled between and inside neurons, in order to understand the gene–environment interaction between manganese and mutant *HTT*.

Manganese transporters: The mechanism by which expression of mutant *HTT* reduces neuronal manganese uptake is unknown. Several distinct metal transporters have been implicated in the movement of manganese and other metals across cell membranes and within cell organelles and intracellular compartments. Surprisingly, alterations in many of these transporters have been definitively linked to other neurodegenerative disorders [257], but not specifically to HD [194]. Among these transporters, PARK9 (also known as ATP13A2) and huntingtin-interacting protein 14 (HIP14) were thought to be especially interesting in the context of HD. Wild-type PARK9 (also known as ATP13a2) has been shown to prevent manganese toxicity in neuronal cells and yeast, but mutations in this gene cause Kufor–Rakeb syndrome, a form of parkinsonism [258]. PARK9 primarily transports zinc into lysosomes and primary cultures of neurons, and fibroblasts from patients with mutations in PARK9 display increased sensitivity to Zn^{2+} that can be rescued with iodochlorhydroxyquin, a chelator of both copper and zinc [259]; however, no difference has been found in zinc levels in HD. HIP14 and its fellow palmitoyl acyltransferase, HIP14L, are associated with the huntingtin protein and known to have at least two functions relevant to molecular derangements in HD, palmitoylation and metal transport [260]. HIP14 must palmitoylate huntingtin for proper functioning of the protein, and the expanded mutant polyglutamine tract reduces the ability of huntingtin to be palmitoylated [260]. In addition, palmitoylation of huntingtin inhibits aggregation [261]. Mice with disrupted expression of HIP14 have reduced palmitoylation and display many of the manifestations seen in HD model mice [262]. Both of these palmitoyl acyltransferases are shown to mediate Mg^{2+} transport, with some selectivity for other cations as well [263]. For example, HIP14 transports Fe^{2+}, but not Cu^{2+}, Co^{2+} nor Ca^{2+}, while HIP14L is impermeable to Fe^{2+} but transports Cu^{2+}, and both HIP14 and HIP14L are permeable to Sr^{2+} Ni^{2+}, Mn^{2+}, Ba^{2+}, and Zn^{2+}[263]. The lack of selectivity for manganese by both PARK9 and HIP14/L indicates unlikeliness that either transporter will explain the reductions in striatal manganese seen in HD models.

Iron and manganese are often transported through the same channels, and iron is found to accumulate in HD neurons, suggesting that lack of manganese may be associated with the excess iron of HD. Ferroportin, which exports iron from cells, is a manganese-responsive protein that has also been shown to export manganese, and it is increased in HD model neurons, while the TfR is decreased [264,265]. The primary manganese transporter, divalent metal transporter 1 (DMT1), is known to transport Fe^{2+} across cell membranes and into organelles such as the early endosome, and to play a role in the absorption of Mn^{2+} [266]. The TfR also transports both iron and manganese, but only in their trivalent forms. Though iron and manganese appear to be linked through these iron-transport mechanisms, iron-transport systems were excluded as the drivers of the manganese-uptake deficit in HD in an experiment measuring manganese toxicity. In this experiment, saturation of Fe^{3+}

levels blocked manganese toxicity, but the effect was identical in both HD mutant and wild-type cell lines [203], ruling out iron transporters as sources of manganese deficiency in these cells.

Solute carrier (SLC) transporters comprise a broad range of 52 families containing over 300 members [267]. This diverse group contains many transporters that both directly and indirectly control intracellular manganese movement. The SLC family 11 is known as a group of divalent transition metal transporters that deal primarily with iron and manganese. SLC11A1 shares a similar homology with SLC11A2 (also known as the divalent metal transporter 1 or DMT1) and has been shown to use manganese as a substrate [268]. Additionally, the SLC39 family includes three divalent metal/bicarbonate ion symporters that transport manganese, known as zinc-interacting protein 7 (ZIP7), zinc-interacting protein 8 (ZIP8), and zinc-interacting protein 14 (ZIP14) [194,269].

ZIP8 and ZIP14 are the most closely related members of the ZIP family by amino acid sequence, although ZIP8, unlike ZIP14, was recently shown to be incapable of transporting non-transferrin-bound iron [270]. One last SLC member, SLC30A10 (also known as ZNT10), was shown to be involved in manganese efflux, and exposure to interleukin-6 increased manganese uptake by downregulating ZnT10 and upregulating ZIP14 [271]. This assortment of SLC transporters may provide insight into the mechanism behind the manganese handling deficit seen in HD, along with identification of new, noncanonical manganese transporters.

Other metal transporters that may have a role in the reduction of striatal manganese accumulation include TRPM3 [272], TRPM7 [273,274], the sodium–calcium exchanger (NCX) [275,276], and metallothioneins [277], which need to be explored further.

CONCLUSIONS

The expanded polyglutamine series on mutant *HTT* has an impact on myriad region-specific neuronal functions, including transcriptional regulation, protein interactions, neurogenesis, cell death, and glucose metabolism [278]. These functions are further impacted both positively and negatively by environmental exposures. Diet alterations can prevent or hasten HD pathology, and environmental enrichment can delay the onset of symptoms, perhaps by increasing BDNF and protecting neurons from huntingtin-driven degeneration. Other environmental exposures, such as metals, appear to have large roles in neurodegenerative diseases, although the exact mechanisms connecting metal exposure and pathology have not been confirmed, and some gene–environment interactions reveal a deficit of essential metals rather than toxic excess. In HD, copper and iron build up in excess in affected brain regions, whereas a deficit of manganese is found in the striata of HD models. A lack or excess of manganese may potentially be involved with other neurodegenerative disorders including Parkinson's disease, Alzheimer's disease, amyotrophic lateral sclerosis (ALS), and prion diseases, in addition to HD. All of these neurodegenerative diseases share alterations in protein aggregation, mitochondrial damage, and oxidative stress, and all are improved with interventions that increase BDNF. In addition, it is possible that mutant *HTT* dysregulates both intracellular neurotrophins and metal transport dynamics, causing impaired downstream signaling cascades that have been implicated in HD and other neurodegenerative diseases.

References

[1] Wexler NS. Venezuelan kindreds reveal that genetic and environmental factors modulate Huntington's disease age of onset. Proc Natl Acad Sci USA 2004;101(10):3498–503.

[2] Gusella JF, MacDonald ME. Huntington's disease: seeing the pathogenic process through a genetic lens. Trends Biochem Sci 2006;31(9):533–40.

[3] Ross CA, Aylward EH, Wild EJ, Langbehn DR, Long JD, Warner JH, et al. Huntington disease: natural history, biomarkers and prospects for therapeutics. Nat Rev Neurol 2014;10:204–16.

[4] Rubinsztein DC, Leggo J, Coles R, Almqvist E, Biancalana V, Cassiman J-J, et al. Phenotypic characterization of individuals with 30-40 CAG repeats in the Huntington disease (HD) gene reveals HD cases with 36 repeats and apparently normal elderly individuals with 36-39 repeats. Am J Hum Genet 1996;59(1):16–22.

[5] Kaltenbach LS, Romero E, Becklin RR, Chettier R, Bell R, Phansalkar A, et al. Huntingtin interacting proteins are genetic modifiers of neurodegeneration. PLoS Genet 2007;3(5):e82.

[6] Gusella JF, MacDonald ME. Huntington's disease: the case for genetic modifiers. Genome Med 2009;1(8):80.

[7] Taherzadeh-Fard E, Saft C, Andrich J, Wieczorek S, Arning L. PGC-1alpha as modifier of onset age in Hunting-ton disease. Mol Neurodegener 2009;4(10):8.

[8] Rosenblatt A, Brinkman R, Liang K, Alqvist E, Margolis R, Huang C, et al. Familial influence on age of onset among siblings with Huntington disease*. Am J Med Genet 2001;105(5):399–403.

[9] Novak M, Tabrizi SJ. Huntington's disease. BMJ 2010;340(4):c3109.

[10] Squitieri F, Jankovic J. Huntington's disease: how intermediate are intermediate repeat lengths? Mov Disord 2012;27(14):1714–7.

[11] Friedman JH, Trieschmann ME, Myers RH, Fernandez HH. Monozygotic twins discordant for Huntington disease after 7 years. Arch Neurol 2005;62(6):995–7.

[12] McNeil SM, Novelletto A, Srinidhi J, Barnes G, Kornbluth I, Altherr MR, et al. Reduced penetrance of the huntington's disease mutation. Hum Mol Genet 1997;6(5):775–9.

[13] Zuccato C, Cattaneo E. Role of brain-derived neurotrophic factor in Huntington's disease. Prog Neurobiol 2007;81(5):294–330.

[14] Gauthier LR, Charrin BC, Borrell-Pagès M, Dompierre JP, Rangone H, Cordelieres FP, et al. Huntingtin controls neurotrophic support and survival of neurons by enhancing BDNF vesicular transport along microtubules. Cell 2004;118(1):127–38.

[15] Duan W, Guo Z, Jiang H, Ware M, Li X-J, Mattson MP. Dietary restriction normalizes glucose metabolism and BDNF levels, slows disease progression, and increases survival in huntingtin mutant mice. Proc Natl Acad Sci USA 2003;100(5):2911–6.

[16] Ciammola A, Sassone J, Cannella M, Calza S, Poletti B, Frati L, et al. Low brain derived neurotrophic fac-tor (BDNF) levels in serum of Huntington's disease patients. Am J Med Genet Part B: Neuropsychiatr Genet 2007;144(4):574–7.

[17] Trembath MK, Horton ZA, Tippett L, Hogg V, Collins VR, Churchyard A, et al. A retrospective study of the impact of lifestyle on age at onset of Huntington disease. Mov Disord 2010;25(10):1444–50.

[18] Tabrizi SJ, Scahill RI, Owen G, Durr A, Leavitt BR, Roos RA, et al. Predictors of phenotypic progression and dis-ease onset in premanifest and early-stage Huntington's disease in the TRACK-HD study: analysis of 36-month observational data. Lancet Neurol 2013;12(7):637–49.

[19] Martinez-Finley EJ, Avila DS, Chakraborty S, Aschner M. Insights from *Caenorhabditis elegans* on the role of metals in neurodegenerative diseases. Metallomics 2011;3(3):271–9.

[20] Dexter D, Carayon A, Javoy-Agid F, Agid Y, Wells F, Daniel S, et al. Alterations in the levels of iron, ferritin and other trace metals in Parkinson's disease and other neurodegenerative diseases affecting the basal ganglia. Brain 1991;114(4):1953–75.

[21] Park J-S, Koentjoro B, Veivers D, Mackay-Sim A, Sue CM. Parkinson's disease-associated human ATP13A2 (PARK9) deficiency causes zinc dyshomeostasis and mitochondrial dysfunction. Hum Mol Genet 2014;34(46):1581–287.

[22] Zoroddu MA, Peana MF, Medici S, Solinas C, Juliano CCA, Remelli M. Interaction of divalent cations with protein PARK9; *International symposium on metal ions in biology and medicine: abstracts and proceedings*, March 11–13, 2013, Punta del Este, Uruguay. Montevideo, [s.n.]. p. 39. ISBN 978-9974-0-0911-0.

[23] Schmidt K, Wolfe DM, Stiller B, Pearce DA. Cd^{2+}, Mn^{2+}, Ni^{2+} and Se^{2+} toxicity to *Saccharomyces cerevisiae* lacking YPK9p the orthologue of human ATP13A2. Biochem Biophys Res Commun 2009;383(2):198–202.

[24] Chesi A, Kilaru A, Fang X, Cooper AA, Gitler AD. The role of the Parkinson's disease gene PARK9 in essential cellular pathways and the manganese homeostasis network in yeast. PloS One 2012;7(3):e34178.

[25] MacDonald ME, Ambrose CM, Duyao MP, Myers RH, Lin C, Srinidhi L, et al. A novel gene containing a trinucleotide repeat that is expanded and unstable on Huntington's disease chromosomes. Cell 1993;72(6):971–83.

[26] Kremer B, Goldberg P, Andrew SE, Theilmann J, Telenius H, Zeisler J, et al. A worldwide study of the Huntington's disease mutation: the sensitivity and specificity of measuring CAG repeats. N. Engl J Med 1994;330(20):1401–6.

[27] Schmitt I, Bächner D, Megow D, Henkiein P, Hamelster H, Epplen JT, et al. Expression of the Huntington disease gene in rodents: cloning the rat homologue and evidence for downregulation in non-neuronal tissues during development. Hum Mol Genet 1995;4(7):1173–82.

[28] Barnes GT, Duyao MP, Ambrose CM, McNeil S, Persichetti F, Srinidhi J, et al. Mouse Huntington's disease gene homolog (Hdh). Somat Cell Mol Genet 1994;20(2):87–97.

[29] Cattaneo E, Zuccato C, Tartari M. Normal huntingtin function: an alternative approach to Huntington's disease. Nat Rev Neurosci 2005;6(12):919–30.

[30] Cattaneo E, Rigamonti D, Goffredo D, Zuccato C, Squitieri F, Sipione S. Loss of normal huntingtin function: new developments in Huntington's disease research. Trends Neurosci 2001;24(3):182–8.

[31] Zuccato C, Valenza M, Cattaneo E. Molecular mechanisms and potential therapeutical targets in Huntington's disease. Physiol Rev 2010;90(3):905–81.

[32] Squitieri F, Gellera C, Cannella M, Mariotti C, Cislaghi G, Rubinsztein DC, et al. Homozygosity for CAG mutation in Huntington disease is associated with a more severe clinical course. Brain 2003;126(4):946–55.

[33] Poirier MA, Jiang H, Ross CA. A structure-based analysis of huntingtin mutant polyglutamine aggregation and toxicity: evidence for a compact beta-sheet structure. Hum Mol Genet 2005;14(6):765–74.

[34] Ehrnhoefer DE, Sutton L, Hayden MR. Small changes, big impact: posttranslational modifications and function of huntingtin in Huntington disease. Neurosci 2011 Oct;17(5):475–92. 1073858410390378.

[35] Fox JH, Connor T, Stiles M, Kama J, Lu Z, Dorsey K, et al. Cysteine oxidation within N-terminal mutant huntingtin promotes oligomerization and delays clearance of soluble protein. J Biol Chem 2011;286(20):18320–30.

[36] Xiao G, Fan Q, Wang X, Zhou B. Huntington disease arises from a combinatory toxicity of polyglutamine and copper binding. Proc Natl Acad Sci USA 2013;110(37):14995–5000.

[37] Firdaus WJ, Wyttenbach A, Giuliano P, Kretz-Remy C, Currie RW, Arrigo AP. Huntingtin inclusion bodies are iron–dependent centers of oxidative events. FEBS J 2006;273(23):5428–41.

[38] Wang J, Wang C-E, Orr A, Tydlacka S, Li S-H, Li X-J. Impaired ubiquitin–proteasome system activity in the synapses of Huntington's disease mice. J Cell Biol 2008;180(6):1177–89.

[39] Ferrante RJ, Kubilus JK, Lee J, Ryu H, Beesen A, Zucker B, et al. Histone deacetylase inhibition by sodium butyrate chemotherapy ameliorates the neurodegenerative phenotype in Huntington's disease mice. J Neurosci 2003;23(28):9418–27.

[40] Ryu H, Lee J, Hagerty SW, Soh BY, McAlpin SE, Cormier KA, et al. ESET/SETDB1 gene expression and histone H3 (K9) trimethylation in Huntington's disease. Proc Natl Acad Sci USA 2006;103(50):19176–81.

[41] Seong IS, Woda JM, Song J-J, Lloret A, Abeyrathne PD, Woo CJ, et al. Huntingtin facilitates polycomb repressive complex 2. Hum Mol Genet 2010;19(4):573–83.

[42] Browne SE, Bowling AC, Macgarvey U, Baik MJ, Berger SC, Muquit MM, et al. Oxidative damage and metabolic dysfunction in Huntington's disease: selective vulnerability of the basal ganglia. Ann Neurol 1997;41(5):646–53.

[43] Huang CC, Faber PW, Persichetti F, Mittal V, Vonsattel J-P, MacDonald ME, et al. Amyloid formation by mutant huntingtin: threshold, progressivity and recruitment of normal polyglutamine proteins. Somat cell Mol Genet 1998;24(4):217–33.

[44] Kazantsev A, Preisinger E, Dranovsky A, Goldgaber D, Housman D. Insoluble detergent-resistant aggregates form between pathological and nonpathological lengths of polyglutamine in mammalian cells. Proc Natl Acad Sci USA 1999;96(20):11404–9.

[45] Wheeler VC, White JK, Gutekunst C-A, Vrbanac V, Weaver M, Li X-J, et al. Long glutamine tracts cause nuclear localization of a novel form of huntingtin in medium spiny striatal neurons in HdhQ92 and HdhQ111 knock-in mice. Hum Mol Genet 2000;9(4):503–13.

[46] Lo DC, Hughes RE, Hersch SM, Rosas HD. Biomarkers to enable the development of neuroprotective therapies for Huntington's disease; 2011.

[47] Guidetti P, Charles V, Chen E-Y, Reddy PH, Kordower JH, Whetsell Jr WO, et al. Early degenerative changes in transgenic mice expressing mutant huntingtin involve dendritic abnormalities but no impairment of mitochondrial energy production. Exp neurol 2001;169(2):340–50.

[48] Younes L, Ratnanather JT, Brown T, Aylward E, Nopoulos P, Johnson H, et al. Regionally selective atrophy of subcortical structures in prodromal HD as revealed by statistical shape analysis. Hum Brain Mapp 2014 Mar;35(3):792–809.

[49] Vonsattel JPG, DiFiglia M. Huntington disease. J Neuropathol Exp Neurol 1998;57(5):369–84.

[50] Stack EC, Dedeoglu A, Smith KM, Cormier K, Kubilus JK, Bogdanov M, et al. Neuroprotective effects of synaptic modulation in Huntington's disease R6/2 mice. J Neurosci 2007;27(47):12908–15.

[51] Kim J, Bordiuk OL, Ferrante RJ. Experimental models of HD and reflection on therapeutic strategies. Int Rev Neurobiol 2011;98:419–81.

[52] Orr HT, Zoghbi HY. Trinucleotide repeat disorders. Annu Rev Neurosci 2007;30:575–621.

[53] Wolfe KJ, Cyr DM. Amyloid in neurodegenerative diseases: friend or foe? Semin Cell Dev Biol 2011;2011: 476–81. Elsevier.

[54] Hannan AJ. Review: environmental enrichment and brain repair: harnessing the therapeutic effects of cognitive stimulation and physical activity to enhance experience–dependent plasticity. Neuropathol Appl Neurobiol 2014;40(1):13–25.

[55] Babenko O, Kovalchuk I, Metz GA. Epigenetic programming of neurodegenerative diseases by an adverse environment. Brain Res 2012;1444:96–111.

[56] Pouladi MA, Morton AJ, Hayden MR. Choosing an animal model for the study of Huntington's disease. Nat Rev Neurosci 2013;14(10):708–21.

[57] Menalled L, Lutz C, Ramboz S, Brunner D, Lager B, Noble S, et al. A field guide to working with mouse models of Huntingtons disease. Bar Harbor, Tarrytown, Princeton: The Jackson Laboratory, PsychoGenics, CHDI Foundation; 2014. 2014.

[58] Menalled LB, Chesselet M-F. Mouse models of Huntington's disease. Trends Pharmacol Sci 2002;23(1):32–9.

[59] Menalled L, El-Khodor BF, Patry M, Suarez-Farinas M, Orenstein SJ, Zahasky B, et al. Systematic behavioral evaluation of Huntington's disease transgenic and knock-in mouse models. Neurobiol Dis 2009;35(3):319–36.

[60] Cowin R-M, Bui N, Graham D, Green JR, Grueninger S, Yuva-Paylor LA, et al. Onset and progression of behavioral and molecular phenotypes in a novel congenic R6/2 line exhibiting intergenerational CAG repeat stability. PLoS One 2011;6(12):e28409.

[61] Cummings DM, Alaghband Y, Hickey MA, Joshi PR, Hong SC, Zhu C, et al. A critical window of CAG repeat-length correlates with phenotype severity in the R6/2 mouse model of Huntington's disease. J Neurophysiol 2012;107(2):677–91.

[62] Dragatsis I, Goldowitz D, Del Mar N, Deng Y, Meade C, Liu L, et al. CAG repeat lengths ≥335 attenuate the phenotype in the R6/2 Huntington's disease transgenic mouse. Neurobiol Dis 2009;33(3):315–30.

[63] Morton AJ, Glynn D, Leavens W, Zheng Z, Faull RL, Skepper JN, et al. Paradoxical delay in the onset of disease caused by super-long CAG repeat expansions in R6/2 mice. Neurobiol Dis 2009;33(3):331–41.

[64] Hodgson JG, Agopyan N, Gutekunst C-A, Leavitt BR, LePiane F, Singaraja R, et al. A YAC mouse model for Huntington's disease with full-length mutant huntingtin, cytoplasmic toxicity, and selective striatal neurodegeneration. Neuron 1999;23(1):181–92.

[65] Slow EJ, van Raamsdonk J, Rogers D, Coleman SH, Graham RK, Deng Y, et al. Selective striatal neuronal loss in a YAC128 mouse model of Huntington disease. Hum Mol Genet 2003;12(13):1555–67.

[66] Gray M, Shirasaki DI, Cepeda C, Andre VM, Wilburn B, Lu X-H, et al. Full-length human mutant huntingtin with a stable polyglutamine repeat can elicit progressive and selective neuropathogenesis in BACHD mice. J Neurosci 2008;28(24):6182–95.

[67] Benn CL, Slow EJ, Farrell LA, Graham R, Deng Y, Hayden MR, et al. Glutamate receptor abnormalities in the YAC128 transgenic mouse model of Huntington's disease. Neuroscience 2007;147(2):354–72.

[68] Lin C-H, Tallaksen-Greene S, Chien W-M, Cearley JA, Jackson WS, Crouse AB, et al. Neurological abnormalities in a knock-in mouse model of Huntington's disease. Hum Mol Genet 2001;10(2):137–44.

[69] Brooks SP, Janghra N, Workman VL, Bayram-Weston Z, Jones L, Dunnett SB. Longitudinal analysis of the behavioural phenotype in R6/1 (C57BL/6J) Huntington's disease transgenic mice. Brain Res Bull 2012;88(2):94–103.

[70] Heng MY, Duong DK, Albin RL, Tallaksen-Greene SJ, Hunter JM, Lesort MJ, et al. Early autophagic response in a novel knock-in model of Huntington disease. Hum Mol Genet 2010 Oct 1;19(19):3702–20.

[71] Hickey MA, Kosmalska A, Enayati J, Cohen R, Zeitlin S, Levine MS, et al. Extensive early motor and non-motor behavioral deficits are followed by striatal neuronal loss in knock-in Huntington's disease mice. Neuroscience 2008;157(1):280–95.

[72] Menalled LB, Kudwa AE, Miller S, Fitzpatrick J, Watson-Johnson J, Keating N, et al. Comprehensive behavioral and molecular characterization of a new knock-in mouse model of Huntington's disease: zQ175. PloS One 2012;7(12):e49838.

[73] Kazemi-Esfarjani P, Benzer S. Genetic suppression of polyglutamine toxicity in *Drosophila*. Science 2000;287(5459):1837–40.

[74] Fernandez-Funez P, Nino-Rosales ML, de Gouyon B, She W-C, Luchak JM, Martinez P, et al. Identification of genes that modify ataxin-1-induced neurodegeneration. Nature 2000;408(6808):101–6.

[75] Marsh JL, Walker H, Theisen H, Zhu Y-Z, Fielder T, Purcell J, et al. Expanded polyglutamine peptides alone are intrinsically cytotoxic and cause neurodegeneration in *Drosophila*. Hum Mol Genet 2000;9(1):13–25.

[76] Warrick JM, Chan HE, Gray-Board GL, Chai Y, Paulson HL, Bonini NM. Suppression of polyglutamine-mediated neurodegeneration in *Drosophila* by the molecular chaperone HSP70. Nat Genet 1999;23(4):425–8.

[77] Brand AH, Perrimon N. Targeted gene expression as a means of altering cell fates and generating dominant phenotypes. Development 1993;118(2):401–15.

[78] Lee W-CM, Yoshihara M, Littleton JT. Cytoplasmic aggregates trap polyglutamine-containing proteins and block axonal transport in a *Drosophila* model of Huntington's disease. Proc Natl Acad Sci USA 2004;101(9):3224–9.

[79] Pallos J, Bodai L, Lukacsovich T, Purcell JM, Steffan JS, Thompson LM, et al. Inhibition of specific HDACs and sirtuins suppresses pathogenesis in a *Drosophila* model of Huntington's disease. Hum Mol Genet 2008;17(23):3767–75.

[80] Lu B, Al-Ramahi I, Valencia A, Wang Q, Berenshteyn F, Yang H, et al. Identification of NUB1 as a suppressor of mutant Huntingtin toxicity via enhanced protein clearance. Nat Neurosci 2013;16(5):562–70.

[81] Faber PW, Alter JR, MacDonald ME, Hart AC. Polyglutamine-mediated dysfunction and apoptotic death of a *Caenorhabditis elegans* sensory neuron. Proc Natl Acad Sci USA 1999;96(1):179–84.

[82] Parker JA, Connolly JB, Wellington C, Hayden M, Dausset J, Neri C. Expanded polyglutamines in *Caenorhabditis elegans* cause axonal abnormalities and severe dysfunction of PLM mechanosensory neurons without cell death. Proc Natl Acad Sci USA 2001;98(23):13318–23.

[83] Gidalevitz T, Wang N, Deravaj T, Alexander-Floyd J, Morimoto RI. Natural genetic variation determines susceptibility to aggregation or toxicity in a *C. elegans* model for polyglutamine disease. BMC Biol 2013;11(1):100.

[84] Cherny RA, Ayton S, Finkelstein DI, Bush AI, McColl G, Massa SM. PBT2 reduces toxicity in a *C. elegans* model of polyQ aggregation and extends lifespan, reduces striatal atrophy and improves motor performance in the R6/2 mouse model of Huntington's disease. J Huntingt Dis 2012;1(2):211–9.

[85] Prana. Prana announces successful phase 2 results in Huntington disease trial. Melbourne & New York: Prana Biotechnology; 2014.

[86] Chan AW, Xu Y, Jiang J, Rahim T, Zhao D, Kocerha J, et al. A two years longitudinal study of a transgenic Huntington disease monkey. BMC Neurosci 2014;15(1):36.

[87] Ferrante RJ, Kowall NW, Cipolloni P, Storey E, Beal MF. Excitotoxin lesions in primates as a model for Huntington's disease: histopathologic and neurochemical characterization. Exp Neurol 1993;119(1):46–71.

[88] Brouillet E, Hantraye P, Ferrante RJ, Dolan R, Leroy-Willig A, Kowall NW, et al. Chronic mitochondrial energy impairment produces selective striatal degeneration and abnormal choreiform movements in primates. Proc Natl Acad Sci USA 1995;92(15):7105–9.

[89] Dal-Pan A, Pifferi F, Marchal J, Picq J-L, Aujard F, Consortium R. Cognitive performances are selectively enhanced during chronic caloric restriction or resveratrol supplementation in a primate. PloS One 2011;6(1):e16581.

[90] Verina T, Kiihl SF, Schneider JS, Guilarte TR. Manganese exposure induces microglia activation and dystrophy in the substantia nigra of non-human primates. Neurotoxicology 2011;32(2):215–26.

[91] Morton AJ, Rudiger SR, Wood NI, Sawiak SJ, Brown GC, McLaughlan CJ, et al. Early and progressive circadian abnormalities in Huntington's disease sheep are unmasked by social environment. Hum Mol Genet 2014 Jul 1; 23(13):3375–83.

[92] Caron NS, Desmond CR, Xia J, Truant R. Polyglutamine domain flexibility mediates the proximity between flanking sequences in huntingtin. Proc Natl Acad Sci USA 2013;110(36):14610–5.

[93] Harjes P, Wanker EE. The hunt for huntingtin function: interaction partners tell many different stories. Trends Biochem Sci 2003;28(8):425–33.

[94] Kim MW, Chelliah Y, Kim SW, Otwinowski Z, Bezprozvanny I. Secondary structure of Huntingtin amino-terminal region. Structure 2009;17(9):1205–12.

[95] Warby SC, Doty CN, Graham RK, Shively J, Singaraja RR, Hayden MR. Phosphorylation of huntingtin reduces the accumulation of its nuclear fragments. Mol Cell Neurosci 2009;40(2):121–7.

[96] Steffan JS, Agrawal N, Pallos J, Rockabrand E, Trotman LC, Slepko N, et al. SUMO modification of Huntingtin and Huntington's disease pathology. Science 2004;304(5667):100–4.

[97] Kalchman MA, Graham RK, Xia G, Koide HB, Hodgson JG, Graham KC, et al. Huntingtin is ubiquitinated and interacts with a specific ubiquitin-conjugating enzyme. J Biol Chem 1996;271(32):19385–94.

[98] Steffan JS, Bodai L, Pallos J, Poelman M, McCampbell A, Apostol BL, et al. Histone deacetylase inhibitors arrest polyglutamine-dependent neurodegeneration in *Drosophila*. Nature 2001;413(6857):739–43.

[99] Jeong H, Then F, Melia Jr TJ, Mazzulli JR, Cui L, Savas JN, et al. Acetylation targets mutant huntingtin to autophagosomes for degradation. Cell 2009;137(1):60–72.

[100] Jones L, Hughes A. Pathogenic mechanisms in Huntington's disease. Int Rev Neurobiol 2011;98:373–418.

[101] Ratovitski T, Gucek M, Jiang H, Chighladze E, Waldren E, D'Ambola J, et al. Mutant huntingtin N-terminal fragments of specific size mediate aggregation and toxicity in neuronal cells. J Biol Chem 2009;284(16):10855–67.

[102] Yanai A, Huang K, Kang R, Singaraja RR, Arstikaitis P, Gan L, et al. Palmitoylation of huntingtin by HIP14is essential for its trafficking and function. Nat Neurosci 2006;9(6):824–31.

[103] Landles C, Sathasivam K, Weiss A, Woodman B, MOffitt H, Finkbeiner S, et al. Proteolysis of mutant huntingtin produces an exon 1 fragment that accumulates as an aggregated protein in neuronal nuclei in Huntington disease. J Biol Chem 2010;285(12):8808–23.

[104] Wellington CL, Hayden MR. Caspases and neurodegeneration: on the cutting edge of new therapeutic approaches. Clin Genet 2000;57(1):1–10.

[105] Sathasivam K, Neueder A, Gipson TA, Landles C, Benjamin AC, Bondulich MK, et al. Aberrant splicing of HTT generates the pathogenic exon 1 protein in Huntington disease. Proc Natl Acad Sci USA 2013;110(6):2366–70.

[106] Nasir J, Floresco SB, O'Kusky JR, Diewert VM, Richman JM, Zeisler J, et al. Targeted disruption of the Huntington's disease gene results in embryonic lethality and behavioral and morphological changes in heterozygotes. Cell 1995;81(5):811–23.

[107] Zeitlin S, Liu J-P, Chapman DL, Papaioannou VE, Efstratiadis A. Increased apoptosis and early embryonic lethality in mice nullizygous for the Huntington's disease gene homologue. Nat Genet 1995;11(2):155–63.

[108] Dragatsis I, Dietrich P, Zeitlin S. Expression of the Huntingtin-associated protein 1 gene in the developing and adult mouse. Neurosci Lett 2000;282(1):37–40.

[109] Ismailoglu I, Chen Q, Popowski M, Yang L, Gross SS, Brivanlou AH. Huntingtin protein is essential for mitochondrial metabolism, bioenergetics and structure in murine embryonic stem cells. Dev Biol 2014;391(2):230–40.

[110] Waelter S, Boeddrich A, Lurz R, Scherzinger E, Lueder G, Lehrach H, et al. Accumulation of mutant huntingtin fragments in aggresome-like inclusion bodies as a result of insufficient protein degradation. Mol Biol Cell 2001;12(5):1393–407.

[111] Hoffner G, Kahlem P, Djian P. Perinuclear localization of huntingtin as a consequence of its binding to microtubules through an interaction with β-tubulin: relevance to Huntington's disease. J Cell Sci 2002;115(5):941–8.

[112] Trushina E, Dyer RB, Badger JD, Ure D, Eide L, Tran DD, et al. Mutant huntingtin impairs axonal trafficking in mammalian neurons in vivo and in vitro. Mol Cell Biol 2004;24(18):8195–209.

[113] Schulte J, Littleton JT. The biological function of the Huntingtin protein and its relevance to Huntington's disease pathology. Curr Trends Neurol 2011;5:65.

[114] Cornett J, Cao F, Wang C-E, Ross CA, Bates GP, Li S-H, et al. Polyglutamine expansion of huntingtin impairs its nuclear export. Nat Genet 2005;37(2):198–204.

[115] Jucker M, Walker LC. Pathogenic protein seeding in Alzheimer disease and other neurodegenerative disorders. Ann Neurol 2011;70(4):532–40.

[116] Mangiarini L, Sathasivam K, Seller M, Cozens B, Harper A, Hetherington C, et al. Exon 1 of the *HD* gene with an expanded CAG repeat is sufficient to cause a progressive neurological phenotype in transgenic mice. Cell 1996;87(3):493–506.

[117] van Dellen A, Blakemore C, Deacon R, York D, Hannan AJ. Delaying the onset of Huntington's in mice. Nature 2000;404(6779):721–2.

[118] Carter RJ, Hunt MJ, Morton AJ. Environmental stimulation increases survival in mice transgenic for exon 1 of the Huntington's disease gene. Mov Disord 2000;15(5):925–37.

[119] Hockly E, Cordery PM, Woodman B, Mahal A, Van Dellen A, Blakemore C, et al. Environmental enrichment slows disease progression in R6/2 Huntington's disease mice. Ann Neurol 2002;51(2):235–42.

[120] Van Dellen A, Cordery PM, Spires TL, Blakemore C, Hannan AJ. Wheel running from a juvenile age delays onset of specific motor deficits but does not alter protein aggregate density in a mouse model of Huntington's disease. BMC Neurosci 2008;9(1):34.

[121] Wood NI, Carta V, Milde S, Skillings EA, McAllister CJ, Ang YM, et al. Responses to environmental enrichment differ with sex and genotype in a transgenic mouse model of Huntington's disease. PLoS One 2010;5(2):e9077.

[122] Harrison DJ, Busse M, Openshaw R, Rosser AE, Dunnett SB, Brooks SP. Exercise attenuates neuropathology and has greater benefit on cognitive than motor deficits in the R6/1 Huntington's disease mouse model. Exp Neurol 2013;248:457–69.

[123] Potter MC, Yuan C, Ottenritter C, Mughal M, van Praag H. Exercise is not beneficial and may accelerate symptom onset in a mouse model of Huntington's disease. PLoS Curr 2010;2.

[124] Mo C, Pang TY, Ransome MI, Hill RA, Renoir T, Hannan AJ. High stress hormone levels accelerate the onset of memory deficits in male Huntington's disease mice. Neurobiol Dis 2014;69:248–62.

[125] Turner CA, Lewis MH. Environmental enrichment: effects on stereotyped behavior and neurotrophin levels. Physiol Behav 2003;80(2):259–66.

[126] Nofuji Y, Suwa M, Sasaki H, Ichimiya A, Nishichi R, Kumagai S. Different circulating brain-derived neurotrophic factor responses to acute exercise between physically active and sedentary subjects. J Sports Sci Med 2012;11:83–8.

[127] Schmolesky MT, Webb DL, Hansen RA. The effects of aerobic exercise intensity and duration on levels of brain-derived neurotrophic factor in healthy men. J Sports Sci Med 2013;12(3):502.

[128] Frazzitta G, Maestri R, Ghilardi MF, Riboldazzi G, Perini M, Bertotti G, et al. Intensive rehabilitation increases BDNF serum levels in parkinsonian patients a randomized study. Neurorehabil Neural Repair 2014;28(2):163–8.

[129] van Praag H, Kempermann G, Gage FH. Running increases cell proliferation and neurogenesis in the adult mouse dentate gyrus. Nat Neurosci 1999;2(3):266–70.

[130] Laviola G, Hannan AJ, Macrì S, Solinas M, Jaber M. Effects of enriched environment on animal models of neurodegenerative diseases and psychiatric disorders. Neurobiol Dis 2008;31(2):159–68.

[131] Zuccato C, Ciammola A, Rigamonti D, Leavitt BR, Goffredo D, Conti L, et al. Loss of huntingtin-mediated BDNF gene transcription in Huntington's disease. Science 2001;293(5529):493–8.

[132] Spires TL, Grote HE, Varshney NK, Cordery PM, van Dellen A, Blakemore C, et al. Environmental enrichment rescues protein deficits in a mouse model of Huntington's disease, indicating a possible disease mechanism. J Neurosci 2004;24(9):2270–6.

[133] Huang EJ, Reichardt LF. Neurotrophins: roles in neuronal development and function. Annu Rev Neurosci 2001;24:677.

[134] Saylor AJ, McGinty JF. An intrastriatal brain-derived neurotrophic factor infusion restores striatal gene expression in Bdnf heterozygous mice. Brain Struct Funct 2010;215(2):97–104.

[135] Altar CA, Cai N, Bliven T, Juhasz M, Conner JM, Acheson AL, et al. Anterograde transport of brain-derived neurotrophic factor and its role in the brain. Nature 1997;389(6653):856–60.

[136] Shimojo M. Huntingtin regulates RE1-silencing transcription factor (REST/NRSF) nuclear trafficking indirectly through a complex with REST/NRSF-interacting LIM domain protein (RILP) and dynactin p150Glued. J Biol Chem 2008 Dec 12;283(50):34880–86.

[137] Zuccato C, Tartari M, Crotti A, Goffredo D, Valenza M, Conti L, et al. Huntingtin interacts with REST/NRSF to modulate the transcription of NRSE-controlled neuronal genes. Nat Genet 2003;35(1):76–83.

[138] Samadi P, Boutet A, Rymar V, Rawal K, Maheux J, Kvann JC, et al. Relationship between BDNF expression in major striatal afferents, striatum morphology and motor behavior in the R6/2 mouse model of Huntington's disease. Genes, Brain Behav 2013;12(1):108–24.

[139] Canals JM, Pineda JR, Torres-Peraza JF, Bosch M, Martin-Ibanez R, Munoz MT, et al. Brain-derived neurotrophic factor regulates the onset and severity of motor dysfunction associated with enkephalinergic neuronal degeneration in Huntington's disease. J Neurosci 2004;24(35):7727–39.

[140] Gharami K, Xie Y, An JJ, Tonegawa S, Xu B. Brain-derived neurotrophic factor over-expression in the forebrain ameliorates Huntington's disease phenotypes in mice. J Neurochem 2008;105(2):369–79.

[141] Wu CL, Hwang CS, Chen SD, Yin JH, Yang DI. Neuroprotective mechanisms of brain–derived neurotrophic factor against 3–nitropropionic acid toxicity: therapeutic implications for Huntington's disease. Ann N Y Acad Sci 2010;1201(1):8–12.

[142] Hurlbert MS, Zhou W, Wasmeier C, Kaddis F, Hutton J, Freed C. Mice transgenic for an expanded CAG repeat in the Huntington's disease gene develop diabetes. Diabetes 1999;48(3):649–51.

[143] Farrer LA. Diabetes mellitus in Huntington disease. Clin Genet 1985;27(1):62–7.

[144] Pratley RE, Salbe AD, Ravussin E, Caviness JN. Higher sedentary energy expenditure in patients with Huntington's disease. Ann Neurol 2000;47(1):64–70.

[145] Ciarmiello A, Cannella M, Lastoria S, Simonelli M, Frati L, Rubinsztein DC, et al. Brain white-matter volume loss and glucose hypometabolism precede the clinical symptoms of Huntington's disease. J Nucl Med 2006;47(2):215–22.

[146] Vittori A, Breda C, Repici M, Orth M, Roos RA, Outeiro TF, et al. Copy-number variation of the neuronal glucose transporter gene SLC2A3 and age of onset in Huntington's disease. Hum Mol Genet 2014;23(12):3129–37.

[147] Goodman AO, Barker RA. Body composition in premanifest Huntington's disease reveals lower bone density compared to controls. PLoS Curr 2011;3.

[148] Goodman AO, Rogers L, Pilsworth S, McAllister CJ, Shneerson JM, Morton AJ, et al. Asymptomatic sleep abnormalities are a common early feature in patients with Huntington's disease. Curr Neurol Neurosci Rep 2011;11(2):211–7.

[149] Ahmad Aziz N, Pijl H, Frölich M, Maurits van der Graaf A, Roelfsema F, Roos RA. Leptin secretion rate increases with higher CAG repeat number in Huntington's disease patients. Clin Endocrinol 2010;73(2):206–11.

[150] Hult S, Soylu R, Petersén Å. B30 Expression of mutant huntingtin in leptin receptor-expressing neurons does not influence the metabolic phenotype and psychiatric-like features in Huntington's disease. J Neurol, Neurosurg Psychiatry 2012;83(Suppl. 1):A14–5.

[151] Björkqvist M, Fex M, Renström E, Wierup N, Petersen A, Gil J, et al. The R6/2 transgenic mouse model of Huntington's disease develops diabetes due to deficient β-cell mass and exocytosis. Hum Mol Genet 2005;14(5):565–74.

[152] Clifford J, Drago J, Natoli A, Wong J, Kinsella A, Waddington J, et al. Essential fatty acids given from conception prevent topographies of motor deficit in a transgenic model of Huntington's disease. Neuroscience 2002;109(1):81–8.

[153] Vaddadi K, Soosai E, Chiu E, Dingjan P. A randomised, placebo-controlled, double blind study of treatment of Huntington's disease with unsaturated fatty acids. Neuroreport 2002;13(1):29–33.

[154] Miyazawa D, Yasui Y, Yamada K, Ohara N, Okuyama H. Regional differences of the mouse brain in response to an α-linolenic acid-restricted diet: neurotrophin content and protein kinase activity. Life Sci 2010;87(15):490–4.

[155] Weindruch R. The retardation of aging by caloric restriction: studies in rodents and primates. Toxicol pathol 1996;24(6):742–5.

[156] Duan W, Guo Z, Jiang H, Ware M, Mattson MP. Reversal of behavioral and metabolic abnormalities, and insulin resistance syndrome, by dietary restriction in mice deficient in brain-derived neurotrophic factor. Endocrinology 2003;144(6):2446–53.

[157] Steinkraus KA, Smith ED, Davis C, Carr D, Pendergrass WR, Sutphin GL, et al. Dietary restriction suppresses proteotoxicity and enhances longevity by an hsf–1–dependent mechanism in Caenorhabditis elegans. Aging Cell 2008;7(3):394–404.

[158] Lin MT, Beal MF. Mitochondrial dysfunction and oxidative stress in neurodegenerative diseases. Nature 2006;443(7113):787–95.

[159] Damiano M, Galvan L, Déglon N, Brouillet E. Mitochondria in Huntington's disease. Biochim Biophys Acta (BBA)-Mol Basis Dis 2010;1802(1):52–61.

[160] Beal MF, Brouillet E, Jenkins BG, Ferrante RJ, Kowall NW, Miller J, et al. Neurochemical and histologic characterization of striatal excitotoxic lesions produced by the mitochondrial toxin 3-nitropropionic acid. J Neurosci 1993;13(10):4181–92.

[161] Napolitano M, Centonze D, Calce A, Picconi B, Spiezia S, Gulino A, et al. Experimental parkinsonism modulates multiple genes involved in the transduction of dopaminergic signals in the striatum. Neurobiol Dis 2002;10(3):387–95.

[162] Walaas SI, Hemmings Jr HC, Greengard P, Nairn AC. Beyond the dopamine receptor: regulation and roles of serine/threonine protein phosphatases. Front Neuroanat 2011;5.

[163] Cowan CM, Raymond LA. Selective neuronal degeneration in Huntington's disease. Curr Top Dev Biol 2006;75:25–71.

[164] Beal MF. Mitochondria take center stage in aging and neurodegeneration. Ann Neurol 2005;58(4):495–505.

[165] Goswami A, Dikshit P, Mishra A, Mulherkar S, Nukina N, Jana NR. Oxidative stress promotes mutant huntingtin aggregation and mutant huntingtin-dependent cell death by mimicking proteasomal malfunction. Biochem Biophys Res Commun 2006;342(1):184–90.

[166] Yang L, Calingasan NY, Wille EJ, Cormier K, Smith K, Ferrante RJ, et al. Combination therapy with coenzyme Q10 and creatine produces additive neuroprotective effects in models of Parkinson's and Huntington's diseases. J Neurochem 2009;109(5):1427–39.

[167] Dedeoglu A, Kubilus JK, Yang L, Ferrante KL, Hersch SM, Beal MF. Creatine therapy provides neuroprotection after onset of clinical symptoms in Huntington's disease transgenic mice. J Neurochem 2003;85(6):1359–67.

[168] Wright RO, Baccarelli A. Metals and neurotoxicology. J Nutr 2007;137(12):2809–13.

[169] Migliore L, Coppedè F. Environmental-induced oxidative stress in neurodegenerative disorders and aging. Mutat Research/Genetic Toxicol Environ Mutagen 2009;674(1):73–84.

[170] Hermosura MC, Nayakanti H, Dorovkov MV, Calderon FR, Ryazanov AG, Haymer DS, et al. A TRPM7 variant shows altered sensitivity to magnesium that may contribute to the pathogenesis of two Guamanian neurodegenerative disorders. Proc Natl Acad Sci USA 2005;102(32):11510–5.

[171] Beal MF. Mitochondria in neurodegeneration. Mitochondrial disorders. Springer; 2002. p. 17–35.

[172] Brustovetsky T, Purl K, Young A, Shimizu K, Dubinsky JM. Dearth of glutamate transporters contributes to striatal excitotoxicity. Exp Neurol 2004;189(2):222–30.

[173] Brustovetsky N, LaFrance R, Purl K, Brustovetsky T, Keene C, Low W, et al. Age–Dependent changes in the calcium sensitivity of striatal mitochondria in mouse models of Huntington's disease. J Neurochem 2005;93(6):1361–70.

[174] Grubman A, White A, Liddell J. Mitochondrial metals as a potential therapeutic target in neurodegeneration. Br J Pharmacol 2014;171(8):2159–73.

[175] Tarohda T, Yamamoto M, Amamo R. Regional distribution of manganese, iron, copper, and zinc in the rat brain during development. Anal Bioanal Chem 2004;380(2):240–6.

[176] Krebs N, Langkammer C, Goessler W, Ropele S, Fazekas F, Yen K, et al. Assessment of trace elements in human brain using inductively coupled plasma mass spectrometry. J Trace Elem Med Biol 2014;28(1):1–7.

[177] Bonilla E, Estevez J, Suarez H, Morales L, de Bonilla LC, Villalobos R, et al. Serum ferritin deficiency in Huntington's disease patients. Neurosci Lett 1991;129(1):22–4.

[178] Fox JH, Kama JA, Lieberman G, Chopra R, Dorsey K, Chopra V, et al. Mechanisms of copper ion mediated Huntington's disease progression. PLoS One 2007;2(3):e334.

[179] Tao T, Tartakoff AM. Nuclear relocation of normal huntingtin. Traffic 2001;2(6):385–94.

[180] Gutekunst C-A, Levey AI, Heilman CJ, et al. Identification and localization of huntingtin in brain and human lymphoblastoid cell lines with anti-fusion protein antibodies. Proc Natl Acad Sci USA 1995;92(19):8710–4.

[181] DiFiglia M, Sapp E, Chase K, Schwarz C, Meloni A, Young C, et al. Huntingtin is a cytoplasmic protein associated with vesicles in human and rat brain neurons. Neuron 1995;14(5):1075–81.

[182] Velier J, Kim M, Schwarz C, Kim TW, Sapp E, Chase K, et al. Wild-type and mutant huntingtins function in vesicle trafficking in the secretory and endocytic pathways. Exp Neurol 1998;152(1):34–40.

[183] Kegel KB, Sapp E, Yoder J, Cuiffo B, Sobin L, Kim YJ, et al. Huntingtin associates with acidic phospholipids at the plasma membrane. J Biol Chem 2005;280(43):36464–73.

[184] Engelender S, Sharp AH, Colomer V, Tokito MK, Lanahan A, Worley P, et al. Huntingtin-associated protein 1 (HAP1) interacts with the p150Glued bubunit of dynactin. Hum Mol Genet 1997;6(13):2205–12.

[185] Li S-H, Gutekunst C-A, Hersch SM, Li X-J. Interaction of huntingtin-associated protein with dynactin P150Glued. J Neurosci 1998;18(4):1261–9.

[186] Qin Z-H, Wang Y, Sapp E, Cuiffo B, Wanker E, Hayden MR, et al. Huntingtin bodies sequester vesicle-associated proteins by a polyproline-dependent interaction. J Neurosci 2004;24(1):269–81.

[187] Hilditch-Maguire P, Trettel F, Passani LA, Auerbach A, Persichetti F, MacDonald ME. Huntingtin: an iron-regulated protein essential for normal nuclear and perinuclear organelles. Hum Mol Genet 2000;9(19):2789–97.

[188] Lumsden AL, Henshall TL, Dayan S, Lardelli MT, Richards RI. Huntingtin-deficient zebrafish exhibit defects in iron utilization and development. Hum Mol Genet 2007;16(16):1905–20.

[189] Gorrell JM, DiMonte D, Graham D. The role of the environment in Parkinson's disease. Environ Health Perspect 1996;104(6):652.

[190] Racette BA, McGee-Minnich L, Moerlein S, Mink J, Videen T, Perlmutter J. Welding-related parkinsonism clinical features, treatment, and pathophysiology. Neurology 2001;56(1):8–13.

[191] Saxena S, Caroni P. Selective neuronal vulnerability in neurodegenerative diseases: from stressor thresholds to degeneration. Neuron 2011;71(1):35–48.

[192] Mela L, Chance B. Spectrophotometric measurements of the kinetics of Ca^{2+} and Mn^{2+} accumulation in mitochondria. Biochemistry 1968;7(11):4059–63.

[193] Gavin CE, Gunter KK, Gunter TE. Manganese and calcium efflux kinetics in brain mitochondria. Relevance to manganese toxicity. Biochem J 1990;266:329–34.

[194] Bowman AB, Kwakye GF, Herrero Hernández E, Aschner M. Role of manganese in neurodegenerative diseases. J Trace Elem Med Biol 2011;25(4):191–203.

[195] Prohaska JR. Functions of trace elements in brain metabolism. Physiol Rev 1987;67(3):858–901.

[196] Prabhakaran K, Ghosh D, Chapman G, Gunasekar P. Molecular mechanism of manganese exposure-induced dopaminergic toxicity. Brain Res Bull 2008;76(4):361–7.

[197] Choo YS, Johnson GV, MacDonald M, Detloff PJ, Lesort M. Mutant huntingtin directly increases susceptibility of mitochondria to the calcium-induced permeability transition and cytochrome c release. Hum Mol Genet 2004;13(14):1407–20.

[198] Tamm C, Sabri F, Ceccatelli S. Mitochondrial-mediated apoptosis in neural stem cells exposed to manganese. Toxicol Sci 2008;101(2):310–20.

[199] Brazier MW, Davies P, Player E, Marken F, Viles JH, Brown DR. Manganese binding to the prion protein. J Biol Chem 2008;283(19):12831–9.

[200] Martin DP, Anantharam V, Jin H, Witte T, Houk R, Kanthasamy A, et al. Infectious prion protein alters manganese transport and neurotoxicity in a cell culture model of prion disease. Neurotoxicology 2011;32(5):554–62.

[201] Wang R-G, Zhu X-Z. Subtoxic concentration of manganese synergistically potentiates 1-methyl-4-phenylpyridinium-induced neurotoxicity in PC12 cells. Brain Res 2003;961(1):131–8.

[202] Williams BB, Li D, Wegrzynowicz M, Vadodaria BK, Anderson JG, Kwakye GF, et al. Disease–toxicant screen reveals a neuroprotective interaction between Huntington's disease and manganese exposure. J Neurochem 2010;112(1):227–37.

[203] Williams BB, Kwakye GF, Wegrzynowicz M, et al. Altered manganese homeostasis and manganese toxicity in a Huntington's disease striatal cell model are not due to defects in the iron transport system. Toxicol Sci 2010 Sep;117(1):169–79.

[204] Madison JL, Wegrzynowicz M, Aschner M, Bowman AB. Disease-toxicant interactions in manganese exposed Huntington disease mice: early changes in striatal neuron morphology and dopamine metabolism. PLoS One 2012;7(2):e31024.

[205] Kwakye GF, Li D, Bowman AB. Novel high-throughput assay to assess cellular manganese levels in a striatal cell line model of Huntington's disease confirms a deficit in manganese accumulation. Neurotoxicology 2011;32(5):630–9.

[206] Maciejewski PK, Rothman DL. Proposed cycles for functional glutamate trafficking in synaptic neurotransmission. Neurochem Int 2008;52(4):809–25.

[207] Sidoryk–Węgrzynowicz M, Lee E, Albrecht J, Aschner M. Manganese disrupts astrocyte glutamine transporter expression and function. J Neurochem 2009;110(3):822–30.

[208] Lobsiger CS, Cleveland DW. Glial cells as intrinsic components of non-cell-autonomous neurodegenerative disease. Nat Neurosci 2007;10(11):1355–60.

[209] Behrens P, Franz P, Woodman B, Lindenberg K, Landwehrmeyer G. Impaired glutamate transport and glutamate–glutamine cycling: downstream effects of the Huntington mutation. Brain 2002;125(8):1908–22.

[210] Butterworth J, Yates CM, Reynolds GP. Distribution of phosphate-activated glutaminase, succinic dehydrogenase, pyruvate dehydrogenase and γ-glutamyl transpeptidase in post-mortem brain from Huntington's disease and agonal cases. J Neurol Sci 1985;67(2):161–71.

[211] Carter C. Glutamine synthetase activity in Huntington's disease. Life Sci 1982;31(11):1151–9.

[212] Zelko IN, Mariani TJ, Folz RJ. Superoxide dismutase multigene family: a comparison of the CuZn-SOD (SOD1), Mn-SOD (SOD2), and EC-SOD (SOD3) gene structures, evolution, and expression. Free Radic Biol Med 2002;33(3):337–49.

[213] Mizuno K, Whittaker MM, Bächinger HP, Whittaker JW. Calorimetric studies on the tight binding metal interactions of *Escherichia coli* manganese superoxide dismutase. J Biol Chem 2004;279(26):27339–44.

[214] Aguirre JD, Culotta VC. Battles with iron: manganese in oxidative stress protection. J Biol Chem 2012;287(17):13541–8.

[215] Whittaker J. The irony of manganese superoxide dismutase. Biochem Soc Trans 2003;31(6):1318–21.

[216] Ash DE. Structure and function of arginases. J Nutr 2004;134(10):2760S–4S.

[217] Morris Jr SM. Recent advances in arginine metabolism: roles and regulation of the arginases. Br J Pharmacol 2009;157(6):922–30.

[218] Deckel AW, Tang V, Nuttal D, Gary K, Elder R. Altered neuronal nitric oxide synthase expression contributes to disease progression in Huntington's disease transgenic mice. Brain Res 2002;939(1):76–86.

[219] Deckel AW, Gordinier A, Nuttal D, Tang V, Kuwada C, Freitas R, et al. Reduced activity and protein expression of NOS in R6/2 HD transgenic mice: effects of L-NAME on symptom progression. Brain Res 2001;919(1):70–81.

[220] Nakamura T, Wang L, Wong CC, Scott FL, Eckelman BP, Han X, et al. Transnitrosylation of XIAP regulates caspase-dependent neuronal cell death. Mol Cell 2010;39(2):184–95.

[221] Chiang M-C, Chen H-M, Lee Y-H, Chang H-H, Wu Y-C, Soong B-W, et al. Dysregulation of C/EBPα by mutant Huntingtin causes the urea cycle deficiency in Huntington's disease. Hum Mol Genet 2007;16(5):483–98.

[222] Leopold NA, Podolsky S. Exaggerated growth hormone response to arginine infusion in Huntington's disease 1. J Clin Endocrinol Metab 1975;41(1):160–3.

[223] Deckel AW, Volmer P, Weiner R, Gary KA, Covault J, Sasso D, et al. Dietary arginine alters time of symptom onset in Huntington's disease transgenic mice. Brain Res 2000;875(1):187–95.

[224] Maywood ES, Fraenkel E, McAllister CJ, Wood N, Reddy AB, Hastings MH, et al. Disruption of peripheral circadian timekeeping in a mouse model of Huntington's disease and its restoration by temporally scheduled feeding. J Neurosci 2010;30(30):10199–204.

[225] Estévez AG, Sahawneh MA, Lange PS, Bae N, Egea M, Ratan RR. Arginase 1 regulation of nitric oxide production is key to survival of trophic factor-deprived motor neurons. J Neurosci 2006;26(33):8512–6.

[226] Scrutton MC, Utter MF, Mildvan AS. Pyruvate carboxylase VI. The presence of tightly bound manganese. J Biol Chem 1966;241(15):3480–7.

[227] Scrutton MC, Utter MF. Pyruvate carboxylase III. Some physical and chemical properties of the highly purified enzyme. J Biol Chem 1965;240(1):1–9.

[228] Mildvan AS, Scrutton MC, Utter MF. Pyruvate carboxylase VII. A possible role for tightly bound manganese. J Biol Chem 1966;241(15):3488–98.

[229] Mochel F, DeLonlay P, Touati G, Brunengraber H, Kinman RP, Rabier D, et al. Pyruvate carboxylase deficiency: clinical and biochemical response to anaplerotic diet therapy. Mol Genet Metab 2005;84(4):305–12.

[230] Gines S, Seong IS, Fossale E, Ivanova E, Trettel F, Gusella JF, et al. Specific progressive cAMP reduction implicates energy deficit in presymptomatic Huntington's disease knock-in mice. Hum Mol Genet 2003;12(5):497–508.

[231] Weydt P, Pineda VV, Torrence AE, Libby RT, Satterfield TF, Lazarowski ER, et al. Thermoregulatory and metabolic defects in Huntington's disease transgenic mice implicate PGC-1α in Huntington's disease neurodegeneration. Cell Metab 2006;4(5):349–62.

[232] Mochel F, Durant B, Meng X, O'Callaghan J, Yu H, Brouillet E, et al. Early alterations of brain cellular energy homeostasis in Huntington disease models. J Biol Chem 2012;287(2):1361–70.

[233] Seong IS, Ivanova E, Lee J-M, Choo YS, Fossale E, Anderson M, et al. HD CAG repeat implicates a dominant property of huntingtin in mitochondrial energy metabolism. Hum Mol Genet 2005;14(19):2871–80.

[234] Harms L, Meierkord H, Timm G, Pfeiffer L, Ludolph A. Decreased N-acetyl-aspartate/choline ratio and increased lactate in the frontal lobe of patients with Huntington's disease: a proton magnetic resonance spectroscopy study. J Neurol, Neurosurg Psychiatry 1997;62(1):27–30.

[235] Jenkins BG, Koroshetz WJ, Beal MF, Rosen BR. Evidence for irnnairment of energy metabofism in vivo in Huntington's disease using localized 1H NMR spectroscopy. Neurology 1993;43(12):2689.

[236] Ortez C, Jou C, Cortès-Saladelafont E, Moreno J, Perez A, Ormazabal A, et al. Infantile parkinsonism and gabaergic hypotransmission in a patient with pyruvate carboxylase deficiency. Gene 2013;532(2):302–6.

[237] Woźniak-Celmer E, Ołdziej S, Ciarkowski J. Theoretical models of catalytic domains of protein phosphatases 1 and 2A with Zn^{2+} and Mn^{2+} metal dications and putative bioligands in their catalytic centers. Acta Biochim Pol 2000;48(1):35–52.

[238] Xiao L, Gong L, Yuan D, Deng M, Zeng X, Chen L, et al. Protein phosphatase-1 regulates Akt1 signal transduction pathway to control gene expression, cell survival and differentiation. Cell Death Differ 2010;17(9):1448–62.

[239] Li DW, Liu J, Schmid P, Schlosser R, Feng H, Liu W, et al. Protein serine/threonine phosphatase-1 dephosphorylates p53 at Ser-15 and Ser-37 to modulate its transcriptional and apoptotic activities. Oncogene 2006;25(21):3006–22.

[240] Marion S, Urs NM, Peterson SM, Sotnikova TD, Beaulieu J-M, Gainetdinov RR, et al. Dopamine D2 receptor relies upon PPM/PP2C protein phosphatases to dephosphorylate huntingtin protein. J Biol Chem 2014;289(17):11715–24.

[241] Napolitano M, Centonze D, Gubellini P, Rossi S, Spiezia S, Bernardi G, et al. Inhibition of mitochondrial complex II alters striatal expression of genes involved in glutamatergic and dopaminergic signaling: possible implications for Huntington's disease. Neurobiol Dis 2004;15(2):407–14.

[242] Canman CE, Lim D-S. The role of ATM in DNA damage responses and cancer. Oncogene 1998;17(25):3301–8.

[243] Banin S, Moyal L, Shieh S-Y, Taya Y, Anderson C, Chessa L, et al. Enhanced phosphorylation of p53 by ATM in response to DNA damage. Science 1998;281(5383):1674–7.

[244] Guo Z, Kozlov S, Lavin MF, Person MD, Paull TT. ATM activation by oxidative stress. Science 2010;330(6003):517–21.

[245] Chan DW, Son S-C, Block W, Ye R, Khanna KK, Wold MS, et al. Purification and characterization of ATM from human placenta a manganese-dependent, wortmannin-sensitive serine/threonine protein kinase. J Biol Chem 2000;275(11):7803–10.

[246] Ferlazzo ML, Sonzogni L, Granzotto A, Bodgi L, Lartin O, Devic C, et al. Mutations of the Huntington's disease protein impact on the ATM-dependent signaling and repair pathways of the radiation-induced DNA double-strand breaks: corrective effect of statins and bisphosphonates. Mol Neurobiol 2014;49(3):1200–11.

[247] Trujillo KM, Yuan S-SF, Eva Y-HL, Sung P. Nuclease activities in a complex of human recombination and DNA repair factors Rad50, Mre11, and p95. J Biol Chem 1998;273(34):21447–50.

[248] Hopkins BB, Paull TT. The P. furiosus Mre11/Rad50 complex promotes 5′ strand resection at a DNA double-strand break. Cell 2008;135(2):250–60.

[249] Paull TT, Gellert M. The 3′ to 5′ exonuclease activity of Mre11 facilitates repair of DNA double-strand breaks. Mol Cell 1998;1(7):969–79.

[250] Buis J, Wu Y, Deng Y, Leddon J, Westfield G, Eckersdorff M, et al. Mre11 nuclease activity has essential roles in DNA repair and genomic stability distinct from ATM activation. Cell 2008;135(1):85–96.

[251] Lee J-H, Paull TT. ATM activation by DNA double-strand breaks through the Mre11-Rad50-Nbs1 complex. Science 2005;308(5721):551–4.

[252] Hopfner K-P, Karcher A, Shin DS, Craig L, Arthur LM, Carney JP, et al. Structural biology of Rad50 ATPase: ATP-driven conformational control in DNA double-strand break repair and the ABC-ATPase superfamily. Cell 2000;101(7):789–800.

[253] Waters LS, Minesinger BK, Wiltrout ME, D'Souza S, Woodruff RV, Walker GC. Eukaryotic translesion polymerases and their roles and regulation in DNA damage tolerance. Microbiol Mol Biol Rev 2009;73(1):134–54.

[254] Frank EG, Woodgate R. Increased catalytic activity and altered fidelity of human DNA polymerase ι in the presence of manganese. J Biol Chem 2007;282(34):24689–96.

[255] Hays H, Berdis AJ. Manganese substantially alters the dynamics of translesion DNA synthesis. Biochemistry 2002;41(15):4771–8.

[256] Wegrzynowicz M, Holt HK, Friedman DB, Bowman AB. Changes in the striatal proteome of YAC128Q mice exhibit gene–environment interactions between mutant huntingtin and manganese. J Proteome Res 2012;11(2):1118–32.

[257] Ugolino J, Fang S, Kubisch C, Monteiro MJ. Mutant Atp13a2 proteins involved in parkinsonism are degraded by ER-associated degradation and sensitize cells to ER-stress induced cell death. Hum Mol Genet 2011 Sep 15;20(18):3565–77.

[258] Tan J, Zhang T, Jiang L, Chi J, Hu D, Pan Q, et al. Regulation of intracellular manganese homeostasis by Kufor-Rakeb syndrome-associated ATP13A2 protein. J Biol Chem 2011;286(34):29654–62.

[259] Tsunemi T, Krainc D. Zn²⁺ dyshomeostasis caused by loss of ATP13A2/PARK9 leads to lysosomal dysfunction and alpha-synuclein accumulation. Hum Mol Genet 2014;23(11):2791–801.

[260] Yanai A, Huang K, Kang R, Singaraja RR, Arstikaitis P, Gan L, et al. Palmitoylation of huntingtin by HIP14 is essential for its trafficking and function. Nat Neurosci 2006;9(6):824–31.

[261] Linder ME, Deschenes RJ. Palmitoylation: policing protein stability and traffic. Nat Rev Mol Cell Biol 2007;8(1):74–84.

[262] Singaraja RR, Huang K, Sanders SS, Milnerwood AJ, Hines R, Lerch JP, et al. Altered palmitoylation and neuropathological deficits in mice lacking HIP14. Hum Mol Genet 2011;20(20):3899–909.

[263] Goytain A, Hines RM, Quamme GA. Huntingtin-interacting proteins, HIP14 and HIP14L, mediate dual functions, palmitoyl acyltransferase and Mg^{2+} transport. J Biol Chem 2008;283(48):33365–74.

[264] Yin Z, Jiang H, Lee ESY, Ni M, Erikson KM, Milatovic D, et al. Ferroportin is a manganese-responsive protein that decreases manganese cytotoxicity and accumulation. J Neurochem 2010;112(5):1190–8.

[265] Chen J, Marks E, Lai B, Zhang Z, Duce JA, Lam LQ, et al. Iron accumulates in Huntington's disease neurons: protection by deferoxamine. PLoS One 2013;8(10):e77023.

[266] Au C, Benedetto A, Aschner M. Manganese transport in eukaryotes: the role of DMT1. Neurotoxicology 2008;29(4):569–76.

[267] Hediger MA, Clémençon B, Burrier RE, Bruford EA. The ABCs of membrane transporters in health and disease (SLC series): introduction. Mol Aspects Med 2013;34(2):95–107.

[268] Forbes JR, Gros P. Iron, manganese, and cobalt transport by Nramp1 (Slc11a1) and Nramp2 (Slc11a2) expressed at the plasma membrane. Blood 2003;102(5):1884–92.

[269] Jeong J, Eide DJ. The SLC39 family of zinc transporters. Mol Aspects Med 2013;34(2):612–9.

[270] Himeno S, Yanagiya T, Fujishiro H. The role of zinc transporters in cadmium and manganese transport in mammalian cells. Biochimie 2009;91(10):1218–22.

[271] Fujishiro H, Yoshida M, Nakano Y, Himeno S. Interleukin-6 enhances manganese accumulation in SH-SY5Y cells: implications of the up-regulation of ZIP14 and the down-regulation of ZnT10. Metallomics 2014;6(4):944–9.

[272] Xu SZ, Zeng F, Boulay G, Grimm C, Harteneck C, Beech DJ. Block of TRPC5 channels by 2-aminoethoxydiphenyl borate: a differential, extracellular and voltage-dependent effect. Br J Pharmacol 2005;145(4):405–14.

[273] Guilbert A, Gautier M, Dhennin-Duthille I, Haren N, Sevestre H, Ouadid-Ahidouch H. Evidence that TRPM7 is required for breast cancer cell proliferation. Am J Physiol-Cell Physiol 2009;297(3):C493–502.

[274] Riccio A, Mattei C, Kelsell RE, Medhurst AD, Calver AR, Randall AD, et al. Cloning and functional expression of human short TRP7, a candidate protein for store-operated Ca^{2+} influx. J Biol Chem 2002;277(14):12302–9.

[275] Chen Y, Payne K, Perara VS, Huang S, Baba A, Matsuda T, et al. Inhibition of the sodium–calcium exchanger via SEA0400 altered manganese-induced T1 changes in isolated perfused rat hearts. NMR Biomed 2012;25(11):1280–5.

[276] Waghorn B, Yang Y, Baba A, Matsuda T, Schumacher A, Yanasak N, et al. Assessing manganese efflux using SEA0400 and cardiac T1-mapping manganese-enhanced MRI in a murine model. NMR Biomed 2009;22(8):874–81.

[277] Kobayashi K, Kuroda J, Shibata N, Hasegawa T, Seko Y, Satoh M, et al. Induction of metallothionein by manganese is completely dependent on interleukin-6 production. J Pharmacol Exp Ther 2007;320(2):721–7.

[278] van Dellen A, Hannan AJ. Genetic and environmental factors in the pathogenesis of Huntington's disease. Neurogenetics 2004;5(1):9–17.

Neuroinflammation in Neurological Dysfunction and Degeneration

Shih-Heng Chen, Esteban A. Oyarzabal, Janine Santos,
Qingshan Wang, Lulu Jiang, Jau-Shyong Hong

INFLAMMATION AND NEURODEGENERATION: CLUES FROM EPIDEMIOLOGICAL STUDIES

Inflammation of the central nervous system (CNS) is a hallmark pathological feature of several neurodegenerative [1,2] and psychiatric disorders [3]. Due to the limited regenerative capacities of postmitotic neural tissue, *neuroinflammation* is a highly regulated and essential process for restoring CNS homeostasis after pathological events such as infections, toxicant exposures, and physical trauma. In the brain, inflammatory responses are induced and coordinated by resident innate immune cells called microglia, and are further shaped by reactive astrocytes and infiltrating leukocytes. Neuroinflammatory responses are typically transient and sequentially characterized by the recruitment of additional immune cells to the afflicted site, clearance of the injurious stimulant, tissue repair or scarring, and immune resolution. Together, these overlapping events are termed as *acute neuroinflammation*.

Most neurodegenerative and psychiatric disorders display *low-grade chronic neuroinflammation*, whereas traumatic brain injuries, autoimmune disorders, and encephalopathies typically display *severe chronic neuroinflammation*—both of these forms of neuroinflammation cause collateral damage worse than that caused by the original insult [4]. Beyond inducing sustained neurocircuit disruptions that result in perpetual behavioral phenotypes, many of these conditions progress to cause irreparable neuronal loss and brain atrophy. These effects have been linked to the chronic release of cytotoxic factors by activated microglia and other immune cells, resulting in neurodegeneration by oxidative stress, bioenergetic failure, and excitotoxicity.

Although the root of chronic neuroinflammation in neurodegenerative diseases remains highly disputed, several epidemiological studies have confirmed links between environmental factors capable of inducing neurotoxicity directly and/or indirectly, and the etiology and accelerated progression of neurodegenerative diseases (Table 18.1). Among these factors, the association between infection and neurodegenerative disease is one of the most well studied (see review 5). For instance, infections with systemic inflammation increase the risk for Alzheimer's disease in the elderly by two-fold [6], and are three times as likely to cause relapse in patients with multiple sclerosis [7].

The greatest challenge in understanding the link between neurodegenerative disease and environmental triggers has been sorting through the heterogeneity of a lifetime of exposures—particularly since the average age of onset of these diseases is late adulthood. Interpretations of epidemiological studies are further complicated by considering the developmental stages of exposure (e.g., in utero, infancy, adolescence, and adulthood), whether exposures consist of a single large hit or chronic smaller hits, and whether they require multiple hits (e.g., exposure to other xenobiotics, aging, and genetic predisposition) in order to initiate neurodegeneration. Furthermore, most associations linking environmental exposures with idiopathic neurodegenerative diseases have been rather weak in establishing probabilistic causality. For this reason, many studies have shifted from the determination of causal relationships of environmental factors to the linking of the severity

TABLE 18.1   Association of Environmental Exposures and Neurodegenerative Diseases

Disease	Environmental exposure	References
Amyotrophic lateral sclerosis	Microbial infection, pesticides, heavy metals	[5,8,9]
Alzheimer's disease	Microbial infection, pesticides, heavy metals	[5,9]
Huntington's disease	Microbial infection, heavy metals	[5,10]
Multiple sclerosis	Microbial infection, industrial solvents	[5,9]
Parkinson's disease	Microbial infection, pesticides, heavy metals, industrial solvents, air pollution	[5,11–15]

of existing neuroinflammation, via direct measurement of inflammatory factors in blood/ cerebral spinal fluid and inflammation-related targets in brain tissue [16–18] as biomarker profiles for different stages of neurodegenerative diseases (Table 18.2). These inflammatory biomarkers are detectable during prodromal stages [19], allowing for presymptomatic diagnoses of neurodegenerative diseases.

To avoid confounding effects from preexisting systemic inflammatory conditions outside of the brain, many epidemiological studies account for the presence of autoimmune diseases, atherosclerosis, diabetes, obesity, rheumatoid arthritis, and lifestyle factors such as smoking. Interestingly, links between peripheral chronic inflammatory diseases and neurodegenerative disorders (Table 18.3) are becoming evident, suggesting that chronic systemic inflammation may also play a role in the etiology or accelerated progression of neurodegenerative diseases.

The strongest evidence supporting the inflammatory hypothesis originates from unbiased, hypothesis-free genome-wide association (GWA) studies that uncovered genetic mutations in inflammation-related genes; these mutations may be involved in the pathogenesis of several neurodegenerative diseases. GWA studies of Alzheimer's disease patients found an association with mutations in the inflammation-related genes *TREM2, CLU, CD31, EPHA1*, and *CR1* [52–54]—all thought to impair anti-inflammatory functions in microglia. Mutations in *LRRK2* and *HLA* [55,56] suggest impairment of anti-inflammatory functions in microglia and the involvement of leukocyte recruitment in Parkinson's disease. Lastly, gain-of-function mutations in the genes that encode the proinflammatory cytokines TNF-α, IL-1α, IL-1β, and IL-6—and loss-of-function mutations for the anti-inflammatory cytokine IL-10—have been associated with Alzheimer's disease [57,58]. Similarly, gain-of-function mutations in TNF-α and IL-1β are associated with Parkinson's disease [59].

TABLE 18.2   Expression of Inflammatory Marker Profiles Detected through Noninvasive Sampling in Neurodegenerative Diseases

Disease	Detectable inflammatory markers	Sources
Amyotrophic lateral sclerosis	Serum: IL-6, IL-13, IL-17, TNF-α, MCP-1, CRP	[20–24]
	CSF: IFNγ	[25]
	Tissue: Translocator protein upregulation	[26]
Alzheimer's disease	Serum: IL-6, IL-1β, TNF-α, CRP, TGF-β, IL-12, IL-18	[27,28]
	CSF: TGF-β	[27]
	Tissue: Translocator protein upregulation	[17]
Huntington's disease	Serum: IL-6, IL-8, IL-13, TNF-α	[29]
	CSF: IL-6, IL-8	[30]
	Tissue: Translocator protein upregulation	[31]
Multiple sclerosis	Serum: TNF-α, IFNγ, IL-1β, IL-6, IL-10, CXCL8, TGF-β	[32,33]
	CSF: TNF-α, IL-1β	[34]
	Tissue: Translocator protein upregulation	[16]
Parkinson's disease	Serum: MCP-1, RANTES, MIP-1α, IL-8, IFNγ, IL-1β, TNF-α	[35]
	CSF: IL-1β, IL-6	[36]
	Tissue: Translocator protein upregulation	[18]

TABLE 18.3   Association of Chronic Inflammatory Diseases and Neurodegenerative Diseases

Disease	Linked chronic inflammatory disease	Sources
Amyotrophic lateral sclerosis	Asthma, celiac disease, diabetes, myasthenia gravis, myxedema, polymyositis, Sjögren syndrome, systemic lupus erythematosus, ulcerative colitis	[37]
Alzheimer's disease	Atherosclerosis, diabetes, obesity, periodontitis	[38–42]
Huntington's disease	Diabetes, periodontitis	[43,44]
Multiple sclerosis	Diabetes, periodontitis, obesity, autoimmune diseases	[45–48]
Parkinson's disease	Diabetes, periodontitis, ulcerative colitis	[49–51]

DIRECT AND INDIRECT INDUCTION OF CHRONIC NEUROINFLAMMATION

The best way to illustrate how chronic inflammatory diseases or exposures to environmental factors can generate chronic neuroinflammation is by describing a few models of neuroinflammation. For the sake of consistency, we will focus on models using exposures to the gram-negative bacterial endotoxin lipopolysaccharide (LPS). These models vary only in the dose, frequency, and route of exposure to the same endotoxin, but result in distinct neuroinflammatory phenotypes.

LPS is innocuous to all cells except those equipped with the correct surface pattern recognition receptors (e.g., TLR-4, β2-integrin, and scavenger receptors) that detect the endotoxin. In the body, LPS is primarily detected by tissue macrophages, which in turn become activated to release proinflammatory factors such as TNF-α, IL-1β, IL-6, IL-18, nitric acid, superoxide, and prostaglandins [60]. When LPS is administered directly into the brain, it is detected by microglia and induces their activation. This in turn results in neuroinflammation followed by subsequent nuclei-specific pathophysiological and behavioral changes, which typically occur within a week (Table 18.4). Induction of neuroinflammation through this method reflects the actions of selective toxicants that can penetrate the blood–brain barrier and damage neurons in a confined brain nuclei (e.g., MPP+, trimethyltin) through sustained microglial activation. Acute parkinsonian pathologies and behavioral changes occur when neuroinflammation is experimentally induced at any of the three nuclei thought to be impaired in the motor deficits of Parkinson's disease: the substantia nigra, the globus pallidus, and the dorsal striatum. Interestingly, neuroinflammation and disease-specific protein lesions (e.g., tau phosphorylation or the formation of Lewy body-like inclusions) travel along the neural circuits that interconnect these nuclei, suggesting that the neuroinflammatory hypothesis can account for nuclei-specific progression across long distances [61]. Likewise, inflammation restricted to the hippocampus and cortex results in an acute model of Alzheimer's disease, with similar transmission of neuroinflammation along these nuclei and the thalamus. Together, these models implicate a causal role for nuclei-specific neuroinflammation in the etiology of acute neurodegenerative-like diseases.

When LPS is injected in the periphery, it produces far different neurological dysfunctions and pathologies, which vary greatly in a dose-dependent manner. LPS enters the bloodstream

TABLE 18.4   Pathophysiological and Behavioral Effects Detected in Rodents Stereotaxically Injected with LPS into a Defined Brain Region

Injection site	Pathophysiological and behavioral effects	Sources
Substantia nigra	Loss of substantia nigra dopaminergic neurons; microgliosis with increased levels of IL-1β, TNF-α, IL-6, and NO; motor deficits.	[62–66]
Globus pallidus	Loss of substantia nigra dopaminergic neurons; microgliosis with increased levels of IL-1β, TNF-α, IL-6, and NO in the globus pallidus and substantia nigra; enhanced α-synuclein nitration and oligomerization; increased iron accumulation; motor deficits.	[67,68]
Dorsal striatum	Loss of substantia nigra dopaminergic neurons; microgliosis; depletion of striatal dopamine; Lewy body-like accumulations of α-synuclein in dopaminergic neurons; motor deficits.	[69–72]
Cortex	Enhanced neuronal apoptosis; microgliosis with increased levels of TNF-α, iNOS, IL-4, MIP-2, and BDNF; enhanced neurofibrillary tangles and β-amyloid clearance in APP transgenic mice.	[73–76]
Hippocampus	Sustained microgliosis in the hippocampus, cerebral cortex, and thalamus; accumulated ubiquitinated proteins with impaired proteasome activity; enhanced β-amyloid clearance in APP transgenic mice; deficits in social behavior, learning, and prepulse inhibition.	[77–81]

when administered peripherally (e.g., by subcutaneous, intravenous, or intraperitoneal injection) and induces systemic endotoxemia. A limited amount of LPS penetrates the blood–brain barrier [75], thus inducing neuroinflammation indirectly through a *humoral route* via peripheral cytokines generated by macrophages in the liver and spleen, and in endothelial cells in the cerebrovasculature. Although cytokines are too large to diffuse through the blood–brain barrier, humoral TNF-α, IL-1β, and to a lesser extent IL-6 can rapidly enter the brain through paracellular diffusion in regions with a discontinuous blood–brain barrier such as circumventricular organs and fenestrated capillaries, as well as through active endothelial transporters across the intact blood–brain barrier into the brain parenchyma [82]. These secondary messengers induce microglial activation to produce inflammatory factors centrally, exacerbating the neuroinflammatory response. Furthermore, these secondary messengers are detected in the periphery via a *neurogenic route* (i.e., the afferent arc of the inflammatory reflex) by innervating receptors on parasympathetic nerve endings that trigger neuroinflammation along the hypothalamic–pituitary–adrenal axis [83].

When a subfebrile dose of LPS (25 μg/kg) is administered into the peritoneal cavity of rodents, it induces a transient sickness behavior that lasts up to 24 h with no detectable long-term pathologies in the brain. These effects are solely restricted to hypothalamic–pituitary–adrenal axis activation through a neurogenic route, with no contribution from humoral cytokines. When the dose of LPS injected into the peritoneal cavity is increased to a febrile, subseptic dose (100–500 μg/kg), it induces sickness behavior followed by depression-like behavior that may last several weeks with no detectable long-term pathologies in the brain. This dose of LPS induces acute neuroinflammation in two waves—one through an instantaneous neurogenic route in the hypothalamic–pituitary–adrenal axis, and the other through a delayed humoral route with limited tissue diffusion from endothelial and epithelial barriers. Together, these findings implicate that most mild infections may not play a role in triggering a sufficient neuro-inflammatory response to drive neurodegenerative diseases.

In contrast, when septic doses of LPS (>1 mg/kg) are administered into the peritoneal cavity of mice, neuroinflammation is induced primarily through the humoral route, resulting in permanent depression-like behavior in the mice that survive. Greater concentrations of circulating cytokines diffuse into and throughout the brain to induce global neuroinflammation via volume transmission and glia-mediated propagation. Interestingly, sublethal endotoxemia (5 mg/kg of LPS) can generate chronic neuroinflammation throughout the entire brain for the remaining lifetime of the mouse [84]. Indeed, the animals exhibit parkinsonian pathologies with a selective loss of 40–50% of the dopaminergic neurons of the substantia nigra by 10 months after exposure, loss of striatal dopaminergic fibers, formation of Lewy body-like aggregates, and the presence of motor deficits. The induction of a chronic neuro-inflammation is thought to occur through a positive feedback loop, wherein the release of cytotoxic factors by activated microglia during neuroinflammation results in collateral damage to neighboring neurons that in turn release paracrine factors, which serve as endogenous danger-associated molecular patterns (DAMPs) that sustain the neuroinflammation [4,85]. These findings suggest that severe infections in the peripheral system may trigger a chronic inflammatory response in the CNS to drive neuronal dysfunction and degeneration. Furthermore, the mechanism that drives chronic neuroinflammation is also thought to be inducible by chronic inflammatory diseases and direct neurotoxicants (Figure 18.1)—since stressed neurons shed membrane breakdown products (e.g., laminin) and leak cytosolic compounds (e.g., α-synuclein, β-amyloid, μ-calpain, neuromelanin, and HMGB-1).

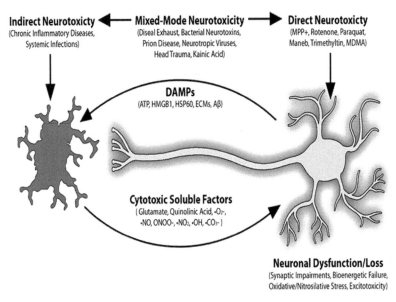

Indirect Neurotoxicty (Chronic Inflammatory Diseases, Systemic Infections)

Mixed-Mode Neurotoxicity (Diseal Exhaust, Bacterial Neurotoxins, Prion Disease, Neurotropic Viruses, Head Trauma, Kainic Acid)

Direct Neurotoxicty (MPP+, Rotenone, Paraquat, Maneb, Trimethyltin, MDMA)

DAMPs (ATP, HMGB1, HSP60, ECMs, Aβ)

Cytotoxic Soluble Factors (Glutamate, Quinolinic Acid, $\cdot O_2^-$, $\cdot NO$, $ONOO-$, $\cdot NO_2$, $\cdot OH$, $\cdot CO_3^-$)

Neuronal Dysfunction/Loss (Synaptic Impairments, Bioenergetic Failure, Oxidative/Nitrosilative Stress, Excitotoxicity)

FIGURE 18.1   Reactive microgliosis drives chronic and progressive neurotoxicity through two mechanisms. Microglia can become activated by recognizing cytokines and microbial pathogens, producing cytotoxic factors that damage neurons (indirect neurotoxicity); alternatively, neurons damaged by neurotoxicants (direct neurotoxicity) can release danger-associated molecular patterns (DAMPs) that induce microglial activation. Sufficient neuroinflammation can drive a feed-forward cycle, known as reactive microgliosis, that further damages neurons.

The mismatch between the prevalence of neurodegenerative diseases and the incidence of individual histories of systemic inflammation is largely attributed to variability in susceptibility risk factors (Figure 18.2). Age can be a risk factor for both systemic inflammation and neurotoxicant exposures, since microglia are less quiescent and thus more easily

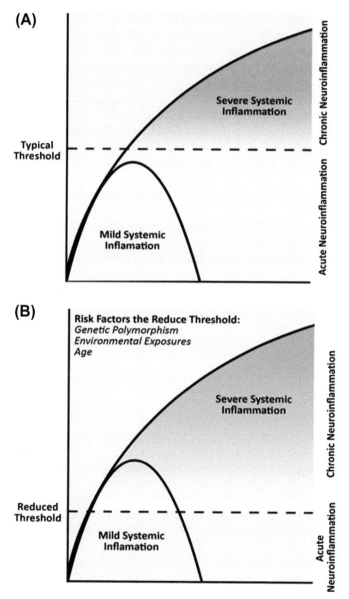

FIGURE 18.2 Severe systemic inflammation is typically associated with developing chronic neuroinflammation that drives neurodegeneration. (A) In order to generate chronic neuroinflammation, a threshold level of inflammation that drives reactive microgliosis must be achieved. (B) The threshold level of inflammation is highly variable among individuals, as it can be reduced by differential susceptibility risk factors such as genetic polymorphisms, environmental exposures, and age.

primed during neurodevelopment [86] and aging [87]. This raises the interesting possibility that infections or toxicant exposures that occur in utero during brain development, in addition to those occurring in late adulthood, may contribute to the etiology of neurodegenerative diseases. For instance, systemic maternal inflammation in response to bacterial vaginosis can enter the chorioamniotic compartment and induce neuroinflammation in the fetus through a humoral route. Pregnant rat dams exposed to subseptic doses of LPS at gestation day 10.5 induced neuroinflammation in their offspring, an effect lasting up to 16 months [88–90]— indicating that age-related risk factors increase susceptibility by priming microglia to reduce the humoral cytokine threshold required for induction of chronic neuroinflammation. More importantly, the data obtained with these exposed litters further support the inflammatory hypothesis, as the animals developed parkinsonian symptoms, including a 30% loss of dopaminergic neurons in the substantia nigra, reduction in striatal dopamine levels with increased compensatory dopamine synthesis in the remaining projection neurons, and formation of Lewy body-like aggregates [88–90], in addition to the loss of pyramidal cells in the hippocampus [91].

Genetic risk factors (e.g., loss-of-function, gain-of-function, and null mutations) can also reduce the humoral cytokine threshold required for induction of chronic neuroinflammation, highlighting the importance of gene–environment interactions in neurodegenerative disease etiology. LPS treatment on transgenic mice expressing familial mutations associated with Parkinson's disease in the genes *PARK2* [92], *SNCA* [93], and *LRRK2* [56] resulted in accelerated progression of parkinsonian pathologies and motor deficits. Although mutations in *PARK2* and *SNCA* are themselves insufficient to cause Parkinson's disease and are indirectly related to the modulation of neuroinflammation, accumulating evidence supports a *multiple-hit hypothesis* in the pathogenesis of Parkinson's and other neurodegenerative diseases. Similarly, systemic inflammation in the Tg2567 [94], *PSEN1* [95], and 3xTg [96] mouse models of Alzheimer's disease, the SOD1^{G37R} [97] mouse model of amyotrophic lateral sclerosis (ALS), and the Crry$^{-/-}$ [98] mouse model of multiple sclerosis resulted in the presentation of accelerated pathologies and neurodegeneration.

Lastly, exposures to multiple environmental risk factors and their simultaneous interactions may also reduce the humoral cytokine threshold required for induction of chronic neuroinflammation. Subseptic doses of LPS generated cytotoxic chronic neuroinflammatory response with neurodegeneration in mice with prion disease [99]. Repetitive daily administration of subseptic doses of LPS in the peritoneal cavity of mice (five consecutive days every month for four consecutive months) induced chronic depression-like behavior [100]. Subfebrile doses of LPS administered intranasally in mice with a nebulizer every other day for five months showed parkinsonian pathologies with an 80% reduction in dopaminergic neurons of the substantia nigra, reduced striatal dopamine content, Lewy body-like aggregates, and progressive hypokinesia [101].

Furthermore, preexisting systemic inflammatory events, such as in a dextran sulfate sodium-induced ulcerative colitis mouse model [102] and in a Wallerian degeneration mouse model via the crushing of the optic nerve [103], showed exacerbated damage when coupled with a single administration of subfebrile doses of LPS. Together, these findings suggest that an additional environmental hit coupled with a preexisting systemic inflammation (e.g., chronic infections, chronic disease, or chronic exposures to environmental contaminants) could also play a major role in the etiology of neurodegenerative diseases.

NEUROINFLAMMATION AND NEUROLOGICAL DYSFUNCTION: MODULATING BEHAVIOR

Behavioral changes associated with peripheral and central inflammation are linked to proinflammatory cytokines in the brain. In fact, humans exposed to LPS [104], *Salmonella typhi* titers [105], or recombinant interleukin-2 [106] or interferon-α [107] showed cognitive deficits, lethargy, locomotor retardation, reduced appetite, loss of pleasure from rewarding activities, sleep disturbances, and increased pain sensitivity—all of which overlap with symptoms of major depression disorder (MDD). Interestingly, these inflammation-driven depressive-like behaviors are a common denominator linking neurodegenerative diseases like Alzheimer's, multiple sclerosis, and Parkinson's with systemic inflammatory conditions such as psoriasis [108], rheumatoid arthritis [109], inflammatory bowel disease [110], cancer [111], and coronary artery disease [111]. Moreover, depression-like behaviors in neurodegenerative diseases appear during prodromal stages, suggesting that they may initially be attributed to neurocircuitry dysfunction prior to the onset of neurodegeneration.

Repeated LPS exposures [100] or chronic infection with bacilli Calmette–Guerin [112] caused depression-like behavior that permanently disrupted food intake, hindered operant response to food rewards, retarded exploration and social interactions, increased pain sensitivity, altered cognition, and increased slow-wave sleep, when administered to the peritoneal cavity of mice. Interestingly, severing the vagus nerve below the diaphragm prior to administering LPS or IL-1β into the peritoneal cavity prevented the aforementioned behavior changes, implying that cytokine receptors expressed on afferent nerves in the peripheral system can modulate the behavioral changes observed in LPS-treated mice. Among the many proinflammatory cytokines generated by LPS, only the administration of recombinant IL-1β, both systemically and centrally, has been shown to induce depression-like behavior. Furthermore, this effect can be reversed in caspase 1-deficient mice incapable of processing inactive pro-IL-1β to its active form, by centrally administering IL-1 receptor antagonist.

Relatively little is known about how IL-1β may interfere with neurocircuits and neurotransmission to induce depression-like behavior. Neuronal activity following LPS stimulation in the peritoneal cavity showed pronounced activity of the nucleus of the solitary tract, locus coeruleus, dorsal raphe, paraventricular hypothalamus, bed nucleus of the stria terminalis, nucleus accumbens, extended amygdala and motor, and cingulate and piriform cortices, as measured by *c-fos* expression [113]. Most prominently, the activation of the hypothalamic–pituitary–adrenal relay is involved in modulatory changes of glucocorticoid levels in the blood that can suppress systemic inflammation [114]. Furthermore, both LPS stimulation in the peritoneal cavity and direct electrical stimulation of the vagus nerve result in the production of IL-1β in regions such as the locus coeruleus [115,116] and dorsal raphe [115], suggesting that neuroinflammation can be induced through direct afferent stimulation.

The vagus nerve travels to the nucleus of the solitary tract where it projects to the locus coeruleus. Projection neurons from the locus coeruleus account for approximately 70% of the brain's production of the neuromodulator norepinephrine—believed to be important in modulating attention, fear–response, anxiety, cognitive functions, depression, REM sleep, and gastrocolic *reflex* [117]. Upon detecting systemic inflammation along the neurogenic route, IL-1β is released in the locus coeruleus, resulting in a doubling of the firing rate of noradrenergic neurons for prolonged periods (confirmed up to three weeks after onset) [118]. Interestingly, similar changes to firing rates were observed in noradrenergic neurons of the locus coeruleus, and serotonergic neurons of the dorsal raphe, following two weeks of vagus nerve stimulation without any desensitization [119]—whereby monosynaptic projections from the locus coeruleus to the dorsal raphe connected these nuclei [117]. The increased levels of norepinephrine and serotonin released throughout the brain during the initial phase of chronic neuroinflammation are thought to provide antiepileptic and antidepressant effects that may be partly attributed to the perceived modulation of neuroinflammation. Yet, sustained overexcitation and vesicular dumping of norepinephrine and serotonin can generate H_2O_2 when catabolized by monoamine oxidases and through nonenzymatic autoxidation, thereby forming reactive quinones that contribute to oxidative stress [120,121]—partly explaining the inevitable and highly pronounced neuronal loss in both the locus coeruleus and dorsal raphe in Alzheimer's disease and Parkinson's disease [122].

As central norepinephrine declines with the loss of noradrenergic neurons of the locus coeruleus, the severity of cognitive impairment in Alzheimer's disease increases [123]. One report has linked this association with increased neuroinflammation due to the loss of the tonic anti-inflammatory actions of norepinephrine on the hippocampus [124]—more specifically to higher levels of microglial-derived IL-1β [125]. Additionally, noradrenergic neurons of the locus coeruleus seem to modulate neurotransmission in the nigrostriatal dopaminergic system and activity of the basalocortical cholinergic system during neuroinflammation [126]—which are thought to modulate fatigue, locomotor retardation and loss of pleasure from rewarding activities, and anti-inflammatory effects on neuroinflammation.

Chronic neuroinflammation induces a significant reduction in cortical acetyl cholinesterase activity, which is involved in the degradation of the anti-inflammatory acetylcholine that typically attenuates microglial activation through actions on nicotinic receptors [127]. The degradation of acetylcholine can be prevented with acetyl cholinesterase inhibitors (e.g., donepezil, tacrine, and rivastigmine), which have been used to attenuate neuroinflammation [128] and slow the progression of mild and moderate dementia in Alzheimer's disease [129]. Conversely, tricyclic antidepressants, which show anticholinergic properties, generate delirious states with hallucinations and excessive anxiety [128].

Serotonin is thought to control affective and cognitive functions and modulate circadian rhythms, and declines in serotonin are directly linked with depression. Humans with genetic mutations that impair serotonin transporter activity and a subtype of serotonin receptor are at higher risk of developing MDD [130]. Conversely, the use of selective serotonin reuptake inhibitors such as fluoxetine, paroxetine, and citalopram successfully treat depression, and have been shown to improve cognitive functions and reduce the accumulation of β-amyloid and tau fibrillary tangles in individuals with Alzheimer's disease [131,132]. Loss of serotonin observed during inflammation-induced depressive-like behavior and in Alzheimer's disease results from TNF-α-mediated loss of serotonergic neurons [133] and a drastic decline in the blood tryptophan required for serotonin production in the brain [134]. Blood tryptophan reductions are associated with increased enzymatic activity by indoleamine 2,3-dioxygenase (IDO), generated by TNF-α-activated microglia to reduce the bioavailability of serotonin in the brain [135]. Inhibition of IDO was able to suppress the formation of depressive-like behaviors in a mouse model of chronic neuroinflammation [136].

Some behavioral deficits induced by chronic neuroinflammation in rodents are likely due to inflammation-driven synaptic [137] and dendritic [138] stripping of stressed and damaged neurons. Furthermore, oligodendrocytes are highly vulnerable to inflammation-mediated oxidative stress, and their loss demyelinates synaptic projects, thus interrupting nerve conduction [139]. Furthermore, chronic neuroinflammation impairs synaptoplasticity by inhibiting long-term potentiation [140] through AMPA receptor endocytosis [141]. These pathological changes are mediated by activated microglia [141–143], and occur concurrently and progressively with aging and neurodegenerative disorders in humans [144,145]. Cognitive deficits are associated with inflammation-driven alterations of synaptic integrity and function in the hippocampus, resulting in impaired spatial learning and memory [137], which can be attenuated by suppressing microglial activation [146–149]. Furthermore, chronic neuroinflammation decreases basal expression of neurotrophins [150] and alters adult neurogenesis [151,152], further contributing to deficits in learning and memory formation.

Lastly, chronic neuroinflammation results in increased glutamate neurotransmission, resulting in increased spontaneous excitatory postsynaptic current [153–155] and a lower seizure threshold, which can be reversed with the microglial-activation inhibitor minocycline. This hyperexcitation is also present in several neurodegenerative disorders [156,157], and is primarily attributed to activated microglia that release glutamate and activated astrocytes that no longer buffer glutamate [155]. High levels of glutamate release due to chronic neuroinflammation can drive neuronal loss via excitotoxicity [158].

NEUROINFLAMMATION IN NEURODEGENERATION: RECAPITULATING PROGRESSIVE NEURONAL DEATH

Oxidative stress is a state where intracellular production or accumulation of reactive oxygen species (ROS) supersedes the antioxidant capacity of the cell. Neurons are highly susceptible to damage by oxidative stress due to their rich polyunsaturated fatty acids composition (a major site of oxidation) and minimal antioxidant capacity. Furthermore, the oxygen-rich environment of the brain, containing many transition metals needed to maintain basal neuronal function, can facilitate the escalation of ROS production. Neurodegenerative diseases are linked to oxidative stress, whereby high levels of ROS are generated extracellularly by activated microglial NADPH oxidase, and intracellularly by aerobic oxidative phosphorylation (OXPHOS) in mitochondria, and are thought to drive mitochondrial dysfunction as well as inevitable neuronal loss.

Phagocytic NADPH oxidase, located on the surface of activated microglia during neuroinflammation, can catalytically generate superoxide anions ($O_2^{\bullet-}$) that rapidly undergo spontaneous dismutation to hydrogen peroxide (H_2O_2) in the interstitial space. This H_2O_2 can easily diffuse through biological membranes and participate in Fenton chemistry with transition metals, generating the very reactive hydroxyl radicals ($^{\bullet}OH$) that together can damage cellular components such as proteins, lipids, carbohydrates, and nucleic acids. Neuronal populations differ in their vulnerability to these ROS due to their ability to enzymatically (e.g., glutathione peroxidase, glutathione reductase, superoxide dismutases, and catalase) and nonenzymatically (e.g., glutathione, ubiquinol, and uric acid) neutralize ROS. For instance, the selective degeneration of dopaminergic neurons in Parkinson's disease is partly attributed to the lower constitutive levels of superoxide dismutase, glutathione, and catalase in these neurons [159]. Recent findings suggest that NADPH oxidase isoforms are also present on neurons, and may directly generate intracellular ROS during neuroinflammatory conditions [160].

Superoxide generated as a by-product of complexes I and III of the electron transport chain (ETC) during aerobic OXPHOS in mitochondria is another major contributor to oxidative stress. Mitochondrial-derived $O_2^{\bullet-}$ is converted to H_2O_2 in the mitochondria by superoxide dismutases (Mn–SOD within the matrix and Cu, Zn–SOD within the intermembrane space). Typically, most of this H_2O_2 diffuses out of the mitochondria where it is neutralized by glutathione peroxidase and peroxiredoxins. Nondismutated superoxide does not diffuse from the mitochondria due to its charge, and thus these radicals can react with nearby biomolecules such as the citric acid cycle enzyme aconitase. Once this reaction occurs, aconitase releases iron from its 4Fe–S cluster, not only inactivating the enzyme but also freeing iron to undergo Fenton chemistry [161]. Together, these intramitochondrial ROS can inhibit components of the mitochondrial respiratory chain, resulting in mitochondrial dysfunction and neuronal bioenergetic failure that can further drive neurodegeneration.

Activated microglia also overexpress inducible nitric oxide synthase (iNOS) to produce nitric oxide (NO) through the catalytic reduction of L-arginine. NO readily binds to O_2^- in the interstitial space to produce the extremely active oxidizing peroxynitrite anions ($ONOO^-$) [162]. Peroxynitrite can penetrate into neurons through anion exchanges where they oxidize, peroxidize, nitrosylate, and nitrate lipids, proteins, and nucleic acids [163]. Treatment with L-N(G)-nitroarginine, a selective inhibitor of nitric oxide synthase, rescued dopaminergic neurons from inflammation-mediated neurodegeneration—implicating that NO also plays a fundamental role in neurodegeneration [164,165]. Furthermore, neurons are thought to generate additional intracellular NO through the oxidative stress-mediated activation of neuronal nitric oxide synthase (nNOS) and through the reduction of NO_2^- by complex IV of the ETC. Increased expressions of both iNOS and nNOS have been observed in neurodegenerative disorders such as Alzheimer's disease and Parkinson's disease. Together, this increase in intracellular reactive nitrogen species (RNS) can modify a discrete set of mitochondrial proteins, and inhibit the mitochondrial respiratory chain, thus causing mitochondrial dysfunction and neuronal bioenergetic failure. In Parkinson's disease, RNS are thought to nitrosylate α-synuclein to form the pathogenic aggregates found in Lewy bodies [166]. Additionally, RNS simultaneously suppress lysosomal degradation by nitrosylating ubiquitin E3 ligase, parkin, and protein–disulfide isomerase [167]. These posttranslational protein adducts not only inhibit protein function, but are important pathophysiological mechanisms associated with the progression of neurodegeneration, by promoting protein misfolding into insoluble fibrils.

Mitochondrial dysfunction is present in neurodegenerative diseases, whereby metabolic changes are among the earliest clinically detectable hallmarks of neurodegeneration along with neuroinflammation [168]. Mitochondria seem to have two major roles in neurodegeneration, one that involves chronic mitochondrial dysfunction as a driver of progressive altered metabolism (and perhaps progressive degeneration), and another associated with apoptosis (and as such, the loss of cells). In terms of mitochondrial dysfunction, inhibition of the ETC is achieved using the neurotoxicants malonate and 3-nitropropionic acid (inhibiting complex II), and 1-methyl-4-phenyl-1,2,3,6-tetrahydropyridine (MPTP) and rotenone (inhibiting complex I) are widely used in animal models to induce huntingtonian and parkinsonian pathologies, respectively. As the acute nature of the mitochondrial failure achieved in these models does not recapitulate the progressive nature of these diseases, it is feasible to consider that mild but perhaps long-lasting and chronic changes in mitochondrial function are more relevant to the progression of neurodegeneration. Better animal models, in which mitochondria are not affected as drastically, could be a more informative means of addressing this question. It is important to keep in mind that as of today, the extent of mitochondrial dysfunction necessary to affect neuronal function and drive progressive neurodegeneration is unclear, and whether mitochondrial dysfunction in neurodegenerative diseases is a cause, consequence, or both is unclear as well. Irrespective of these open questions, current models of mitochondrial dysfunction point to the important role of energy homeostasis in the brain, and allow for the posing of questions about the long-term effects of changes in metabolism and bioenergetics on cellular function and brain health.

In trying to understand the broad impact of mitochondrial dysfunction on cellular homeostasis, one may consider a simple scenario in which the mitochondrial genome is damaged. As the mitochondrial DNA encodes 13 critical protein components of the ETC, genetic mutations or oxidative damage to mitochondrial DNA can decrease ETC activity, affecting the ability of mitochondria to maintain their membrane potential ($\Delta\Psi$m). Since the $\Delta\Psi$m is established by pumping out protons at complexes I, III, and IV of the ETC, direct oxidative damage or inhibition to these complexes can result in the loss of $\Delta\Psi$m. Decreased $\Delta\Psi$m can further affect mitochondrial function, by impairing the import of more than 1500 nuclear DNA-encoded proteins to mitochondria. As mentioned above, the mitochondrial genome encodes components of the ETC, and thus proteins encoded in the nucleus have functions varying from mtDNA metabolism, to antioxidant defenses (MnSOD), iron homeostasis (frataxin), and fatty acid oxidation (acyl-CoA dehydrogenase). Thus, the loss of $\Delta\Psi$m can sustain a cycle of dysfunction. In addition, decreased $\Delta\Psi$m impairs calcium buffering in mitochondria, further impacting neuronal bioenergetics because the citric acid cycle enzymes pyruvate dehydrogenase phosphatase, isocitrate dehydrogenase, and α-ketoglutarate dehydrogenase are regulated by calcium [169]. Loss of $\Delta\Psi$m may further impair autophagy and clearance of dysfunctional mitochondria [170], which are required for maintenance of the mitochondrial network function within cells. Maintenance of the homeostasis of the mitochondrial network within cells is especially important in neurons, given their size and the length of axons. The loss of Parkin or Pink-1, two proteins intimately involved in mitochondrial dynamics and clearance, is a genetic risk factor for Parkinson's disease [171]. Thus, it is easy to see how perturbation of one aspect of mitochondrial function may have a broader impact on cellular homeostasis.

Severe mitochondrial abnormalities are incompatible with life [172]. Thus, human and animal studies of mitochondrial diseases may give the greatest understanding of the pathophysiological impacts of chronic mitochondrial dysfunction on human health. In this context, studies on mitochondrial disorders in which OXPHOS activity is diminished have highlighted the progressive development of many symptoms shared with neurodegenerative disease, such as the loss of neurons and neuronal function, deficits in muscle coordination and strength, sensory impairments, learning disabilities, gastrointestinal problems, diabetes, autonomic dysfunction, and dementia [172]. Although many of these studies involved mutations that altered the integrity or function of the OXPHOS complexes during infancy [173], one can envision that de novo genetic mutations or oxidative stress-derived adducts, perhaps resulting from chronic inflammation, can produce mild but chronic mitochondrial dysfunction that could theoretically give rise to delayed and progressive neurodegeneration in adulthood, such as Alzheimer's disease and Parkinson's disease. Hence, studies focusing on models in which mitochondrial dysfunction is secondary, but chronically affected, may be more relevant in understanding the role of mitochondria in the progressive nature of neurodegeneration. Models of chronic neuroinflammation are available and have demonstrated mitochondrial dysfunction, and thus could serve as starting points for these analyses.

As mentioned above, while the impact of mild but chronic mitochondrial dysfunction on neuronal physiology is still poorly understood, the role that mitochondria have in promoting cell loss is well established. Neuronal death due to oxidative and nitrosylative stress is driven by apoptosis triggered from or mediated by mitrochondrial dysfunction, bioenergetic failure, and excitotoxicity, which become progressively worse with age. These apoptotic driving factors are well established in all neurodegenerative diseases.

Accumulating evidence reveals that increased bioenergetic failure is an early hallmark of neurodegenerative diseases. Neurons are particularly vulnerable to damage from hypometabolic stress due to their high energetic demands and limited energetic reserves. Glucose and ketone bodies represent the primary and secondary energy compounds that fuel the brain. Bioenergetic failure is thought to occur during chronic neuroinflammation associated with neurodegenerative diseases that are caused by regional changes in protein expression patterns in mitochondria that prevent pyruvate and eventually ketone metabolism, and inevitably lead to apoptosis. Glucose hypometabolism is thought to occur early in Alzheimer's, Parkinson's, and Huntington's diseases—shifting to ketosis as the primary metabolic pathway to fuel the brain through lipid oxidation and mitochondrial respiration. In fact, shifting mice to a ketogenic diet resulted in significant pathophysiological improvements in experimental models of Alzheimer's disease [174], multiple sclerosis [175], and ALS [176]. Whether the shift to ketone body metabolism also modulates the level of neuroinflammation or oxidative stress is still controversial.

Inflammation coupled with nitrosylative stress is thought to drive glutamatergic excitotoxicity in neurodegeneration. The inability of activated astrocytes, during inflammation, to uptake and convert glutamate to glutamine at tripartite synapse is thought to result in the hyperaccumulation of glutamate at synapses resulting in excessive activation of glutamate receptors. Although typically these receptors can become desensitized or undergo endocytosis, the inotropic glutamate receptor N-methyl-D-aspartate (NMDA) can undergo a conformational shift through the nitrosylation of an intracellular modulatory domain by NO, allowing for free entry of extracellular calcium to continuously enter the neuron. In order to maintain low concentrations of cytosolic calcium, neurons must expend a great deal of energy to pump calcium out of the cell or into endoplasmic reticulum and mitochondrial stores, eventually leading to collapse of the $\Delta\Psi m$, and the opening of the mitochondrial permeability transition pore, resulting in apoptosis [177].

THERAPEUTIC IMPLICATIONS: BREAKING THE CYCLE

The discovery that greater intake of dietary antioxidants such as vitamins C and E, β-carotene, and coenzyme Q, as well as the long-term administration of nonsteroidal anti-inflammatory drugs, reduced the risk of developing neurodegenerative diseases prompted the investigation of anti-inflammatory therapies as potential treatments [178–180]. However, when these strategies were implemented in human clinical settings they overwhelmingly ended in failure due to their inability to halt the progression of neurodegeneration, and increased risks of hemorrhagic stroke and mortality. The failure of these anti-inflammatory therapies is likely attributable to their inability to interrupt the positive feedback loop associated with chronic neuroinflammation. This is primarily because these strategies prevent or sequester downstream proinflammatory compounds without addressing the source of the inflammation [181]. For this reason, many laboratories have been screening compounds to target surface pattern recognition receptors such as TLR-4 in order to modulate chronic neuroinflammation [182]—yet this strategy may also be ineffective due to the redundancy of surface pattern recognition receptors that can detect the same DAMPs.

The inflammatory hypothesis assumes that effective strategies targeting the sources of chronic neuroinflammation and breaking the positive feedback loop associated with inflammation-driven neurodegeneration can halt the progression of neurodegenerative diseases. If this is indeed the case, then by definition neurodegenerative risk factors that reduce the humoral cytokine threshold and thereby induce chronic neuroinflammation may be therapeutically modifiable. Both NADPH oxidase and iNOS are necessary to generate an oxidative burst of diffusible $O_2^{\bullet-}$ and NO that generates the oxygen radicals, nitrogen radicals, and peroxynitrite species directly involved in promoting neuronal oxidative and nitrosylative stress. Targeting the catalytic production of $O_2^{\bullet-}$ and/or NO can effectively reduce the toxicity generated by chronic neuroinflammation. Mounting evidence suggests that inhibition of the catalytic subunit of NADPH oxidase is also important in reducing proinflammatory signal transduction by autocrine signaling of lipid-diffusible H_2O_2 into microglia—effectively blocking the release of proinflammatory factors. Furthermore, NADPH oxidase interacts directly with most surface pattern recognition receptors associated with DAMP recognition—thus effectively blocking microglial reactivation by endogenous neuronal distress signals.

Our laboratory has shown that selectively inhibiting the catalytic subunit of NADPH oxidase (gp91^{phox}) prevents neuroinflammation-driven degeneration of dopaminergic neurons of the substantia nigra in several mouse models of Parkinson's disease. Furthermore, we have shown that suppressing superoxide production significantly reduces neuronal oxidative stress as detected with a loss of dihydroethidium staining in neurons—including punctate localization of ROS in neuronal mitochondria [183]. Together, these findings indicate that suppression of NADPH oxidase-generated ROS can reduce the oxidative stress and mitochondrial dysfunction that drive neuronal loss in neurodegenerative diseases. Among the greatest obstacles yet to be overcome is developing NADPH oxidase inhibitors that can cross the blood–brain barrier, are highly efficacious and specific, and are incapable of permanently inhibiting the respiratory burst associated with typical immune responses to pathogen detection. Our laboratory is currently in the process of screening several promising compounds that fit these exclusion criteria.

CONCLUSIONS

The inflammatory hypothesis of neurodegenerative diseases continues to be strengthened by accumulating research and clinical evidence. Beyond having a key role in accelerating the progression of neurodegenerative diseases, we have shown that global and nuclei-specific chronic neuroinflammation can generate disease-specific protein lesions that can travel long distances in the brain. The prevalence of neurodegenerative disease onset is disproportionately attributed to infection or chronic inflammatory disorders due to variable susceptibility risk factors that range from age to genetic polymorphisms. Lastly, the failure of current anti-inflammatory therapeutics to halt the progression of neurodegenerative diseases is due to their inability to suppress the source of inflammation. Evidence strongly supports the hypothesis that neuroinflammation in neurodegenerative diseases is not only pathogenic but also involved in the etiology of these diseases.

References

[1] Gao HM, Hong JS. Why neurodegenerative diseases are progressive: uncontrolled inflammation drives disease progression. Trends Immunol 2008;29(8):357–65.

[2] Liu B, Hong JS. Role of microglia in inflammation-mediated neurodegenerative diseases: mechanisms and strategies for therapeutic intervention. J Pharmacol Exp Ther 2003;304(1):1–7.

[3] Prinz M, Priller J. Microglia and brain macrophages in the molecular age: from origin to neuropsychiatric disease. Nat Rev Neurosci 2014;15(5):300–12.

[4] Block ML, Zecca L, Hong J-S. Microglia-mediated neurotoxicity: uncovering the molecular mechanisms. Nat Rev Neurosci 2007;8(1):57–69.

[5] Nicolson GL, Haier J. Role of chronic bacterial and viral infections in neurodegenerative, neurobehavioral, psychiatric, autoimmune and fatiguing illnesses: Part 1. BJMP 2009;2(4):20–8.

[6] Dunn N, Mullee M, Perry VH, Holmes C. Association between dementia and infectious disease: evidence from a case-control study. Alzheimer Dis Assoc Disord 2005;19(2):91–4.

[7] Sibley WA, Bamford CR, Clark K. Clinical viral infections and multiple sclerosis. Lancet 1985;1(8441):1313–5.

[8] Johnson FO, Atchison WD. The role of environmental mercury, lead and pesticide exposure in development of amyotrophic lateral sclerosis. Neurotoxicology 2009;30(5):761–5.

[9] Cannon JR, Greenamyre JT. The role of environmental exposures in neurodegeneration and neurodegenerative diseases. Toxicol Sci Off J Soc Toxicol 2011;124(2):225–50.

[10] Fox JH, Kama JA, Lieberman G, Chopra R, Dorsey K, Chopra V, et al. Mechanisms of copper ion mediated Huntington's disease progression. PloS One 2007;2(3):e334.

[11] Alasia DD, Asekomeh GA, Unachuku CN. Parkinsonism induced by sepsis: a case report. Niger J Med J Natl Assoc Resid Dr Niger 2006;15(3):333–6.

[12] Nobrega AC, Rodrigues B, Melo A. Is silent aspiration a risk factor for respiratory infection in Parkinson's disease patients? Park Relat Disord 2008;14(8):646–8.

[13] Arai H, Furuya T, Mizuno Y, Mochizuki H. Inflammation and infection in Parkinson's disease. Histol Histopathol 2006;21(6):673–8.

[14] Harris MA, Tsui JK, Marion SA, Shen H, Teschke K. Association of Parkinson's disease with infections and occupational exposure to possible vectors. Mov Disord Off J Mov Disord Soc 2012;27(9):1111–7.

[15] Vlajinac H, Dzoljic E, Maksimovic J, Marinkovic J, Sipetic S, Kostic V, . Infections as a risk factor for Parkinson's disease: a case–control study. Int J Neurosci 2013;123(5):329–332.

[16] Giannetti P, Politis M, Su P, Turkheimer F, Malik O, Keihaninejad S, et al. Microglia activation in multiple sclerosis black holes predicts outcome in progressive patients: an in vivo [(11)C](R)-PK11195-PET pilot study. Neurobiol Dis 2014;65:203–10.

[17] Venneti S, Lopresti BJ, Wang G, Hamilton RL, Mathis CA, Klunk WE, et al. PK11195 labels activated microglia in Alzheimer's disease and in vivo in a mouse model using PET. Neurobiol Aging 2009;30(8):1217–26.

[18] Maia S, Arlicot N, Vierron E, Bodard S, Vergote J, Guilloteau D, et al. Longitudinal and parallel monitoring of neuroinflammation and neurodegeneration in a 6-hydroxydopamine rat model of Parkinson's disease. Synapse 2012;66(7):573–83.

[19] Gimenez-Llort L, Torres-Lista V, la Fuente MD. Crosstalk between behavior and immune system during the prodromal stages of Alzheimer's disease. Curr Pharm Des 2014;20(29):4723–4732.

[20] Fiala M, Chattopadhay M, La Cava A, Tse E, Liu G, Lourenco E, et al. IL-17A is increased in the serum and in spinal cord CD8 and mast cells of ALS patients. J Neuroinflammation 2010;7:76.

[21] Shi N, Kawano Y, Tateishi T, Kikuchi H, Osoegawa M, Ohyagi Y, et al. Increased IL-13-producing T cells in ALS: positive correlations with disease severity and progression rate. J Neuroimmunol 2007;182(1–2):232–5.

[22] Kuhle J, Lindberg RL, Regeniter A, Mehling M, Steck AJ, Kappos L, et al. Increased levels of inflammatory chemokines in amyotrophic lateral sclerosis. Eur J Neurol Off J Eur Fed Neurol Soc 2009;16(6):771–4.

[23] Moreau C, Devos D, Brunaud-Danel V, Defebvre L, Perez T, Destée A, et al. Elevated IL-6 and TNF-α levels in patients with ALS: inflammation or hypoxia? Neurology 2005;65(12):1958–60.

[24] Keizman D, Rogowski O, Berliner S, Ish-Shalom M, Maimon N, Nefussy B, et al. Low-grade systemic inflammation in patients with amyotrophic lateral sclerosis. Acta Neurol Scand 2009;119(6):383–9.

[25] Holmoy T, Roos PM, Kvale EO. ALS: cytokine profile in cerebrospinal fluid T-cell clones. Amyotroph lateral Scler Off Publ World Fed Neurol Res Group Mot Neuron Dis 2006;7(3):183–6.

[26] Turner MR, Cagnin A, Turkheimer FE, Miller CC, Shaw CE, Brooks DJ, et al. Evidence of widespread cerebral microglial activation in amyotrophic lateral sclerosis: an [11C](R)-PK11195 positron emission tomography study. Neurobiol Dis 2004;15(3):601–9.

[27] Swardfager W, Lanctot K, Rothenburg L, Wong A, Cappell J, Herrmann N. A meta-analysis of cytokines in Alzheimer's disease. Biol Psychiatry 2010;68(10):930–41.

[28] Kreisl WC, Lyoo CH, McGwier M, Snow J, Jenko KJ, Kimura N, et al. In vivo radioligand binding to translocator protein correlates with severity of Alzheimer's disease. Brain J Neurol 2013;136(Pt 7):2228–38.

[29] Nguyen HP, Bjorkqvist M, Bode FJ, Stephan M, von Horsten S. Serum levels of a subset of cytokines show high interindividual variability and are not altered in rats transgenic for Huntington's disease. PLoS Curr 2010;2:RRN1190.

[30] Gaus SE, Lin L, Mignot E. CSF hypocretin levels are normal in Huntington's disease patients. Sleep 2005;28(12):1607–8.

[31] Pavese N, Gerhard A, Tai YF, Ho AK, Turkheimer F, Barker RA, et al. Microglial activation correlates with severity in Huntington disease: a clinical and PET study. Neurology 2006;66(11):1638–43.

[32] Vrethem M, Kvarnstrom M, Stenstam J, Cassel P, Gustafsson M, Landtblom AM, et al. Cytokine mapping in cerebrospinal fluid and blood in multiple sclerosis patients without oligoclonal bands. Mult Scler 2012;18(5):669–73.

[33] Hollifield RD, Harbige LS, Pham-Dinh D, Sharief MK. Evidence for cytokine dysregulation in multiple sclerosis: peripheral blood mononuclear cell production of pro-inflammatory and anti-inflammatory cytokines during relapse and remission. Autoimmunity 2003;36(3):133–41.

[34] Hauser SL, Doolittle TH, Lincoln R, Brown RH, Dinarello CA. Cytokine accumulations in CSF of multiple sclerosis patients: frequent detection of interleukin-1 and tumor necrosis factor but not interleukin-6. Neurology 1990;40(11):1735–9.

[35] Reale M, Iarlori C, Thomas A, Gambi D, Perfetti B, Di Nicola M, et al. Peripheral cytokines profile in Parkinson's disease. Brain Behav Immun 2009;23(1):55–63.

[36] Blum-Degen D, Muller T, Kuhn W, Gerlach M, Przuntek H, Riederer P. Interleukin-1 beta and interleukin-6 are elevated in the cerebrospinal fluid of Alzheimer's and de novo Parkinson's disease patients. Neurosci Lett 1995;202(1–2):17–20.

[37] Turner MR, Goldacre R, Ramagopalan S, Talbot K, Goldacre MJ. Autoimmune disease preceding amyotrophic lateral sclerosis: an epidemiologic study. Neurology 2013;81(14):1222–5.

[38] Kamer AR, Craig RG, Dasanayake AP, Brys M, Glodzik-Sobanska L, de Leon MJ. Inflammation and Alzheimer's disease: possible role of periodontal diseases. Alzheimer's Dementia J Alzheimer's Assoc 2008;4(4):242–50.

[39] Luchsinger JA. Adiposity, hyperinsulinemia, diabetes and Alzheimer's disease: an epidemiological perspective. Eur J Pharmacol 2008;585(1):119–29.

[40] Breteler MM. Vascular risk factors for Alzheimer's disease: an epidemiologic perspective. Neurobiol Aging 2000;21(2):153–60.

[41] Kodl CT, Seaquist ER. Cognitive dysfunction and diabetes mellitus. Endocr Rev 2008;29(4):494–511.

[42] Qiu C, De Ronchi D, Fratiglioni L. The epidemiology of the dementias: an update. Curr Opin Psychiatry 2007;20(4):380–5.

[43] Farrer LA. Diabetes mellitus in Huntington disease. Clin Genet 1985;27(1):62–7.

[44] Saft C, Andrich JE, Muller T, Becker J, Jackowski J. Oral and dental health in Huntington's disease – an observational study. BMC Neurol 2013;13(1):114.

[45] Sheu JJ, Lin HC. Association between multiple sclerosis and chronic periodontitis: a population-based pilot study. Eur J Neurol Off J Eur Fed Neurol Soc 2013;20(7):1053–9.

[46] Ramagopalan SV, Dyment DA, Ebers GC, Sadovnick AD. Gestational diabetes and multiple sclerosis. Epidemiology 2009;20(5):783–4.

[47] Hedstrom AK, Lima Bomfim I, Barcellos L, Gianfrancesco M, Schaefer C, Kockum I, et al. Interaction between adolescent obesity and HLA risk genes in the etiology of multiple sclerosis. Neurology 2014;82(10):865–72.

[48] International Multiple Sclerosis Genetics C, Welcome Trust Case Control C, Sawcer S, Hellenthal G, Pirinen M, Spencer CC, et al. Genetic risk and a primary role for cell-mediated immune mechanisms in multiple sclerosis. Nature 2011;476(7359):214–9.

[49] Hu G, Jousilahti P, Bidel S, Antikainen R, Tuomilehto J. Type 2 diabetes and the risk of Parkinson's disease. Diabetes Care 2007;30(4):842–7.

[50] Weller C, Oxlade N, Dobbs SM, Dobbs RJ, Charlett A, Bjarnason IT. Role of inflammation in gastrointestinal tract in aetiology and pathogenesis of idiopathic parkinsonism. FEMS Immunol Med Microbiol 2005;44(2):129–35.

[51] Cicciu M, Risitano G, Lo Giudice G, Bramanti E. Periodontal health and caries prevalence evaluation in patients affected by Parkinson's disease. Parkinson's Dis 2012;2012:541908.

[52] Carter C. Alzheimer's disease: APP, gamma secretase, APOE, CLU, CR1, PICALM, ABCA7, BIN1, CD2AP, CD33, EPHA1, and MS4A2, and their relationships with herpes simplex, C. Pneumoniae, other suspect pathogens, and the immune system. Int J Alzheimer's Dis 2011;2011:501862.

[53] Hazrati LN, Van Cauwenberghe C, Brooks PL, Brouwers N, Ghani M, Sato C, et al. Genetic association of CR1 with Alzheimer's disease: a tentative disease mechanism. Neurobiol Aging 2012;33(12):2949, e5–e12.

[54] Neumann H, Daly MJ. Variant TREM2 as risk factor for Alzheimer's disease. N Engl J Med 2013;368(2):182–4.

[55] Hamza TH, Zabetian CP, Tenesa A, Laederach A, Montimurro J, Yearout D, et al. Common genetic variation in the HLA region is associated with late-onset sporadic Parkinson's disease. Nat Genet 2010;42(9):781–5.

[56] Moehle MS, Webber PJ, Tse T, Sukar N, Standaert DG, DeSilva TM, et al. LRRK2 inhibition attenuates microglial inflammatory responses. J Neurosci Off J Soc Neurosci 2012;32(5):1602–11.

[57] McGeer PL, McGeer EG. Polymorphisms in inflammatory genes and the risk of Alzheimer disease. Archives Neurol 2001;58(11):1790–2.

[58] Ramos EM, Lin MT, Larson EB, Maezawa I, Tseng LH, Edwards KL, et al. Tumor necrosis factor alpha and interleukin 10 promoter region polymorphisms and risk of late-onset Alzheimer disease. Archives Neurol 2006;63(8):1165–9.

[59] Wahner AD, Sinsheimer JS, Bronstein JM, Ritz B. Inflammatory cytokine gene polymorphisms and increased risk of Parkinson disease. Arch Neurol 2007;64(6):836–40.

[60] Van Amersfoort ES, Van Berkel TJ, Kuiper J. Receptors, mediators, and mechanisms involved in bacterial sepsis and septic shock. Clin Microbiol Rev 2003;16(3):379–414.

[61] Guo JL, Lee VM. Cell-to-cell transmission of pathogenic proteins in neurodegenerative diseases. Nat Med 2014;20(2):130–8.

[62] Castano A, Herrera AJ, Cano J, Machado A. The degenerative effect of a single intranigral injection of LPS on the dopaminergic system is prevented by dexamethasone, and not mimicked by rh-TNF-alpha, IL-1beta and IFN-gamma. J Neurochem 2002;81(1):150–7.

[63] Herrera AJ, Castano A, Venero JL, Cano J, Machado A. The single intranigral injection of LPS as a new model for studying the selective effects of inflammatory reactions on dopaminergic system. Neurobiol Dis 2000;7(4):429–47.

[64] Machado A, Herrera AJ, Venero JL, Santiago M, de Pablos RM, Villarán RF, et al. Inflammatory animal model for Parkinson's disease: the intranigral injection of LPS induced the inflammatory process along with the selective degeneration of nigrostriatal dopaminergic neurons. ISRN Neurol 2011;2011:476158.

[65] Zhou HF, Liu XY, Niu DB, Li FQ, He QH, Wang XM. Triptolide protects dopaminergic neurons from inflammation-mediated damage induced by lipopolysaccharide intranigral injection. Neurobiol Dis 2005;18(3):441–9.

[66] Lu X, Bing G, Hagg T. Naloxone prevents microglia-induced degeneration of dopaminergic substantia nigra neurons in adult rats. Neuroscience 2000;97(2):285–91.

[67] Zhang J, Stanton DM, Nguyen XV, Liu M, Zhang Z, Gash D, et al. Intrapallidal lipopolysaccharide injection increases iron and ferritin levels in glia of the rat substantia nigra and induces locomotor deficits. Neuroscience 2005;135(3):829–38.

[68] Choi DY, Zhang J, Bing G. Aging enhances the neuroinflammatory response and alpha-synuclein nitration in rats. Neurobiol Aging 2010;31(9):1649–53.

[69] Choi DY, Liu M, Hunter RL, Cass WA, Pandya JD, Sullivan PG, et al. Striatal neuroinflammation promotes Parkinsonism in rats. PloS One 2009;4(5):e5482.

[70] Hunter RL, Cheng B, Choi DY, Liu M, Liu S, Cass WA, et al. Intrastriatal lipopolysaccharide injection induces parkinsonism in C57/B6 mice. J Neurosci Res 2009;87(8):1913–21.

[71] Hunter RL, Choi DY, Kincer JF, Cass WA, Bing G, Gash DM. Fenbendazole treatment may influence lipopolysaccharide effects in rat brain. Comp Med 2007;57(5):487–92.

[72] Hunter RL, Dragicevic N, Seifert K, Choi DY, Liu M, Kim HC, et al. Inflammation induces mitochondrial dysfunction and dopaminergic neurodegeneration in the nigrostriatal system. J Neurochem 2007;100(5): 1375–86.

[73] Nimmervoll B, White R, Yang JW, An S, Henn C, Sun JJ, et al. LPS-induced microglial secretion of TNFα increases activity-dependent neuronal apoptosis in the neonatal cerebral cortex. Cereb Cortex 2013;23(7):1742–55.

[74] de Pablos RM, Villaran RF, Arguelles S, Herrera AJ, Venero JL, Ayala A, et al. Stress increases vulnerability to inflammation in the rat prefrontal cortex. J Neurosci Off J Soc Neurosci 2006;26(21):5709–19.

[75] Lee DC, Rizer J, Selenica ML, Reid P, Kraft C, Johnson A, et al. LPS-induced inflammation exacerbates phospho-tau pathology in rTg4510 mice. J Neuroinflammation 2010;7:56.

[76] Park KW, Lee DY, Joe EH, Kim SU, Jin BK. Neuroprotective role of microglia expressing interleukin-4. J Neurosci Res 2005;81(3):397–402.

[77] Zhu F, Zhang L, Ding YQ, Zhao J, Zheng Y. Neonatal intrahippocampal injection of lipopolysaccharide induces deficits in social behavior and prepulse inhibition and microglial activation in rats: implication for a new schizophrenia animal model. Brain Behav Immun 2014;38:166–74.

[78] Pintado C, Gavilan MP, Gavilan E, García-Cuervo L, Gutiérrez A, Vitorica J, et al. Lipopolysaccharide-induced neuroinflammation leads to the accumulation of ubiquitinated proteins and increases susceptibility to neurodegeneration induced by proteasome inhibition in rat hippocampus. J Neuroinflammation 2012;9(1):87.

[79] Herber DL, Maloney JL, Roth LM, Freeman MJ, Morgan D, Gordon MN. Diverse microglial responses after intrahippocampal administration of lipopolysaccharide. Glia 2006;53(4):382–91.

[80] DiCarlo G, Wilcock D, Henderson D, Gordon M, Morgan D. Intrahippocampal LPS injections reduce Aβ load in APP+PS1 transgenic mice. Neurobiol Aging 2001;22(6):1007–12.

[81] Ma TC, Zhu XZ. Suppression of lipopolysaccharide-induced impairment of active avoidance and interleukin-6-induced increase of prostaglandin E2 release in rats by indometacin. Arzneim 1997;47(5):595–7.

[82] Pan W, Kastin AJ. TNFalpha transport across the blood–brain barrier is abolished in receptor knockout mice. Exp Neurol 2002;174(2):193–200.

[83] Kessler W, Diedrich S, Menges P, Ebker T, Nielson M, Partecke LI, et al. The role of the vagus nerve: modulation of the inflammatory reaction in murine polymicrobial sepsis. Mediat Inflamm 2012;2012:467620.

[84] Qin L, Wu X, Block ML, Liu Y, Breese GR, Hong HS, et al. Systemic LPS causes chronic neuroinflammation and progressive neurodegeneration. Glia 2007;55(5):453–62.

[85] Block ML, Hong JS. Chronic microglial activation and progressive dopaminergic neurotoxicity. Biochem Soc Trans 2007;35(Pt 5):1127–32.

[86] Bilbo SD, Schwarz JM. Early-life programming of later-life brain and behavior: a critical role for the immune system. Front Behav Neurosci 2009;3:14.

[87] Wong WT. Microglial aging in the healthy CNS: phenotypes, drivers, and rejuvenation. Front Cell Neurosci 2013:7.

[88] Ling Z, Gayle DA, Ma SY, Lipton JW, Tong CW, Hong JS, et al. In utero bacterial endotoxin exposure causes loss of tyrosine hydroxylase neurons in the postnatal rat midbrain. Mov Disord Off J Mov Disord Soc 2002;17(1):116–24.

[89] Ling Z, Chang QA, Tong CW, Leurgans SE, Lipton JW, Carvey PM. Rotenone potentiates dopamine neuron loss in animals exposed to lipopolysaccharide prenatally. Exp Neurol 2004;190(2):373–83.

[90] Ling ZD, Chang Q, Lipton JW, Tong CW, Landers TM, Carvey PM. Combined toxicity of prenatal bacterial endotoxin exposure and postnatal 6-hydroxydopamine in the adult rat midbrain. Neuroscience 2004;124(3):619–28.

[91] Golan HM, Lev V, Hallak M, Sorokin Y, Huleihel M. Specific neurodevelopmental damage in mice offspring following maternal inflammation during pregnancy. Neuropharmacology 2005;48(6):903–17.

[92] Frank-Cannon TC, Tran T, Ruhn KA, Martinez TN, Hong J, Marvin M, et al. Parkin deficiency increases vulnerability to inflammation-related nigral degeneration. J Neurosci Off J Soc Neurosci 2008;28(43):10825–34.

[93] Gao HM, Zhang F, Zhou H, Kam W, Wilson B, Hong JS. Neuroinflammation and alpha-synuclein dysfunction potentiate each other, driving chronic progression of neurodegeneration in a mouse model of Parkinson's disease. Environ Health Perspect 2011;119(6):807–14.

[94] Sly LM, Krzesicki RF, Brashler JR, Buhl AE, McKinley DD, Carter DB, et al. Endogenous brain cytokine mRNA and inflammatory responses to lipopolysaccharide are elevated in the Tg2576 transgenic mouse model of Alzheimer's disease. Brain Res Bull 2001;56(6):581–8.

[95] Lee J, Chan SL, Mattson MP. Adverse effect of a presenilin-1 mutation in microglia results in enhanced nitric oxide and inflammatory cytokine responses to immune challenge in the brain. Neuromolecular Med 2002;2(1):29–45.

[96] Kitazawa M, Oddo S, Yamasaki TR, Green KN, LaFerla FM. Lipopolysaccharide-induced inflammation exacerbates tau pathology by a cyclin-dependent kinase 5-mediated pathway in a transgenic model of Alzheimer's disease. J Neurosci Off J Soc Neurosci 2005;25(39):8843–53.

[97] Nguyen MD, D'Aigle T, Gowing G, Julien JP, Rivest S. Exacerbation of motor neuron disease by chronic stimulation of innate immunity in a mouse model of amyotrophic lateral sclerosis. J Neurosci 2004;24(6):1340–9.

[98] Ramaglia V, Hughes TR, Donev RM, Ruseva MM, Wu X, Huitinga I, et al. C3-dependent mechanism of microglial priming relevant to multiple sclerosis. Proc Natl Acad Sci USA 2012;109(3):965–70.

[99] Cunningham C, Wilcockson DC, Campion S, Lunnon K, Perry VH. Central and systemic endotoxin challenges exacerbate the local inflammatory response and increase neuronal death during chronic neurodegeneration. J Neurosci 2005;25(40):9275–84.

II. NEURODEGENERATIVE DISORDERS

[100] Kubera M, Curzytek K, Duda W, Leskiewicz M, Basta-Kaim A, Budziszewska B, et al. A new animal model of (chronic) depression induced by repeated and intermittent lipopolysaccharide administration for 4 months. Brain Behav Immun 2013;31:96–104.

[101] He Q, Yu W, Wu J, Chen C, Lou Z, Zhang Q, et al. Intranasal LPS-mediated Parkinson's model challenges the pathogenesis of nasal cavity and environmental toxins. PloS One 2013;8(11):e78418.

[102] Villaran RF, Espinosa-Oliva AM, Sarmiento M, De Pablos RM, Argüelles S, Delgado-Cortés MJ, et al. Ulcerative colitis exacerbates lipopolysaccharide-induced damage to the nigral dopaminergic system: potential risk factor in Parkinson's disease. J Neurochem 2010;114(6):1687–700.

[103] Palin K, Cunningham C, Forse P, Perry VH, Platt N. Systemic inflammation switches the inflammatory cytokine profile in CNS Wallerian degeneration. Neurobiol Dis 2008;30(1):19–29.

[104] Reichenberg A, Yirmiya R, Schuld A, Kraus T, Haack M, Morag A, et al. Cytokine-associated emotional and cognitive disturbances in humans. Archives General Psychiatry 2001;58(5):445–52.

[105] Brydon L, Harrison NA, Walker C, Steptoe A, Critchley HD. Peripheral inflammation is associated with altered substantia nigra activity and psychomotor slowing in humans. Biol Psychiatry 2008;63(11):1022–9.

[106] Kammula US, White DE, Rosenberg SA. Trends in the safety of high dose bolus interleukin-2 administration in patients with metastatic cancer. Cancer 1998;83(4):797–805.

[107] Capuron L, Gumnick JF, Musselman DLLawson DH, Reemsnyder A, Nemeroff CB, et al. Neurobehavioral effects of interferon-alpha in cancer patients: phenomenology and paroxetine responsiveness of symptom dimensions. Neuropsychopharmacol Off Publ Am Coll Neuropsychopharmacol 2002;26(5):643–52.

[108] Tyring S, Gottlieb A, Papp K, Gordon K, Leonardi C, Wang A, et al. Etanercept and clinical outcomes, fatigue, and depression in psoriasis: double-blind placebo-controlled randomised phase III trial. Lancet 2006;367(9504):29–35.

[109] Dickens C, McGowan L, Clark-Carter D, Creed F. Depression in rheumatoid arthritis: a systematic review of the literature with meta-analysis. Psychosom Med 2002;64(1):52–60.

[110] Graff LA, Walker JR, Bernstein CN. Depression and anxiety in inflammatory bowel disease: a review of comorbidity and management. Inflamm Bowel Dis 2009;15(7):1105–18.

[111] Celano CM, Huffman JC. Depression and cardiac disease: a review. Cardiol Rev 2011;19(3):130–42.

[112] Moreau M, Andre C, O'Connor JC, Dumich SA, Woods JA, Kelley KW, et al. Inoculation of Bacillus Calmette-Guerin to mice induces an acute episode of sickness behavior followed by chronic depressive-like behavior. Brain Behav Immun 2008;22(7):1087–95.

[113] Valles A, Marti O, Armario A. Mapping the areas sensitive to long-term endotoxin tolerance in the rat brain: a c-fos mRNA study. J Neurochem 2005;93(5):1177–88.

[114] Reyes EP, Abarzua S, Martin A, Rodriguez J, Cortes PP, Fernandez R. LPS-induced c-Fos activation in NTS neurons and plasmatic cortisol increases in septic rats are suppressed by bilateral carotid chemodenervation. Adv Exp Med Biol 2012;758:185–90.

[115] Dunn AJ. Effects of cytokines and infections on brain neurochemistry. Clin Neurosci Res 2006;6(1–2):52–68.

[116] Borsody MK, Weiss JM. Alteration of locus coeruleus neuronal activity by interleukin-1 and the involvement of endogenous corticotropin-releasing hormone. Neuroimmunomodulation 2002;10(2):101–21.

[117] Manta S, Dong J, Debonnel G, Blier P. Enhancement of the function of rat serotonin and norepinephrine neurons by sustained vagus nerve stimulation. J Psychiatry Neurosci 2009;34(4):272–80.

[118] Borsody MK, Weiss JM. Peripheral endotoxin causes long-lasting changes in locus coeruleus activity via IL-1 in the brain. Acta Neuropsychiatr 2002;14(6):303–21.

[119] Dorr AE, Debonnel G. Effect of vagus nerve stimulation on serotonergic and noradrenergic transmission. J Pharmacol Exp Ther 2006;318(2):890–8.

[120] Asanuma M, Miyazaki I, Diaz-Corrales FJ, Ogawa N. Quinone formation as dopaminergic neuron-specific oxidative stress in the pathogenesis of sporadic Parkinson's disease and neurotoxin-induced parkinsonism. Acta Medica Okayama 2004;58(5):221–33.

[121] Miyazaki I, Asanuma M. Dopaminergic neuron-specific oxidative stress caused by dopamine itself. Acta Medica Okayama 2008;62(3):141–50.

[122] Zarow C, Lyness SA, Mortimer JA, Chui HC. Neuronal loss is greater in the locus coeruleus than nucleus basalis and substantia nigra in Alzheimer and Parkinson diseases. Archives Neurol 2003;60(3):337–41.

[123] Grudzien A, Shaw P, Weintraub S, Bigio E, Mash DC, Mesulam MM. Locus coeruleus neurofibrillary degeneration in aging, mild cognitive impairment and early Alzheimer's disease. Neurobiol Aging 2007;28(3):327–35.

[124] Heneka MT, Nadrigny F, Regen T, Martinez-Hernandez F, Dumitrescu-Ozimek F, Terwel F, et al. Locus ceruleus controls Alzheimer's disease pathology by modulating microglial functions through norepinephrine. Proc Natl Acad Sci USA 2010;107(13):6058–63.

[125] Holmes C, El-Okl M, Williams AL, Cunningham C, Wilcockson D, Perry VH. Systemic infection, interleukin 1beta, and cognitive decline in Alzheimer's disease. J Neurol Neurosurg Psychiatry 2003;74(6):788–9.

[126] Marien MR, Colpaert FC, Rosenquist AC. Noradrenergic mechanisms in neurodegenerative diseases: a theory. Brain Res Brain Res Rev 2004;45(1):38–78.

[127] Tyagi E, Agrawal R, Nath C, Shukla R. Influence of LPS-induced neuroinflammation on acetylcholinesterase activity in rat brain. J Neuroimmunol 2008;205(1–2):51–6.

[128] Tyagi E, Agrawal R, Nath C, Shukla R. Effect of anti-dementia drugs on LPS induced neuroinflammation in mice. Life Sci 2007;80(21):1977–83.

[129] Rountree SD, Chan W, Pavlik VN, Darby EJ, Siddiqui S, Doody RS. Persistent treatment with cholinesterase inhibitors and/or memantine slows clinical progression of Alzheimer disease. Alzheimer's Res Ther 2009;1(2):7.

[130] Bull SJ, Huezo-Diaz P, Binder EB, Cubells JF, Ranjith G, Maddock C, et al. Functional polymorphisms in the interleukin-6 and serotonin transporter genes, and depression and fatigue induced by interferon-alpha and ribavirin treatment. Mol Psychiatry 2009;14(12):1095–104.

[131] Nelson RL, Guo Z, Halagappa VM, Pearson M, Gray AJ, Matsuoka Y, et al. Prophylactic treatment with paroxetine ameliorates behavioral deficits and retards the development of amyloid and tau pathologies in 3xTgAD mice. Exp Neurol 2007;205(1):166–76.

[132] Sheline YI, West T, Yarasheski K, Swarm R, Jasielec MS, Fisher JR, et al. An antidepressant decreases CSF Aβ production in healthy individuals and in transgenic AD mice. Sci Transl Med 2014;6(236):236re4.

[133] Hochstrasser T, Ullrich C, Sperner-Unterweger B, Humpel C. Inflammatory stimuli reduce survival of serotonergic neurons and induce neuronal expression of indoleamine 2,3-dioxygenase in rat dorsal raphe nucleus organotypic brain slices. Neuroscience 2011;184:128–38.

[134] Capuron L, Ravaud A, Neveu PJ, Miller AH, Maes M, Dantzer R. Association between decreased serum tryptophan concentrations and depressive symptoms in cancer patients undergoing cytokine therapy. Mol Psychiatry 2002;7(5):468–73.

[135] Andre C, O'Connor JC, Kelley KW, Lestage J, Dantzer R, Castanon N. Spatio-temporal differences in the profile of murine brain expression of proinflammatory cytokines and indoleamine 2,3-dioxygenase in response to peripheral lipopolysaccharide administration. J Neuroimmunol 2008;200(1–2):90–9.

[136] O'Connor JC, Lawson MA, Andre C, Moreau M, Lestage J, Castanon N, et al. Lipopolysaccharide-induced depressive-like behavior is mediated by indoleamine 2,3-dioxygenase activation in mice. Mol Psychiatry 2009;14(5):511–22.

[137] Deng XH, Ai WM, Lei DL, Luo XG, Yan XX, Li Z. Lipopolysaccharide induces paired immunoglobulinlike receptor B (PirB) expression, synaptic alteration, and learning-memory deficit in rats. Neuroscience 2012;209:161–70.

[138] Richwine AF, Parkin AO, Buchanan JB, Chen J, Markham JA, Juraska JM, et al. Architectural changes to CA1 pyramidal neurons in adult and aged mice after peripheral immune stimulation. Psychoneuroendocrinology 2008;33(10):1369–77.

[139] Pang Y, Cai Z, Rhodes PG. Effects of lipopolysaccharide on oligodendrocyte progenitor cells are mediated by astrocytes and microglia. J Neurosci Res 2000;62(4):510–20.

[140] Hennigan A, Trotter C, Kelly AM. Lipopolysaccharide impairs long-term potentiation and recognition memory and increases p75NTR expression in the rat dentate gyrus. Brain Res 2007;1130(1):158–66.

[141] Zhang J, Malik A, Choi HB, Ko RW, Dissing-Olesen L, Macvicar BA. Microglial CR3 activation triggers long-term synaptic depression in the Hippocampus via NADPH oxidase. Neuron 2014;82(1):195–207.

[142] Kettenmann H, Kirchhoff F, Verkhratsky A. Microglia: new roles for the synaptic stripper. Neuron 2013;77(1):10–8.

[143] Stellwagen D, Malenka RC. Synaptic scaling mediated by glial TNF-alpha. Nature 2006;440(7087):1054–9.

[144] Heinonen O, Soininen H, Sorvari H, Kosunen O, Paljärvi L, Koivisto E, et al. Loss of synaptophysin-like immunoreactivity in the hippocampal formation is an early phenomenon in Alzheimer's disease. Neuroscience 1995;64(2):375–84.

[145] DeKosky ST, Scheff SW. Synapse loss in frontal cortex biopsies in Alzheimer's disease: correlation with cognitive severity. Ann Neurol 1990;27(5):457–64.

[146] Lawson MA, Parrott JM, McCusker RH, Dantzer R, Kelley KW, OC JC. Intracerebroventricular administration of lipopolysaccharide induces indoleamine-2,3-dioxygenase-dependent depression-like behaviors. J Neuroinflammation 2013;10(1):87.

[147] Choi DY, Lee JW, Lin G, Lee YK, Lee YH, Choi IS, et al. Obovatol attenuates LPS-induced memory impairments in mice via inhibition of NF-κB signaling pathway. Neurochem Int 2012;60(1):68–77.

[148] Rosi S, Vazdarjanova A, Ramirez-Amaya V, Worley PF, Barnes CA, Wenk GL. Memantine protects against LPS-induced neuroinflammation, restores behaviorally-induced gene expression and spatial learning in the rat. Neuroscience 2006;142(4):1303–15.

[149] Zarifkar A, Choopani S, Ghasemi R, Naghdi N, Maghsoudi AH, Maghsoudi N, et al. Agmatine prevents LPS-induced spatial memory impairment and hippocampal apoptosis. Eur J Pharmacol 2010;634(1–3):84–8.

[150] Guan Z, Fang J. Peripheral immune activation by lipopolysaccharide decreases neurotrophins in the cortex and hippocampus in rats. Brain Behav Immun 2006;20(1):64–71.

[151] Belarbi K, Arellano C, Ferguson R, Jopson T, Rosi S. Chronic neuroinflammation impacts the recruitment of adult-born neurons into behaviorally relevant hippocampal networks. Brain Behav Immun 2012;26(1):18–23.

[152] Jakubs K, Bonde S, Iosif RE, Ekdahl CT, Kokaia Z, Kokaia M, et al. Inflammation regulates functional integration of neurons born in adult brain. J Neurosci Off J Soc Neurosci 2008;28(47):12477–88.

[153] Espey MG, Kustova Y, Sei Y, Basile AS. Extracellular glutamate levels are chronically elevated in the brains of LP-BM5-infected mice: a mechanism of retrovirus-induced encephalopathy. J Neurochem 1998;71(5):2079–87.

[154] Fine SM, Angel RA, Perry SW, Epstein LG, Rothstein JD, Dewhurst S, et al. Tumor necrosis factor alpha inhibits glutamate uptake by primary human astrocytes. Implications for pathogenesis of HIV-1 dementia. J Biol Chem 1996;271(26):15303–6.

[155] Pascual O, Ben Achour S, Rostaing P, Triller A, Bessis A. Microglia activation triggers astrocyte-mediated modulation of excitatory neurotransmission. Proc Natl Acad Sci USA 2012;109(4):E197–205.

[156] Lauderback CM, Harris-White ME, Wang Y, Pedigo Jr NW, Carney JM, Butterfield DA. Amyloid beta-peptide inhibits Na^+-dependent glutamate uptake. Life Sci 1999;65(18–19):1977–81.

[157] Rothstein JD. Excitotoxicity and neurodegeneration in amyotrophic lateral sclerosis. Clin Neurosci 1995;3(6):348–59.

[158] Kigerl KA, Ankeny DP, Garg SK, Wei P, Guan Z, Lai W, et al. System x(c)(-) regulates microglia and macrophage glutamate excitotoxicity in vivo. Exp Neurol 2012;233(1):333–41.

[159] Floyd RA. Antioxidants, oxidative stress, and degenerative neurological disorders. Proc Soc Exp Biol Med Soc Exp Biol Med 1999;222(3):236–45.

[160] Choi DH, Cristovao AC, Guhathakurta S, Lee J, Joh TH, Beal MF, et al. NADPH oxidase 1-mediated oxidative stress leads to dopamine neuron death in Parkinson's disease. Antioxid Redox Signal 2012;16(10):1033–45.

[161] Vasquez-Vivar J, Kalyanaraman B, Kennedy MC. Mitochondrial aconitase is a source of hydroxyl radical. An electron spin resonance investigation. J Biol Chem 2000;275(19):14064–9.

[162] Chinta SJ, Andersen JK. Nitrosylation and nitration of mitochondrial complex I in Parkinson's disease. Free Radic Res 2011;45(1):53–8.

[163] Szabo C, Ischiropoulos H, Radi R. Peroxynitrite: biochemistry, pathophysiology and development of therapeutics. Nat Rev Drug Discov 2007;6(8):662–80.

[164] Arimoto T, Bing G. Up-regulation of inducible nitric oxide synthase in the substantia nigra by lipopolysaccharide causes microglial activation and neurodegeneration. Neurobiol Dis 2003;12(1):35–45.

[165] Bal-Price A, Brown GC. Inflammatory neurodegeneration mediated by nitric oxide from activated glia-inhibiting neuronal respiration, causing glutamate release and excitotoxicity. J Neurosci Off J Soc Neurosci 2001;21(17):6480–91.

[166] Uversky VN, Yamin G, Munishkina LA, Karymov MA, Millett IS, Doniach S, et al. Effects of nitration on the structure and aggregation of alpha-synuclein. Brain Res Mol Brain Res 2005;134(1):84–102.

[167] Nakamura T, Cho DH, Lipton SA. Redox regulation of protein misfolding, mitochondrial dysfunction, synaptic damage, and cell death in neurodegenerative diseases. Exp Neurol 2012;238(1):12–21.

[168] Gibson GE, Starkov A, Blass JP, Ratan RR, Beal MF. Cause and consequence: mitochondrial dysfunction initiates and propagates neuronal dysfunction, neuronal death and behavioral abnormalities in age-associated neurodegenerative diseases. Biochim Biophys Acta 2010;1802(1):122–34.

[169] Carafoli E. The fateful encounter of mitochondria with calcium: how did it happen? Biochim Biophys Acta 2010;1797(6–7):595–606.

[170] Cardenas C, Miller RA, Smith I, Bui T, Molgó J, Müller M, et al. Essential regulation of cell bioenergetics by constitutive InsP3 receptor Ca^{2+} transfer to mitochondria. Cell 2010;142(2):270–83.

[171] Corti O, Brice A. Mitochondrial quality control turns out to be the principal suspect in parkin and PINK1-related autosomal recessive Parkinson's disease. Curr Opin Neurobiol 2013;23(1):100–8.

[172] Pinto M, Moraes CT. Mitochondrial genome changes and neurodegenerative diseases. Biochim Biophys Acta 2014;1842(8):1198–1207.

[173] Schon EA, Przedborski S. Mitochondria: the next (neurode)generation. Neuron 2011;70(6):1033–53.

[174] Brownlow ML, Benner L, D'Agostino D, Gordon MN, Morgan D. Ketogenic diet improves motor performance but not cognition in two mouse models of Alzheimer's pathology. PLoS One 2013;8(9):e75713.

[175] Kim do Y, Hao J, Liu R, Turner G, Shi FD, Rho JM. Inflammation-mediated memory dysfunction and effects of a ketogenic diet in a murine model of multiple sclerosis. PLoS One 2012;7(5):e35476.

[176] Zhao W, Varghese M, Vempati P, Dzhun A, Cheng A, Wang J, et al. Caprylic triglyceride as a novel therapeutic approach to effectively improve the performance and attenuate the symptoms due to the motor neuron loss in ALS disease. PLoS One 2012;7(11):e49191.

[177] Beal MF. Energetics in the pathogenesis of neurodegenerative diseases. Trends Neurosci 2000;23(7):298–304.

[178] Kamat CD, Gadal S, Mhatre M, Williamson KS, Pye QN, Hensley K. Antioxidants in central nervous system diseases: preclinical promise and translational challenges. J Alzheimer's Dis 2008;15(3):473–93.

[179] Golden TR, Patel M. Catalytic antioxidants and neurodegeneration. Antioxidants Redox Signal 2009;11(3):555–70.

[180] Ajmone-Cat MA, Bernardo A, Greco A, Minghetti L. Non-steroidal anti-inflammatory drugs and brain inflammation: effects on microglial functions. Pharmaceuticals 2010;3(6):1949–65.

[181] Uttara B, Singh AV, Zamboni P, Mahajan RT. Oxidative stress and neurodegenerative diseases: a review of upstream and downstream antioxidant therapeutic options. Curr Neuropharmacol 2009;7(1):65–74.

[182] Hutchinson MR, Northcutt AL, Hiranita T, Wang X, Lewis SS, Thomas J, et al. Opioid activation of toll-like receptor 4 contributes to drug reinforcement. J Neurosci Off J Soc Neurosci 2012;32(33):11187–200.

[183] Qin L, Liu Y, Hong JS, Crews FT. NADPH oxidase and aging drive microglial activation, oxidative stress, and dopaminergic neurodegeneration following systemic LPS administration. Glia 2013;61(6):855–68.

Late Neurological Effects of Early Environmental Exposures

David C. Bellinger, Maitreyi Mazumdar

The hypothesis that early-life exposures and health events play roles in laying the foundation for adult neurological diseases and degenerative processes has come to be called the "Developmental Origins of Health and Disease" (DOHaD) [1]. Such developmental programming has been investigated most actively with regard to early nutrition and later cardiovascular and metabolic disease [2], but has been also extended to noncommunicable diseases including cancer [3] and mental disorders [4].

Within this framework, the late neurotoxic effects of early-life exposures to chemicals are described as "delayed" or initially "silent," becoming manifest only after some interval since the exposure occurred. Such effects could, in principle, be explained by a diversity of (nonmutually exclusive) mechanisms (Table 19.1). For example, impairment might be

Environmental Factors in Neurodevelopmental and Neurodegenerative Disorders
http://dx.doi.org/10.1016/B978-0-12-800228-5.00019-4

409

TABLE 19.1   Alternative Models to Account for an Association between an Early-Life Exposure to a Chemical and a Late Neurological Effect

- Late maturation of the neural substrate underlying a function
- Chemical-induced epigenetic modification causing delayed gene expression
- Chemical-induced loss of capacity for plasticity or acceleration of neurodegenerative processes associated with aging
- Methodological artifact due to age-related differences in sensitivity of measurement tools

expressed only after the central nervous system (CNS) substrate underlying a function has matured sufficiently that the adverse impact of an earlier event on that substrate can be expressed. This has been suggested in the case of preterm infants, whereby the likelihood of detecting associations between neonatal brain abnormalities and neurological outcomes becomes greater as the children get older [5,6]. Alternatively, an emerging impairment could reflect an exposure-related process, initiated early in life, involving epigenetic modification that is delayed in expression, as has been shown for lead [7] and manganese [8]. This hypothesis is discussed in detail in a later section. The appearance of impairment could also reflect the waning efficacy of protective factors that previously helped to prevent an impairment from being expressed. Or it could be the expression of a progressive process, such as a toxicant-induced acceleration of the rate of the neuronal loss associated with normal aging, which has reached the point at which compensatory processes fail, causing the onset of functional limitations [9]. For example, a 0.5% increase per year, after age 25 years, in the rate of functional loss would result in a reduction of functional capacity to 70% by age 55 (vs >95 years under conditions of "normal" aging). Weiss and colleagues [10] speculated that neurons differ in their vulnerability to neurotoxicants, with cells that are more susceptible dying first, and, "surviving cells assume their functions but eventually because of increased functional load and metabolic stress, these cells also succumb. At some point, the neuronal population has exhausted its capacity to compensate for the cell less and clinical signs rapidly erupt" (p. 852). In this latter view, the neurotoxicant would not be the primary cause of the degenerative process but a risk modifier, influencing the timing or severity of the expression of genetic predispositions germane to aging [11]. It is also now well established that neurogenesis in rodents and nonhuman primates continues into adulthood, so an early life exposure might interfere with the integrity of the processes that underlie this plasticity. For example, Gilbert and colleagues [12] showed that chronic lead exposure reduced neurogenesis in the dentate granule layer of the hippocampus in adult rats. Finally, the appearance of delayed neurotoxicity could be explained by a methodological artifact, namely differences in the sensitivity of the tools available for detecting neurotoxicity at different life stages.

Two historical episodes provide "proof-of-concept" for the impact of early-life exposures on adult neurological function. First, men born to women who contracted influenza during the pandemic of 1918 enjoyed considerably less economic success as adults compared to men whose mothers did not, as reflected in literacy, educational attainment, employment, income, and socioeconomic status [13]. Second, individuals who were in utero, particularly early in gestation, at the time of the Dutch famine of 1944–1945 were at greater risk of a variety of adult health problems, including neurological diseases [14]. The early-life malnutrition associated with this episode also produced transgenerational effects, as the offspring of women who themselves were born during the famine were at increased risk for neurological diseases in adulthood [15].

Three examples will be used to illustrate the potential lifelong nature of neurological effects of early-life exposure to a chemical. These examples do not necessarily illustrate delayed neurotoxicity, however, as adverse effects might well have been evident soon after the onset of exposure. First, White and colleagues [16] reported a study of adults who had been hospitalized 48–60 years before for treatment of severe clinical lead poisoning in childhood. Compared to controls matched for age, race, sex, and neighborhood, the former patients scored significantly worse on a variety of neuropsychological tests, showing global impairment that is more typical of the sequelae of childhood lead exposure than of adult lead exposure [17]. Second, individuals who developed congenital Minamata disease because of high dose in utero exposure to methylmercury suffered permanent disabilities, including mental retardation, cerebellar ataxia, chorea and athetosis, dysarthria, sensory impairments, and other signs of neurological dysfunction that required them to reside in an institution that provided lifelong care [18]. Even among community members who did not meet criteria for Minamata disease, the prevalence odds ratios of mood and behavioral disorders and of psychiatric symptoms were higher than in a control region [19]. In nonhuman primates, Rice [20] demonstrated a variety of late somatosensory effects of early-life methylmercury exposure that had not been detectable in assessments conducted at younger ages. Third, Japanese infants who, in 1955, consumed milk powder highly contaminated with arsenic suffered adverse sequelae that persisted into adulthood, including intellectual disability and other neurological disorders [21].

In this chapter, we discuss the evidence for the long-term effects of early-life chemical exposures on neurological status, alternative models of these associations, and, to a limited extent, mechanisms, most notably the putative role of epigenetic modifications.

METHODOLOGICAL CHALLENGES

Numerous obstacles face an investigator who seeks to clarify the natural histories of early-life exposure–adult outcome relationships. Some of the challenges are generic to the DOHaD field, with the most difficult one being the long interval between the hypothesized antecedent (early-life exposure to a chemical) and effect (an endpoint that provides information about a neurodegenerative process). It is difficult, both logistically and financially, to keep a cohort intact over 10 years, much less over the human lifespan. Given this, it is not surprising that studies that assess outcomes in the first few years after an exposure occurred are legion, but studies of later outcomes are rare. In part, this might be due to an assumption that the impairments detected in children at, say, 10 years of age, are likely to be irreversible, vitiating the need for further follow-up assessments. The viability of this assumption with regard to neurodegeneration is questionable, however, because, as far as is known, prodromal signs of such a process are not present in childhood. Therefore, to address this hypothesis adequately in humans, prospective studies sustained over a period of decades are needed. In fact, the length of the follow-up interval is likely to exceed the duration of an investigator's research career, requiring the use of creative strategies to organize a study. In addition to a long follow-up interval, it is also necessary to have access to high-quality data on the exposures that occurred during critical windows of vulnerability, which might be the prenatal period, as well as on the neurological outcomes of interest. Under some circumstances, owing to the foresight of an earlier investigator, archived biological samples (e.g., blood, serum, teeth, hair, nails) are available, permitting valid estimation of exposure to chemicals that do not degrade over time, such as heavy metal elements. More often, however, in studying a very late outcome such as neurodegeneration, investigators have to rely on proxy measures of exposure (e.g., residential history, parental occupation), which will likely introduce misclassification that is difficult to characterize in terms of its direction and severity.

Other common epidemiologic challenges are faced, including loss to follow-up of individuals in the cohort, which, except under the highly unlikely assumption that loss is random, will introduce one or more selection biases that could threaten the validity of any associations that appear to exist. For example, greater early-life exposure to lead might predispose an individual to develop both chronic kidney disease and neurodegeneration in later adulthood [22]. If the kidney disease results in a premature death from end-stage renal disease, before neurodegeneration is apparent, the resulting censoring of the data would produce an underestimate of the relationship between early-life exposure to lead and neurodegeneration late in life. Finally, confounding can be a serious problem when follow-up occurs after an interval of many decades owing to the potential influences on the outcomes of interest of the myriad of events and exposures that occur in the interval. Unless the investigator who set up a cohort developed plans to collect data on potential confounding variables across the entire period between exposure and outcome assessment, the critical data needed to address this possibility are likely to be unavailable. This can introduce either positive or negative bias into the effect estimates, both of which can threaten the validity of the inferences drawn.

The idea that neurotoxicity remains "silent," only to be unmasked at a time distant from the original event, also poses a logical challenge, as it requires proof that subtle prodromal impairment does not precede the eventual expression of neurotoxicity in adulthood. Proving a negative proposition is always problematic, especially when it might be that the form of a prodrome differs from the form in which the late effect is expressed. For example, in one study it was found that the density of ideas in autobiographies written 60 years prior predicted cognitive function and the presence of neurofibrillary tangles in the brains of participants diagnosed with Alzheimer's disease (AD) [23]. Whether the process that eventually resulted in the late expressions was "silent" in the preceding decades thus depends on how narrowly the outcome is defined and on the disease model that is constructed to describe the association.

Therefore, it is necessary that anyone seeking to demonstrate delayed or silent neurotoxicity make explicit the model of the association of interest. Recent literature on childhood lead exposure and adult mental health provides a useful example. Studies using a variety of designs (ecologic, case–control, cross-sectional, prospective) have reported significant associations between early-life lead exposure, as reflected by biomarkers such as blood or bone lead levels, and delinquency, antisocial behavior, and homicide in late childhood or early adulthood [24–29]. Is this an example of delayed neurotoxicity or does this association reflect a process, which, initiated by early-life exposure, is expressed in different forms throughout the follow-up interval, including as antisocial behavior in adulthood? The latter seems more plausible. Early-life lead exposure has been shown to reduce children's cognitive abilities, impair executive functions and attention, and reduce impulse control and the ability to delay gratification. Those impairments, in turn, place a child at risk of educational failure, behavioral problems such as Attention Deficit Hyperactivity Disorder and Conduct Disorder [30], all of which are known to be risk factors for antisocial behavior [31]. Thus, early-life lead exposure might set in motion a process that renders a child poorly equipped to make responsible decisions, with criminality an eventual downstream result for some. This example highlights the importance of articulating clearly the model motivating a hypothesis and clearly describing the approach taken to data analyses. In analyses exploring the associations between prenatal and early-life factors and late-life cognition, Zhang and colleagues [32] found significant crude associations between prenatal factors and adult cognition, but these associations were attenuated to nonsignificance once adjustments were made for early childhood events and characteristics (e.g., nutritional, social, and environmental factors). If these early childhood factors were, in part, the result of the prenatal factors, this analytical approach would represent overcontrol, that is, control for mediating variables. To apply this concept to the lead example, significant associations between childhood lead exposure and adult criminality might not have been found if the models were adjusted for child IQ, executive functions, impulse control, and so forth. These factors themselves reflect an impact of lead and so would "take up" variance in the outcome variable that should more appropriately be attributed to lead.

Another conceptual issue requires mention. It pertains to effect modification, but not in the way that this is usually discussed. Typically, one is concerned with clarifying ways in which genotype, medical comorbidity, or the social environment modifies the neurotoxicity of early-life exposures [33]. The converse is possible as well, however, insofar as one effect of early-life exposure to a chemical might be to alter the individual's responses to CNS insults that occur later. Such exposures might reduce resilience (or "cognitive reserve"), resulting in the subsequent insults being more neurotoxic. This has been shown in some animal studies, whereby rats exposed early in life to lead show slower recovery of function following an induced ischemic stroke in the somatosensory cortex [34] and show less efficient recovery of organization in the neurons of the barrel field cortex following whisker ablation [35].

EARLY-LIFE EXPOSURE AND LATER COGNITIVE STATUS

Lead is the chemical for which the most data are available in the context of a longitudinal study from early life, although the follow-up intervals usually cover less than half of the average lifespan. A pooled analysis that included the data from seven prospective studies, with a total sample size of 1333 children, showed that an increase in childhood blood lead level from 2.4 to 10 μg/dL was associated with a decline of 3.9 points in Full-Scale IQ measured at preschool or school-age (Figure 19.1) [36]. In one of the studies that contributed to these analyses, the Full-Scale IQ scores at 10 years of age of children who

FIGURE 19.1 Two models of the relationship between concurrent blood lead level and IQ score, pooling the data from seven prospective studies (N=1333). The dotted lines are the 95% confidence interval for the restricted cubic spline model. *Lanphear et al. Environmental Health Perspectives 2005;113:894–9.*

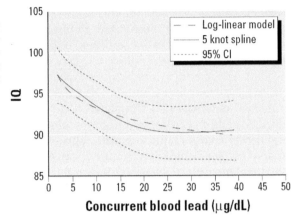

had been enrolled in a prospective study at birth were significantly associated with their blood lead levels at 2 years of age [37]. For 90% of the children, this level was less than 10 μg/dL, and IQ score declined 0.58 points for each μg/dL increase. Mazumdar and colleagues [38] followed up this cohort at approximately 30 years of age and, although only about one-third of the participants could be assessed, the estimated Full-Scale IQ score (assessed using the Wechsler Abbreviated Scale of Intelligence) was inversely and significantly associated with umbilical cord blood lead level. Needleman and colleagues [39] measured deciduous tooth dentine lead levels in first- and second-graders and found the concentration to be inversely associated with scores on a variety of neuropsychological tests, including Full-Scale IQ, as well as with teachers' ratings of classroom behavior. A subset of the children was followed up in late adolescence [40,41], at which time children with higher dentine lead levels continued to score lower on neuropsychological tests and were at significantly higher risk of failing to show grade-appropriate progress in reading, and at higher risk of having dropped out of school. Similar findings were reported for children from New Zealand, as higher deciduous tooth dentine lead levels were significantly associated with the achievement of fewer academic qualifications at age 18 years [42].

Another prospective study, conducted in Port Pirie, South Australia, found evidence of persistent associations between blood lead levels in childhood and tests of intellectual capacity to age 11–13 years [43]. This study provides persuasive evidence that the deficits associated with early lead exposure are nonprogressive but persistent. Children were categorized into tertiles based on their blood lead levels between birth and 2 years of age, and the scores of the three groups compared on standardized neurodevelopmental tests administered at age 2, 4, 7, and 11–13. The group with blood lead levels that placed them in the highest tertile had mean deficits of four to five points at all ages. In other words, the deficit remained constant, neither attenuating nor increasing. Interestingly, the blood lead levels of the children who were in the highest tertile, based on levels measured between birth and 2 years, were also higher in the period after 2 years. This suggests that having a persistently higher blood lead level after age 2 years did not cause further harm.

Analyzing historical data for several cohorts, Kaufman and colleagues [44] found that the mean IQ of US adults, adjusted for demographic characteristics, is highly correlated with the decline in population blood lead level, representing a four to five point increase in score since the 1970s.

EARLY-LIFE EXPOSURE AND LATER PSYCHIATRIC STATUS

Several studies have reported significant cross-sectional associations between a lead biomarker level and psychiatric status in adulthood [45,46], but what is not clear is whether this reflects concurrent exposure or exposure in early life. This is true even when bone lead level is used. The concentration of lead in the bone compartment is considered to be an index of cumulative exposure over a period of decades (cortical bone) or a decade (trabecular bone), but it appears that these indices do not provide information about lead exposure in childhood. In a follow-up study of children treated for clinical lead poisoning in early life, Nie and colleagues [47] found that none of the children had a bone lead concentration above the limit of detection. This is perhaps because the rapid remodeling of bone that occurs as a child grows prevents the accumulation of lead in bone until the pace of remodeling slows.

In the Port Pirie prospective study, participants were followed to a mean age of 26 years, permitting assessment of the relationship between early-life lead exposure and adult psychiatric status. At age 11–13 years, the investigators had reported that children with greater childhood blood lead levels showed more externalizing problems (both boys and girls), and greater internalizing problems (girls only) [48]. At age 26 years, lifetime prevalence of DSM-IV disorders was determined using a standardized diagnostic interview [49]. Unadjusted analyses showed sex differences, in that no associations were observed in boys, but girls with higher early lead exposure had higher odds of social phobia and substance abuse, as well as higher levels of anxiety, somatic problems, and antisocial personality problems. These associations were attenuated to nonsignificance, however, when adjusted for covariates such as characteristics of the home environment and socioeconomic status. Only about one-third of the original cohort participated in this follow-up, however, and the evidence shows that the loss to follow-up did not occur at random. However, the effect sizes observed were similar in magnitude to those observed at the 11–13-year follow-up, though they did not reach statistical significance due to the reduced statistical power.

Utilizing archived biological samples available for a subset of children enrolled in the Collaborative Perinatal Project cohort, Opler and colleagues [50,51] investigated the association between prenatal lead exposure and the risk of schizophrenia in young adulthood. Because only serum was available, a proxy for lead was measured, δ-amino levulinic acid (ALA), an endogenous metabolite in the heme pathway. Based on the known relationship between ALA and lead, women were categorized in groups based on whether the second trimester blood lead level was likely to have been greater or less than 15 μg/dL. Adults in the group suspected to have had greater in utero lead exposure were at twofold increased risk (OR=2.3, 95% CI: 1.0–4.3) of having a disorder on the schizophrenia spectrum. Guilarte and colleagues [52] summarized a wealth of anatomical, biochemical, and neuropathological evidence supporting the hypothesis that lead-induced hypoactivity of the N-methyl-D-aspartate (NMDA) subtype of excitatory amino acid receptors, in combination with a genetic susceptibility to schizophrenia, is a possible mechanism underlying this association. They speculated that other environmental pollutants that function as N-methyl-D-aspartate receptor (NMDAR) antagonists, including manganese and polycyclic aromatic hydrocarbons (PAHs), should also be evaluated as contributors to this disorder.

EARLY-LIFE EXPOSURES AND NEUROIMAGING IN ADOLESCENCE AND ADULTHOOD

Increasing evidence is being generated showing that a variety of early-life exposures, including lead, methylmercury, air pollution, pesticides, and alcohol, have discernable effects on later brain structure and organization. In the Cincinnati Prospective Lead Study, a variety of neuroimaging modalities were applied to a subset of the cohort at age 19–24 years, including volumetric MRI [53,54], diffusion tensor MRI [55], nuclear magnetic resonance spectroscopy [56], and functional MRI [57]. For each modality, children with higher early-life blood lead levels showed evidence of altered brain structure (e.g., volume loss in the prefrontal cortex, altered white matter organization) and function (e.g., altered metabolite concentrations and ratios, altered patterns of activation during task performance).

Calderón-Garcidueñas et al. [58,59] reported that a higher percentage of children from Mexico City (high air pollution) than control children had white matter hyperintensities in the prefrontal cortex and that children with higher exposures showed greater decrement in performance on neuropsychological tests over a one-year follow-up period. They also showed that, compared to controls, more highly exposed children had more frontal tau hyperphosphorylation and amyloid-β diffuse plaques, suggesting processes associated with neurodegeneration [60].

In a cohort of adolescent boys enrolled in the Faroe Islands birth cohort, White and colleagues [61] found that those with high mixed exposures to methylmercury and PCBs had activation patterns on functional MRI during visual and motor tasks that differed from those of boys with low mixed exposures, specifically activation in more brain areas and wider patterns of activation.

Rauh and colleagues [62] conducted morphometric MRI studies of 5–11-year-old children with varying levels of prenatal exposure to chlorpyrifos, an organophosphate pesticide. Higher exposures were associated with significant enlargements in several brain areas bilaterally (superior temporal, posterior middle temporal, inferior postcentral gyri) and unilaterally (superior frontal gyrus, gyrus rectus, cuneus, and precuneus) unilaterally (right hemisphere), as well as frontal and parietal cortical thinning, and the absence of expected sexual dimorphisms.

In children and adolescents with fetal alcohol syndrome, longitudinal imaging using various modalities indicated delayed white matter development as well as reduced total brain, white, cortical gray, and deep gray matter volumes [63].

EPIGENETIC PROCESSES AS A MECHANISM OF DEVELOPMENTAL PROGRAMMING

In this section, we discuss in some detail the importance of epigenetic processes as one possible basis for developmental programming. In the last decade, our understanding of epigenetics has advanced considerably, revealing that gene expression can be influenced by factors other than DNA sequence. The best-studied epigenetic regulators are DNA methylation, histone modifications, and noncoding RNAs. Together, these processes affect patterns of gene expression and transcript stability, influence DNA accessibility and chromatin compaction, regulate the integrity and function of the genome, and maintain higher-order nuclear organization in a manner that determines the disease risk of the cell or tissue [64–66]. Epidemiological studies have most frequently focused on DNA methylation due to its relative stability with storage and the many technical platforms available for analysis. Although epigenetics refers to the study of single genes or gene sets, epigenomics refers to analyses that are more global, reflecting epigenetic changes across the entire genome.

Epigenetics may provide a missing mechanistic link underlying the latent effects of adverse fetal, infant, and childhood environments on chronic disease later in life. In animal and human models, many chemicals and pollutants can alter epigenetic programs and thereby heighten or lessen the risk of disease development [67]. In some cases, the impacts of an epigenetic disruption only become manifest at a later life stage [68,69]. An extreme case is the Dutch Famine (1944–1945). Six decades later, individuals who were prenatally exposed to famine had less DNA methylation of the *IGF2* gene compared with their unexposed, same-sex siblings [70].

EARLY-LIFE LEAD EXPOSURE AND AD

Human epidemiological research on early-life environmental exposures, neurological disease, and epigenetics faces many challenges. Animal models, therefore, have been most widely used to study the hypothesis that early-life exposures cause neurological disease later in life through the epigenome. Much of this work has focused on the toxic effects of lead. A series of animal and cell culture studies conducted by Zawia and colleagues collectively demonstrate that early-life lead exposure reduces DNA methyltransferase activity (a family of enzymes instrumental in DNA methylation), and specifically alters the regulation of many genes implicated in AD. Basha et al. [71] exposed newborn rats to lead. At 20 months of age, the gene encoding the β-amyloid precursor protein (APP) exhibited delayed overexpression. The increase in *APP* gene expression in old age was accompanied by an elevation in β-amyloid protein (Aβ) in brain tissue. These changes were not seen in rats exposed to lead as adults, suggesting that early timing of lead exposure is an important determinant of gene expression and protein production. Cynomolgus monkeys exposed to lead as infants had more plaques that were neuritic in their brains and exhibited higher levels of expression of AD-related genes in the brain than did monkeys not exposed to lead [7]. In human cell culture studies, the administration of lead caused a significantly decreased expression of epigenetic intermediates, including DNA methyltransferase enzymes [72]. This model has some support in human studies. We recently showed that an umbilical cord blood lead concentration greater than 10 μg/dL was associated with lower expression in young adults of several genes involved in the clearance of amyloid proteins from the brain as well as with changes in the expression of numerous genes not involved in amyloid pathways but implicated in neurogenesis, growth, and general cell development [73] (Table 19.2).

TABLE 19.2   Alzheimer's Disease-Related Genes and other Genes with Altered Expression at Umbilical Cord Blood Lead Levels >10 µg/dL

Genes encoding proteins that affect Aβ production and deposition
- ADAM metallopeptidase domain 9
- Low-density lipoprotein receptor-related protein associated protein 1
- Reticulon 4

Network analysis representing functional association of 39 genes with altered expression
Gene Ontology Category of biological process
- Neuron development
  - Negative regulation of axonogenesis
  - Negative regulation of cell projection organization
  - Negative regulation of neurogenesis
  - Nerve growth factor receptor signaling pathway
  - Regulation of axonogenesis
  - Regulation of neuron projection development
- Cell development
  - Negative regulation of cell development
  - Negative regulation of cell differentiation
- Immune response
  - Negative regulation of cell differentiation
  - T-cell receptor signaling pathway

Mazumdar et al. Environmental Health Perspectives 2012;120:702–7.

EPIGENETIC MARKERS AS BIOMARKERS OF ENVIRONMENTAL EXPOSURES

Given the challenges of linking early-life exposures to late-life outcomes, most epidemiological studies currently use epigenetics as a marker of exposure. Epigenetic markers may represent "tombstones" of previous exposures to chemicals that have since been excreted from the body [74]. For example, researchers found DNA methylation sites in newborn cord blood that correlated to maternal smoking behavior in pregnancy [75]. Among 103 umbilical cord samples, prenatal lead exposure measured by X-ray fluorescence was inversely associated with genomic methylation [76]. Similar associations have been reported for cord blood and prenatal arsenic exposure, though only among males [77].

CONCLUSION

Considerable progress has been made in recent decades in our understanding the effects on human health of early-life exposures to environmental chemicals, though only the near-term consequences have been the primary focus of this work. Rapid progress in the relatively new field of epigenetics has increased interest in the longer-term effects, including those not expressed until late adulthood. At the same time, studies have demonstrated that the manner and severity in which a chemical's neurotoxicity is expressed depends, to some extent, on contextual factors such as medical co-morbidities, stress, aspects of the social environment, and other chemical exposures [78]. What all of these developments are pointing to is the need to adopt a lifespan approach, involving birth cohort studies, that involve careful assessment, over a sustained period of time, of a myriad of factors, including chemical exposures, medical history, family history, and social environment, that influence human neurodevelopment [79]. It will be necessary to use the methods developed in the social science literature to construct complex models of developmental trajectories [80,81]. The phenomena that we seek to explain are complex. To date, the methodological approaches we have pursued have not been commensurate in their sophistication. We have no alternative but to rectify this if we wish to achieve a full understanding of the import of chemical exposures for human health.

References

[1] Barouki R, Gluckman PD, Grandjean P, Hanson M, Heindel JJ. Developmental origins of non-communicable disease: implications for research and public health. Environ Health 2012;11:42.

[2] Benyshek DC. The "early life" origins of obesity-related health disorders: new discoveries regarding the intergenerational transmission of developmentally programmed traits in the global cardiometabolic health crisis. Am J Phys Anthropol 2013;152(Suppl. 57):79093.

[3] Walker CL, Ho SM. Developmental reprogramming of cancer susceptibility. Nat Rev Cancer 2012;12:479–86.

[4] Van den Bergh BR. Developmental programming of early brain and behavior development and mental health: a conceptual framework. Dev Med Child Neurol 2011;53(Suppl. 4):19–23.

[5] Rogers CE, Anderson PJ, Thompson DK, Kidokoro H, Wallendorf M, Treyvaud K, et al. Regional cerebral development at term relates to school-age social-emotional development in very preterm children. J Am Acad Child Adolesc Psychiatry 2012;51:181–91.

[6] Woodward LJ, Clark CA, Bora S, Inder TE. Neonatal white matter abnormalities an important predictor of neurocognitive outcome for very preterm children. PLoS One 2012;7:e51879.

[7] Wu J, Basha R, Brock B, Cox DP, Cardozo-Pelaez F, McPherson CA, et al. Alzheimer's disease (AD)-like pathology in aged monkeys after infantile exposure to environmental metal lead (Pb): evidence for a developmental origin and environmental link for AD. J Neurosci 2008;28:3–9.

[8] Guilarte TR. APLP.1, Alzheimer's-like pathology and neurodegeneration in the frontal cortex of manganese-exposed non-human primates. Neurotoxicology 2010;31:572–4.

[9] Weiss B, Simon W. Quantitative perspectives on the long-term toxicity of methylmercury and similar poisons. In: Weiss B, Laties VG, editors. Behavioral toxicology. New York: Plenum Press; 1975. p. 429–38.

[10] Weiss B, Clarkson TW, Simon W. Silent latency periods in methylmercury poisoning and in neurodegenerative disease. Environ Health Perspect 2002;110(Suppl. 5):851–4.

[11] Weiss B. Lead, manganese, and methylmercury as risk factors for neurobehavioral impairment in advanced age. Int J Alzheimer's Dis 2010:607543.

[12] Gilbert ME, Kelly ME, Samsam TE, Goodman JH. Chronic developmental lead exposure reduces neurogenesis in adult rat hippocampus but does not impair spatial learning. Toxicol Sci 2005;86:365–74.

[13] Nelson RE. Testing the fetal origins hypothesis in a developing country: evidence from the 1918 influenza pandemic. Health Econ 2010;19:1181–92.

[14] Roseboom T, de Rooij S, Painter R. The Dutch famine and its long-term consequences for adult health. Early Hum Dev 2006;82:485–91.

[15] Painter RC, Osmond C, Gluckman P, Hanson M, Phillips DI, Roseboom TJ. Transgenerational effects of prenatal exposure to the Dutch famine on neonatal adiposity and health in later life. Br J Obstet Gynecol 2008;115:1243–9.

[16] White RF, Diamond R, Proctor S, Morey C, Hu H. Residual cognitive deficits 50 years after lead poisoning during childhood. Br J Ind Med 1993;50:613–22.

[17] Weisskopf MG, Hu H, Wright RO. In: Bellinger DC, editor. Human developmental neurotoxicology. New York: Taylor & Francis; 2006.

[18] Harada M. Neurotoxicity of methylmercury: Minamata and the Amazon. In: Yasui M, Strong MJ, Ota K, Verity MA, editors. Mineral and metal neurotoxicology. Boca Raton, FL: CRC Press; 1997. pp. 177–88.

[19] Yorifuji T, Tsuda T, Inoue S, Takao S, Harada M. Long-term exposure to methylmercury and psychiatric symptoms in residents of Minamata, Japan. Environ Int 2011;37:907–13.

[20] Rice DC. Evidence for delayed neurotoxicity produced by methylmercury. Neurotoxicology 1996;17:583–96.

[21] Dakeishi M, Murata K, Grandjean P. Long-term consequences of arsenic poisoning during infancy due to contaminated milk powder. Environ Health 2006;5:31.

[22] Brewster UC, Perazella MA. A review of chronic lead intoxication: an unrecognized cause of chronic kidney disease. Am J Med Sci 2004;327:341–7.

[23] Snowdon DA, Kemper SJ, Mortimer JA, Greiner LH, Wekstein DR, Markesbery WR. Linguistic ability in early life and cognitive function and Alzheimer's disease in late life. Findings from the Nun Study. JAMA 1996;275:528–32.

[24] Stretesky PB, Lynch MJ. The relationship between lead and crime. J Health Soc Behav 2004;45:214–29.

[25] Needleman HL, McFarland C, Ness RB, Fienberg SE, Tobin MJ. Bone lead levels in adjudicated delinquents: a case control study. Neurotoxicol Teratol 2002;24:711–7.

[26] Nevin R. How lead exposure relates to temporal changes in IQ, violent crime, and unwed pregnancy. Environ Res 2000;83:1–22.

[27] Nevin R. Understanding international crime trends: the legacy of preschool lead exposure. Environ Res 2007;104:315–36.

[28] Wright JP, Dietrich KN, Ris MD, Hornung RW, Wessel SD, Lanphear BP, et al. Association of prenatal and childhood blood lead concentrations with criminal arrests in early adulthood. PLoS Med 2008;5:e101.

[29] Needleman HL, Riess JA, Tobin MJ, Biesecker GE, Greenhouse JB. Bone lead levels and delinquent behavior. JAMA 1996;275:363–9.

[30] Yolton K, Cornelius M, Ornoy A, McGough J, Makris S, Schantz S. Exposure to neurotoxicants and the development of attention deficit hyperactivity disorder and its related behaviors in childhood. Neurotoxicol Teratol 2014;44C:30–45.

[31] Storebø OJ, Simonsen E. The association between ADHD and antisocial personality disorder (ASPD): a review. J Atten Disord (in press).

[32] Zhang ZX, Plassman BL, Xu Q, Zahner GE, Wu B, Gai MY, et al. Lifespan influences on mid- to late-life cognitive function in a Chinese birth cohort. Neurology 2009;73:186–94.

[33] Weston HI, Sobolewski ME, Allen JL, Weston D, Conrad K, Pelkowski S, et al. Sex-dependent and nonmonotonic enhancement and unmasking of methylmercury neurotoxicity by prenatal stress. Neurotoxicology 2014;41:123–40.

[34] Schneider JS, Decamp E. Postnatal lead poisoning impairs behavioral recovery following brain damage. Neurotoxicology 2007;28:1153–7.

[35] Wilson MA, Johnston MV, Goldstein GW, Blue ME. Neonatal lead exposure impairs development of rodent barrel field cortex. Proc Natl Acad Sci USA 2000;97:5540–5.

[36] Lanphear BP, Hornung R, Khoury J, Yolton K, Baghurst P, Bellinger DC, et al. Low-level environmental lead exposure and children's intellectual function: an international pooled analysis. Environ Health Perspect 2005;113:894–9.

[37] Bellinger DC, Stiles KM, Needleman HL. Low-level lead exposure, intelligence and academic achievement: a long-term follow-up study. Pediatrics 1992;90:855–61.

[38] Mazumdar M, Bellinger DC, Gregas M, Abanilla K, Bacic J, Needleman HL. Low-level environmental lead exposure in childhood and adult intellectual function: a follow-up study. Environ Health 2011;10:24.

[39] Needleman HL, Gunnoe C, Leviton A, Reed R, Peresie H, Maher C, et al. Deficits in psychologic and classroom performance of children with elevated dentine lead levels. N Engl J Med 1979;300:689–95.

[40] Needleman HL, Schell A, Bellinger D, Leviton A, Allred EN. The long-term effects of exposure to low doses of lead in childhood. An 11-year follow-up report. N Engl J Med 1990;322:83–8.

[41] Bellinger D, Hu H, Titlebaum L, Needleman HL. Attentional correlates of dentin and bone lead levels in adolescents. Arch Environ Health 1994;49:98–105.

[42] Fergusson DM, Horwood LJ, Lynskey MT. Early dentine lead levels and educational outcomes at 18 years. J Child Psychol Psychiatry 1997;38:471–8.

[43] Tong S, Baghurst P, McMichael A, Sawyer M, Mudge J. Lifetime exposure to environmental lead and children's intelligence at 11–13 years: the Port Pirie cohort study. Br Med J 1996;312:1569–75.

[44] Kaufman AS, Zhou X, Reynolds MR, Kaufman NL, Green GP, Weiss LG. The possible societal impact of the decrease in U.S. blood lead levels on adult IQ. Environ Res 2014;132:413–20.

[45] Bouchard MF, Bellinger DC, Weuve J, Matthews-Bellinger J, Gilman SE, Wright RO, et al. Blood lead levels and major depressive disorder, panic disorder, and generalized anxiety disorder in US young adults. Arch General Psychiatry 2009;66:1313–9.

[46] Rajan P, Kelsey KT, Schwartz JD, Bellinger DC, Weuve J, Sparrow D, et al. Lead burden and psychiatric symptoms and the modifying influence of the delta-aminolevulinic acid dehydratase (ALAD) polymorphism: the VA normative aging Study. Am J Epidemiol 2007;166:1400–8.

[47] Nie LH, Wright RO, Bellinger DC, Hussain J, Amarasiriwardena C, Chettle DR, et al. Blood lead levels and cumulative blood lead index (CBLI) as predictors of late neurodevelopment in lead poisoned children. Biomarkers 2011;16:517–24.

[48] Burns JM, Baghurst PA, Sawyer MG, McMichael AJ, Tong SL. Lifetime low-level exposure to environmental lead and children's emotional and behavioral development at ages 11–13 years. The Port Pirie cohort study. Am J Epidemiol 1999;149:740–9.

[49] McFarlane AC, Searle AK, Van Hooff M, Baghurst PA, Sawyer MG, Galletly C, et al. Prospective associations between childhood low-level lead exposure and adult mental health problems: the Port Pirie cohort study. Neurotoxicology 2013;39:11–7.

[50] Opler MG, Brown AS, Graziano J, Desai M, Zheng W, Schaefer C, et al. Prenatal lead exposure, delta-aminolevulinic acid, and schizophrenia. Environ Health Perspect 2004;112:548–52.

[51] Opler MG, Buka SL, Groeger J, McKeague I, Wei C, Factor-Litvak P, et al. Prenatal exposure to lead, delta-aminolevulinic acid, and schizophrenia: further evidence. Environ Health Perspect 2008;116:1586–90.

[52] Guilarte TR, Opler M, Pletnikov M. Is lead exposure in early life an environmental risk factor for schizophrenia? Neurobiological connections and testable hypotheses. Neurotoxicology 2012;33:560–74.

[53] Cecil KM, Brubaker CJ, Adler CM, Dietrich KN, Altaye M, Egelhoff JC, et al. Decreased brain volume in adults with childhood lead exposure. PLoS Med 2008;5:e112.

[54] Brubaker CJ, Dietrich KN, Lanphear BP, Cecil KM. The influence of age of lead exposure on adult gray matter volume. Neurotoxicology 2010;31:259–66.

[55] Brubaker CJ, Schmithorst VJ, Haynes EN, Dietrich KN, Egelhoff JC, Lindquist DM, et al. Altered myelination and axonal integrity in adults with childhood lead exposure: a diffusion tensor imaging study. Neurotoxicology 2009;30:867–75.

[56] Cecil KM, Dietrich KN, Altaye M, Egelhoff JC, Lindquist DM, Brubaker CJ, et al. Proton magnetic resonance spectroscopy in adults with childhood lead exposure. Environ Health Perspect 2011;119:403–8.

[57] Yuan W, Holland SK, Cecil KM, Dietrich KN, Wessel SD, Altaye M, et al. The impact of early childhood lead exposure on brain organization: a functional magnetic resonance imaging study of language function. Pediatrics 2006;118:971–7.

[58] Calderón-Garcidueñas L, Mora-Tiscareño A, Ontiveros E, Gómez-Garza G, Barragán-Mejía G, Broadway J, et al. Air pollution, cognitive deficits and brain abnormalities: a pilot study with children and dogs. Brain Cogn 2008;68:117–27.

[59] Calderón-Garcidueñas L, Mora-Tiscareño A, Styner M, Gómez-Garza G, Zhu H, Torres-Jardón R, et al. White matter hyperintensities, systemic inflammation, brain growth, and cognitive functions in children exposed to air pollution. J Alzheimer's Dis 2012;31:183–91.

[60] Calderón-Garcidueñas L, Franco-Lira M, Mora-Tiscareño A, Medina-Cortina H, Torres-Jardón R, Kavanaugh M. Early Alzheimer's and Parkinson's disease pathology in urban children: friend versus foe responses – it is time to face the evidence. Biomed Res Int 2013;2013:161687.

[61] White RF, Palumbo CL, Yurgelun-Todd DA, Heaton KJ, Weihe P, Debes F, et al. Functional MRI approach to developmental methylmercury and polychlorinated biphenyl neurotoxicity. Neurotoxicology 2011;32:975–80.

[62] Rauh VA, Perera FP, Horton MK, Whyatt RM, Bansal R, Hao X, et al. Brain anomalies in children exposed prenatally to a common organophosphate pesticide. Proc Natl Acad Sci USA 2012;109:7871–6.

[63] Treit S, Lebel C, Baugh L, Rasmussen C, Andrew G, Beaulieu C. Longitudinal MRI reveals altered trajectory of brain development during childhood and adolescence in fetal alcohol spectrum disorders. J Neurosci 2013;33:10098–109.

[64] Ho SM, Johnson A, Tarapore P, Janakiram V, Zhang X, Leung YK. Environmental epigenetics and its implication on disease risk and health outcomes. ILAR J/Natl Res Counc Inst Lab Anim Resour 2012;53:289–305.

[65] Nyon MP, Kirkpatrick J, Cabrita LD, Christodoulou J, Gooptu B. 1H, 15N and 13C backbone resonance assignments of the archetypal serpin alpha1-antitrypsin. Biomol NMR Assignments 2012;6:153–6.

[66] Calvanese V, Lara E, Fraga MF. Epigenetic code and self-identity. Adv Exp Med Biol 2012;738:236–55.

[67] Feil R, Fraga MF. Epigenetics and the environment: emerging patterns and implications. Nat Rev Genet 2011;13:97–109.

[68] Tang WY, Ho SM. Epigenetic reprogramming and imprinting in origins of disease. Rev Endocr Metab Disord 2007;8:173–82.

[69] Tang WY, Morey LM, Cheung YY, Birch L, Prins GS, Ho SM. Neonatal exposure to estradiol/bisphenol A alters promoter methylation and expression of Nsbp1 and Hpcal1 genes and transcriptional programs of Dnmt3a/b and Mbd2/4 in the rat prostate gland throughout life. Endocrinology 2012;153:42–55.

[70] Heijmans BT, Tobi EW, Stein AD, Putter H, Blauw GJ, Susser ES, et al. Persistent epigenetic differences associated with prenatal exposure to famine in humans. Proc Natl Acad Sci USA 2008;105:17046–9.

[71] Basha MR, Wei W, Bakheet SA, Benitez N, Siddiqi HK, Ge YW, et al. The fetal basis of amyloidogenesis: exposure to lead and latent overexpression of amyloid precursor protein and beta-amyloid in the aging brain. J Neurosci 2005;25:823–9.

[72] Bihaqi SW, Zawia NH. Alzheimer's disease biomarkers and epigenetic intermediates following exposure to Pb in vitro. Curr Alzheimer Res 2012;9:555–62.

[73] Mazumdar M, Xia W, Hofmann O, Gregas M, Ho S, Hide W, et al. Prenatal lead levels, plasma amyloid beta levels, and gene expression in young adulthood. Environ Health Perspect 2012;120:702–7.

[74] Bakulski KM, Fallin MD. Epigenetic epidemiology: promises for public health research. Environ Mol Mutagen 2014;55:171–83.

[75] Joubert BR, Håberg SE, Nilsen RM, Wang X, Vollset SE, Murphy SK, et al. 450K epigenome-wide scan identifies differential DNA methylation in newborns related to maternal smoking during pregnancy. Environ Health Perspect 2012;120:1425–31.

[76] Pilsner JR, Hu H, Ettinger A, Sánchez BN, Wright RO, Cantonwine D, et al. Influence of prenatal lead exposure on genomic methylation of cord blood DNA. Environ Health Perspect 2009;117:1466–71.

[77] Pilsner JR, Hall MN, Liu X, Ilievski V, Slavkovich V, Levy D, et al. Influence of prenatal arsenic exposure and newborn sex on global methylation of cord blood DNA. PloS One 2012;7:e37147.

[78] Bellinger DC. Interpreting epidemiologic studies of developmental neurotoxicity: conceptual and analytic issues. Neurotoxicol Teratol 2009;31:267–74.

[79] Power C, Kuh D, Morton S. From developmental origins of adult disease to life course research on adult disease and aging: insights from birth cohort studies. Annu Rev Public Health 2013;34:7–28.

[80] Salum GA, Polanczyk GV, Miguel EC, Rohde LA. Effects of childhood development on late-life mental disorders. Curr Opin Psychiatry 2010;23:498–503.

[81] Kundakovic M, Champagne FA. Early life experience, epigenetics, and the developing brain. Neuropsychopharmacology 2015;40:141–53.

Environmental Factors in Neurodegenerative Disorders: Summary and Perspectives

Michael Aschner, Lucio G. Costa

As noted in a recent issue in Science [1], brain trajectory is affected by numerous factors, such as genetic, physical, and psychological factors to name a few. With age, the human brain constantly reorganizes, to accommodate and compensate, enabling us to remain vigilant and interactive. As health care around the globe has improved, demographics have changed and the median age of life expectancy has drastically increased. Today, there are more people over 60 than under 15 in Europe [2], and the USA will follow this trend by as early as 2030. Most striking, those aged 80 and over will constitute ~12% of Europe's population by 2060.

These aforementioned demographic changes have financially strained health systems around the globe, reflecting in part the increase of financial investment in combating the increased incidence of late-life onset diseases, such as Parkinson's disease, Alzheimer's disease, multiple sclerosis, amyotrophic lateral sclerosis (ALS), and Huntington's disease, to name a few. It is estimated that, today, ~5 million people in the USA suffer from Alzheimer's disease, ~1 million from Parkinson's disease, ~400,000 from multiple sclerosis, ~30,000 from ALS, and ~30,000 from Huntington's disease. Further, it is estimated that, by 2045, as many as 12 million people in the USA will suffer from neurodegenerative diseases, for which effective treatments have yet to be developed.

Great strides have been made over the last several decades in understanding the etiology of neurodegenerative diseases. Although of multifaceted nature, neurodegenerative diseases are rarely characterized solely by genetic predisposition, and many have been linked to environmental exposures to pesticides, metals, and other xenobiotics.

The second part of this book addresses the role of these compounds as environmental risk factors in neurodegenerative diseases. Leading experts in the field discuss the epidemiology of the aforementioned diseases, highlighting relevant epidemiological and experimental data, and molecular triggers of neurodegeneration. The role of both prenatal and postnatal exposure to metals and pesticides in late-life onset neurodegenerative diseases is also

Environmental Factors in Neurodevelopmental and Neurodegenerative Disorders
http://dx.doi.org/10.1016/B978-0-12-800228-5.00020-0

discussed, attesting to the fact that intrauterine or early childhood exposure may play a role in neurodegenerative changes that unmask themselves much later in life.

Animal and human studies have provided a wealth of results on the role of environmental exposures to neurodegenerative disorders. Yet, much more is to be learned and done. We must turn our attention to remediating exposures to neurotoxins, as disease prevention and early interventions should not only diminish the incidence of neurodegenerative disorders, but also afford a sound financial approach to an aging population that strains economic resources. In addition, more attention should be directed at social policies that avoid exposures, as scientific research continues to decipher the relationship between neurodegeneration and its environmental links.

References

[1] Stern P, Hines PJ, Travis J. The aging brain. Science 2014;346:567.
[2] Harper S. Economic and social implications of aging societies. Science 2014;346:587–91.

Index

Note: Page numbers followed by "f" and "t" indicate figures and tables respectively.